D1803784

Wind
In Architectural
And Environmental Design

Wind In Architectural And Environmental Design

Michele Melaragno
Professor of Architecture
University of North Carolina at Charlotte

VNR VAN NOSTRAND REINHOLD COMPANY
NEW YORK CINCINNATI TORONTO LONDON MELBOURNE

Van Nostrand Reinhold Company Regional Offices:
New York Cincinnati

Van Nostrand Reinhold Company International Offices:
London Toronto Melbourne

Library of Congress Catalog Card Number: 80-29613
ISBN: 0-442-25130-0

Manufactured in the United States of America

Published by Van Nostrand Reinhold Company
135 West 50th Street, New York, N.Y. 10020

Published simultaneously in Canada by Van Nostrand Reinhold Ltd.

15 14 13 12 11 10 9 8 7 6 5 4 3 2 1

Library of Congress Cataloging in Publication Data

Melaragno, Michele G.
Wind in architecture and environmental design.

Includes index.
1. Wind-pressure. 2. Tornadoes. 3. Wind power.
4. Environmental engineering. 5. Architectural design.
I. Title.
TA654.5.M46 624.1'75 80-29613
ISBN 0-442-25130-0

To
Helen, Patricia, Elizabeth, and Mark

Preface

The intent of *Wind in Architectural and Environmental Design* is to study the wind, not just as a natural phenomenon, but also in terms of its influence on the human environment, i.e., structures, the urban agglomerates, and the landscape. Therefore, it is addressed to those who are interested in exploring the potential applications of the wind in design and in finding innovative ways to improve the quality of life.

My interest in this topic originated from an investigation of the effects of tornado forces on buildings, which I conducted in the 1960s while teaching at Kansas State University. My first bibliographical search showed that the literature on tornadoes was scarce and confined mostly to the meteorological aspects of the problem rather than extending to engineering implications. Among structural engineers, the consensus was that tornado forces were too great to allow any design criterion to be economically feasible. Their only comfort was the low probability of any particular point being struck by a tornado. Thus, it was obvious that an urgent need existed for gathering pertinent data on the subject and organizing them in a format that could be easily used by designers, from both an engineering and architectural point of view. In addition, the proliferation of nuclear power plants had created a need for design criteria in order to protect such potentially dangerous structures from the action of tornadoes.

Influenced by these and other factors, I expanded my interest in the wind to include a wider range of wind effects. The ecological movement of the 1960s, for instance, made me realize the importance of many wind applications in the design of the human environment—for example, the ventilation problems in and around buildings that influence their thermal control, the reduction of wind-induced waves for the expansion of the human habitat on water, and the realization that wind in urban centers is a subtle energizing force. In addition, I became aware of the wind as a logical solution to the demand for new energy sources.

Wind in Architectural and Environmental Design presents the material in a historical context, tapping the mythologies of past civilizations. It also presents a detailed bibliography of papers presented at recent international conferences, thus offering a state of the art of wind engineering.

With the clear goal of stimulating the interest of designers and guiding their attention in this relatively new area, it is my hope that *Wind in Architectural and Environmental Design* will be of practical use to the reader.

MICHELE MELARAGNO
Charlotte, North Carolina

Acknowledgments

With deep appreciation, the author wishes to acknowledge the contribution of Rahim Banizaman, who prepared the ink drawings, and the editorial assistance of Patricia Mansfield. The author also extends his sincere gratitude to Dr. and Mrs. Paul A. Saman for their help in proofreading, to the reference librarians at the University of North Carolina at Charlotte for their help, and to all those who furnished the photographs used in the book.

Contents

Wind
In Architectural
And Environmental Design

1
Winds

SYMBOLISM AND MYTHOLOGY

Wind is one of the manifestations of nature that catches the imagination of humans, often inspiring feelings of awe and fear that can be linked with primitive religious experiences. Associated with the primary elements of earth, air, fire, and water, wind is also related to the idea of creative breath or human exhalation. In Arabic, the word *ruh* means "breath" and "spirit," as well as "wind." In Hebrew, the word *ruach,* which has a similar meaning, is an example of the association of wind with divinity.

Among the first graphic symbols used by humans, especially in North and South America, were those representing the hurricane; these included the sigma, double sigma, and swastika. American Indians recognized the supreme force of the hurricane and were deeply influenced by its terrible power. Worshipping the hurricane as a force of divine origin, they cowered in fear before its destructive fury.

Hurricanes have also been symbolically linked with fecundation and regeneration. Furthermore, they share with the planetary systems the dual motions of rotation and revolution, and thus planets and hurricanes have often been connected in primitive mythologies. This connection is further supported by the feeling of absolute calm represented by the harmonious motion of celestial bodies in the universe and by the calm at the center of a hurricane.

During the Assyro-Babylonian civilization beginning after 3000 B.C., a rich mythology developed. The original god of the hurricane and lord of the wind was Enlil, whose characteristics gradually changed throughout history. By 2000 B.C., he was replaced by Adad, the god of lightning and tempest, and lord of the storms. He was usually represented holding thunderbolts in his hands and standing on a bull. Nevertheless, Adad was also the god of beneficial, rain-bearing winds that produced the fertilizing floods necessary for agriculture. (See Figure 1-1.)

In Phoenician mythology, there were five winds: Aether and Air, which were thought to have existed at the beginning of time, followed by Wind and the two other winds, Lips and Notus. In the classical Greek culture, winds had a multiple significance, which included their beneficial effects on the climate and the fertility of plants and

Figure 1-1. Teshup, the Babylonian storm god, with an axe in one hand and lightning in the other, in a style of the eighth century B.C. (Courtesy of Arts Council of Great Britain)

animals, as well as their violent, storm-producing nature. Depending upon their benevolent or malevolent nature, winds were represented by gods or monsters. For instance, demonic figures called *Harpies* symbolized storms (Figure 1-2). Respected and feared by sea travelers, they were the goddesses of tempests.

Figure 1-2. According to Virgil, the Harpies, the Roman-age Greek goddesses of storms, each had the body of a bird and the face of a woman. Their names were Aello, Ocypete, and Celaeno (or Podarge). Their secondary function was to inflict pain and punishment according to the will of the gods. Two of the Harpies are shown here.

The ancient Greeks also associated winds with the souls of the dead, and human sacrifices were offered to calm the violent winds. The sacrifice of Iphigenia, Agamemnon's daughter, as the Greeks were leaving for Troy, and the sacrifice offered by Menelaus as he departed for Egypt, are examples of this practice. Another ancient rite related to the winds was the annual offering of a horse, which was burned on the Taygetus Mountains.* The ashes were dispersed in the wind. In the first century B.C., an octagonal structure—the Tower of the Winds—was built in Athens, with each side representing one of the eight winds (see Figure 1-3).

Figure 1-3. The Tower of the Winds, built in Athens in the first century B.C. by Andronicus of Syros, is an octagonal building, 40 feet high. On each side, a winged figure represents the wind blowing from the directions along which the eight sidewalls are oriented.

*Mountain range near the border of Laconia and Messenia.

In Greek mythology, the major wind gods were: Boreas (the north wind), Zephyrus or Zephyr (the west wind), Eurus (the east wind), and Notus (the south wind). They were born from Eos, the beautiful goddess of the dawn, and the Titan, Astraeus, the starry sky. Boreas, represented as a mature man with wings, lived in the caves in the mountains in Thrace. Strong and violent, he produced storms and heavy seas. The Athenians built a temple in Boreas' honor in gratitude for his courage in dispersing the attacking enemy fleet of Xerxes. Among several children Boreas had with Oreithyia (Figure 1-4) were the twins, Zetes and Calais, also called the *Boreades.* They were killed by Hercules, who changed them into the mild northeasterly winds called the *Prodromes* ("runners"). In fact, they preceded the rise of the Dog Star.

Figure 1-4. Drawing of Boreas, the north wind, showing the strong winged god carrying Oreithyia, the beautiful young daughter of Cecrops, the king of Attica. For his help in defeating the attacking Persian fleet, a temple in Athens was dedicated to Boreas, who received sacrifices for the protection of the city.

Zephyrus was, in the beginning, a violent wind who lived with his brother Boreas. With the Harpy Podarge, he fathered Xanthos and Balios, the two horses that pulled Achilles' chariot. He later became a mild wind, carrying sweet scents over the Elysium, the fields on the banks of the Oceanus River where the holy lived after death.

Echidna and Typhon, spirit of typhoons and hurricanes, were, respectively, the mother and father of various monsters that produced storm winds and tempests. Echidna was represented as a monster with the upper body of a young nymph and the lower body of a serpent with scales. Among their children was the Chimera, representing a storm cloud, who was sometimes depicted with a lion's head, a goat's body, and a dragon's tail.

Although demonic in nature, the Harpies were also goddesses like their divine parents. According to Hesiod, there were two Harpies; other authors list more. The Harpies combined their power over the tempest with the role of carrying human souls in the journey after death. This dual function, which associates storms and death, does not appear to be accidental but rather a significant theme. According to some sources, the Harpies were the daughters of Taumantes and Electra. They inspired many poets, including Hesiod, Homer, Virgil, and Dante, with their symbolism.

Much of Greek mythology passed on into the Roman myths. Aeolus, who in Greek legend was the guardian of the winds, as described in the *Odyssey,* is one example of this carry-over. Living in the Lipari (or Aeolian) Islands, Aeolus was befriended by the gods, who gave him the power to excite and sooth the winds at his pleasure. When Ulysses visited the Lipari Islands, Aeolus gave him a wineskin containing all of the adverse winds and none of the favorable ones. In the Roman legend, Aeolus is the god of winds who lived in the Lipari Islands and kept the winds chained in deep caves.

In the Italic regions, before the advent of Roman culture, the winds were worshipped as frightening demoniac entities and were often symbolized by fierce predatory birds. They later assumed more serene representations.

In Rome, winds were worshipped as gods. *Tempestates* were the goddesses of storms to whom a temple was dedicated, circa 259 B.C., by L. Cornelius Scipio. The winds and the god Neptune were worshipped together because they both controlled navigation. Even Caesar Augustus sacrificed to both gods when preparing for the expedition against Sextus Pompeius Magnus.

In Chinese mythology, the winds were contained in a goatskin bottle carried by the Earl of the Winds (Feng-p'o). He was replaced by Mrs. Wind (Feng-p'o-p'o), a goddess in the shape of an old woman riding a tiger through the clouds.

In Japanese mythology, in a complicated series of myths explaining the birth of the gods and the creation of the world, the god of wind, Susa-no-o, was among the first gods born. He sprang from the nose of the god Izanagi. Izanagi and his wife, the goddess Izanami, were both very important in Japanese legends. Another name associated with the god of winds at a lower scale in Japanese mythology is Fujin, who is often confused with Raiden, the thunder god.

For the Hindus, the god of storms is Rudra. As described in one of the Vedas, four ancient sacred books of the Hindus, the god was both benevolent and destructive. He was the divine healer, as well as a demon who killed men and beasts with his bow and arrow. An eighth-century sculpture of Rudra can be found in one of the 34 cave temples at Ellora, a village in the Maharashtra state in India (Figure 1-5). Also mentioned in the Vedas is the god of the wind, Vāyu, who rides in the sky in the same chariot with Indra. Vāyu, often confused with Indra, is also known by other names that are associated with various characteristics of the wind; for example, Gandha-vaha ("bearer of perfumes") and Satata-ga ("ever moving") (Dawson 1968).

Among the North American Indians, the Iroquois and the Huron tribes have colorful mythologies. In their legends explaining the genesis of the cosmos, the master of winds has a substantial role. He is the father of the two major participants in the creation: Ioskeha, the creator, and Tawiscara, the evil power.

For the Indians of the Great Plains, the legends of creation are different. Tirawa is the main figure in this genesis. The bright star or evening star, mother of all things, was given control over the winds and the clouds, as well as lightning and thunder. Various parts of the creation, including the birth of the earth, mountains, valleys, forests, prairies, and rivers, came from several storms. Such storms were produced by the singing of the gods of winds and other elements.

With the Dakota Indian tribes, the whirlwinds, with their mysterious spiraling motion, have the greatest impact, demonstrated by several graphical representations of the natural phenomenon that they have seen so often (Figure 1-6).

Figure 1-5. Rudra, the storm god, is represented in this sculpture of the eighth century A.D., which is found in one of the cave temples at Ellora in the Maharashtra state in India.

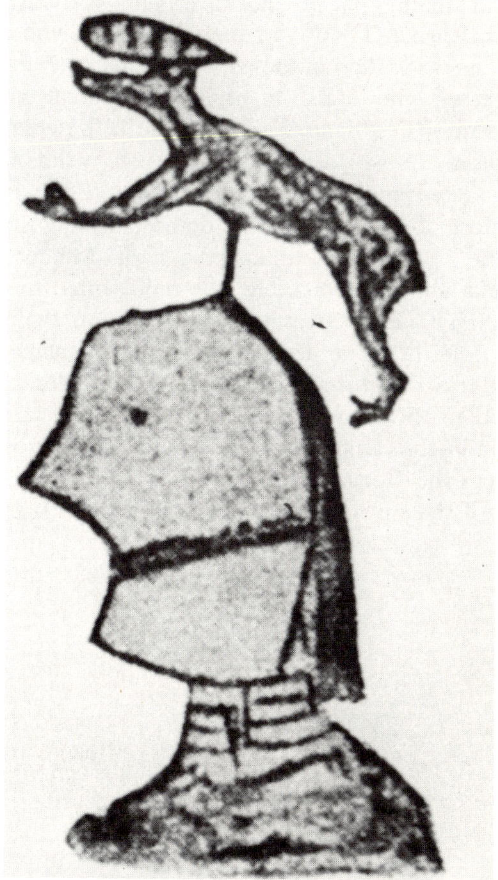

Figure 1-6. *Mato-wamniyomni* (bear-whirlwind) of the Oglala Dakota Indians in North America. (From *Picture-Writing of the American Indians,* Vol. 2, by G. Mallery, Dover Publications, New York, 1972.)

For the Haida Indians of the Queen Charlotte Islands, British Columbia, and the Prince of Wales Island, Alaska, T'Kul was the wind spirit who controlled the winds and brought the rain. The deification of the winds was also popular among the Aztecs. Such gods include Ehecatl, who is also identified as Quetzalcoatl; Tezcatlipoca, the god of wind; and the son of Coatlicue, Huitzilopochtli, the great storm god.

Other tribes throughout the world paid tribute to the winds in their mythologies. The tribes of Nicaragua recognized Ecalchot as the god of winds. A god of great importance, he was close to the two major deities, Tamagostad and the goddess Zipaltonal. In Chile, the Araucanian Indians recognized Meuler as the god of winds, whirlwinds, waterspouts, and typhoons. He was represented by a lizard. The Taino tribe of the Bahamas, Haiti, and Greater Antilles had in their language the word *huracan,* meaning "evil spirit." The Tainos worshipped Guabancex, represented by a stone idol, as a goddess of storms, winds, and water. *Hunraken* was the name given to the storm god in the Mayan dialect, and *hyoracan,* meaning "devil," was the word for *hurricane* in the language of the Galibi Indians of Dutch and French Guiana. Even in Africa, the Yoruba, inhabitants of southwest Nigeria, worshipped wind deities. The god of the storms was Shango (Figure 1-7).

Figure 1-7. Ceremonial axe of the Shango cult. The double axe-blade represents the magic thunderbolt of Shango, the storm god of the Yoruba people of southwest Nigeria. The kneeling figure represents one of the followers of the cult. The piece is now in the British Museum.

Several gods of winds were included in the mythology of the pagan Slavs. One legend spoke of three brothers: the north wind, the east wind, and the west wind. The west wind, Dogoda, was a mild, gentle wind. In other Slavic myths, there were seven wind gods. Another legend mentions wind gods named Stribog and Varpulis. Erisvorsh, god of tempest, was another wind god in Slavic mythology. In Iceland and Norway, prior to the advent of Christianity (circa 1000), the Teutonic cults worshipped a powerful god of thunder and storms, Thor (Figure 1-8).

Figure 1-8. Thor, the thunder and storm god of the Teutons, particularly popular in Norway and Iceland until the advent of Christianity (circa 950–1000 A.D.). The bronze statuette, found in North Iceland in the second half of the tenth century A.D., is now in Reykjavik.

Remnants of pagan and fetish legends extend into the twentieth century in some parts of the world; for example, some rites are still practiced in remote villages of Southeast Asia. Some of these include the wind as the main subject of their ceremonies. Typical examples of this are found in northeastern Thailand, in the Ubon Province. Sometime in spring, the villagers celebrate the visitation of the dead. Tables are set in the open with food offerings, and mats are placed on the ground. In the evening darkness, the murmuring crowd suddenly becomes silent as gusts of wind begin: the spirits have arrived. Lying on the mats, female mediums go into spasms and scream; then, sitting up, they start speaking in the voices of the ghosts that have possessed them. Through the mediums, the dead are now speaking to their relatives and beloved ones, in a suggestive but also frightening atmosphere. At the end, they depart as they arrived, and the gustiness of the wind suddenly picks up.

The human need to replace abstractions and intangible entities with more concrete forms has led to the creation of symbolism, its purpose being to reproduce in a sensorial way the same feelings produced by the original source. Because it is intangible and invisible, the wind required symbolic representation, which in fact it attained (Figure 1-9). From the ancient expressions of religious significance that had been inspired by its mysterious power, to the current symbols for the wind used in the meteorological science, the applications of symbols to the various aspects of the wind are numerous. As a benevolent energy-giver, as an indispensable agent in the perennial changing of the weather, and as a destructive and powerful force, the wind has played many roles and has inspired many symbolic interpretations in the arts, including literature, painting, sculpture, and music.

In those Oriental languages in which words are written with symbolic images, the wind and its characteristics have a pictorial representation that is intriguing to the Western observer. For example, see the Chinese expressions illustrated in Figure 1-10.

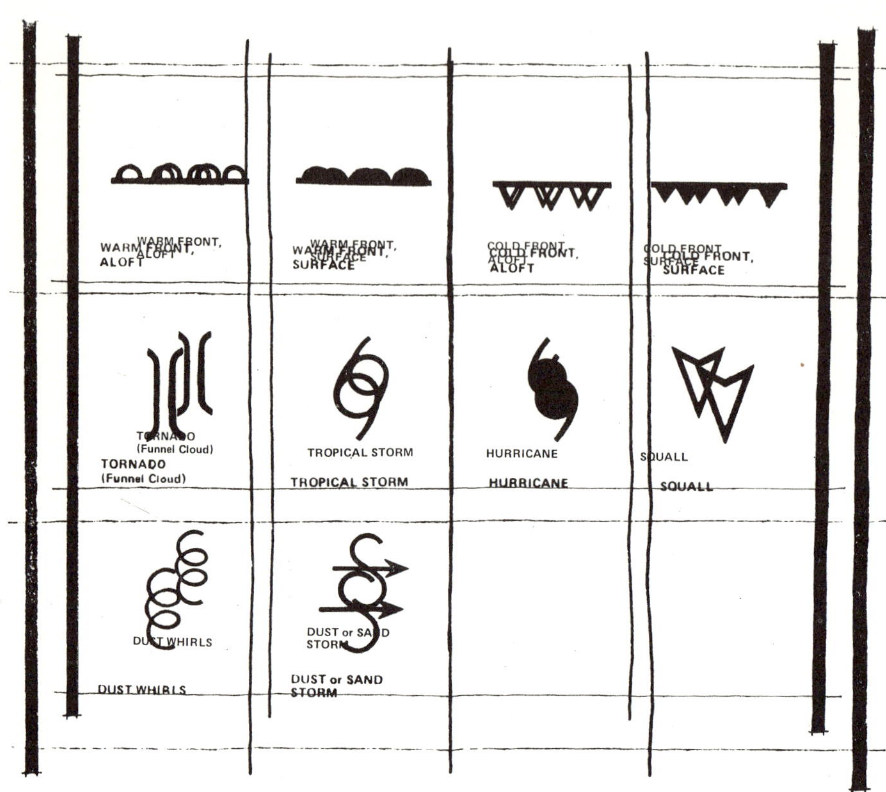

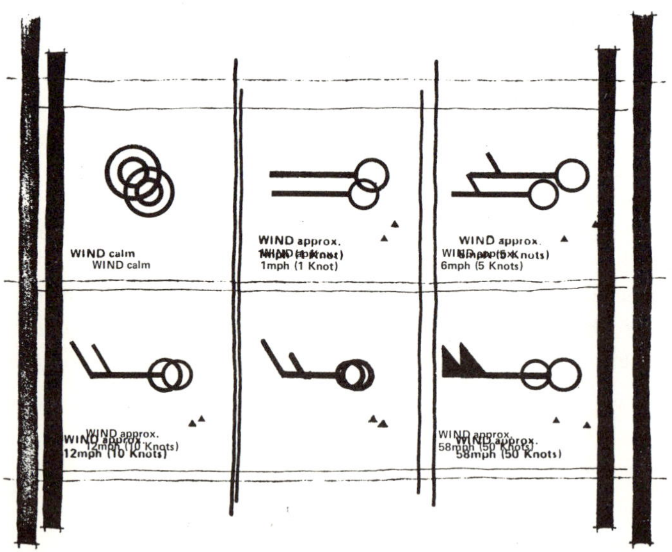

Figure 1-9. Symbols used in meteorology to describe wind characteristics and associated phenomena.

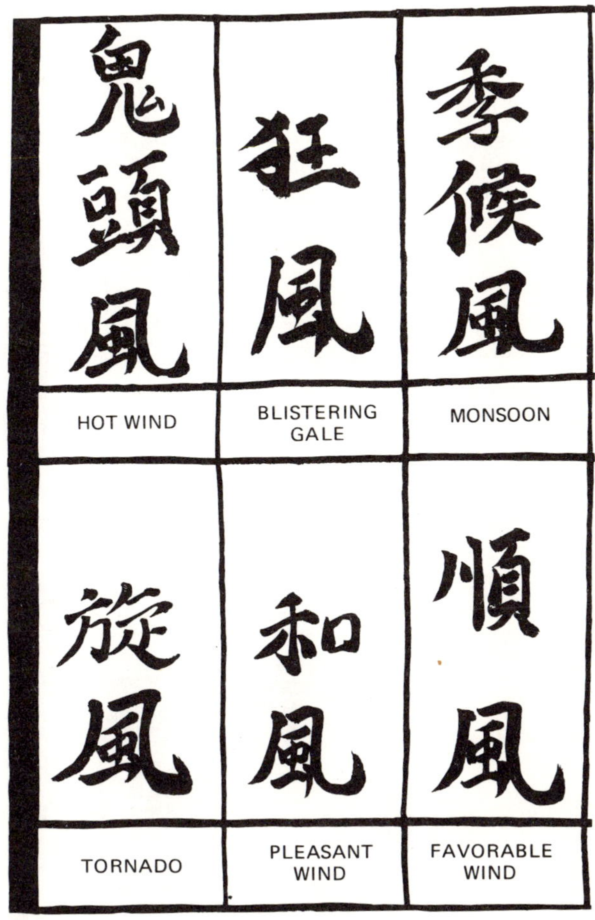

Figure 1-10. The pictorial representations of Chinese writing illustrate the meaning, the symbolism, and other qualities of the word. The words describing several wind conditions are shown as symbols of the natural phenomenon comparable to other visual illustrations.

Another symbolic representation of the wind, which was popular in Chinese houses and often exposed over entry doors, is shown in Figure 1-11. The figure illustrates an octagon with a central piece surrounded by eight groups of bars. The central part represents the Chinese concept of universalism, while in one of the eight groups the wind is symbolized. More precisely, the meaning expressed is not of the wind in general, but of the soft wind with its pleasant, peaceful characteristics of gentleness and mildness.

Anemone

Greek mythology tells us that Anemone, the windflower, was a beautiful nymph, dearly beloved by Zephyr, the gentle wind of the southwest. Jealous of such love, a goddess changed Anemone into a flower—the windflower that bears her name. However, another legend tells a different story. Young Adonis, beloved by Venus, was attacked by a ferocious boar while hunting. Responding to his groans, Venus rushed down to earth. Too late, she found him dead. She sprinkled nectar on Adonis' blood, from which a flower bloomed. It had a red hue, and its name was *anemone.* Each year the flower would bloom to renew the memory of Adonis for the inconsolable Venus.

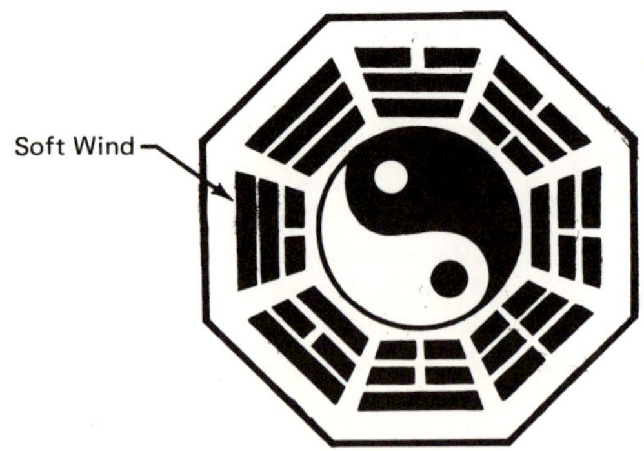

Figure 1-11. Famous symbol of universalism often displayed in Chinese houses. Includes the symbol of the soft wind in the trigram at the 9:00 o'clock position.

Passing from myth to early science, according to Pliny the Elder (23–79) the anemone opens its blossom when the spring breezes blow over it. At the end of the flower's short life, the wind takes its petals away. Thus, the anemone is the windflower. See Figure 1-12.

Aeolian Harp

Inspired probably by the sound of gentle winds, the *aeolian harp* appeared in history since remote times, not as a musical instrument per se (there is no human participation in the sequence of the tunes), but as a magic object made alive by the invisible power of the wind. The aeolian harp, with 10 or 12 strings of catgut stretched over it, was usually hung across an open window to catch the wind. The sounds of the harp have been described in literature with such vivid terms as "drowsy and lulling" and "ethereal and ghostly." Aeolus, the wind god of the Greeks and the Romans, was the divine player in the placid and serene setting of the pagan world. Even before the Roman and Greek cultures, the harp of Aeolus existed with different

Figure 1-12. Anemone lancifolia.

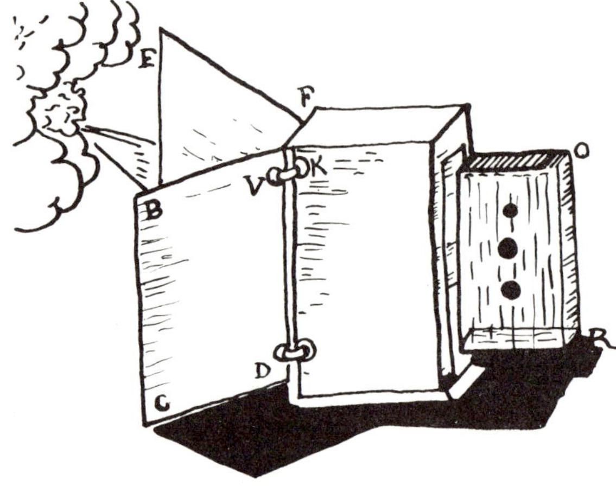

Figure 1-13. The aeolian harp, a stringed instrument played by the wind. This illustration is from a treatise on sound by Athanasius Kircher, circa 1650.

names in Hebrew history. The Rabbinic record, in fact, tells us that King David had over his bed a harp hung in the wind. At night, the gentle breeze played harmonic tunes. In the legends of the Middle Ages, Saint Dunstan (925–988) of Glastonbury, England, was credited with inventing the aeolian harp. This was probably because he restored monastic life, and its spirit of tranquility and serenity was associated with the instrument. Interest in the instrument continued for centuries. As late as 1650, Athanasius Kircher described the aeolian harp he built in his *Misurgia Universalis* (Figure 1-13).

The magic instrument is not only a Western invention. We find a similar example of the aeolian harp in the construction of paper kites in Asia. At each end of a kite, strings stretched across an opening are set in vibration by the wind, producing harmonic sounds.

Kinetic Art

As will be discussed in Chapter 6, the energizing action of the wind has recently been applied in sculpture by a new movement referred to as *kinetic art.* Light, movable compositions either suspended from wires or hinged with various mechanical devices acquire motion as air moves through them. Gentle air currents or actual winds give life to either indoor or outdoor pieces of sculpture.

HISTORICAL PROGRESS IN THE KNOWLEDGE OF WIND

Humanity's exploration of wind began with mythology and continued with the beginning of the sciences, but a real understanding of the wind did not begin before the eighteenth century. Early civilizations saw in the wind the manifestation of a divine force used by benevolent or malevolent gods. With the beginning of the Western school of philosophy in Greece, the scientific investigation of wind began. Anaximander, a philosopher of the Milesian school, in the sixth century B.C., offered definitions of the nature and origin of wind. Empedocles, in the fifth century B.C., proved the physical and material consistency of air and wind. About 340 B.C., in his treatise, *Meteorologica,* Aristotle proposed a speculative theory on wind and the complex whole of atmospheric phenomena. Of course, without the basic knowledge of physics, the Aristotelian theory was imaginative but quite far from the truth, yet it lasted undisputed through the sixteenth century. In fact, the *ipse dixit* that carried the authority of Aristotelian dogmatic theories had stagnated progress in many of the arts and sciences.

The absence of tools for measuring temperature, pressure, and so on impaired real progress for many centuries. About 1600, however, Galileo invented the thermometer, and, in the 1640s, his pupil, Torricelli, invented the barometer. In 1625, Francis Bacon published a treatise in which he tried to pursue a scientific understanding of wind apart from Aristotle's theories. Despite the knowledge of physics in existence at that time, his *Historia Ventorum* (*History of Winds*) is still far from scientific reality.

In 1680, Edmund Halley, an English astronomer wrote about trade winds and monsoons. He recognized that solar heating of the atmosphere over the equator was the energy source for tropical winds. Halley's treatise was the beginning of the scientific approach to the study of wind. George Hardley continued to work on trade winds and explained their direction along the equatorial belt. In 1835, the French engineer Gaspard Gustav de Coriolis (1792–1843) explained the deflection of masses that rotate around the earth, which was later applied to winds. In 1856, a French-American meterologist, William Ferrell (1817–1891), applied Coriolis' principles to the wind paths and presented his findings in the *Essay on the Winds and Currents of the Ocean.*

Wind studies continue, of course, to the present day and listing all the scientists who have contributed to them would take too much time and space. However, specific contributions to the knowledge of winds in the low part of the atmosphere near the earth's surface have been made by Ekman and Prandtl of Germany and Taylor and Sutton of England. In the field of planetary winds, we can remember the Norwegians, Bjerknes and Bergeron, Palmen of Finland, and the Swedish-American, Rossby.

WINDS AROUND THE WORLD

Around the world, naming the winds seems to be a practical necessity, allowing people to discuss them without confusion. Identification of winds using only the cardinal points of wind direction is an unimaginative, although more practical, form of reference. Several lists of wind names have been compiled by various authors, including Murchie (1954), De la Rue (1955), and Forrester (1959). These identifications are not scientific, and can be confusing, especially when one is translating names from one language into another. The same wind is called by different names as it blows through different geographical areas and countries; for example, over 50 names are used to identify the sirocco.

The common practice of identifying individual winds with specific names rather than distinguishing them by means of their direction was obviously derived from the need to simplify verbal communication. Moreover, the literary needs of prose and poetry reinforced this practice. For instance, in Greek and Latin, the personification of winds with mythological characters was a more vivid and effective identification than merely naming them. The wind names give a total picture of the weather. For instance, a day of sirocco in Europe implies not just a windy day when the wind comes from the south, but a day with hot humid weather that prevents the evaporation of perspiration; low barometric pressure that makes breathing difficult; feelings of irritability and depression; and overall discomfort. Literary descriptions of the winds can at times be so colorful that they convey more about the weather and environmental conditions than do modern scientific weather reports. The name *zephyr,* for instance, brings to mind the fragrances of flowers, herbs, and fruit trees that grow along the Mediterranean.

The following compendium of wind names includes not only the specific names commonly used, but also the nomenclature used for general wind categories. The *bora,* for example, is a specific name for a wind in Trieste, whereas *hurricane* is a general term for a category of winds. The list is by no means all-inclusive, but rather a list of the most commonly found names in literature.

Wind Names

African monsoon. A monsoon of Africa.

Africo. A southwesterly wind over the Mediterranean, equivalent to the Italian libeccio.

Altanus. A southwesterly wind of the ancient world (Vitruvius*).

Anticyclone (or *high*). A system of spiraling winds moving downward and outward from a central area of high pressure. Its mechanism is the reverse of a cyclone and often causes good weather. Anticyclones may be either cold or warm. Forming in polar areas, the cold type moves in a southerly direction and is generally short-lived. Forming in warm areas, the warm anticyclone extends into the upper atmosphere and moves slowly, sometimes even remaining stationary. Generally long-lasting, the warm type influences the weather to a large extent. Typical zones where anticyclones are formed include the Azores and Bermuda.

Aquilo. A northwesterly wind of the ancient world (Vitruvius).

Argestes. A southwesterly wind of the ancient world (Vitruvius).

Asian monsoon. A monsoon of Asia.

Asifa-t. A violent Arabian wind of hurricane force.

Aura. A gentle breeze.

Auster. A southerly wind of the ancient world (Vitruvius).

Australian monsoon. A monsoon of Australia.

Baguio. A wind of hurricane force in the Philippines.

* Vitruvius Pollio, Roman architect and engineer of first century B.C.

Belat. A northern wind blowing in winter on the coast of Arabia.

Berg. A wind, of the same category as the föhn, which blows on the southern coast of Africa.

Bise. A cold stormy wind from the northeast, although sometimes blowing from the north. It is associated with low pressure in France.

Blast. A squally wind in Scotland.

Black roller. A dust storm of the western United States.

Blizzard. A cold, northerly gale occurring in the northern part of the midwestern United States, especially in North and South Dakota. Bringing rapidly falling temperatures and fine crystallized snow, this suffocating wind often kills animals that are stranded out in the open.

Boekifu. A trade wind in Japan.

Bofu. A wind of high gale force in Japan.

Bohorok. A wind in the region of Sumatra belonging to the föhn category.

Bora. A cold, descending, mountain wind blowing over the Adriatic, Aegean, and Caspian Seas. Typical in Trieste, its speed can reach 100 mph.

Breeze. The general name given to light winds blowing along seashores and lakes. During the daylight hours, when the land temperature is warmer than that of the water, the air over the land rises, creating a low-pressure area. Cold air from the sea or lake then blows toward the land, beginning very gradually after sunrise, increasing to a peak in the afternoon and diminishing in the evening. At night, when the land temperature drops below that of the water, the process is reversed, and there is a flow of air from the land toward the water. This seaward breeze starts gently in the evening, increases to a peak during the night, and is still by morning.

Breva. A gentle breeze on Lake Como in Italy.

Brickfielder. A desert wind of Australia.

Brize carabinere. A wind typical in Spain.

Bull's-eye squall. A squally wind off the Cape of Good Hope.

Buran. A stormy, snow-bearing northeasterly wind blowing over Russia.

Caecias. A southeasterly wind of the ancient world (Vitruvius).

California northern. A wind, generated by a low pressure, originating in Texas or the Gulf of Mexico. It attracts cold, polar air from the north and lasts for about one day.

Canadian north wind. A very cold wind of Canada.

Canterbury föhn. A prevailing westerly wind of New Zealand, occurring frequently on the Canterbury Plains during the colder half of the year and bringing clear skies and fog.

Carbas. A northeasterly wind of the ancient world (Vitruvius).

Caspian monsoon. A monsoon of the Caspian Sea.

Cat's paw. A light breeze in the United States.

Caurus. A northwesterly wind of the ancient world (Vitruvius).

Chamsin. A southerly wind in Arabia.

Chergui. A Moroccan wind of the sirocco family.

Chichili. An Algerian wind of the sirocco family.

Chihili (or *khamsin*). A hot, dry, dust-laden, south or southeasterly wind of the sirocco family. Blowing north from the interior of the Sahara Desert, it occurs in winter and in spring.

Chili. A Tunisian wind of the sirocco family.

Ching fung. A light breeze of China.

Chinook. A wind of the föhn family blowing over the United States and Canada. Although basically a winter wind, it can occur at any time of year. It blows from the west and, upon reaching the Rockies, leaves most of its moisture on the west side of the mountains, becoming a dry wind. Temperatures produced by this wind can rise or fall as much as 30°F in 3 minutes. Sometimes it melts the snow, thus allowing cattle to pasture.

Chocolatero. A squally wind in the Gulf of Mexico.

Chubasco. A violent squall with thunderstorms, lasting only a few hours.

Chwa. A light breeze of Wales.

Cierco. Similar to the mistral, it is a cold, westerly wind of the Ebro Valley of Spain.

Colla. A light gale in the Philippines.

Coronazo de San Francisco. A gale of up to 75 knots that is typical in the Gulf of Mexico between May and November.

Creithleag. A light breeze in Ireland.

Crivetz. A northeastern wind of Rumania, similar to a blizzard.

Cyclone. A rotating air mass formed by winds that converge from all directions and spiral toward a central point. From this calm central zone of lower barometric pressure, the converging winds rise, forming clouds and eventually producing precipitation. Encompassing areas from 50 to 1000 miles in diameter, cyclones are nearly circular. The rotation of the spiraling winds is counterclockwise in the Northern Hemisphere and clockwise in the Southern Hemisphere; the intensity of these winds ranges from high to extremely low. Cyclones are divided into tropical and extratropical categories; hurricanes are tropical cyclones.

Datoo. A western sea breeze over Gibraltar.

Doctor. A breeze of Australia and Africa, so named for its beneficial effects on people.

Doinionn. A strong gale in Ireland.

Dust-devil (also *dust whirl* and *desert devil*). A whirlwind rotating around a vertical axis in a clockwise or counterclockwise direction. No more than a few yards in diameter, it can reach a height of 100 to 300 feet—and up to 2000 feet in the desert. Usually carrying dust, sand, and debris, which form a dark cloud, a dust-devil is much less violent than a tornado and usually forms on calm, hot, sunny afternoons.

Elephanta. A strong wind of gale force on the coast of Malabar, in southwest India.

Eliseos. A trade wind of Spain.

Elvegast (or *Sno*). A cold, dry, offshore wind blowing in winter toward the west coast of Norway from the interior.

Erh chi chi fung. A northerly wind of northern China.

Etesian. A northerly wind in Greece and the general region of the Eastern Mediterranean. It is cool and blows during the summer.

Eurocircias. A southeasterly wind of the ancient world (Vitruvius).

Euroclydon. A northeasterly wind associated with storm activity in the eastern Mediterranean.

Eurus. A southeasterly wind of the ancient world (Vitruvius).

Favonius. A warm westerly wind of the ancient world (Vitruvius), descending the Alps; equivalent to the föhn.

Feh. A breeze typical of Shanghai.

Flakt. A breeze typical of Sweden.

Flauwewind. A breeze typical of Holland.

Föhn (or *foehn*). The general name given to all winds descending from the tops of mountains and moving down along their sides. This air, usually warm and dry, is compressed, and its temperature rises. A distinction is made between north and south föhns. Typical of the northern side of the Alps, the south föhn is strong and gusty, bringing warm, dry air that melts the snow and evaporates any clouds on the mountain top. The north föhn blows over the southern side of the Alps and has the same characteristics as the south föhn, although it is not as strong or as warm. Many winds of different names throughout the world have characteristics similar to the föhn and can be included in the same category.

Frisk wind. A high gale wind typical of Sweden.

Fuga. A wind typical of Crimea.

Fung chiao hsueh. A northeasterly wind associated with storm activity; typical of China.

Gale. A wind measuring 32 to 63 mph on the Beaufort scale.*

Galerna. A squally wind in the Bay of Biscay.

Gallego. A true north wind typical of Spain.

* Scale, invented by a British admiral, Sir Francis Beaufort (1774–1857), which indicates the force of the wind.

Gallicus. A northeasterly wind of the ancient world (Vivutrius).

Garvi. A modified sirocco of the Aegean.

General wind. Wind that stretches for thousands of miles across the surface of the earth.

Geostrophic wind. Wind characterized by a state of equilibrium between pressure, gravity, and the Coriolis force.

Ghibli. A hot, dry wind typical of Libya, which may acquire föhn characteristics when descending a mountain slope. Its temperature will sometimes increase well over 100°F due to air compression.

Greenland double föhn. A strong, warm, seaward wind blowing over the east and west coasts of Greenland in the winter; originating in the high, ice-covered plateaus of the country's interior.

Gregale. A true north wind typical of Spain.

Gust. The sudden increase in a wind's speed, lasting only a few seconds.

Haboob. A desert wind of Sudan.

Harmattan. Dry, desert wind typical of Algeria and Morocco.

Hawa janubi. A south wind of Arabia.

Hawa shimali. A true north wind of Arabia.

Helm. A northeasterly wind blowing over the northern and central parts of England. Usually lasting over a week, it is quite strong, although not as forceful as the French mistral.

Hokuto no kaze. A northeasterly wind associated with storm activity; typical of Japan.

Hurricane (or *tropical cyclone*). A system of spiraling winds converging with increasing speed toward a center where they rise vertically around an area of relative calm. Spreading over an area between 50 and 600 miles in diameter, the hurricane travels over the ocean at speeds from 10 to 25 mph while tangential wind speed varies from 40 to 200 mph. A hurricane is formed over the ocean by the rising of a hot, humid column of air that rotates in a counterclockwise direction in the

Northern Hemisphere. As the warm air rises and cools, the vapor is condensed into rain, and the latent heat that is released is the energy source that feeds the hurricane system. Associated with the storm is a tide from 10 to 25 ft high that is produced by the low pressure of the air above the ocean waters. Usually lasting from 8 to 12 days, hurricanes occur at the end of summer.

Imbat. A cool breeze on the coast of northern Africa.

Indian summer monsoon. A southwesterly wind blowing in summer from the Indian Ocean, where it absorbs moisture and brings beneficial rain over central Asia. Asian agriculture is heavily dependent on the rain brought by this wind.

Indian winter monsoon. A dry, northeasterly wind blowing over India in the winter.

Inverna. A southerly afternoon breeze over Lake Maggiore in Italy.

I tien tien fung. A light breeze of China.

Kadja. An ascending mountain wind of Bali.

Kai. A balmy southerly wind in China.

Kapalilua. A wind of Hawaii.

Katabatic. General name for a wind that follows an incline. It includes the föhn and the gravity winds. The former is characterized by the fact that the wind descending a mountain slope is warmer than the air in the valley, and the latter are characterized by the fact that the descending wind is colder than the air in the valley.

Kaus. A wind of the Persian Gulf.

Kawaihae. A squally wind in Hawaii.

Kessava. A cold, gusty wind from the south that blows over Hungary and Yugoslavia in the winter. It has the same origin as the bise.

Khamsin. A desert wind of the sirocco family blowing over Egypt and Arabia.

Klod. A wind of Bali.

Kohala. A gale in Hawaii.

Kohilo. A gentle breeze in Hawaii.

Kona. A southerly wind in Hawaii.

Koshava. A stormy, northeasterly wind of Yugoslavia, usually carrying snow.

Krivetz. A cold, gusty, northeasterly wind in Rumania and Hungary.

Landlash. A strong wind of gale force in Scotland.

Leste. A warm, easterly desert wind blowing in January over the Madeira Islands.

Leuconotus. A southeasterly wind of the ancient world (Vitruvius).

Leung. A true north wind along the coast of China.

Levante. A sea breeze of the Mediterranean blowing from the east over the Balearic Islands.

Leveche. A wind of Spain that is the Spanish equivalent of the sirocco.

Libeccio. A strong, southwesterly wind typical of Italy.

Libonotus. A southwesterly wind of the ancient world (Vitruvius).

Line squall. A squall typical of the United States.

Local winds. Small-scale phenomena due to local orographic conditions.

Maestral. A cold wind of gale force blowing from the mountains behind the city of Genoa into the Gulf of Genoa.

Maloja. A föhn wind blowing through the Maloja Pass in Switzerland.

Maoi fung. A trade wind of China.

Mauritius hurricane. A hurricane typical of the Indian Ocean.

Meltemia. A persistent summer wind lasing for several months over Greece and the Middle East.

Mistral. A violent, cold, dry, northerly wind typical of France, particularly in Marseilles, Avignon Province, and the Gulf of Lyons. In winter and spring, it usually blows steadily for weeks at a time, and Marseilles is affected by this wind for an average of 100 days per year. Navigation in the Gulf of Lyons is made very hazardous by the mistral, and when channeled in the Rhone Valley, it can reach speeds of up to 90 mph.

Moncao. A trade wind of Portugal.

Monsoon. A seasonal wind, named from the Arabic *mansin* ("season"). In the summer, it blows in one direction; in the winter, the direction is reversed. The difference in temperature between land and water is the main cause for this wind, which resembles the breezes on a larger scale.

Mountain wind (also *downslope winds* or *drainage winds*). Descending mountain winds that form the föhn family, which includes the Swiss and Austrian föhns, the chinook of the Rocky Mountains, the zonda of Argentina, the puelche of the Andes, the Canterbury northwester of New Zealand, and others. As the wind rises over mountain slopes, it expands, cools, and loses moisture through condensation. As the wind descends the other side of the mountain, it is dry and becomes hot by compression.

Myatel. A stormy, northeasterly wind of northern Russia.

Naalehu. A desert wind blowing from the interior of Hawaii.

Naf hat. A squally wind typical of Arabia.

Narai. A true north wind of Japan.

Nasim. A breeze typical of Arabia.

Nemere. A cold, northeasterly wind combined with blizzards. Occurring on the Transylvanian Alps, it has its origin with the bise wind.

Nor'easter. A cold, violent wind of the same family as the blizzard. It is typical of New England.

North American monsoon. A monsoon of North America.

Norther. A wind generated by a low pressure in Texas or in the Gulf of Mexico. Pulling cold polar air from the north, the norther lasts for about one day.

Om. A squally wind typical of Canton, China.

Ora. A breeze on Lake Garda in Italy.

Orkan. A strong wind of gale force in Norway.

Ornithiae. A southeasterly wind of the ancient world (Vitruvius).

Pampero. A violent, northerly wind typical of the Argentine pampas.

Papagayos. A mountain descending wind typical of Costa Rica.

Pei fung. A true north wind in China.

Ponente. A western breeze of Italy, typical of Rome.

Puelche. A descending mountain wind of the Andes, belonging to the föhn family.

Purga. A very violent, stormy, snow-bearing wind typical of Siberia. It may blow for up to a week.

Reffoli. A gusty wind of up to 100 mph in the Adriatic Sea. It is associated with the bora.

Reppu. A very violent wind in Japan.

Reshabar. A descending wind in the Caucasus Mountains.

Samiel. A desert wind, similar to the simoom in Turkey.

Samum. A desert wind of Egypt, belonging to the sirocco family.

Santa Ana. A dry, hot, northerly or northwesterly descending wind, blowing into the Los Angeles basin from the Mojave Desert in southern California. It is the cause of the very warm winters in this area.

Seistan. A northwesterly wind in eastern Iran. Often called the *land wind* or *wind of 120 days.* It blows from May through September with speeds from 70 to 120 mph.

Septentrio. A northerly wind of the ancient world (Vitruvius).

Shamal. A summer wind, typical of southern Iraq and Iran, that blows dust. Its origin is comon to that of monsoons, and its speed is less than 30 mph.

Shi ling fung. A trade wind in China.

Siffanto. A warm wind in Italy.

Simoom. A desert wind of north Africa, Arabia, and Syria.

Sirocco (or *scirocco*). An extremely hot desert wind from Africa, carrying unpleasant and oppressive weather.

Sno (or *Elvegast*). A cold, dry, offshore wind blowing in winter toward the west coast of Norway from the interior.

Solano. A summer wind typical of the southeast coast of Spain. It is similar to the sirocco. Dry and dusty at times, but occasionally it is humid and can carry rain.

Southern buster (or *southeast buster*). A cold, strong, southerly wind of Australia, associated with line squalls, cold fronts, and bad weather.

Soyo Kaze. A breeze typical of Japan.

Squall. A wind arising suddenly from a calm and lasting for only a few minutes.

Steppenwind. A northeasterly wind blowing over the steppes in Germany.

Subvesperus. A southwesterly wind of the ancient world (Vitruvius).

Suestada. A stormy, northeasterly wind typical of the Rio de la Plata, an estuary between Uruguay and Argentina.

Sumatra. A squally, westerly wind in the Indies.

Supernas. A northeasterly wind of the ancient world (Vitruvius).

Sveszhest. A breeze typical of Russia.

Sz. A breeze blowing in China in the fall.

Tapayagua. A squally wind of Central America.

Tegenwind. A stormy, northeasterly wind in Holland.

Tehuantepecer. A cold, northerly wind, similar to the French mistral, that carries cold air to the coast of Mexico.

Thal wind. A breeze typical of Germany.

Thracias. A northwesterly wind of the ancient world (Vitruvius).

Tivano. A breeze on Lake Como in Italy.

Tornado. The most violent storm in nature, consisting of spiraling winds converging with increasing speed toward a central, vertical axis. Winds rotate and move forward (translate) at the same time; latest estimates on maximum wind speeds combining tangential and translation velocities indicate speeds less than 200 mph. The width of the tornado's destructive path varies from a few feet up to 2 miles, with an average width of 400 yards. Average length is about 15 miles, and forward velocity varies from 0 to over 60 mph. Tornadoes occur everywhere in the world, but mostly in the United States, where they also tend to be the most violent. Their frequency is highest during the month of May.

Trade wind. A prevailing wind blowing over the ocean in a belt extending around the world from 30° north latitude to 30° south latitude. In the Northern Hemisphere, this wind blows in a northeasterly direction, while in the Southern Hemisphere, its direction is southeasterly.

Tramontana. A cold, true north wind in Italy.

Tropical cyclone of the Arabian Sea. Comparable to a hurricane.

Tropical cyclone on the Bay of Bengal. Comparable to a hurricane.

Tung shang fung. A trade wind of China.

Typhoon. A hurricane on the Pacific Ocean.

Vendaveles. A typical Spanish wind coming from Morocco.

Ventania. A wind of high gale force, typical of Portugal.

Vento de baixo. A sea breeze of Portugal.

Vind-blaer. A breeze typical of Iceland.

Vinds-gnyr. A squally wind in Iceland.

Virazon. A very strong wind typical of Valparaiso, Chile.

Viuga. A stormy, northeasterly wind typical of the south of Russia.

Volturnus. A southeasterly wind of the ancient world (Vitruvius).

Vyeterok. A breeze typical of Russia.

Waff. A breeze typical of Scotland.

Warm braw. A descending mountain wind of the Schouten Island similar to the föhn.

Waterspout. A phenomenon similar to a tornado only less violent and destructive. Forming over water, it consists of winds spiraling around a freshwater core due to vapor condensation that rises into the funnel-shaped mother cloud. When waterspouts form over the ocean, only the first few feet of the spout contains salt water sucked from the sea's surface. Spouts can occur on lakes as well and can travel on land for short periods of time.

Whirly. A violent whirlwind of Antarctica that occurs during the equinoxes, carrying aloft ice and water.

Williwaw. A squally wind of short duration occurring in Alaska and in the Strait of Magellan.

Willy-willy. A very strong wind of Australia, comparable to a hurricane.

Wisper. A squally wind of the Rhine.

Yamo oroshi. A wind of the föhn family, typical of Japan.

Yuh. A wind of high gale force, typical of Shanghai.

Zephyr. Ancient name for a mild westerly breeze in the intermediate latitudes, equivalent to the ponente of Italy.

Zonda. A desert wind typical of the Argentine pampas and of a föhn in the Andes of Argentina.

Figure 1-14 shows an example of wind classification from Roman times.

Table 1-1 lists the names of the winds of the classic world. Although already known from Latin and Greek literature, they were also mentioned and described in terms of their prevailing direction by Vitruvius in *De Architectura.* The 24 winds listed on the table cover the total 360° spectrum of directions with equal sectors of 15° each.

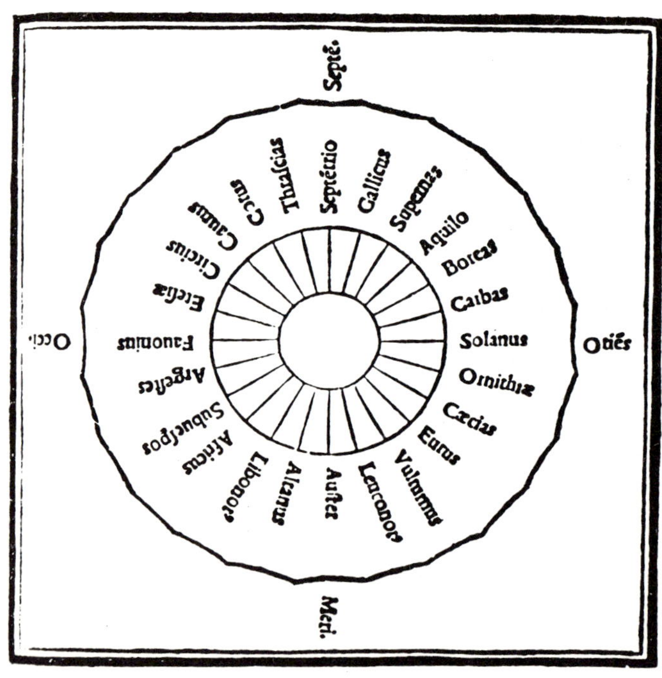

Figure 1-14. An early example of wind classification from Roman times: diagram of the winds. An illustration by Fra Ciocondo (Giovanni da Verona, 1433–1515) to the 1511 Venetian edition of Vitruvius' *De Architectura.*

Table 1-1. Classification of the winds by Vitruvius.

DIRECTION	NAME OF WIND
North	Thracias Septentrio Gallicus
Northeast	Supernas Aquilo Boreas
East	Garbas Solanus Ornithiae
Southeast	Eurocircias Eurus Volturnus
South	Leuconotus Auster Altanus
Southwest	Libonotus Africus Subvesperus
West	Argestes Favonius Etesiae
Northwest	Circias Caurus Corus

REFERENCES

De la Rue, E. A. 1955. *Man and the Winds.* New York: Philosophical Library.

Dowson, J. 1968. *A Classical Dictionary of Hindu Mythology and Religion, Geography, History, and Literature.* London: Routledge & Kegan Paul Ltd.

Forrester, F. H. 1959. *1001 Questions Answered about the Weather.* New York: Dodd, Mead & Co.

Murchie, G. 1954. *Song of the Sky.* Boston: Houghton Mifflin Co.

2
Destructive Winds And Their Effects

ATMOSPHERIC CIRCULATION

Air Motion

Since the sixth century B.C., wind has been defined as air in motion. The energy for wind is provided by the sun. The mechanics of air motion is based on differences in temperature between various points in the atmosphere. In fact, warmer air has less density and thus rises, whereas cold air falls with respect to the adjacent air. In addition to such vertical motion, there is a horizontal one that is due to the movement of cold, dense air toward zones of lower pressure and warmer air.

How does the atmosphere attain different temperatures from one point to another? There are two basic mechanisms. As the sun's radiation enters the atmosphere, part of it is absorbed in different degrees by the air and water vapor and part of it either is reradiated or is absorbed by the earth's surface. Different types of earth surfaces, such as bodies of water, land formations, types of soil, and species of vegetation, reflect heat in different amounts, producing unequal temperatures in the air above. For instance, we can imagine vertical air currents that have different speeds as we go from one type of terrain to another. Under such conditions, a glider flying over these vertical currents would perceive varying lifting forces, while a small plane would experience a series of bumps (Figure 2-1).

The Coriolis Force

The Coriolis force (named after Gaspard Gustav de Coriolis, who discovered it) is a force due to the rotation of the earth that makes the path of moving bodies, and therefore of the winds, deviate toward the right in the Northern Hemisphere and toward the left in the Southern Hemisphere. As the wind speed increases, so does the Coriolis force; it is zero at the equator and maximum at the poles, and it increases proportionally to the sine of the latitude (Figure 2-2).

Figure 2-1. Solar energy striking the earth's surface is reradiated to the sky at a rate that varies according to the type of terrain. Air over hot soil surfaces is hot and rises at a higher speed than does air over colder terrain surfaces. Thus, a small airplane would experience a bumpy flight while passing over various types of earth's surfaces.

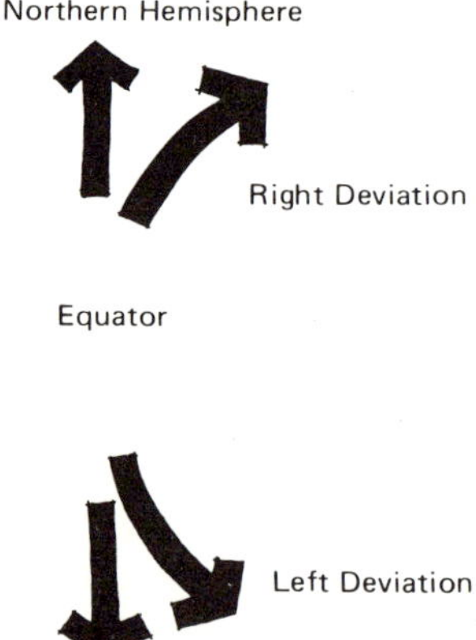

Figure 2-2. Coriolis force.

Geostrophic Winds

Geostrophic winds are controlled by the simultaneous action of pressure, gravity, and the Coriolis force. The speed of the wind is given by the following expression:

$$V = \frac{dp/dn}{2\, w \rho \sin \phi}$$

where

V = velocity of the wind above the gradient height
$\frac{dp}{dn}$ = horizontal pressure gradient along a horizontal line n
ρ = average air density
w = angular velocity of the earth
ϕ = angle of latitude

Buys Ballot's Law

Buys Ballot's law states that in the Northern Hemisphere an observer with his or her back to the geostrophic wind will have low pressure to the left and high pressure to the right. The reverse is true for the Southern Hemisphere (Figure 2-3).

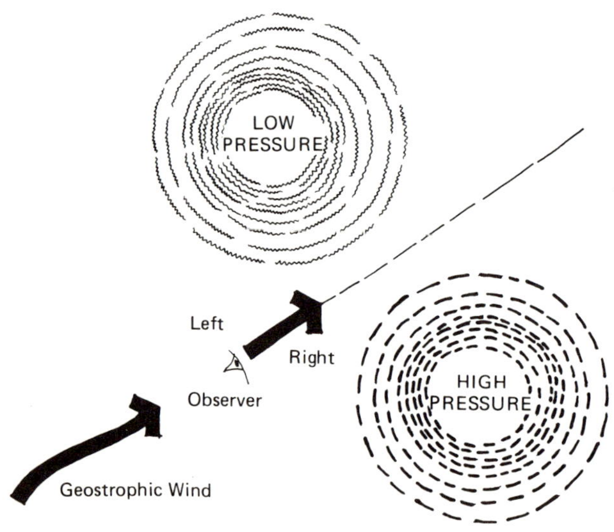

Figure 2-3. Buys Ballot's law.

Theoretical Global Circulation

In theory, the overall circulation of the atmosphere could be visualized as follows: hot air over the tropics rises to the upper atmosphere and moves toward the North and South Poles, deflecting under the Coriolis force; at the poles, the air cools and descends, then moves from the poles toward the tropics with a deflected path because of the Coriolis force. Because of seasonal changes and different heat radiations from continents and ocean surfaces, such a simplistic circulation system does not exist as such but becomes more complex. (See Figure 2-4.)

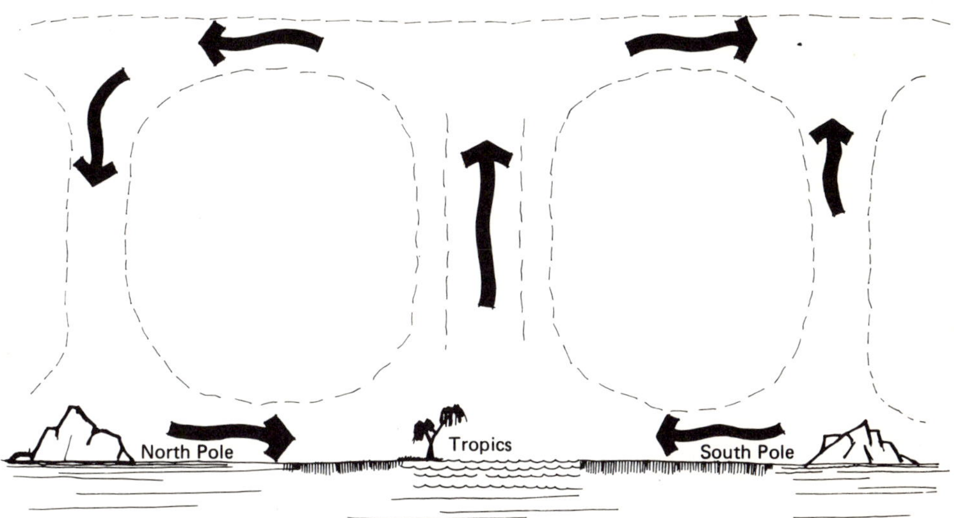

Figure 2-4. Theoretical global circulation.

Global Circulation of Atmosphere

Hot air over the tropics rises to the upper atmosphere, forming jet streams that move toward the poles at speeds over 200 mph (Figure 2-5). Some of these winds descend on the subtropical areas producing anticyclonic systems of high pressure. This air descends farther down to the earth's surface and proceeds partially to the north and south. In the Northern Hemisphere, the northbound air is deflected to the east and forms the middle-latitude westerlies, whereas the southbound air is deflected to the west and forms the northeasterly trade winds. (See Figure 2-6.)

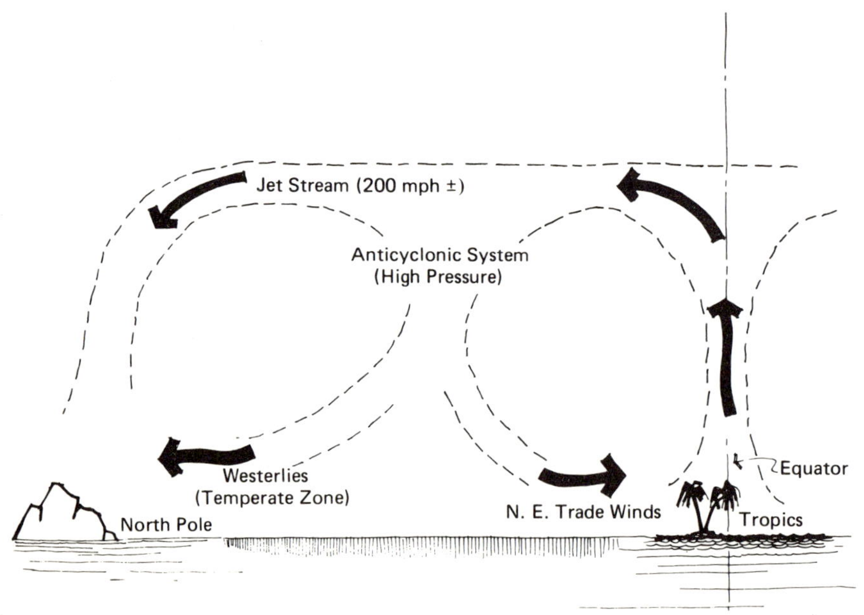

Figure 2-5. Global circulation.

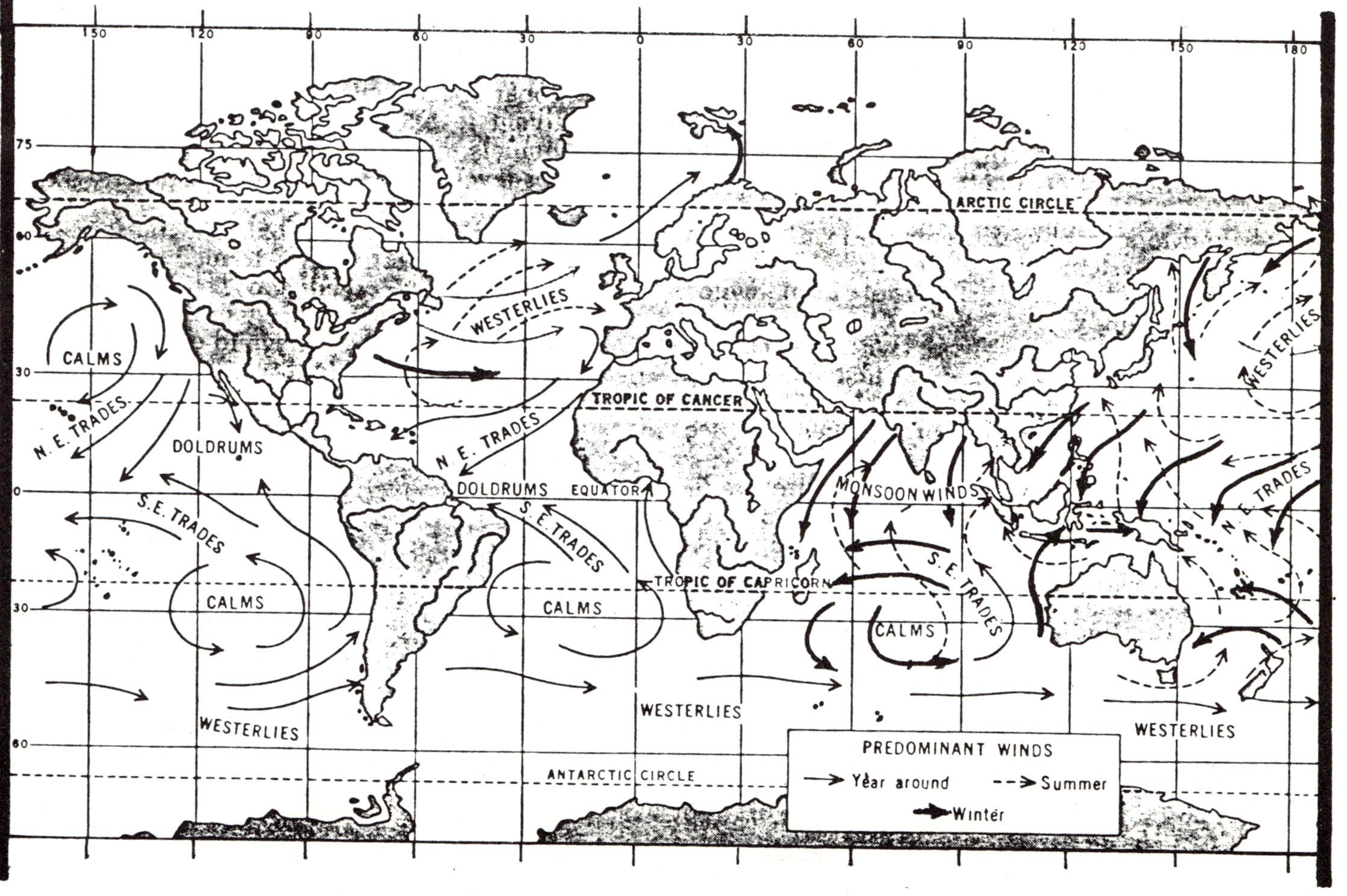

Figure 2-6. Predominant winds of the world.

Gradient Wind

Gradient wind is wind at a certain altitude where the friction from the ground is no longer effective.

Wind Velocity

Wind velocity is a vectorial quantity that includes the magnitude, or speed, and the direction. Wind velocity varies from instant to instant in speed and direction.

Wind Speed

The speed of wind is measured as:

Peak wind speed. Maximum instantancous value.

Fastest-mile wind. The average speed of a column of air 1 mi long that passes over a given point. This is the most common measurement used. (See Figure 2-7.)

Extreme fastest-mile wind. The highest fastest mile wind speed recorded in a certain time period (1, 25, 50, 100 yr). These are the values used in designing structures.

Gradient wind speed. The speed of the wind above the gradient height where the surface friction is no longer effective. Past the gradient speed, the wind velocity stays constant, unless causes other than friction intervene. (See Figure 2-8.)

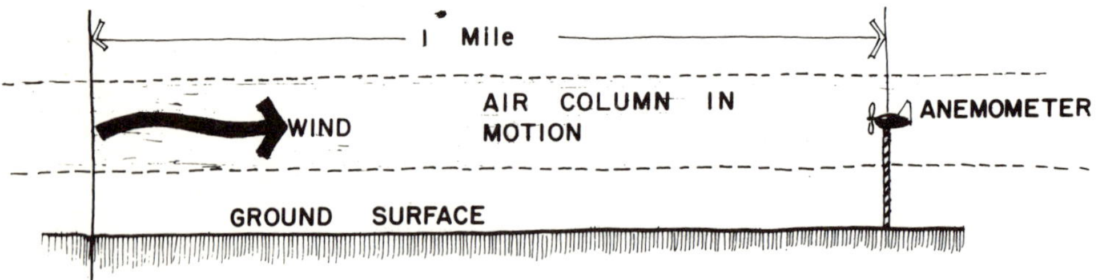

Figure 2-7. Fastest-mile wind speed.

Gradient Height (Z_g)

The gradient height is that height above which the friction from the earth's surface no longer affects the wind speed. Such height varies according to the surface conditions as follows:

Open country: $Z_g = 900$ ft
Suburban areas: $Z_g = 1200$ ft
Metropolitan: $Z_g = 1500$ ft

Wind Speed below Gradient Height

Below the gradient height, the wind speed is affected by the earth's friction and is calculated from the following:

$$V_Z = V_G \left(\frac{Z}{Z_G}\right)^{1/\alpha}$$

where

V_Z = wind speed, mph
Z = height above ground, ft
Z_G = gradient height, ft
$1/\alpha$ = exponent dependent on the terrain (see values below)

Open country (typical farmland with few trees): $1/\alpha = 1/7$
Suburban areas (with buildings from 30 to 50 ft high or wooded areas): $1/\alpha = 1/4.5$
Metropolitan areas (urban centers with tall buildings or terrain with pronounced orography): $1/\alpha = 1/3$

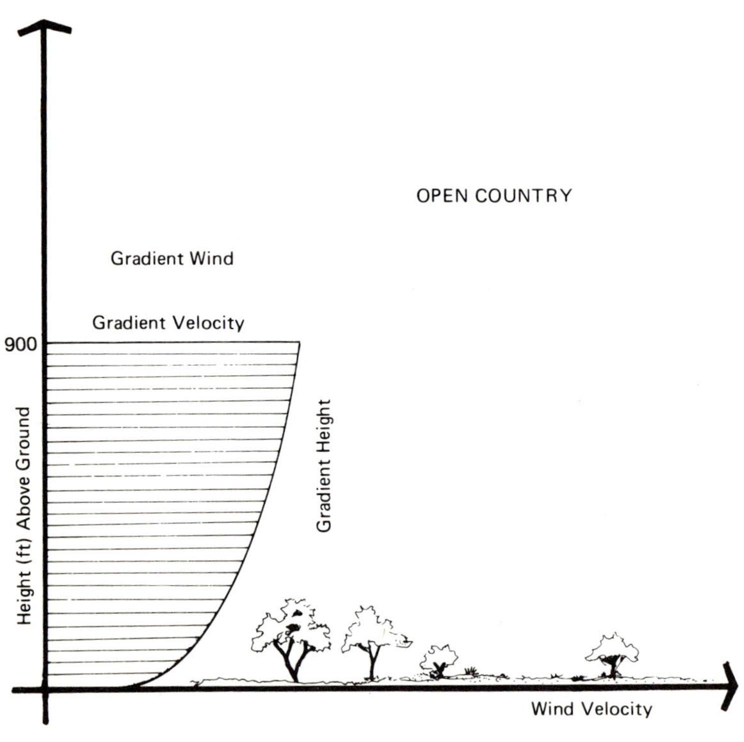

Figure 2-8. Reduction of wind velocity due to ground friction extends vertically through viscosity and turbulence up to the gradient height, where the wind is no longer affected by the ground. The conditions considered here are: *a,* open country; *b,* suburban; and *c,* city.

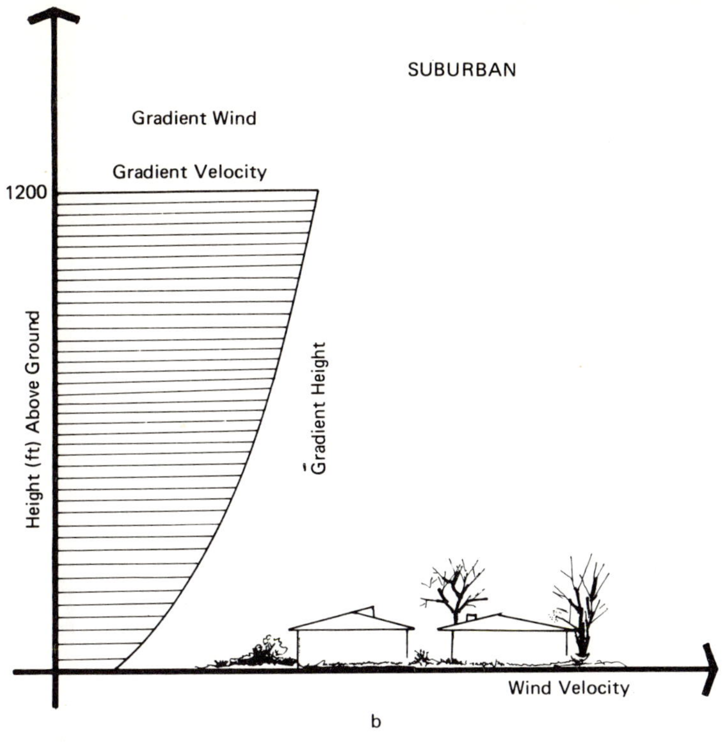

b

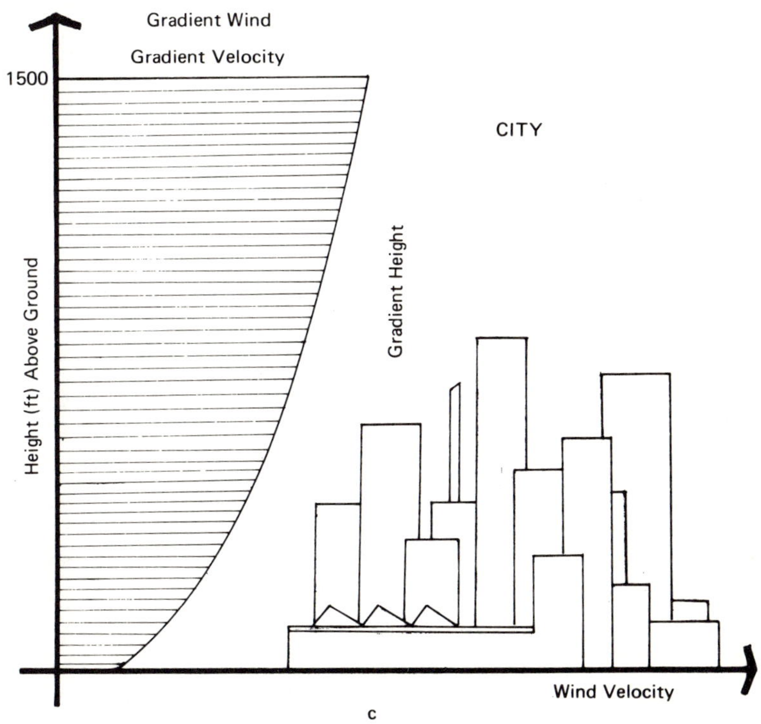

c

Figure 2-8. (Continued)

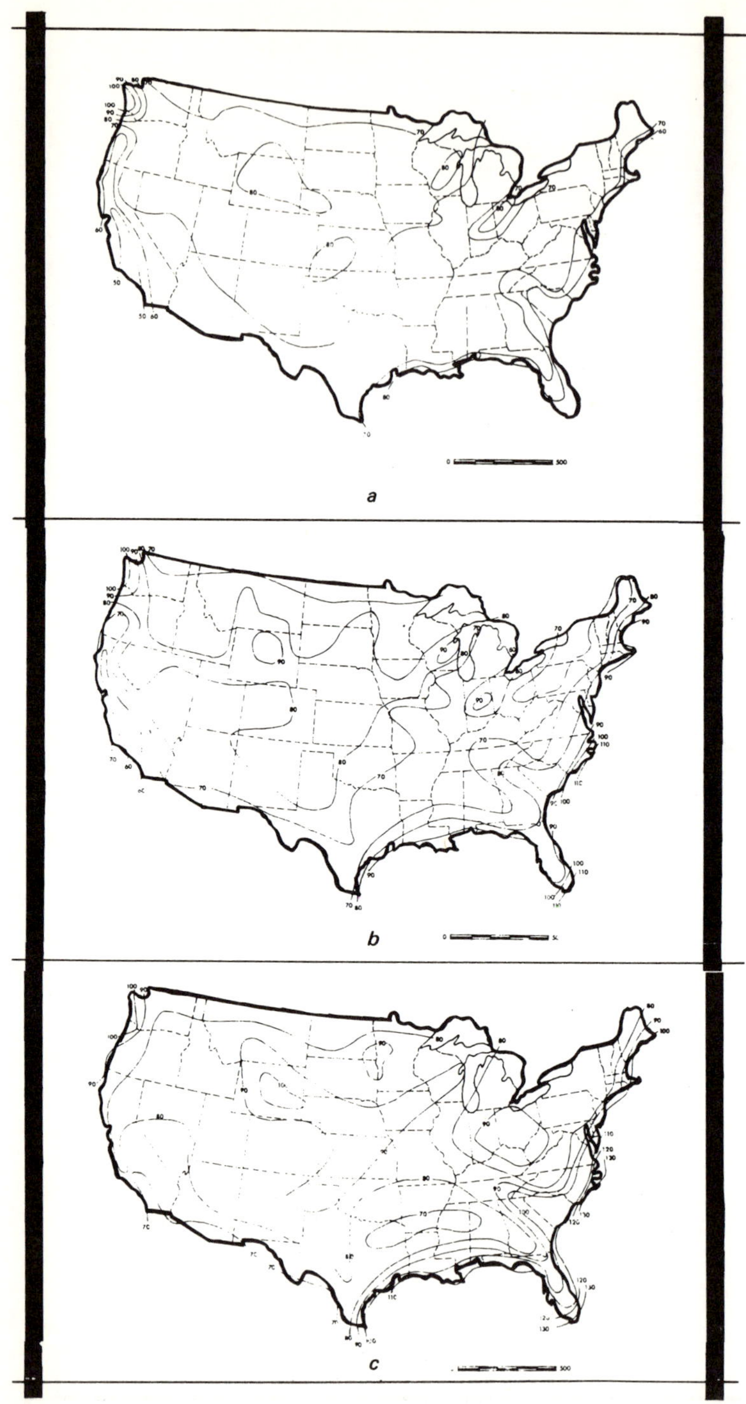

Figure 2-9. *a,* Basic wind speed in miles per hour—25-yr mean recurrence interval. Annual extreme fastest-mile speed 30 ft above ground. *b,* Basic wind speed in miles per hour—50-yr mean recurrence interval. Annual extreme fastest-mile speed 30 ft above ground. *c,* Basic wind speed in miles per hour—100-yr mean recurrence interval. Annual extreme fastest-mile speed 30 ft above ground.

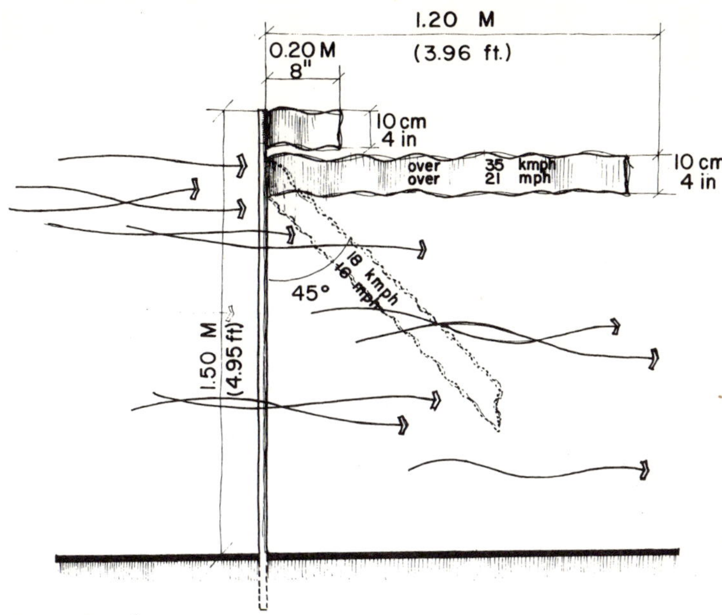

Figure 2-10. A light cloth flag and a pole of certain specified dimensions are sufficient to measure approximate wind speeds up to 16 mph according to the position of the flag.

Wind-Speed Maps

Wind-speed maps are available for most parts of the world. In the United States, they are prepared based on observations recorded by numerous meteorological stations. Annual extreme fastest-mile speeds 30 ft above ground are indicated on charts based on mean recurrence intervals of 25, 50, and 100 yr (Figure 2-9a–c).

Perception of Wind Speed

With the advent of the anemometer and its evolution, wind speed could be measured precisely, eliminating the guesswork of the past. Before the advent of scientific instrumentation, an accepted methodology—still in use today—was the one proposed by Sir Francis Beaufort in 1805. Based on observations of the effect of wind on a sailing ship, the Beaufort scale estimates the wind speed in terms of visual signals commonly observed at sea; these are based on wave height (see Table 2-1). On land, the modified Beaufort scale considers such signs as the drifting of rising smoke and the swaying of trees. On the same basis, one could make one's own wind scale by relating personal experiences to established wind speeds. Practical methods are at times suggested for this purpose. For instance, in hang-gliding, it is vitally important to have some practical means to measure wind speeds before taking off. A little flag, 1.20 m long, on a pole 1.50 m high (see Figure 2-10), can be used to measure speeds up to 16 mph according to the position of the flag in response to the wind.

For the reader interested in the technology of the numerous types of anemometers, an abundant amount of literature on the subject is available. For the reader involved in environmental design, the effect of a certain wind velocity is much more important than the knowledge of the velocity per se. In other words, once we know the existence of a wind of a given speed, how can we associate value with the reality of its effects?

Table 2-1. The Beaufort scale.

BEAUFORT NUMBER	SEAMAN'S DESCRIPTION OF WIND	TERMS USED BY U.S. WEATHER BUREAU	VELOCITY M.P.H.	VELOCITY KNOTS	ESTIMATING VELOCITIES ON LAND	ESTIMATING VELOCITIES ON SEA	PROBABLE MEAN HEIGHT OF WAVES IN FEET	DESCRIPTION OF SEA
0	Calm	Calm	Less than 1	Less than 1	Smoke rises vertically	Sea like a mirror		Calm (glassy)
1	Light air	Light	1–3	1–3	Smoke drifts; wind vanes unmoved.	Ripples with the appearance of scales are formed but without foam crests.	½	Rippled
2	Light breeze		4–7	4–6	Wind felt on face; leaves rustle; ordinary vane moved by wind.	Small wavelets, still short but more pronounced; crests have a glassy appearance and do not break.	1	Smooth
3	Gentle breeze	Gentle	8–12	7–10	Leaves and small twigs in constant motion; wind extends light flag.	Large wavelets. Crests begin to break. Foam of glassy appearance. Perhaps scattered white caps.	2½	
4	Moderate breeze	Moderate	13–18	11–16	Raises dust and loose paper; small branches are moved.	Small waves, becoming longer; fairly frequent white caps.	5	Slight
5	Fresh breeze	Fresh	19–24	17–21	Small trees in leaf begin to sway; crested wavelets form on inland water.	Moderate waves, taking a more pronounced long form; many white caps are formed. (Chance of some spray.)	10	Moderate
6	Strong breeze	Strong	25–31	22–27	Large branches in motion; whistling heard in telegraph wires; umbrellas used with difficulty.	Large waves begin to form; the white foam crests are more extensive everywhere. (Probably some spray.)	15	Rough
7	Moderate gale		32–38	28–33	Whole trees in motion; inconvenience felt in walking against the wind.	Sea heaps up and white foam from breaking waves begins to be blown in streaks along the direction of the wind.	20	Very rough

					generally impedes progress.	greater length; edges of crests break into spindrift. The foam is blown in well-marked streaks along the direction of the wind.		
9	Strong gale		47–54	41–47	Slight structural damage occurs	High waves. Dense streaks of foam along the direction of the wind. Sea begins to roll. Spray may affect visibility.	30	
10	Whole gale	Whole gale	55–63	48–55	Trees uprooted; considerable structural damage occurs.	Very high waves with long, overhanging crests. The resulting foam, in great patches, is blown in dense white streaks along the direction of the wind. On the whole, the surface of the sea takes a white appearance. The rolling of the sea becomes heavy and shocklike. Visibility is affected.	35	Very high
11	Storm		64–75	56–65		Exceptionally high waves. (Small and medium-sized ships might for a long time be lost to view behind the waves.) The sea is completely covered with long white patches of foam lying along the direction of the wind. Everywhere edges of the wave crests are blown into froth. Visibility affected.	40	
12	Hurricane	Hurricane	Above 75	Above 65		The air is filled with foam and spray. Sea completely white with driving spray; visibility very seriously affected.	45 or more	Phenomenal

We should probably relate our own personal experiences to certain wind-speed values. For instance, a 5-mph breeze can carry a kite aloft without breaking the string. Putting your arm out of a car window on a still day gives you an idea of the wind force—it is indicated by the speedometer. The correlation of these experiences with the corresponding air speed will make us aware of the wind speed in concrete terms. To a structural engineer, a certain wind speed implies a given amount of structural damage, based on the predictable strength of the structure.

HURRICANES

Cyclones and Hurricanes

Cyclones are systems of winds rotating counterclockwise in the Northern Hemisphere with a diameter varying from 50 to 1000 mi and a barometric pressure that diminishes toward the center (see Figure 2-11). Other features of cyclones include the vertical ascent of the winds around the center, the formation of a cloud above the center, and condensation with precipitation. They can be so mild that they can pass over inhabited areas without being detected by people in general, although they would be registered by meteorological stations. On the other hand, they can reach very high intensities, such as in the case of hurricanes, which are the most violent of tropical cyclones. Tornadoes are somewhat similar to hurricanes and even more violent. At times, they are produced by hurricanes. However, they are not part of the cyclone family. Furthermore, tornadoes have a diameter much smaller than 50 mi, which is the minimum of cyclones.

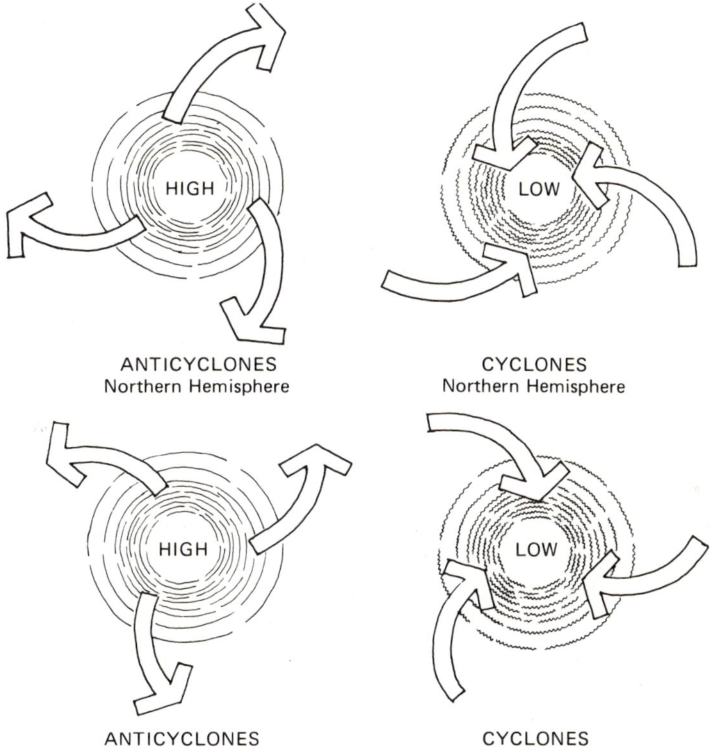

Figure 2-11. Centrifugal anticyclone circulation: clockwise in the Northern Hemisphere and counterclockwise in the Southern Hemisphere. Centripetal cyclone circulation: counterclockwise in the Northern Hemisphere, clockwise in the Southern Hemisphere.

Hurricanes are tropical cyclones of a particularly high intensity. As cyclones, they consist of rotating masses of air that converge spirally toward the center where the atmospheric pressure is low. The winds gradually increase in speed as they approach the center; at the same time, the barometric pressure is dropping. At the periphery of the storm, the winds are relatively horizontal, but they rise gradually as they approach the center that they never reach. At a certain distance, the vertical component of their velocity is at its peak. As the winds rise, they form a cloud that produces heavy torrential rains. The maximum wind speed occurs right at the edge of a central zone called the *eye,* where there is sunshine and relative calm and the fury of the winds ceases.

Most hurricanes are formed in the southwest part of the North Atlantic Ocean including the Caribbean Sea and the Gulf of Mexico. With few exceptions, most hurricanes occur from June through October. The average life-span of a hurricane is about 9 days. Hurricanes forming in August, however, have an average life of 12 days. (See Figures 2-12 and 2-13.)

The annual frequency of tropical cyclones has been slowly increasing since the 1930s. At present, there are from 8 to 10 annually. This increase can be partly attributed to better detection systems and to a gradual warming of the atmosphere. Not all tropical cyclones attain hurricane strength, but approximately 60% do.

Other storms similar to the North Atlantic hurricanes are:

- typhoon in the Pacific
- tropical cyclones of the Bay of Bengal
- tropical cyclones of the Arabic Sea
- willy-willy of Australia

For the average hurricane, the diameter of the area where winds reach gale force (40 mph) is from 350 to 400 mi, but the diameter where winds reach hurricane force (75 mph) is 100 mi. The diameters of some hurricanes have been estimated to be up to 600 mi, and of some typhoons up to 1000 mi.

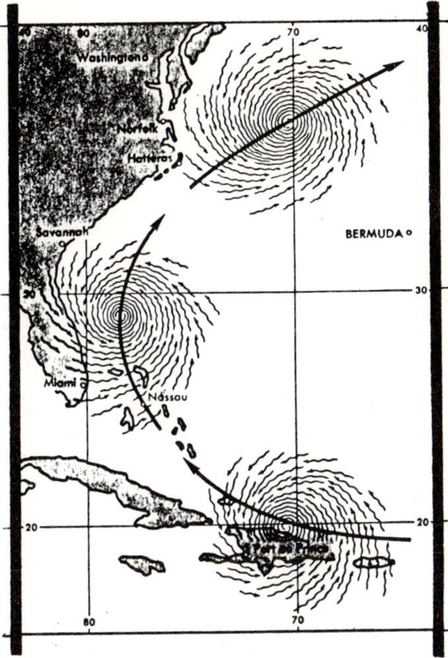

Figure 2-12. Track and wind system of a typical hurricane. Winds are depicted here as they are momentarily directed about the center of a progressing storm. (Source: United States Weather Bureau)

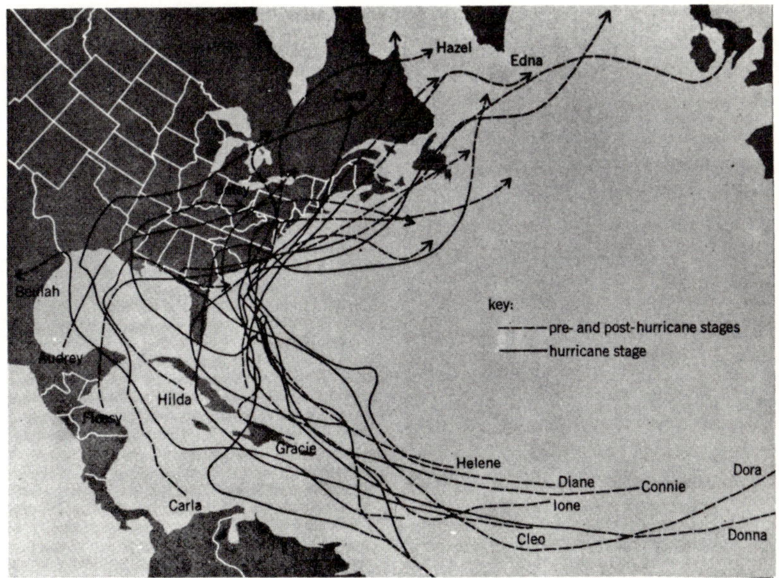

Figure 2-13. Paths of hurricanes that hit the Atlantic coast of the United States from 1954 to 1967. (Courtesy National Oceanic and Atmospheric Administration)

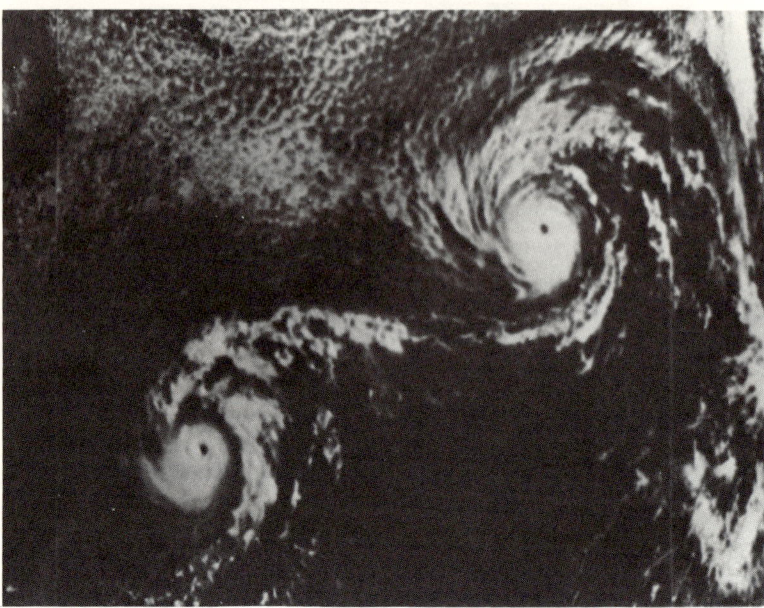

Figure 2-14. Twin Atlantic hurricanes photographed from NOAA-3 artificial satellite on August 24, 1974. Clouds over the eye of each hurricane are higher than the surrounding clouds. (Courtesy National Oceanic and Atmospheric Administration)

The Eye

The central part of a hurricane, the eye, is the area of relative calm. Winds there are not higher than 15 mph, although absolute calm does not exist. In this area, it does not rain, the cloud deck is very thin, and occasionally the sun shines through. Birds are sometimes driven into the eye by the fury of the peripheral winds. The eye of a hurricane is often penetrated by observers, such as ships at sea, unlike the eye of a tornado where penetration is impossible. The low pressure therein, however, gives one a feeling of suffocation, and many observers have described the eye as oppressive. The average diameter is about 14 mi, and the smallest diameter is about 4 mi. The size of the eye tends to increase as hurricanes pass over land. The shape of the eye is usually circular; when the hurricane passes over land, however, the eye tends to become elliptical. The long axis of the ellipse is in the direction of translation (forward motion) of the storm. Elliptically shaped eyes have sometimes reached 90 mi in diameter. As the storm progresses into its life cycle, the size of the eye also varies. The temperature in the eye is much higher than in the outer area, especially at higher altitudes. (See Figure 2-14.)

Winds

The highest wind speed ever recorded with an instrument is 188 mph with gusts of 229 mph. This speed, measured at Mount Washington Observatory in New Hampshire on April 12, 1934, was not related to a hurricane. Such speed seems to be partially caused by the local configuration of the terrain. Other measurements of record wind speeds during hurricanes are:

Speed	*Place*	*Date*
186 mph	Mount Washington, New Hampshire	January 11, 1878
186 mph	Mount Washington, New Hampshire	September 21, 1938
175 mph	Chetumal, Mexico	September 27–28, 1955
163 mph	Havana, Cuba	October 18, 1944
155 mph	Hillsboro Lighthouse, Florida	September 17, 1947
150 mph	San Juan, Puerto Rico	September 13, 1928
132 mph	Jupiter Lighthouse, Florida	August 26, 1949
128 mph	Miami Beach, Florida	September 18, 1926
121 mph	Milton, Massachusetts	September 21, 1938

All of the above measurements were made on land. Wind speeds at sea are appreciably higher because of the reduced surface friction over water.

Precipitation

It is impossible to measure the exact amount of rainfall in a hurricane because the wind does not allow the rain to be collected by the measuring gauge. However, the largest rainfalls are usually associated with hurricanes and tropical cyclones.

The number of inches of rainfall at a given point depends on several factors. The slower the forward motion of the storm, the longer the raining time on that location. If the hurricane is over land, the rate of rainfall increases. Some records of precipitation associated with tropical cyclones, including hurricanes, follow:

Inches	*No. Days*	*Place*	*Date*
96.5	4	Silver Hill, Jamaica	November 1909
88	4	Baguio, Philippines	July 1911
81.5	3	Funkike, Formosa	July 18–20, 1913
63	3	Mount Malloy, Queensland, Australia	Date unknown
47	7	Reunion, Mauritius	February 15–21, 1896
46	–	Baguio, Philippine Islands	July 1911
41	7	La Marie, Tamarind Falls, and L'Etoile, Mauritius	February 15–21, 1896
33.7	5	Crowley, Louisiana	August 6–10, 1940
31.66	5	Abbeville, Louisiana	August 6–10, 1940
29.65	5	Lafayette, Louisiana	August 6–10, 1940
29.60	2	Adjuntas, Puerto Rico	September 13–14, 1928
26	7	Pamplemousse, Mauritius	February 15–21, 1896
23.11	2	Taylor, Texas	September 9–10, 1921
22.22	2	Altapass, North Carolina	July 14–15, 1916
21.4	2	Alexandria, Louisiana	June 15–16, 1886

Hurricane Forward Speed

The average forward speed (translational speed) of hurricanes is about 12 mph.

Hurricane Surge

Hurricane surge is the rapid rise of the water level caused by winds and reduced barometric pressure. For each inch of pressure drop, the water rises approximately 1 ft. Thus, the rise due to the reduction of barometric pressure during a hurricane is usually from 3 to 4 ft. The wave action increases the surge height above these heights, especially near the shoreline where waters are shallow, or in channels, sounds, bays, etc.

Forerunner

A forerunner is a characteristic rise in the water level near the coast, which precedes the actual arrival of the center of a hurricane. This can be explained by the fact that water is actually being transported and accumulated toward the shore. The forerunners extend for long distances over the shores, whereas the surge covers only a small portion of it. The surge is usually located at the right of the storm center where the wind and the forward motion are parallel and additive, thus creating the most severe wind action. The surge adds its action to that of the forerunners in raising the water level.

If the path of the hurricane approaching the shore is at 90° with the coastline, the surge will produce the highest water rise; as the angle diminishes, so will the rise of the water. Waters that are too shallow can reduce the amplification and the surge effects, but both increase when the sea bottom slopes gradually, thus extending the area of shallow waters far into the ocean. Naturally, if the surge coincides with the meteorological high tides, the water level will increase at the turn of the high tide and the surge.

Hurricane Wave

A *hurricane wave,* which is a rare phenomenon, is a very rapid and particularly high surge that comes toward the land. It is usually described as a "wall of water" and its effects are always disastrous. Such a wave was responsible for the deaths of 100,000 people in the Bengal region of India in 1878 during the Backergunger tropical cyclone. On the South Carolina coast, on August 27, 1893, a hurricane wave killed between 1500 and 2000 people. On October 1, 1893, another hurricane wave killed 1800 people near New Orleans.

Wave Action

The dynamic action of the waves produced by hurricanes can cause severe damage. Breaking waves can attain forward speeds of up to 60 mph as they reach land. Buildings and sand dunes can easily be destroyed, not only by the impact of the water but also by the repetitive pounding of heavy debris carried by the waves, such as rocks, trees, and building components (bricks, plaster, etc.).

Hurricane Rating

A rating for hurricanes, similar to the Beaufort scale has been proposed by the National Weather Service. This scale includes five categories, which are defined by wind speed and expected type of damage. As indicated in Table 2-2, the first category is limited to 74 and 95 mph, whereas Category 5 includes wind speeds above 155 mph.

Table 2-2. Hurricane disaster-potential scale (National Weather Service).

CATEGORY	WIND SPEED (MPH)	PROBABLE PROPERTY DAMAGE
5	155	Shrubs and trees blown down; considerable damage to roofs of buildings; all signs down. Very severe and extensive damage to windows and doors. Complete failure of roofs on many residences and industrial buildings. Extensive shattering of glass in windows and doors. Some complete building failures. Complete destruction of mobile homes. Small buildings overturned or blown away.
4	131–155	Shrubs and trees blown down; all signs down. Extensive damage to roofing materials, windows, and doors. Complete failure of roofs on many small residences. Complete destruction of mobile homes.
3	111–130	Foilage torn from trees; large trees blown down. Practically all poorly constructed signs blown down. Some damage to roofing materials of buildings; some window and door damage. Some structural damage to small buildings. Mobile homes destroyed.
2	96–110	Considerable damage to shrubbery and tree foilage; some trees blown down. Major damage to exposed mobile homes. Extensive damage to poorly constructed signs. Some damage to roofing materials of buildings; some window and door damage. No major damage to buildings.
1	74–95	Damage primarily to shrubbery, trees, foilage, and unanchored mobile homes. No real damage to other structures. Some damage to poorly constructed signs.

TORNADOES

Scientific description and analysis of tornado phenomena date back to the latter part of the seventeenth century. In 1689, Francois Lamy published in Paris a work on the physical interpretation of air vortices. Five years later, a posthumous treatise by Geminiano Montanari on the physical and mathematical aspects of tornadoes, was published, describing in particular the storm of July 29, 1686, that struck the territory surrounding Venice in northern Italy. Another early work on tornadoes is the dissertation by Boscovich published in Rome in 1749. A century later, in 1840, Peltier published in Paris his work on tornadoes. Much has been published since these early studies on meteorological and physical aspects of tornadoes, but very little has been written about the forces that produce destructive effects on buildings.

With the approach of each spring in the United States, another so-called tornado season begins, bringing death and damage somewhere in the country. Statistical data collected during a 16-yr period (1953 to 1969) in the United States indicate an average number of 640 tornadoes per year, with a peak of 912 in 1967 and a lower limit of 437 in 1953 (Figure 2-15). Annual averages of losses for this period include $75 million and 125 deaths (Figure 2-16). For many years, scientific information on qualitative and quantitative characteristics of tornadoes was incomplete and inconclusive. Consequently, public opinion (shared also by people in the design profession) concluded that nothing could be done to prevent or alleviate the losses and human suffering.

The present situation, however, is quite changed. Although the problem still persists, scientific knowledge on the subject has increased considerably in the last few years, especially in the fields of meteorology and fluid dynamics. The majority of the research work clearly indicates that forces expected in tornadoes are much lower than previous estimates. In fact, wind velocity estimates, which at one time reached above 300 mph, are now on the order of 200 mph. These new figures put structural design in a completely different perspective where tornado-resistant structures now appear to be feasible, not only from a technical but also from an economic point of view.

Starting in the late 1960s, for the first time in the history of construction, tornado-resistant structures were designed and built to resist tornadic forces on a systematic and scientific approach. These structures included the nuclear power plants of the United States and other countries. It seems feasible to predict that, in the near future, design criteria used by architects and engineers for certain types of other buildings will also include expected forces when the geographic area has a high probability of tornado occurrence.

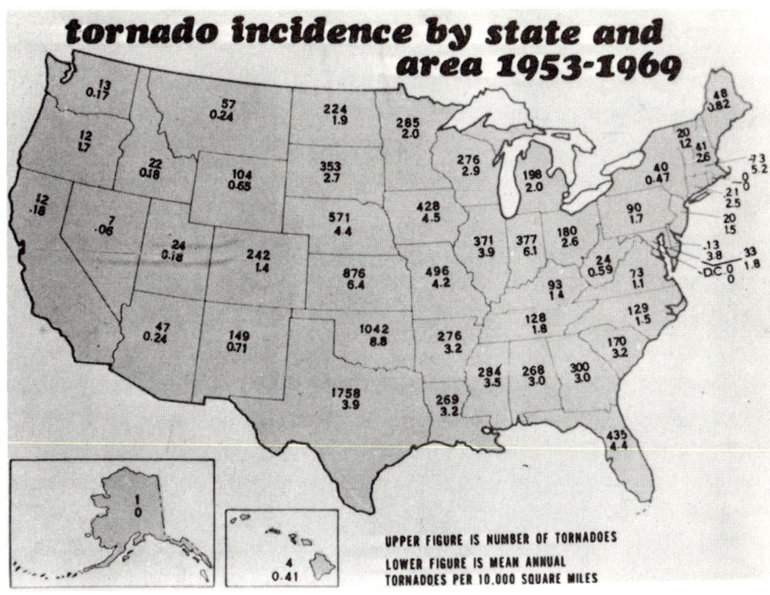

Figure 2-15. Tornado incidence by state and area, 1953–1969. (Courtesy of National Oceanic and Atmospheric Administration)

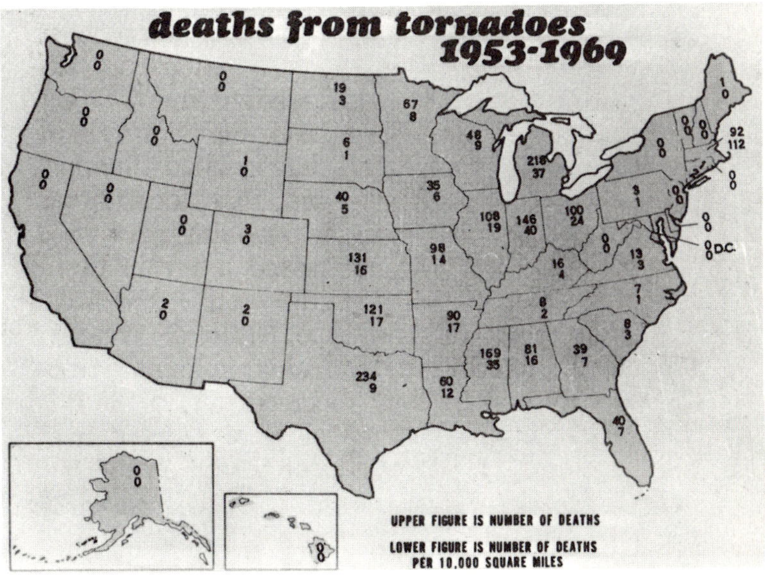

Figure 2-16. Deaths from tornadoes, 1953–1969. (Courtesy of National Oceanic and Atmospheric Administration)

Several considerations corroborate such a prediction. First, the present expansion of housing centers into rural areas is such that the probability of urban centers being struck by tornadoes is much higher than in the past, and this trend is constantly increasing. In addition, the standard of living in the United States and other countries is such that the present toll of lives and destruction becomes increasingly unacceptable.

The present state of the art on tornado engineering includes a general understanding of the atmospheric phenomenon and magnitude of the forces involved, a methodology for computing the probability, and some examples of storm-resistant structures already built. It is essential, therefore, to refine data and to fill gaps when necessary. Moreover, information and often data from other scientific areas must be made available to members of the design profession (architects and engineers) and to the public in general.

Since tornadoes are not limited to the North American continent, tornado engineering is a subject of international concern. The proliferation of nuclear power plants all over the world makes necessary the preparation of tornado probability studies for all proposed sites and the inclusion of specific design criteria. Actually, many countries, including the United Kingdom, do not have policies regarding protection from tornado forces. Since the consequences of radioactive contamination could be catastrophic in the event of a structural failure, the dissemination of information about safety policies is most important.

General Characteristics

Formation

Tornadoes are usually formed during warm and humid weather and quite often are associated with severe thunderstorms. Sometimes, in fact, more than one tornado is formed from the parent thunderstorm.

Theories on the origins of tornadoes vary considerably. Some of them are based on thermic phenomena, others on mechanical or electrical principles or combinations of them. One of the most popular theories is that tornadoes are formed during unstable conditions when layers of cold air lie above warm air near the ground surface. As the cold air at high altitudes drops down, the warm air rises with a spiral motion that initiates the vortex (Figure 2-17). In their relatively short life span, tornado vortices descend from the mother cloud, reaching the ground and lifting up occasionally as they move erratically in their translational motion. (See Figure 2-18.)

Figure 2-17. According to a theory, tornadoes are formed when cold air at higher altitudes drops down while warm air near the ground rises. The warm air acquires a rotary motion as it rises and originates the typical tornado vortex.

Figure 2-18. Sequential evolution of a tornado: (*1*) formation of a thunderstorm mother cloud; (*2*) formation of the tornado funnel; (*3*) extension of the funnel from the cloud down to the ground; (*4*) lifting of the funnel and subsequent descent; (*5*) rising again; (*6*) descent of the funnel and formation of a second funnel (behind the windmill); (*7*) further development of the second funnel; (*8*) regression and extinction of the second funnel while the first is still fully developed.

Figure 2-19. Typical trunk-shaped vortex forming a well-defined curvilinear column stretching continuously from the cloud deck to the ground. (Courtesy of U.S. Department of Commerce, NOAA)

Shape

Tornado vortices extend from the parent cloud of a thunderstorm down to the earth's surface. The dark cloud, several thousand feet high, is a cumulonimbus (i.e., massive cloud, typically shaped like a cauliflower and associated with thundering and heavy precipitation). The rotating air of the vortex as a shape is not equal for all tornadoes. The general funnel shape assumes at times the appearance of a long, slender column, which can either be vertical or form a curve. At other times, for large-diameter vortices, the shape is that of a cone. (See Figures 2-19 through 2-21.)

Figure 2-20. Vortex configuration of the tornado of June 27, 1955, near Scotts Bluff, Nebraska. (Photo by J. E. Jarrell, courtesy of National Weather Service)

Figure 2-21. Conical configuration of tornado vortex. Tornado of June 8, 1953, at Hoytville, Ohio. (Courtesy of National Weather Service)

Fire-Generated Vortices

A convective air column rising over high-intensity fires often causes the formation of air vortices. These are distinguished into two different types: one resembles a whirlwind; the other is very similar to a tornado. This last type originates at about 1000 ft over the fire and descends toward the ground, where it picks up debris, which makes the funnel visible; it would also pick up burning debris if the vortex were to descend within the fire. Such combination of strong winds and burning particles makes these vortices particularly damaging. Although typical of forest fires, tornado-like whirlwinds originating from a long-lasting oil fire have been reported in the literature. In an oil fire that took place in San Luis Obispo, California, in 1925, a great number of vortices of high-intensity speed were observed to move horizontally over the land throughout the several days that the fire burned. By the same token, whirlwinds of this type were generated during the Chicago fire of 1871, with lethal consequences.

Detection

In accordance with the theories on tornado formation and energy mechanism based on electrical energy, it has been found that during electrical discharges in the atmosphere, electromagnetic waves of about 150 kHz are produced (Vonnegut 1960, 1966). Of these, the components within the band between 54 and 60 kHz are captured by television receivers, which show a luminous signal on the screens that lasts a few minutes. This peculiar discovery has practical applications for individual experimentation in tornado detection (Gates 1972). Turn on a television, and set the contrast control at maximum. Turn the dial to the highest numbered channel; then adjust the brightness of the television screen until the contrast is at the threshold of black. Turn the selector back to Channel 2 without resetting the brightness. Lightning will appear on the screen as flashes. Thereafter, if the screen becomes bright or if a picture becomes visible on the darkened screen, this indicates that there is a tornado within a range of 20 mi or less.

Of course, a more dependable technique, which is used by weather stations, is to track tornadoes by radar. In fact, a familiar hook-shaped signal on the screen will indicate the presence of a tornado.

Definitions of Terms

In 1749 in Rome, Boscovich used the word *tromba* (Italian, "trumpet") to describe the shape of the mother cloud of a tornado. The word then spread to other countries, becoming *tromben* in German, *trombe* in French, and so on. The word *tornado* is of Spanish origin. It came into usage in the English language in the nineteenth century.

To avoid misinterpretations, we will here clarify some of the terms used in meteorology. According to the International Meteorological Vocabulary (World Meteorological Organization):

1. *Spout:* phenomenon consisting of an often violent whirlwind, revealed by the presence of a cloud column or inverted cloud cone (funnel cloud), protruding from the base of a Cumulonimbus, and of a "bush" composed of water drops raised from the surface of the sea or of dust, sand, or litter raised from the ground.
2. *Tornadoes:* (a) name given in North America to an intense spout of large diameter; (b) name incorrectly given in West and Equatorial Africa to a violent, thundery squall.
3. *Dust Whirls, Sand Whirls, Dust Devils:* ensemble of particles of dust or sand, sometimes accompanied by small litter, raised from the ground in the form of a whirling column of varying height with a small diameter and an approximately vertical axis.
4. *Waterspouts:* spout above a water surface.

Most European countries follow the preceding definitions and refer to these wind phenomena that occur in Europe as "spouts." However, in the case of an exceptionally large or destructive storm, they also use the term *tornado.* In England and Australia the term *spout* is not used at all in scientific papers; *tornado* is commonly used regardless of the size and location (Evesson 1969, Gilbert and Walker 1963, Kert and Dean 1963, Lacy 1968, Lamb 1957, 1965).

In the United States, as in other English-speaking countries, *tornado* is not limited to storms of the North American territory, nor to large vortices. It includes any phenomenon described as "spout" in the International Meteorological Vocabulary.

Although American tornadoes are often assumed to be much larger than those in other areas, many reports do not validate such a difference. For example, Wegener (1917) reported that the maximum diameter for American tornadoes is about 3 km, compared with 2.3 km for European storms. Therefore, *tornado* is applicable to American as well as to European, Asian, and similar storms in the rest of the world. In fact, the TORRO (*Tor*nado and Storm *R*esearch *O*rganization) scale devised by Meaden (1976) evaluates the average tornado in Great Britain, France, Germany, and the United States with *T* values equal to 8, 9, 9, and 11, respectively.

Waterspouts

Waterspouts are assumed by the International Vocabulary and others to be similar to tornadoes, except that they occur on water surfaces (Figure 2-22). On the contrary, more differences than similarities exist between the two. Waterspouts usually originate in developing cumulus congestus clouds, with maximum top elevations of 12,000 ft; they may also form in shallow stratocumulus clouds. (For explanation of cloud types, see Appendix F.) These phenomena are less intense, shorter, and move at lower speeds than tornadoes. Unlike tornadoes, they do not necessarily develop in severe thunderstorms, which implies that their originating source is of a lower energy level. Also, the tangential velocity seems to increase linearly from the center up to a certain distance, as in the rotation of a solid, and then it decreases. This is in agreement with the latest theories on tornadoes, corroborated by measurements reported by Kessler (1970).

In recent years, several official data have been collected on waterspouts. For instance, Segman (1971) reports on the work conducted by H. Golden and others and supported by several agencies, including the National Severe Storms Laboratory (NSSL). In one study, the temperature in the core was measured directly by passing a towed probe through the vortex of five waterspouts. Other data were obtained by dropping smoke flares on the water surface of and firing smoke flares through a waterspout, while movies and still photos were being taken. In some other experiments, the wing of an airplane was flown through the core, penetrating about 5 ft into it. During the approach, the wing was lifted and during the flyby, it dipped. Such observations seem to imply a rising vertical circulation at the exterior boundary and a downward

Figure 2-22. Waterspout near Florida Keys, September 10, 1969. (Courtesy of National Oceanic and Atmospheric Administration)

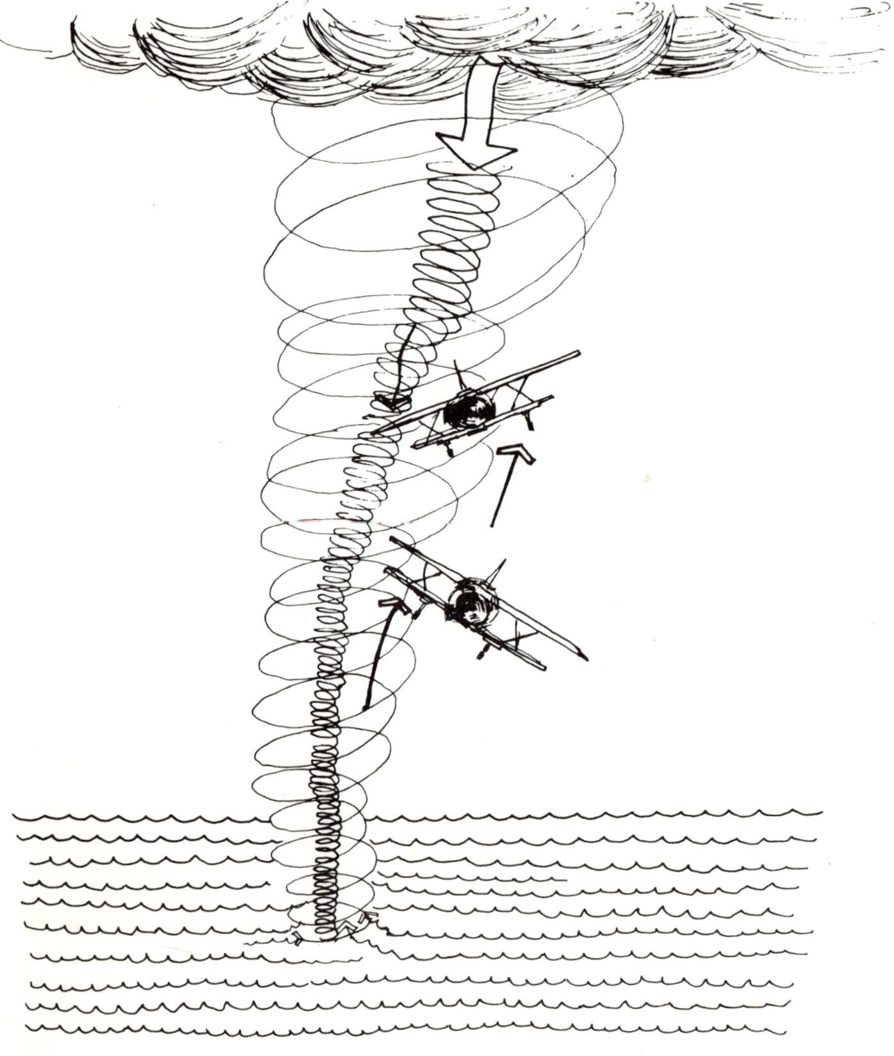

Figure 2-23. Waterspouts differ from tornadoes. They consist of an interior spiraling vortex moving downward and a concentric exterior vortex moving upward. The phenomenon was definitely proved by flying an airplane through the core of a spout. The airplane's wing was lifted as it entered the outside part of the spout, but it was pushed down as the wing penetrated 5 feet into the core.

flow along the axis in the interior. The same circulation is also proposed for tornadoes but has not yet been proved. (See Figure 2-23.)

Another interesting observation that could lead to important conclusions in the future is that the waterspout, in the experiment previously described, dissipated shortly after the plane crossed it. Similarly, on four other occasions when an airplane flew within 50 ft of a waterspout, the vortex dissipated in a short time. There is hope, therefore, that a triggering mechanism can be found to dissipate such phenomena.

Probability of Occurrence

To determine whether tornado forces should be considered in the design of a specific building, it is necessary to know the probability of tornado occurrence in the proposed location. For the United States, it is feasible to use the formula proposed by Thom (1963):

$$P=\frac{\bar{z}\,t}{A}$$

where

P = probability of occurrence (number of probable tornadoes per year) that can strike the point in question.

$\bar{z}$ = average area of damage path. For the United States, Thom calculated it to be 2.8209 mi^2.

t = mean annual frequency of tornadoes in the zone whose area is A.

A = area, 1° latitude and 1° longitude square, that includes the location in question.

The interval of time between two consecutive tornadoes that can strike the location in question is indicated by $1/P$. Thus P is the probable frequency, and $1/P$ is the probable period. Some critics of the formula cite that $\bar{z}$ has been calculated on the basis of Kansas and Iowa tornadoes, whose sizes do not reflect those of the whole country. Still, this formula has been the most practical since 1962, and it is the one that has been used in the design of American nuclear power plants. A tabulation of values for t in any location in the United States can be found in Thom (1963). (See Figures 2-24 and 2-25.)

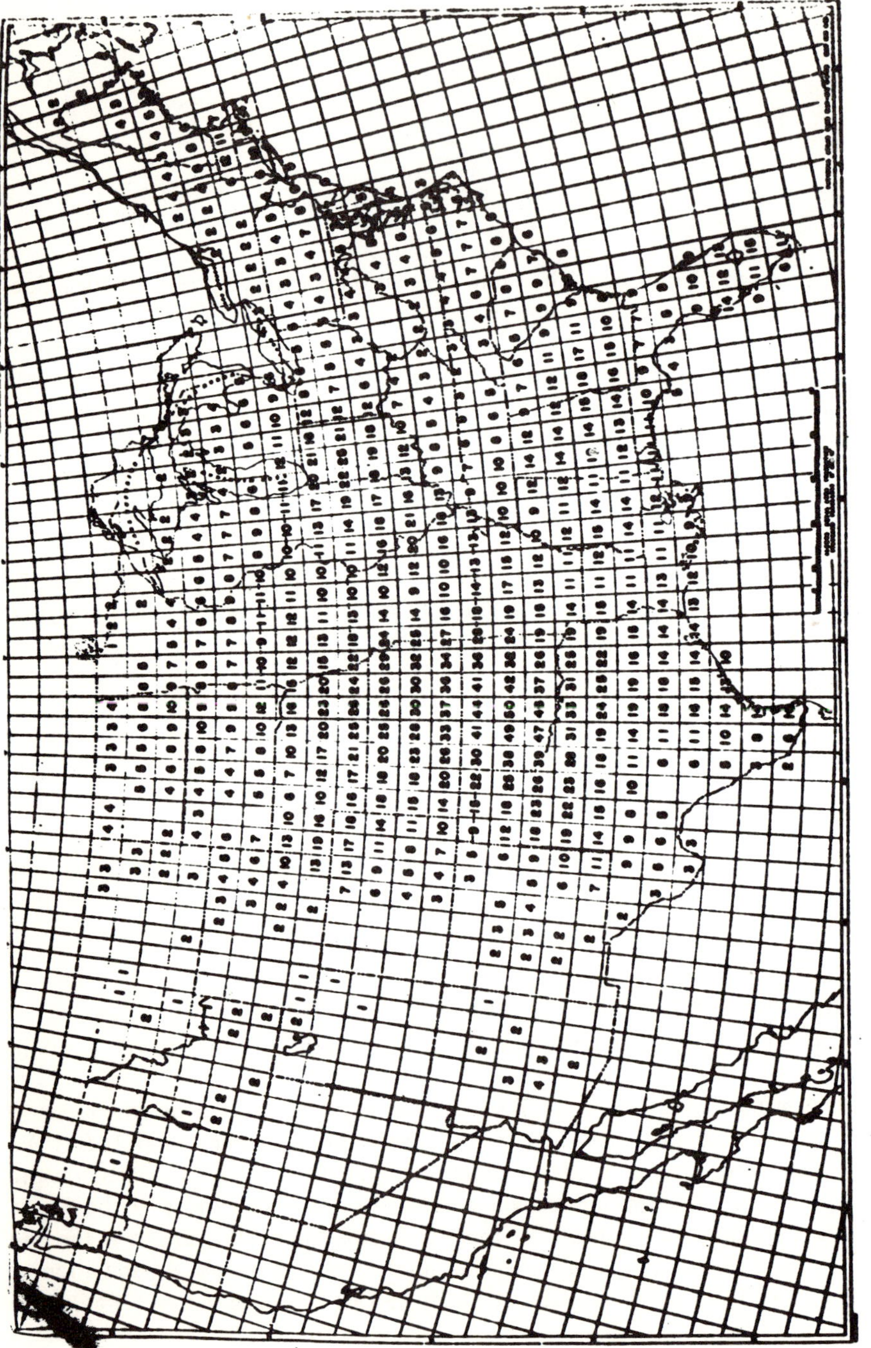

Figure 2-24. Total annual frequency of tornadoes in the United States, 1953–1962.

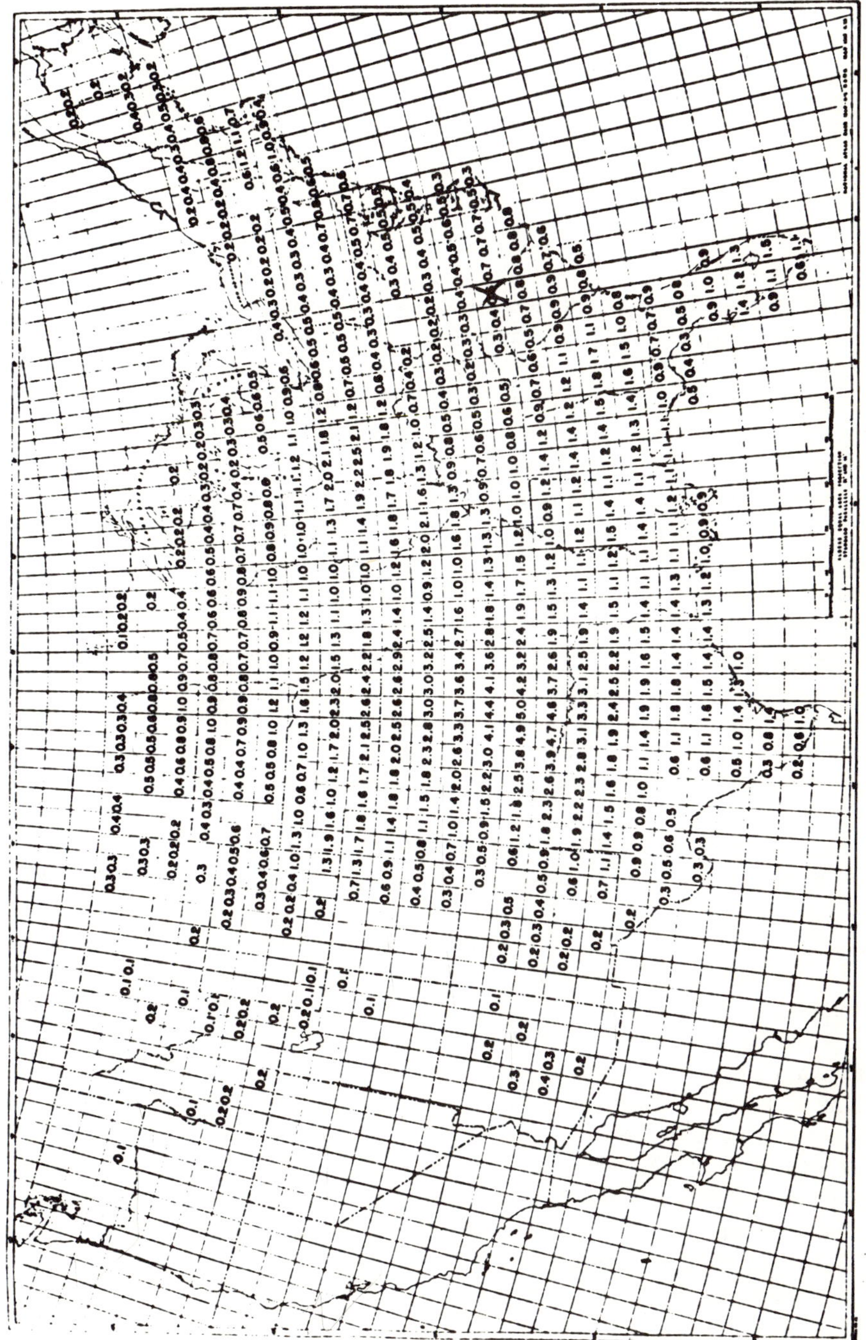

Figure 2-25. Mean annual frequency of tornadoes in the United States, 1953–1962.

Thom (1963) studied several tornadoes that had been recorded in Iowa during the period from 1953 to 1962. Thom's findings, which determined an average area of tornado paths equal to 2.82 mi^2, were also found to be valid for the state of Kansas but not for other parts of the United States, as other researchers later pointed out. In analyzing the data for the determination of the path length, Thom rejected all reports of paths exceeding 100 mi, since these values were doubtful. On the basis of 106 cases, he derived an average length of 3.94 mi. Similarly, for the determination of the average path width, he arrived at an average value of 154.5 yd by rejecting values larger than 1000 yd as unreliable data.

Skaggs (1970) analyzed tornado occurrences in 12 states located in the eastern United States. He indicated that an average tornado area for the United States as a whole would have been an oversimplification since a great variety of tornado sizes exists among various states. For instance, for Texas, Skaggs computed the average tornado path area to be 0.59 mi^2, whereas for Minnesota he found the average path area to be 5.38 mi^2. These two values are the minimum and maximum he found for the United States; for the states in the east coast, he found that the average tornado area was less than 0.75 mi^2. Reevaluating the data for Iowa from 1950 to 1964, he was also able to verify Thom's values of 2.82 mi^2.

Howe (1974) also examined the area of a large number of tornado paths. This work, as well as Skaggs (1970), is quite important because it renders more accurate values, in terms of the area of the tornado paths, when Thom's formula is applied. Howe examined 4400 tornado reports from 1953 to 1972 for all states east of the Rocky Mountains. In doing so, he followed criteria of acceptance that were similar to those proposed by Thom. He rejected values for the width that were less than 4 yd or larger than 1000 yd. He also rejected length values that exceeded 100 mi or were less than 40 yd, and lengths that were shorter than the widths. His calculation showed an average area of 0.96 mi^2 for all states examined; the largest value he found was 1.49 mi^2 for Oklahoma, and the minimum was 0.58 mi^2 for Minnesota (note the discrepancy between Skaggs's values and those of Howe).

The National Severe Storm Forecast Center (NSSFC) also calculated an average tornado area for the United States based on 8900 observations. The average value found was 0.25 mi^2 (Schafer et al. 1979).

As mentioned previously, Thom's formula is still valid, and it can be updated by using the specific values of various locations that were proposed after his work was published. Notice also that Thom's formula was derived as a simplification of a more general formula, which is expressed as follows:

$$P = 1 - \left(1 - \frac{\bar{z}}{A}\right)^t$$

The terms in the above formula are as previously defined.

Further comment on Thom's formula must emphasize the importance of the values assumed for *A*, which influences the final results of the values for *P* (probability of occurrence). Thom specified for *A* to be an area, 1° latitude and 1° longitude square, that includes the point in question. McDonald et al. (1975) pointed out that the probability value could vary considerably in order of magnitude as *A* increases, for example, from 1° to 9° of latitude and longitude.

Tornado Ratings

Similar to the scales used for rating earthquakes, several methods have been proposed for rating the various characteristics of tornadoes. These methods could include the intensity or the overall effects or the size of the storm. In addition, each one of the parameters can be related to the number of casualties or damage magnitude or both. In discussing intensity, the number of deaths and the amount of damage could be referred to a certain specified unit—for instance, a surface area of 1 mi^2. Further specifications may distinguish the unit area in terms of geopolitical conditions, such as open country or built-up areas of various densities. Rating in terms of magnitude, one could consider absolute values of the number of deaths and damage. In conclusion, there are several different ways in which rating can be structured.

Various rating systems have been proposed, including those of Hazen (1890), Seelye (1945), Fujita (1971, modified in 1973), Meaden (1976), and Dames & Moore (1975).

Hazen (1890) proposed the first tornado-rating system, based on the violence of the storms. His study analyzed 2221 tornadoes that occurred from 1873 through 1888. His scale included nine categories, ranging from 3+ down to 1−. The categories are based on the monetary assessment of the tornado's damage.

Seelye (1945) classified tornadoes in terms of intensity using a scale from 0 to 5 to describe the expected type of damage in each class. In class 5, well-constructed buildings are demolished. In class 3, outbuildings, verandas, and roofs are carried away. In class 0, there is no damage because the tornadic funnel cloud does not touch the ground.

Fujita (1971) proposed a rating scale describing the damage to be expected in each of several classes. Each class was defined by the minimum and maximum expected wind speeds and the type of damage produced, described in words and with photographs. In 1973, he modified the scale by substituting the old photographs with new ones. The classes in the scale range from F− to F 12. F− includes wind speeds of less than 40 mph, corresponding to 12 on the Beaufort scale (see Table 2-1). F 12 corresponds to the speed of sound in the atmosphere (Mach 1). In reality, however, class F 6 (maximum speed of 319 mph) is the highest used because it is doubtful that any tornadoes will have a higher speed (see Table 2-3). The equation relating the wind speed to the classes in the Fujita scale is as follows:

$$\text{Wind speed} = 14.1\,(F + 2)^{3/2}\ \text{(mph)}$$

Meaden (1976) proposed a rating system for tornadoes that refines the concepts previously adopted by Fujita. The scale he adopted is the TORRO scale. The various categories in the scale correspond to certain wind speeds and specific types of damage. Each category is identified by a tornado force number: FC, 0, 1, 2, 3, 4, 5, 6, 7, 8, 9, 10, 11, 12 (see Table 2-4). The relationship between the tornado force T and the wind speed v (mph) is derived as follows. Starting with the Beaufort-scale equation:

$$v = 1.870\,(B)^{1.5}$$

substituting B with the following expression:

$$B = 2(T + 4)$$

in the preceding equation, Meaden obtains:

$$v = 5.288\,(T + 4)^{1.5}$$

Table 2-3. Fujita scale classification of tornado wind intensity (1976).[a]

CLASS	CHARACTERISTICS	VERBAL DESCRIPTION OF DAMAGE
F–	Doubtful Tornado less than 40 mph	40 mph speed corresponds to Beaufort 8 or "Fresh Gale." Beaufort specification for use on land is "Breaks twigs off trees." Little damage is expected.
F 0	Very Weak Tornado 40–72 mph	This speed range corresponds to Beaufort 9 through 11. Some damage to chimneys or TV antennae; breaks branches off trees: pushes over shallow-rooted trees; old trees with hollow inside break or fall; sign boards damaged.
F 1	Weak Tornado 73–112 mph	73 mph is the beginning of hurricane windspeed or Beaufort 12. Peels surface off roofs; windows broken; trailer houses pushed or overturned; trees on soft ground uprooted; some trees snapped; moving autos pushed off the road.
F 2	Strong Tornado 113–157 mph	Roof torn off frame houses leaving strong upright walls standing; weak structure or outbuildings demolished; trailer houses demolished; railroad boxcars pushed over, large trees snapped or uprooted; light-object missiles generated; cars blown off highway; block structures and walls badly damaged.
F 3	Severe Tornado 158–206 mph	Roofs and some walls torn off well-constructed frame houses; some rural buildings completely demolished or flattened; trains overturned; steel framed hangar-warehouse type structures torn; cars lifted off the ground and may roll some distance; most trees in a forest uprooted, snapped, or leveled; block structures often leveled.
F 4	Devastating Tornado 207–260 mph	Well-constructed frame houses leveled, leaving piles of debris; structure with weak foundation lifted, torn, and blown off some distance; trees debarked by small flying debris; sandy soil eroded and gravels fly in high winds; cars thrown some distances or rolled considerable distance finally to disintegrate; large missiles generated.
F 5	Incredible Tornado 261–318 mph	Strong frame houses lifted clear off foundation and carried considerable distance to disintegrate; steel-reinforced concrete structures badly damaged; automobile-sized missiles fly through the distance of 100 yds. or more; trees debarked completely; incredible phenomena can occur.
F 6–12	Inconceivable Tornado 319 mph to sonic speed.	Should a tornado with the maximum windspeed in excess of F 6 occur, the extent and types of damage may not be conceived. A number of missiles such as ice boxes, water heaters, storage tanks, automobiles, etc. will fly through a long distance, creating serious secondary damage on structures. Assessment of tornadoes in these categories is feasible only through detailed survey involving engineering and aerodynamical calculations as well as meteorological models of tornadoes.

[a]From assessment of knowledge and implications for man. *Proceedings of the Symposium on Tornadoes.* June 22–24, 1976. Texas Tech University.

Table 2-4. TORRO tornado intensity scale (Meaden 1976).

FORCE (T)	TYPE	VERBAL DESCRIPTION OF DAMAGE[a]
FC	Funnel Cloud or Incipient Tornado	No damage to structures, unless to tops of tallest towers, or to radiosondes, balloons, and aircraft. No damage in the country, except possibly agitation to highest treetops and effect on birds and smoke. Record FC when tornado spout seen aloft but not known to have reached ground level. A whistling or rushing sound may be noticed.
0	Light Tornado 42–58 mph	A. Loose, light litter raised from the ground in spirals. B. Temporary structures such as marquees seriously affected. C. Slight dislodging of the least secure and most exposed tiles, slates, chimney pots, or television aerials may occur. D. Trees severely disturbed, some twigs snapped off. Bushes may be damaged. Hay, straw, and some growing plants and flowers raised in spirals.
1	Mild Tornado 59–77	A. Heavier matter levitated include planks, corrugated iron, deck chairs, light garden furniture, etc. B. Minor to major damage to sheds, outhouses, locksheds, and other wooden structures such as henhouses and outhouses. C. Some dislodging of tiles, slates, and chimney pots. D. Hayricks seriously disarranged; shrubs and trees damage to hedgerows, crops, trees, etc. Trees may be uprooted.
2	Moderate Tornado 78–97 mph	A. Exposed, heavy mobile homes displaced; light caravans damaged. B. Minor to major damage to sheds, outhouses, lockup garages, etc. C. Considerable damage to slates, tiles, and chimney stacks. D. General damage to trees, big branches torn off, some trees uprooted. Tornado track easily followed by damage to hedgerows, crops, trees, etc.
3	Strong Tornado 98–119 mph	A. Mobile home displaced, damaged, or overturned; caravans badly damaged. B. Sheds, lockup garages, outbuildings torn from supports/foundations. C. Severe roof damage to houses, exposing much of roof timbers; thatched roofs stripped. Some serious window and door damage. D. Considerable damage (including twisted tops) to strong trees. A few strong trees uprooted or snapped.

[a] A, B, C apply broadly to urban situations, and D to rural situations.

Table 2-4. (Continued)

FORCE (T)	TYPE	VERBAL DESCRIPTION OF DAMAGE[a]
4	Severe Tornado 120–142 mph	A. Caravans and mobile home destroyed or gravely damaged. C. Entire roofs torn off some frame/wooden houses and small/medium brick or stone houses and light industrial buildings, leaving strong upright walls. D. Large well-rooted trees uprooted, snapped, or twisted apart. The ground possibly furrowed by tornado spout to a depth of about 1 meter.
5	Intense Tornado 143–166 mph	C. More extensive failure of roofs than for force 4, yet with housewalls remaining. Small weak buildings, as in some rural areas (or as existed in medieval towns), may collapse. D. Trees carried through the air.
6	Moderately Devastating Tornado 167–192 mph	A. Motor vehicles over 1 ton lifted well off the ground. C. Most residences lose roofs and some a wall or two; also some heavier roofs torn off (public and industrial buildings, churches). More of the less strong buildings collapse; some totally ruined. D. Across the breadth of the tornado track, every tree in mature woodland or forest uprooted, snapped, twisted, or debranched.
7	Strongly Devastating Tornado 193–219 mph	C. Walls of frame-wooden houses and buildings torn away; some walls of stone or brickhouses and buildings collapsed or are partly beaten down. Steel-framed industrial buildings buckled. Locomotives and trains thrown over.
8	Severely Devastating Tornado 220–247 mph	C. Entire frame houses leveled; most other houses collapse in part or whole. Some steel structures quite badly damaged. Motor cars hurled great distances.
9	248–276 mph	C. Many steel structures badly damaged. Locomotives and trains hurled some distances.
10	Intensely Devastating Tornado 277–306 mph	C. Entire frame/wooden houses hurled from foundations.
11	307–337 mph	C. Steel-reinforced concrete buildings severely damaged.
12	Super Tornado 338+ mph	

Dames & Moore (1972) proposed a tornado-intensity classification that included five categories. Each is defined by wind-velocity limits and expected structural damage. In the first category, wind velocities range from 50 to 90 mph, with structural damage consisting of partial roof removal of weak rural structures and some trees uprooted and blown down. In category 6, wind speeds start at 200 mph, with an upper limit above 300 mph. Structural damage in this category is described as catastrophic. (See Table 2-5.)

Table 2-5. Dames & Moore tornado intensity classification (1972).

CLASS	WIND VELOCITY (mph)	EXPECTED DAMAGE
1	50–90	Partial roof removal of weak rural structures; some trees uprooted and blown down
2	80–120	Total roof removal of rural structures; partial roof removal of single-family residences; house trailers moved or rolled; more extensive tree uprooting
3	100–150	Rural structures heavily damaged; total roof removal of residences; house trailers destroyed; non-reinforced masonry walls overturned; extensive sign damage and tree uprooting
4	120–180	Rural structures demolished; total roof removal of residences and some walls down; partial roof removal of light steel industrial buildings and wood truss commercial buildings
5	150–225	Complete homes destroyed; total roof removal of light industrial buildings and wood truss commercial buildings; partial roof removal of heavy industrial buildings
6	200–300+	Catastrophic destruction; homes off foundations; substantial commercial and industrial buildings destroyed; large steel framed structures heavily damaged.

Tornadoes in Various Countries

The frequency of tornadoes cited in the literature is based on the storms that have been observed and recorded rather than on the absolute number of actual occurrences. It is obvious, therefore, that a large number of factors can affect the number of the recorded observations. The density of the population, the efficiency of the news media, the public interest, the efficiency of the weather stations, and so on, are all factors that influence the number of recorded cases. This is pointed out by comparative analysis, if one considers a certain zone and examines the number of cases reported in different years. In the United States, for instance, a sudden increase beginning in 1953 is clearly noticeable. This date coincides with the beginning of tornado-forecasting programs and the use of radar in weather stations. The number of storms recorded in Europe from 1650 to 1916, as reported by Wegener (1917), varies sharply with political and social factors, such as war (see Figure 2-26). The annual frequency under stable conditions shows considerable variations regardless of the influence of exterior parameters. The graph shown in Figure 2-27 indicates that in the United States, during a period of 16 years, the frequency recorded varies with maximum and minimum values with a ratio greater than 2:1. On the basis of these premises, the meaning of the frequency data results are clearer when such data have been collected under different circumstances in various geographical locations. The European statistics of Figure 2-28, quoted by Wegener (1917), have an informative and historical value, more so than any other meaning. They show very clearly, however, that atmospheric vortices are not limited to the plains of North America, as many believe. The ratio between the number of tornadoes in North America and in Europe, as indicated by records, has been estimated by Wegener (1917) to be about 20:1. The distribution of observations among different European countries, according to Wegener, is illustrated in Figure 2-28. According to data collected by the Swiss Meteorological Institute only three vortices were officially recorded in Switzerland in a period of about 80 yr.* They occurred on August 19, 1890; June 12, 1926; and

* Gensler, G. 1972. Chief Climatological Section, Institut Suisse de Meteorologie. Personal communication, May 2, 1972.

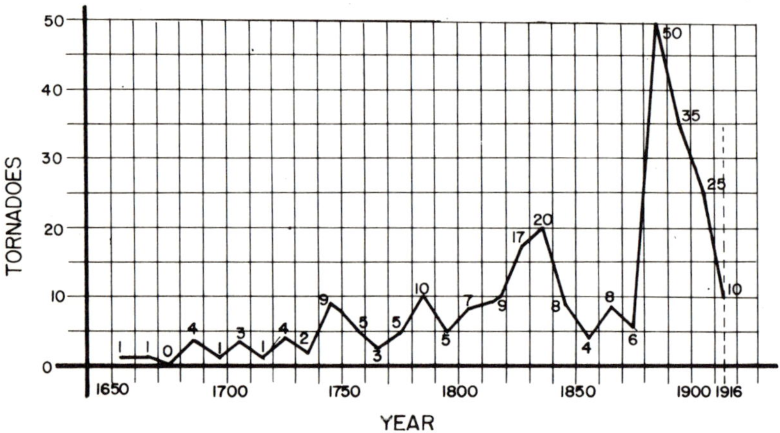

Figure 2-26. Tornadoes in Europe (Wegener).

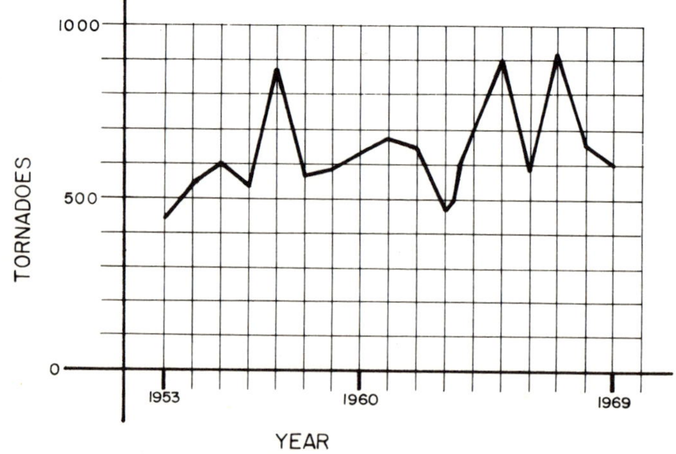

Figure 2-27. Tornadoes in the United States (NOAA).

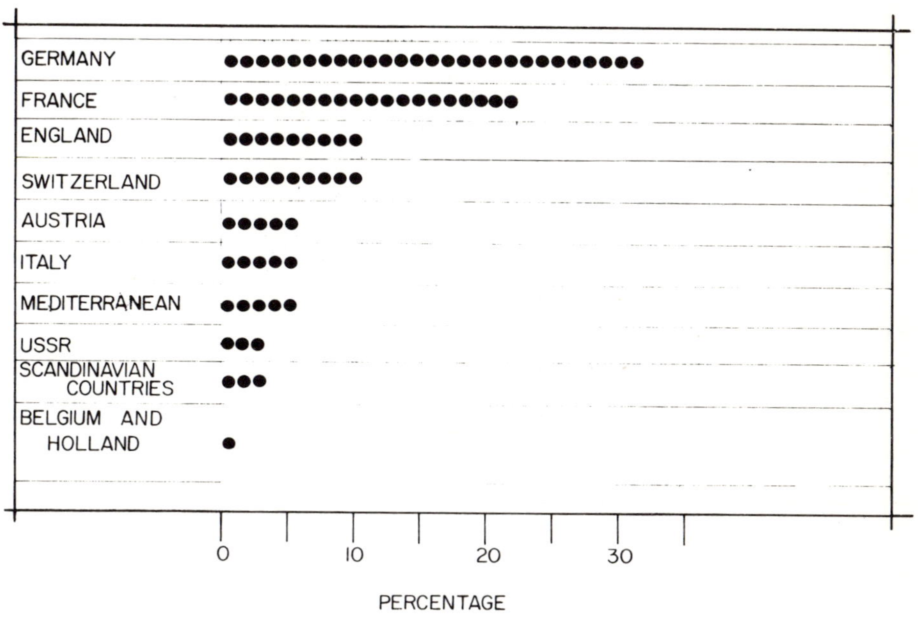

Figure 2-28. Percentage of distribution of tornadoes in Europe (Wegener).

August 26, 1971. Such values are very low compared to those collected in Italy. Furthermore, according to Wegener, they are in contradiction to the percent distribution among European countries. This proves the subjectivity of the various data-collecting sources. Germany seems to have the highest percentage of vortices in comparison to other countries in Europe. Italy has only one-fifth as many cases as reported in Germany; a recent list of storms observed in Italy from 1948 to 1970 indicates a total of 192 reports. The distribution of the annual frequency (Figure 2-29) shows a growing number of cases in Italy, beginning at the end of World War II. This, of course, coincides with a revival of meteorological phenomena. The absence of data

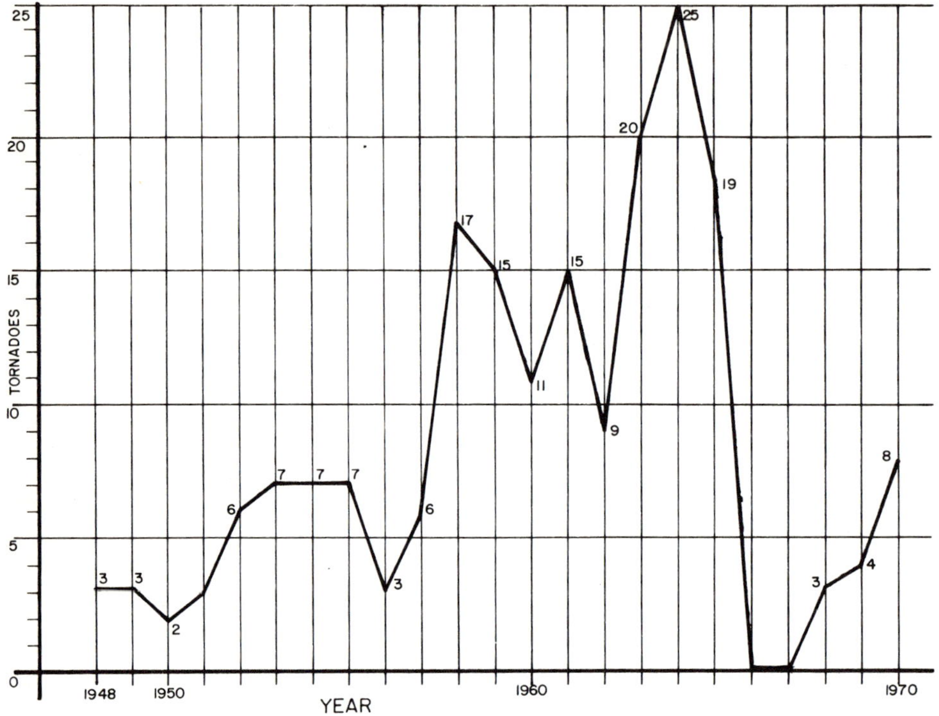

Figure 2-29. Annual tornado frequency in Italy.

in 1966 and 1967 is not clear; it could have been caused by oversight. The frequency data for Italy are also very variable. In 1964, there was a maximum of 25 cases. The monthly distribution in Italy, as for other European countries, indicates that the largest number of cases occur in July (see Figure 2-30). (The highest frequency in the United States occurs in May.) In Italy, the regional distribution is extended to the whole peninsula, including Sicily and Sardinia. Some zones, however, have frequencies that are much higher than others. The province of Rome, for instance, is clearly ahead, with a frequency of 11%, followed by Genoa with 4.6%, Turin and Milan with 4.15%, and Verona and Palermo with 1.56%. It is interesting to note that the first studies on air vortices were initiated in Italy.

Geminiano Montanari, in the posthumous publication of 1694, discussed the theory on air vortices in connection with observations of a vortex reported in the territory of Venice in 1686 (Wegener 1917). A study by Boscovich in 1749 dealt with the

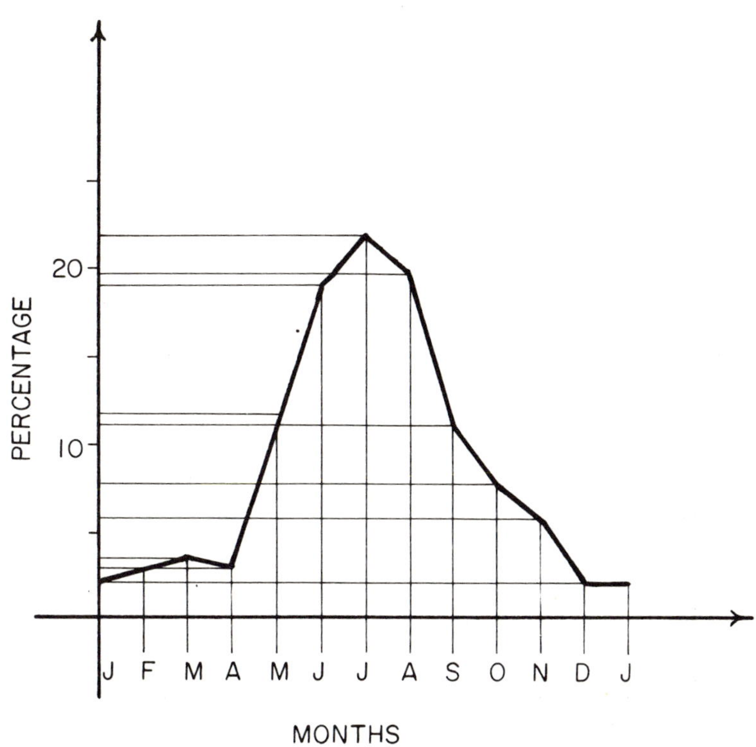

Figure 2-30. Monthly frequency of tornadoes in Europe (Wegener).

action of an air vortex over Rome (Wegener 1917). Other frequency data are illustrated in Figure 2-31. The hourly distribution in Holland and in the United States indicates that the afternoon is the part of the day during which vortices are most likely to occur—more specifically, 3 P.M. in Holland and 5 P.M. in the United States (Wegener 1917).

The first report of tornadoes in Australia seems to be one dated October 20, 1826. More methodical records started in 1887 and included newspaper descriptions given by occasional observers and other nonofficial documents whose reliability is questionable since they were not verified by qualified experts. Apparently, several types of local storms in Australia have similar manifestations that can lead lay observers to erroneous conclusions. In addition, confusion comes from the misuse of such terms as *cyclone, tornadic squall, hurricane, cockeyed bob, willy-willy,* and *tornado.* Clarke (1962) states:

> A search of files on the occurrence of several local windstorms in Australia has made possible some tentative conclusions concerning the frequency, intensity and geographical distribution of these tornado-like storms. They are on the whole about as frequent as, but much less intense than, North American tornadoes.

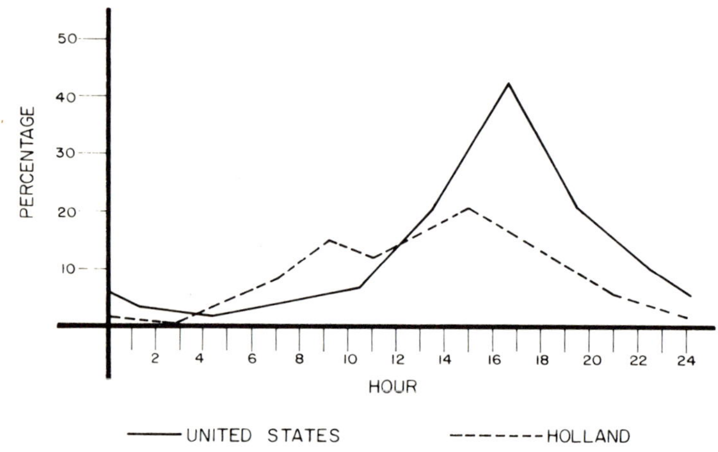

Figure 2-31. Hourly frequency of tornadoes in the United States and Holland (Hann).

Size and violence of tornadoes seem to be related to local geographical conditions. Thus, it is possible to differentiate tornadoes in terms of these parameters. Recent work in this area supports this theory. American tornadoes, for example, can be distinguished from tornadoes in other parts of the world, and, even American tornadoes can be differentiated according to state.

Violence can be measured in terms of wind speed and pressure differentials, or by their manifestations expressed in terms of casualties and damage per unit area of tornado path. It is obvious, however, that other factors such as density of population and effectiveness of warning systems can confuse the analysis for the determination of the storm violence.

The size of the tornado path affects the number of casualties and the amount of damage in absolute values, so that a less violent tornado of larger size could be assessed as being more destructive than a smaller one of higher intensity. (In other words, it is critical not to confuse intensity with overall destructiveness in assessing the effects of tornadoes.) However, new studies support the theory that tornadoes with longer paths are usually not only more destructive but also more violent.

In terms of geographical occurrence, therefore, we can see that tornadoes vary as a function of the previously described characteristics: size, violence, and overall destructiveness. To confirm this differentiation in practical terms, the rating of tornadoes according to Meaden (1976) classifies tornadoes in Great Britain as having an average *T* value of 8. This implies wind velocities from 225 to 247 mph, and the storms are described as "severely devastating tornadoes." Meaden classifies tornadoes in France and Germany as having an average *T* value of 9, which implies wind velocities from 248 to 276 mph and a description similar to the one just mentioned. For the United States instead, the average *T* value is assessed as 11, with wind speeds from 307 to 337 mph. Meaden describes the tornadoes in this category as "intensely devastating."

According to Evesson (1969, 1970), the average number of reported tornado days per year in all of New South Wales in 15.2—11.1 in the inland area and 4.1 in the coastal area. The areas of the United States with the highest frequency are Kansas (24.9), Oklahoma (23.7), and Nebraska (15.4). The average of annually reported tornado days per population density for all New South Wales is 0.16—that is, 0.12 and 0.04, respectively, for the inland and coastal areas. Equivalent data for the United States include: Kansas (1.13), Nebraska (1.10), Oklahoma (0.82), South Dakota (0.72), Arkansas (0.25), and Iowa (0.16).

Tornado sizes in Australia vary from a few feet to over half a mile (Evesson 1969). The percentage distribution of the width of the damage path is indicated in Figure 2-18. There is a theory proposing that a wide path could be produced by two or more tornadoes moving parallel and adjacent to each other, rather than by one of very large diameter (Evesson 1970). However, such theories would apply to the American tornadoes as well, and thus they would not explain the difference in size between the American and the Australian storms.

Maximum tangential velocities are estimated mostly according to minimum force values that could produce observed damage. These values do not imply any upper limits, just lower values; thus, the real wind velocities could have been much higher than that. Such estimates include the following Australian storms: 120 to 150 mph for the Piedmont tornado on July 1, 1861; 140 to 150 mph for the Narrabi tornado on October 13, 1893; and over 110 mph for the Erina tornado on December 5, 1966.

Measurements

A scientific investigation of tornado-induced loads for building design depends largely on the measurement of the physical forces involved in the storms. Here, we are concerned in particular with the destructive forces that cause the damages left in the wake of a tornado. Such forces are the wind pressure itself and the pressure on the exterior walls of structures that derives from the sudden change in atmospheric pressure. Therefore, the measurements discussed herein are those that define these forces: wind velocity, barometric pressure, and pressure gradient.

Measurements can be divided into two categories: direct and indirect. Direct measurements include data obtained by direct readings on calibrated instruments. These are very rare. It is difficult to set up the proper instrumentation for direct measurements. The instruments must be designed and constructed to operate properly under the severe conditions present with tornadoes, and they must be located in a tornado path. The few figures available concern barometric pressure and were recorded by chance on those few occasions when a tornado happened to pass over or near a barometer. However, in recent years, additional direct measurements have been added, as reported by Kessler and Lee (1976). Such measurements were taken by some of the 30 to 60 stations located in zones of high tornado probability. Another series of direct measurements has been taken by penetrating the core of the vortex of waterspouts with probes, as previously mentioned.

Values in the category of indirect measurements are derived from other available data when the mathematical relationships between them and the desired measurements are known. Sometimes, the mathematical relationship is based on an assumption made subjectively, and this, of course, reduces the reliability of the results. In other instances, the results can be considered equivalent to direct measurements because of the direct connection between the data available and the measurements desired.

Most of the values of tornado forces have been computed indirectly by one of the following means:

1. structural analysis of damage
2. analysis of ground marks
3. examination of motion pictures of tornado funnels
4. examination of splinters that have penetrated other objects
5. analysis of funnel shape

The derivation of wind velocity from the analysis of movies (i.e., studying the film frames, pinpointing an airborne particle, observing its change of position from one frame to the next, scaling the distance traveled on the frame, and determining the time elapsed from the frame speed) seems to be a very dependable method. The results of such an analysis could probably be classified as direct measurement because of the precise calculation that is possible. Other indirect measurements, however, must be evaluated according to the reliability of the method used.

Characteristics of the Path

Because the path left behind by tornadoes remains visible for a considerable time, it can be measured without much difficulty. Records of virtually thousands of storms have been collected and have constituted the basis for several studies, as reported by Flora (1954).

Asp (1963) analyzed 924 tornadoes in Oklahoma and concluded that the average length is close to 10 mi and the average width is 440 yd, from which he derived the approximate area of 2.5 mi^2.

Martin (1940) studied 1000 tornadoes chosen at random around the United States and found the average length to be 13.4 mi and the average width 396 yd; consequently, the average area would be approximately 3.0 mi^2.

In a study of 209 tornadoes in 1950, reported by Flora (1954), it was found that the average length was 12 mi and the average width was 438 yd; therefore, the average area was about 3.0 mi^2.

Thom's study (1963), based on tornadoes in Iowa and Kansas, indicates an average path area of 2.8209 mi^2.

Data collected on Australian tornadoes show a considerable variation in size, with maximum values over 2700 ft in diameter. In the category of diameters varying between 2700 ft and over, more than 16% of the storms are included. This seems to disprove the usual belief that tornadoes in other parts of the world are smaller than the typical tornadoes of North America. Figure 2-32 illustrates a distribution of tornadoes in Australia, according to size.

Wolford (1960) indicated that the average path length is 16 mi. Battan (1959) calculated that the mean path length is instead only 2 mi long. To arrive at this conclusion, however, Wolford rejected those accounts of tornadoes that presented discontinuities. He also indicated that in 95% of the cases, the path length was

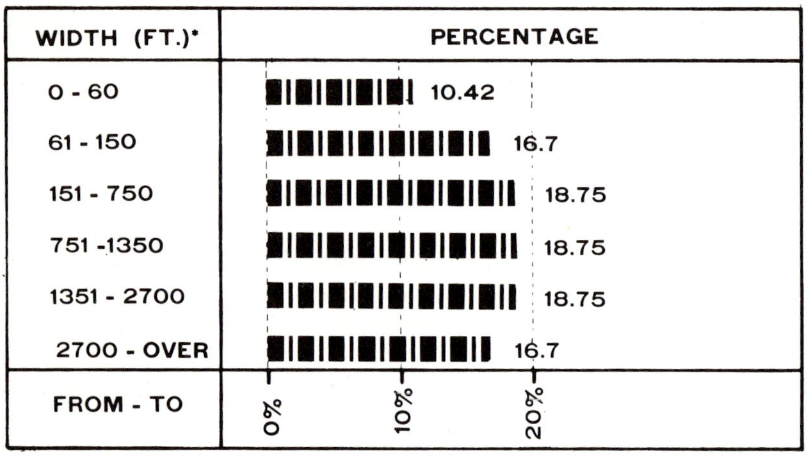

WIDTH (FT.)*	PERCENTAGE
0 - 60	10.42
61 - 150	16.7
151 - 750	18.75
751 - 1350	18.75
1351 - 2700	18.75
2700 - OVER	16.7
FROM - TO	0% 10% 20%

*Damage width does not represent the funnel diameter on the ground, since more than one tornado could translate simultaneously.

Figure 2-32. Distribution of tornadoes in Australia according to size (width of damage path).

found to be less than 8 mi. Thom (1963) concluded that the average path length for tornadoes in Iowa and Kansas was 3.94 mi and the average path width was 154.5 yd. Howe (1974) calculated that the average length and width for tornadoes on the east side of the Rocky Mountains were, respectively, 4.7 mi and 165 yd.

Usually, the prevailing direction of a tornado path is from southwest toward northeast. The paths are mostly curved, and the ends tend to veer toward the north. Wegener (1917) observed that tornadoes in Europe usually turn to the right when emerging from wooded areas into open fields or passing from land to water, and turn left when they pass from water to land. The edges of the path are not continuous smooth lines, but reflect a rather erratic motion with sudden changes of direction and diameter; they also rise at times from the ground and subsequently descend.

It is important to notice that the path indicates the extent of visible damage. This does not necessarily correspond to the edges of the vortex.

There have been many tornadoes with lengths over 100 mi. The longest tornado ever recorded struck Mattoon, Illinois, on May 26, 1917; it was 293 mi long. Widths of up to 2 mi have been measured. The tornado that hit Lubbock, Texas, on May 11, 1970, was estimated to have been 2 mi wide at some points.

Violence as a Function of Path Length

We define the *violence* or *intensity* of a tornado as *the absolute value of the energy involved per unit area of the path.* In this definition, however, we can substitute the wind speed, or the wind pressure, or the destruction (i.e., casualties, fatalities, and damage) per unit area in place of the absolute value of energy, since these are expressions of the energy itself. Since many factors seem to be associated with the intensity level of tornadoes, either as causes or as effects, it should be pointed out that the length of the path is one of the effects that could be used as an index of tornado violence. In other words, tornadoes with longer paths appear to be more violent.

Supporting this theory, we cite the work of several others in recent years. Among these, Wilson and Morgan (1971) indicated that tornadoes with long paths caused more deaths and damage per mile of length than tornadoes with shorter paths. Pearson (1971) also supported this thesis. Fujita and Pearson (1973) indicated, from an analysis of 740 tornadoes that had occurred in 1972, that length and size were proportional to the violence of the storms. More specifically, they calculated that higher intensity was associated with path lengths varying from 3 to 30 mi, and path widths varying from 175 to 556 yd.

Forward Velocity

The tornado system as a whole, from the mother cloud down to the tip of the funnel, moves horizontally over the earth's surface with a velocity that varies both in magnitude and direction from instant to instant and from storm to storm. Although the general direction of the forward motion is usually from southwest to northeast, the path is sometimes very erratic. For instance, there are accounts of storms that have turned 180°. Because the translational velocity is not constant, we must distinguish two different velocities: average and instantaneous.

There are countless records of average velocity because such measurements are easy to make; conversely, little is known of the instantaneous velocity except that it is very erratic, at times dropping to 0 mph and then increasing at an unknown rate.

In terms of average forward velocity, we have minimum and maximum average values computed from recorded cases, and also mean values that have been published by different authors. The minimum average speed recorded is 5 mph for the storm of May 24, 1930, in Pratt, Kansas. Maximum average speed recorded is 65 mph during a storm that occurred in Kansas from Grenola to Uniontown on May 25, 1917. Flora (1954) also mentioned the tornado of March 15, 1938, in Batesville, Illinois, which moved at a speed of 60 to 65 mph, and the tristate tornado of March 18, 1925, that moved at a speed higher than 59 mph in Indiana. The U.S. Department of Commerce (1965) indicated that the highest translational speed ever recorded is 68 mph. Martin (1940) calculated the mean to be 45 mph on the basis of 1000 tornadoes selected by random and reported by Flora (1954). Tepper (1978) assumed the mean to be between 25 and 40 mph, with an average therefore of 32.5 mph. From Kessler (1970), the average path length would be 10 miles and the average duration 20 minutes; thus, the mean would be 33 mph.

Since there are no data indicating what the maximum instantaneous velocity could be, we can only speculate. Naturally, since we know that the forward velocity can at times drop to 0 mph and also change or reverse direction, it is obvious that there are also peak velocity values that are above the average forward velocity. Thus, we know that we cannot use the average values in design problems and be safe—some allowance must be made.

Knowledge of the instantaneous translational velocity is essential in order to compute the following:

1. the total horizontal component of the velocity that is the vectorial sum of forward and tangential velocity.
2. the rate of loading due to wind and to pressure change. A direct consequence of computing (1) and (2) is to establish whether such loads act dynamically or statically.
3. the rate of pressure change in order to design adequate venting systems (see Figure 2-33).

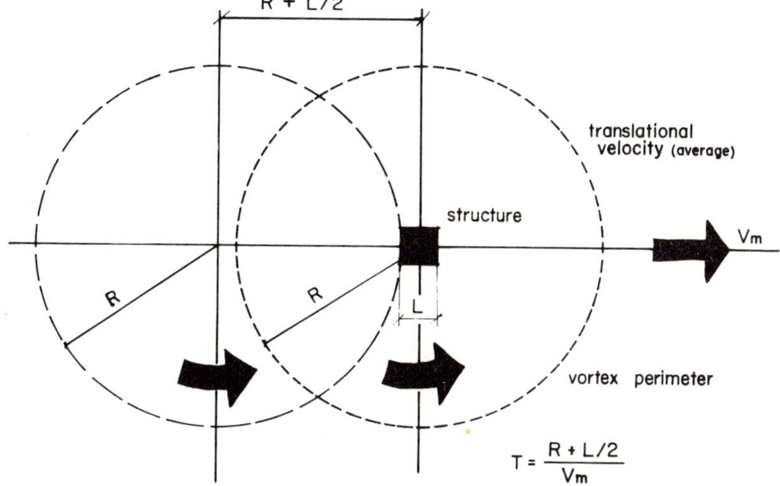

Figure 2-33. Rate of pressure change as a function of the vortex radius.

Tangential Velocity

Tangential air velocity in the funnel is perpendicular to the vertical axis of the tornado and to the radius of the orbits in which the air particles travel. At a given altitude, the tangential velocity varies with the radial distance according to a law that resembles that of the Rankine's vortex. This flow consists of a forced vortex inserted in the core of a free vortex (Figure 2-34). In the forced vortex, which extends from the axis to a distance *R*, the fluid particles have a viscosity that makes the flow rotational;* therefore, the fluid acts as a solid. In fact, as in a solid, velocity varies linearly from zero at the center to a maximum at the edge.

* In the rotational flow, each particle rotates around its own centroidal axis as it revolves around an axis common to all. To visualize the phenomenon, let us imagine that each particle has an arrow marked on it. If at a given moment all arrows are pointing toward the common axis, they will always be pointing toward it. If we compare them with a stationary arrow pointing toward the north, they are constantly changing in direction, which indicates that the particles are rotating. Vice versa, if the arrows at a certain instant are all pointing north and continue to do so constantly during their revolution around the common axis, the flow is called *irrotational* (Figure 2-35).

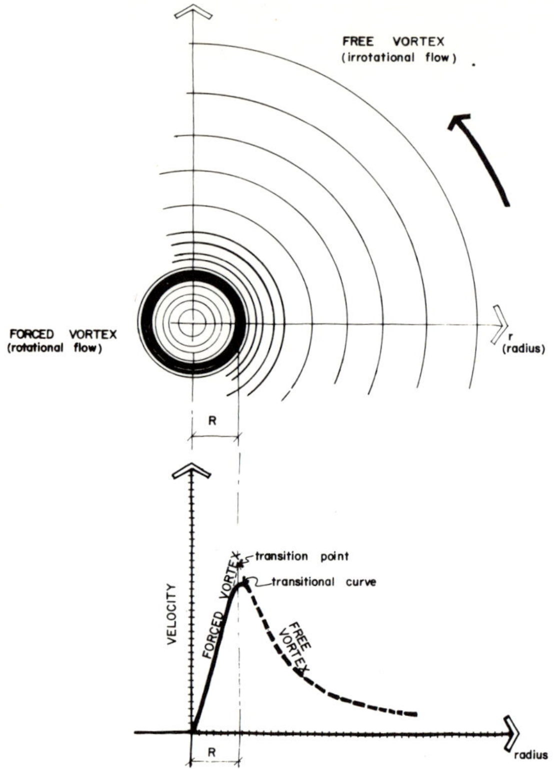

Figure 2-34. Rankine vortex: combination of an interior force vortex and an exterior free vortex.

In the forced vortex, the flow is rotational as in solids. The equation of motion in this case is:

$$V = \xi r/2 = (K/2\pi)(r/R^2) \tag{1}$$

where

ξ = vorticity
r = radius at any point
R = maximum radius at the forced vortex, or minimum radius of the free vortex (critical radius)
$K = \xi \pi R^2$, strength of the forced vortex

In the free vortex, the flow is irrotational, which is typical of ideal fluids. The equation of the flow is:

$$\begin{aligned} V &= K/2\pi r \\ Vr &= K/2\pi = \text{constant} \end{aligned} \tag{2}$$

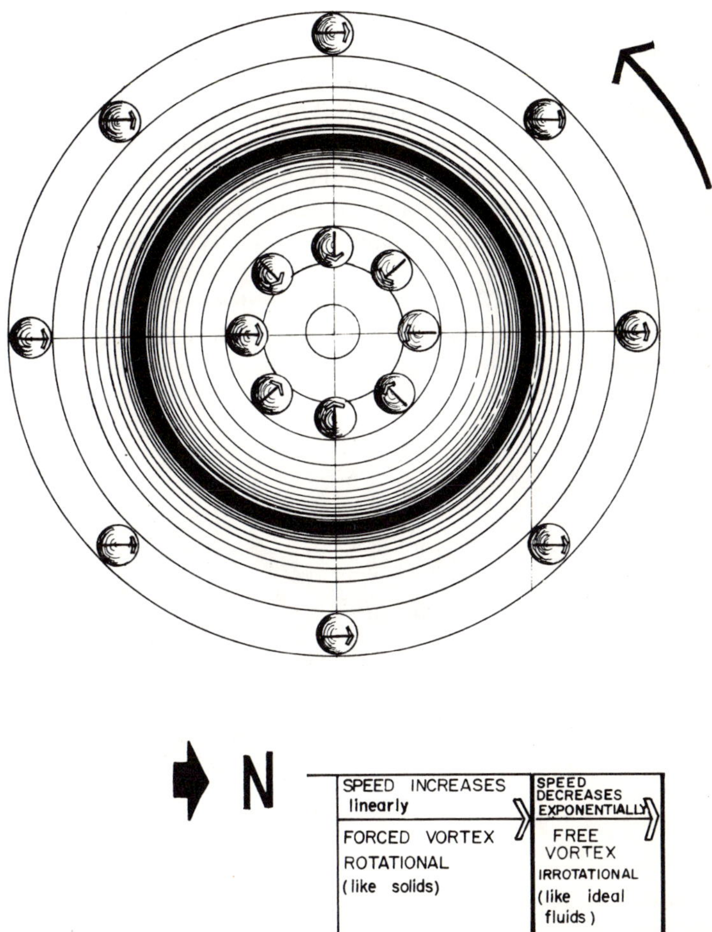

Figure 2-35. Rotational flow in the interior and irrotational flow on the exterior in a combined Rankine vortex.

The equation is represented by an hyperbola where $v = 0$ and $r + 0$ are, respectively, the axis and the asymptotes.

Such a pattern of airflow is corroborated by a study of a tornado in Dallas on April 2, 1957. Hoecker (1960) found that at an altitude of 150 ft, the tangential velocity increased linearly from 0 mph at the axis to a maximum of 170 mph at a radical distance of 125 ft. Then it continued to decrease up to a distance of 250 ft following the equation $Vr^{1.6}$ = constant. Then, from a 250-ft radius outward, the velocity continued to decrease according to the equation $Vr^{0.8}$ = constant. Such a variation from Equation 2 indicates that the flow at this altitude is not irrotational, but, at an elevation of 1200 ft, data do indicate an irrotational flow. Kessler (1970) is in agreement with a distribution of tangential velocity similar to that of the Rankine's combined vortex.

The transition from rotational to irrotational flow at a distance R from the axis is not abrupt, but follows a transitional curve with a consequent reduction of the maximum velocity (see Figure 2-36).

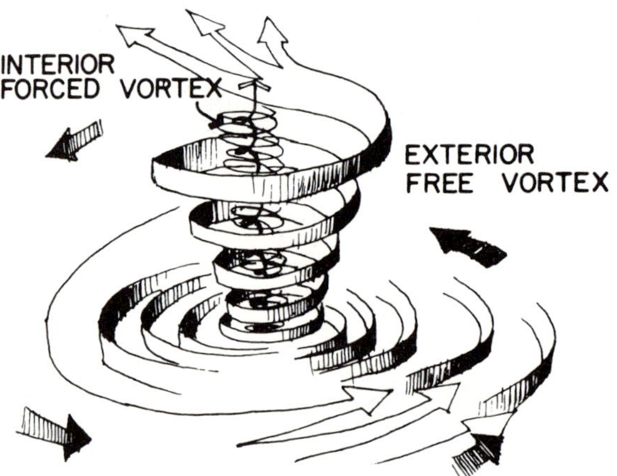

Figure 2-36. There is an almost general consensus that a tornado vortex is a combined Rankine vortex, consisting of a free vortex on the periphery and a forced vortex in the core. In the free vortex, typical of ideal fluids, the velocity increases exponentially as the air particles get closer to the center. In the forced vortex at the core, the velocity decreases linearly as the air particles get closer to the center, just as it does with solid bodies.

Horizontal Component of Resultant Velocity

The vectorial sum of the forward and tangential velocity is the horizontal projection of the resultant velocity. Such a sum is a vector that, for a given air particle, changes continuously in magnitude and direction as the particle rotates and moves forward. For a counterclockwise rotation of the vortex (typical in the Northern Hemisphere), the vector is maximum on the right side of the path, minimum on the left, and intermediate in between. This is illustrated in Figure 2-37, where $\overline{OV}$ represents the magnitude of the vector that varies with the angle θ. Damage on the right side of the path is more severe because of stronger winds on the right (i.e., the observer's right).

General Estimates of Maximum Wind Velocity

Several statements about estimated maximum wind velocity occurring in tornadoes have been made by different authors. These statements are not substantiated directly

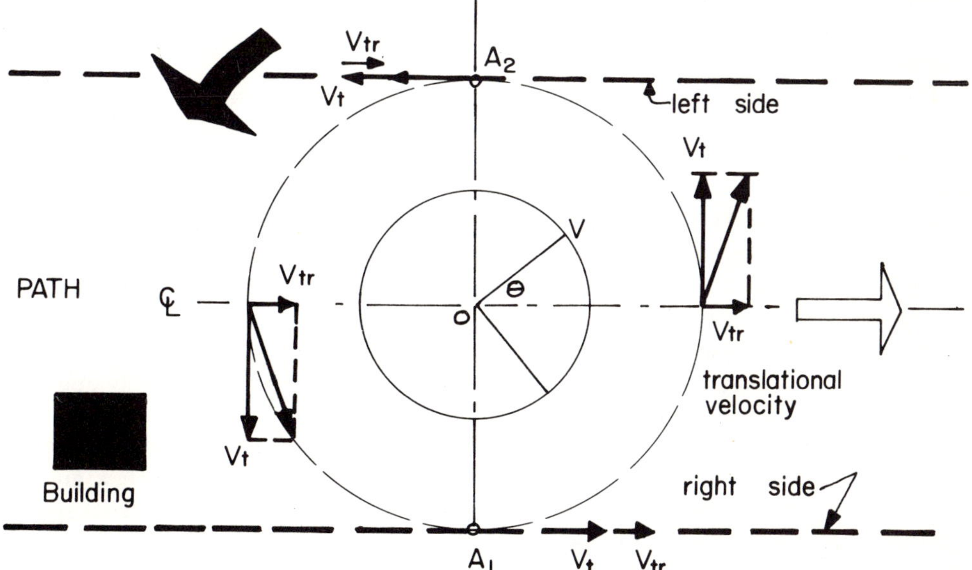

Figure 2-37. Vectorial sum of the tangential and translational velocity.

by any measurement; rather, they represent the opinions of experts. They must be mentioned in order to give a representative illustration of what is known today about tornadoes and to clarify misunderstandings.

In 1961, the American Society of Civil Engineers Task Committee on Wind Forces expressed its conclusions about tornado forces as follows:

> The peripheral wind velocities have been estimated to be in excess of 300 mph, on the basis of damage done; and the dynamic pressure due to the velocity is 230 psf, which corresponds to an average design pressure of 260 psf.

Flora (1954) reported several indicative figures on wind velocities. For example, the estimated maximum rotary wind velocity was 454 mph for the Texas Panhandle tornado of April 6, 1947. Flora also mentioned that, from the observation of damage of many tornadoes, wind velocities ranged from 450 to 500 mph. Moreover, Flora suggested that much evidence indicates that, in some parts of the vortex, the velocity is on the order of magnitude of the speed of sound. A similar conclusion derives from a theory proposed by Abdullah (1955) on supersonic wind speed in the "area limited by the 'limiting circle' and the 'critical circle.' "

Chang (1969) described a tornado as a long vortex core with maximum circumferential velocity ranging from 100 to 400 mph.

Tepper (1978) reported that whirling speeds in tornadoes have been estimated at up to 500 mph.

A report by the U.S. Department of Housing and Urban Development (1966) presented several recommendations for stormproofing houses against hurricanes and tornadoes. The report quotes an engineer who headed a group of technicians in investigating storm damage in Florida and Mississippi: "We know now that the forces of a tornado are not irresistible. We once thought that their velocity was 600–700 miles per hour, but actually it ranges from 100–300 miles per hour and probably approximates 200–250." (See Figure 2-38.)

Edwin Kessler, director of the National Severe Storms Laboratory in Norman, Oklahoma, stated that his personal belief, which was "little better than a guess," was that the damage observed in tornadoes could have been produced by wind speeds of 300 knots (150 m/sec, or 335 mph).*

* Kessler to P. A. Morris of the U.S. Atomic Energy Commission, February 21, 1967.

Figure 2-38. Tornado damage to dwelling in Inverness, Mississippi, in 1971. Notice the garment hanging on the wall *(center of photo)* while the rest of the building has been demolished. (Courtesy of National Oceanic and Atmospheric Administration)

A pamphlet from the U.S. Weather Bureau (1964) presents a concise list of information on tornadoes. It states that the estimated wind velocity within a tornado is more than 300 mph. Estimates of wind velocities for tornadoes, to be used for the design of nuclear power plants, had originally been presented in 1967 by Swanson et al. More specific estimates of these velocities were published in 1974 by the U.S. Atomic Energy Commission.

Walter H. Hoecker of the Atmospheric Trajectory Branch of the Environmental Science Services Administration points out that indirectly measured wind speeds approach 300 mph.*

Swanson et al. reported that wind velocities estimated from structural damages have ranged below 250 mph. They also stated:

* Walter H. Hoecker 1967: personal communication.

A few estimates have been made of speeds-not-exceeded for structures that have withstood the full force of tornadoes. These range from about 200 to 500 mph. It appears that a maximum wind speed of 300 mph is conservative for the "average" tornado, and that a speed of 500 mph has a probability in any given storm that is at or below the level of 2 percent.

Fujita et al. (1967) mention several investigations made by different researchers on structural damages caused by tornadoes and the proposed minimum wind velocities estimated. In conclusion they wrote, "It seems reasonable to assume that minimum speeds ranging between 55 and 217 mph are required to result in typical damage left behind by Midwestern tornadoes."

Maximum Wind Velocity Estimated from Structural Damage

Three transmission towers were destroyed in a tornado in Worcester, Massachusetts, on June 9, 1953. Booker (1963) reported that the towers were previously tested to failure in the factory and that their strength was specifically known. This was quite important because most conclusions based on structural analysis are affected by the validity of the assumptions made; in this case, the test to failure was unequivocally certain.

The towers were identified with the numbers 113, 115, and 116. The direction of the power line supported by the towers was almost at 90° from the direction of the tornado. Tower 113 was to the left of the tornado's centerline path; towers 115 and 116 were to the right. Tower 114, which resisted the tornado winds, was almost in the center.

On the basis of the counterclockwise circular motion of the storm, we can say that the three towers were exposed to transverse winds while tower 114 was subjected to frontal winds. According to Booker (1963), towers 116 and 115 fell forward and tower 113 fell backward, as a result of the counterclockwise direction of the rotating winds.

The minimum wind velocities that would have caused the collapse of the towers are 148 mph for tower 113 and 170 mph for tower 116. Engineering calculations show that the velocity near tower 114 was about 343 mph.

An estimated wind speed was determined by the evaluation of structural damage in connection with small tornadoes that occurred on March 17, 1975, at Avim, Port

Bolivar, and Texas City, Texas (Bigler 1975). The analysis of a damaged water-storage tank in Texas City provided the necessary information for the computation of the minimum wind speed required to produce the damage observed.

The tank was 13.58 ft high and had a diameter of 6.05 ft. In the computations, the tank was assumed to have been completely filled with water. The minimum wind forces required to overturn the tank and to shear the tank off a wooden base were computed. It was found that the minimum wind forces required to produce those effects were 303 mph for overturning and 285 mph for slippage. Considering that at 285 mph the slippage would occur first, this is the value assumed for the wind speed.

Important data on wind velocities can be found in a study by Segner (1960). These data are the results of computations of the minimum wind speed required to produce damage to various structures during the Dallas tornado of April 2, 1957.

The analysis was conducted for each structure, considering several reasonable modes of failure. The results were influenced by so many factors that they have different degrees of dependability. One point, which unfortunately can produce some doubts about the validity of the results, is the impossibility of determining whether the damage was due to wind forces only or to flying objects as well.

The results of the study were as follows:

1. Two 8-in. concrete-block walls in a service station were blown outward. The two walls were 12 ft high and were perpendicular to each other. The minimum wind speeds calculated were 91.6 mph for one wall and 92.2 mph for the other.

2. A small, empty storage tank, 4 ft in diameter and 5.29 ft high, was overturned. The steel tank was 0.125 in. thick, weighed 472 pounds when empty, and was supported on four legs braced horizontally. The supporting structure consisted of 1-½ × 1-½ × 3⁄16-in. steel angles. Eyewitnesses saw the tank rolling for about 100 ft after it had overturned. The author estimated a minimum wind speed of 55 to 65 mph for the failure of the supporting structure and greater speeds to account for the rolling action of the tank after it was overturned.

3. A flat roof covering a two-story structure was lifted. The roof deck, which was estimated to weigh 13 lb/ft^2, was nailed to the supporting 2 × 10 joists. Neither the joists nor the ceiling attached below were damaged, but the deck was blown

off. Segner assumed that only 50% of the common nails anchoring the deck to the joist were effective, and concluded that the lift was caused by a minimum wind speed of 179 mph.

4. Railroad freight cars were overturned. Segner considered eight cars that were overturned and computed the required wind velocity for each car as follows: 128, 144, 144, 217, 83, 144, 105, and 143 mph. The conditions for each car were different because of such factors as weight, amount of load, mode of overturning, and position of doors (open or closed).

5. A 45-ft elevated signboard collapsed. This structure was built with 13 yellow-pine poles, each 45 ft high with an approximate diameter of 15 in. The poles, which were spaced approximately 8 ft on center, were braced together with bracers starting 28 ft from the ground and extending to the top. The signboard itself was 56 ft long and 17 ft high. The minimum velocity of the wind was calculated to be 302 mph. Among the several modes of failure, the most severe that Segner assumed to have been possible was that the billboard failed first and then each pile collapsed individually under the force of the wind acting on the pile surface. The velocity computed in this case was the highest calculated in Segner's analysis. Noticing the contrast of this value to the other wind velocities, Segner suggested that, instead of direct wind action, the impact from flying objects might have caused the structure failure.

6. Perimeter walls of a 200-by-160-ft rectangular building collapsed. The exterior walls were 12 in. thick, consisting of 8-in. lightweight concrete block and 4-in. brick interlaced. The wall-clearance height was 18 ft, 5 in. Of the four exterior walls, two collapsed outward, one collapsed inward, and only one resisted the tornado. For the two walls falling inward, the minimum wind speed was calculated to be 107 mph; for the one falling outward, the velocity was 109 mph.

7. An empty truckbed overturned. Segner assumed that a combination of vertical uplift and horizontal wind action overturned the truckbed and the hydraulic hoist connected to it. He estimated the effects of the shape factor and therefore considers the result to be approximate.

8. A 42-ft-high flagpole, consisting of three steel pipes of different diameters, was

bent. Starting from the base, the first section was 3.5 in. OD with a height of 15.5 ft from the ground. The second section, 17.7 ft long, was 3 in. OD, and the third section, 8.8 ft long, was 2.5 in. OD. The pole, which, extended 6 ft into the ground, was embedded in concrete. The pole remained straight up for its entire length, but it yielded at the connection with the concrete as shown by the permanent deformation confined to that point. Segner considered a minimum yield point of 35,000 psi and estimated the minimum wind speed to produce the yielding at 115 mph, assuming an elastic action, or 133 mph for plastic behavior.

9. A roof section was torn from a warehouse. The portion of the roof considered in the analysis was 25 by 21 ft and consisted of 3-in. longleaf pine sheathing with four-ply tar and gravel roofings. Segner could not determine whether the lifting was due to the typical pressure drop or to suction produced by wind action. The wind velocity estimated was 189 mph; however, the author advised that the accuracy of the result should be assumed to be about 25% because of the approximate assumptions about the joint failure.

Maximum Wind Velocity Estimated from Ground Marks

In a study of the North Platte Valley, Nebraska, tornado of June 27, 1955, Van Tassel (1955) arrived at interesting conclusions and derived a wind velocity of 484 mph in the funnel. On aerial photographs taken about 40 hr after the occurrence of the tornado, ground marks were clearly visible on some cultivated fields. The marks were actually ridges or dikes, from 0.25 to 0.5 in. high, arranged in waves. They were systematic and elliptical in shape, and the dimensions were calculated by scaling the photographs. The major axis was calculated to be 230 ft and the minor axis 152 ft. The forward speed determined by radar observations was estimated at 12 mph; the average distance between rings at the center of the leading edges was determined to be 15 ft, 4 in. The basic assumption was that the marks were produced by some object carried by the tornado within the funnel.

The wind speed was determined by use of the equation:

$$V = eNS$$

where

V = wind speed (mph)
e = approximate perimeter of the ellipse (mi)
N = number of rings per unit length (in 1 mi)
S = forward speed of the tornado (mph)

In the formula that follows, a = one-half of the major axis, and b = one-half of the minor axis.

$$e = 2\pi \sqrt{\frac{(a^2 + b^2)}{2}}$$

$$= 2\pi \sqrt{\frac{(115)^2 + (76)^2}{2}} = 2\pi \sqrt{9500}$$

$$= 612 \text{ ft}$$

$$= 612 \text{ ft}/5280 \text{ ft/mi} = 0.116 \text{ mi}$$

$$\text{Number of orbits in 1 mi: } N = \frac{5280}{15.3} = 345$$

Assumed: $S = 12$ mph
$V = 0.116 \times 345 \times 12 = 480$ mph

Several wind velocities have been estimated by an analysis of ground marks visible along the paths of tornadoes after the occurrence of the storms. These marks are usually of cycloidal shape, indicating two component paths—one rotational and the other rectilinear—representing the rotational and forward motion of a point moving with the tornado.

Interpretations of the mechanics that produce these marks vary among the researchers who investigate the phenomenon. Van Tassel (1955) assumed, for instance, that the marks are produced by an object picked up by tornado winds and moving according to the direction of the wind. Fujita et al. (1967) had a different opinion. They assumed that among several different types of marks left on the ground by tornadoes, one, of cycloidal shape, is produced by a suction force. Around the tornado center, there are several spots in which this suction force is concentrated; these spots are not larger than 50 ft in diameter and rotate around the tornado center. The suction power at these points is less than that of a household vacuum cleaner.

In analyzing the cycloidal suction marks left by a tornado on April 21, 1967, which extended southwest of Belvidere to north of Woodstock, Illinois, Fujita et al. concluded that the rotational speeds on two occasions were 150 to 160 mph. The forward speed of the tornado was measured to be 50 mph. The resultant velocities, found by adding vectorially the rotational and forward speeds, were 210 and 200 mph, respectively.

Using the same method, Fujita et al. computed several velocity values for the suction spots observed after the Fargo, North Dakota, tornadoes of June 20, 1967, and the resultants of rotational and forward velocities were 172, 176, 173, 180, 180, 173, and 166 mph. In this case, the forward speed was 62.5 mph. These values are very close to those calculated by Fujita et al. when they analyzed motion pictures of the Fargo tornadoes. This is, of course, very encouraging because the two completely different methods of analysis agree on relatively low maximum velocities.

Wind Velocity Derived from Scaling Motion Pictures

Hoecker (1960) presents a large number of tornado wind-velocity values for the Dallas tornado of April 2, 1957. High-quality tornado movies were available for study of the Dallas tornado, including some scenes taken with a 10X magnifying telephoto lens. In the funnel, fragments of clouds and debris in a circular motion were clearly visible, and their location at different time intervals could be identified quite accurately because the time periods between pictures were known.

Hoecker (1960) presents a distribution of wind velocities in mph at several points located at different radial distances from the axis of the tornado funnel and at different elevations from the ground. The wind velocities were determined only at those points

Table 2-6. Wind velocity estimates.

a. Maximum speeds (mph)		
Chang (1966)		400
Crompton (1966)		300
Flora (1954)		500
Fujita (1967)		217
Hoecker (1967)		300
Kessler (1967)		335
Swanson et al. (1967)		500
Task Commission on Wind Forces (1961)		300
Tepper (1978)		500
U.S. Weather Bureau (1964)		300 and over
b. Average maximum speeds (mph)		
Flora (1954)		450 to 500
Chang (1966)		100 to 400
Crompton (1966)		200 to 250
Swanson et al. (1967)		200 to 500 300 most probable
c. Maximum speeds estimated from structural damage (mph)*		
Bigler (1957	Water tank	285
Booker (1954)	Transmission tower	148
	Transmission tower	170
	Transmission tower	343
Segner (1960)	Concrete-block wall	91.6
	Concrete-block wall	92.2
	Storage tank	65
	Roof	179
	Railroad freight car	128
	Railroad freight car	144
	Railroad freight car	217
	Railroad freight car	83
	Railroad freight car	144
	Railroad freight car	105

* These values represent speeds required to cause the damage observed. Actual maximum speeds may, of course, be higher.

Table 2-6. (Continued)

Railroad freight car	143
Signboard	302 (doubtful)
Brick-and-block wall	107
Brick-and-block wall	109
Truckbed	133
Roof	189
d. Maximum speeds estimated from cycloidal ground marks (mph)	
Fujita (1967)	176
	173
	180
	180
	173
	166
Fujita et al. (1967)	210
	200
	172
Van Tassel (1955)	484
e. Maximum tangential speeds estimated from scaling motion pictures (mph)	
Fujita (1960)	112
	112
	93
	100
	100
	111
	100
	102
Hoecker (1960)	170
f. Maximum speeds estimated from funnel shape (mph)	
Dergarabedian and Fendell (1971)	208
Glaser (1959)	230

where, by chance, tracer particles were located; but, it is probable that higher wind velocities could have existed at points different from the ones analyzed. Hoecker assumed that the moving tracer elements, cloud fragments, debris, and dust particles were moving at the same speed as the air in which they were imbedded, neglecting aerodynamic shape factors.

The largest velocity value derived was 170 mph, which is a tangential component of the velocity. The point in the tornado where this velocity was calculated, was located 225 ft above the ground and at a 130-ft radial distance from the center of the funnel.

Fujita et al. (1967) derived tangential wind velocities for the Fargo tornadoes of June 20, 1957, from a study made of motion pictures showing the tornado in action. Fujita analyzed seven cloud pendants appearing at the base of a sheared-off funnel just above the ground. These cloud pendants were located at a radial distance from the funnel center of 296, 196, 98, 196, 360, 328, and 296 ft. The velocities were 112, 112, 93, 100, 111, 100, and 102 mph, respectively.

Maximum Wind Velocity Estimated from the Shape of the Funnel

Glaser (1959) reported maximum tangential wind speed computed for the tornado of March 21, 1956, in Texas. The method used was based on the evaluation of the shape of the tornado funnel. Fujita (1967) reported Glaser's calculation of maximum tangential wind speed as 230 mph. This tangential wind speed was computed for an elevation of 820 m above the ground and at a radial distance of 61 m from the center. Table 2-6a–f lists wind-speed values derived by various sources. The values have been grouped according to the criteria that were used to determine them.

Vertical Velocity

Although tornadoes have often been described as nature's giant vacuum cleaners, Fujita et al. (1967) concluded that their suction power near the ground is even less than that of conventional household vacuum cleaners. This is also confirmed by the quick decreasing of the vertical velocity as altitude diminishes. Hoecker's work (1960) on the Dallas tornado includes calculations of vertical wind velocity, which was found to vary with altitude. Along the vertical axis of the tornado, upward velocity was calculated to be 20, 50, 110, 152, and 0 mph at the respective elevations of 0, 6, 50, ·135, and 800 ft above the ground. Such findings indicate a constant increase of speed up to an altitude of 135 ft and then a linear decrease to 0 mph at an elevation of 800 ft. This change in velocity seems to result from the friction effect of the ground that extends upward to a substantial height.

Pressure Change

The relationship between air pressure and velocity in the tornado vortex can be expressed by the cyclostrophic equation:

$$\frac{\partial p}{\partial r} = \frac{\rho V^2}{r}$$

where

P = pressure
r = radial distance
ρ = density
V = velocity

Since the relationship between tangential velocity and radial distance is known, as previously shown, it is possible to integrate the cyclostrophic equation over the radius, starting from a distance where the pressure is undisturbed. From the integration, we obtain the radial distribution of the pressure as theoretically expected in a tornado. Hoecker (1960) used this method in his study of the Dallas tornado. From the velocity field, derived from motion pictures of the vortex, he obtained the pressure field. The pressure distribution in a vertical plane through the tornado axis is represented by plotting the isobar lines as functions of height and radial distance. For the Dallas tornado, the spacing of the isobars diminished up to a certain radius and then increased toward the axis. Obviously, the spacing of the isobars is inversely proportional to the wind velocity. In fact, as the spacing decreases, wind speed increases.

Since the vertical component of the wind force can be estimated from the pressure gradient, the distribution of this component can be derived for any point. It follows that near the axis, there is a zone where the force is directed downward, while in the outer region the force is upward. Note that such a double vertical circulation is in agreement with the latest observations made on waterspouts.

Estimates on Maximum Pressure Differential

Several authors have expressed opinions on the variation of pressure within tornadoes. Finley (1884) considered the possibility that the center of a tornado vortex is a near-vacuum because of the centrifugal force of the tornado cloud, thus implying a pressure drop of about 2000 psf. Teesdale (1928) gave an estimated maximum value of pressure drop of up to 0.5 atm, about 1000 psf. C. F. Brooks (1949) assumed the pressure in the funnel to vary between 0.8 and 0.5 atm (from 1600 to 1000 psf).

Ferrell (1889) assumed the pressure in the funnel to be 0.75 atm, which means a reduction of 0.25 atm (500 psf). Logie (1919) claimed that the reduction of pressure exceeds 50 mb (104 psf) and added that many authorities assume that 500 mb (1040 psf) could be the maximum value. Humphreys (1940) believed the reduction of pressure to be about $\frac{1}{10}$ of an atmosphere (about 200 psf). E. M. Brooks (1951) stated that a probable value could be greater than 200 mb (418 psf).

Abdullah (1955) arrived at a minimum pressure of 500 mb at the limiting circle of the tornado, starting from a pressure of 1000 mb at infinity. The pressure drop is therefore 470 mb at the limiting circle, and it was assumed that this value also could be considered valid for the center of the funnel. A pressure drop of 470 mb is equal to a pressure of 981 psf.

Kessler (1970) suggests a maximum wind speed of 300 knots, or an equivalent pressure drop of about 200 mb, according to Bernoulli's equation:

$$200 \times 2.088 = 418 \text{ psf}$$

Unofficial Pressure Differential Measurements

One of the largest drops of atmospheric pressure recorded during a tornado is an unofficial measurement of questionable reliability. Several authors have mentioned this exceptionally high reading; however, there are some slight differences among them. In reporting on the tornado that passed through St. Louis, Missouri, on the afternoon of May 27, 1896, Frankenfield (1896) explained that he was given an aneroid barometer, with a metric scale to be reset. The barometer belonged to a Mrs. Klemm, and it had been in her house when the tornado struck. Her son had read the barometer at 680 mm (26.78 in.). Frankenfield compared this reading with the official readings of the U.S. Weather Bureau office in St. Louis and corrected the value of 26.78 in. to 27.30 in. to account for the difference in elevation between the Klemm house and the Weather Bureau office. The difference he reported was 2.05 in.

The U.S. Weather Bureau (1965) reported the same differential pressure reading to be 2.42 in. of mercury. Reynolds (1958) indicated that the variance between the two readings was from 2 to 2.5 in. Flora (1954) also mentioned this series of events.

Baier (1896) provided additional information on the subject. He made a complete investigation of the circumstances surrounding the reading of the pressure by Mrs. Klemm's son. He also checked the accuracy of the instrument. Baier concluded that the original value of 27.30 in., representing the corrected reading computed by Frankenfield, should be corrected to 26.94 in. This additional 0.36 in. brings the pressure differential from 2.05 to 2.41 in. of mercury (170.47 psf).

In the report on the tornado of August 20, 1904, in Minneapolis, Minnesota (Outram 1904), an unofficial pressure differential reading is mentioned. The reading is exceptionally high and therefore questionable; however, Outram seems to be convinced of the dependability of the two men who measured the barometric pressure. The men observed an aneroid barometer dropping to 23 in. and then rising almost immediately to the original position. Assuming that the initial reading was the same as that observed by the local forecaster (28.67 in.), the drop was 5.67 in. (401 psf). Outram notes that the aneroid barometer had been checked at the local weather station and compared with the other instruments not very long before, and had been found to be correct.

Official Pressure Measurements

The first recorded official measurement of pressure change in the eye of a tornado is probably the one reported in the *Monthly Weather Review* in reference to the tornado that struck Little Rock, Arkansas, on October 2, 1894. The center of the tornado seemed to have passed right over the telegraph office where the local Weather

Bureau office was also located. George S. Harkness, the observer at the office, made a report of the events. The tornado struck a building located on the south side of the telegraph office, throwing the second story over the office roof. Then the windows in the office were blown in. The wires connecting the instruments on the roof with the recording apparatus inside were broken, and many instruments were damaged. This happened at 8:28 P.M.; it appeared to Harkness that the tornado passed over the office in about 1 min. During this minute, the self-recording Richard barograph recorded a drop of 0.38 in. in the barometer. The diagram that furnishes the record of the variation of the barometer indicates that after a slight oscillation, the pressure dropped in a straight line from 29.31 to 28.93 in. and returned so quickly that it is difficult to tell the time elapsed during the oscillation.

The horizontal movement of the paper in the barograph was such that the recorded oscillation could have occurred in 1 min. or less, but it is not possible to establish the time period exactly. The trace of the pen would have been practically the same whether the time had been 1 min. or 1 sec. Even if the pressure gradient cannot be defined, this datum has great significance because of its dependability.

Another reliable pressure reading was taken during the tornado that passed over Dyersburg, Tennessee, on March 21–22, 1952. A weather station, equipped with a barograph, was operated by the Civil Aeronautic Authority at the local airport. The tornado path at that point was 258 yd wide, and the barograph was in the path located exactly 41 yd from the centerline. The reading indicated a pressure drop of 0.65 in. (45.98 psf). Carr (1952) indicated that the real drop may have been higher than the value indicated on the barogram trace because of a probable lag in the response of the instrument.

On August 20, 1904, a severe storm, which seems to have been a tornado, struck Minneapolis, Minnesota. According to Outram (1904), the official pressure reading was 28.82 in., which gradually fell to 28.67 in. after noon. "Just about the time of the greatest severity of the storm, the barograph dropped suddenly to 28.25, returning immediately and rising to 28.80, then dropping back quickly to 28.70." It then fell gradually until the next day. The jump of 0.55 in. is an official reading.

Pressure Values Estimated from Scaling Motion Pictures

Probably the most interesting and valuable data on pressure change are presented by Hoecker (1960). He estimated wind speeds in the Dallas tornado of April 2, 1957, from high-quality movies that showed the funnel and the movement of cloud fragments and debris circulating around the tornado. The study includes a diagram of the wind velocity in mph at several points located at different elevations above the ground and at different radial distances from the storm center. Curves connecting points of equal velocity (isotachs) were also shown. It should be pointed out that although velocities are indirect measurements, they are logical consequences of the direct measurements made. If the measurements from which the velocities are derived are correct, then the velocities are also correct. In the case of the motion pictures, the interval in time between one frame and the next is definitely known. The distance between two different locations of the same object at two different instants can be accurately scaled from the pictures. We must conclude, therefore, that these measurements are almost direct and are not based just on theoretical assumptions as are many others.

In a second paper, Hoecker (1961) elaborates on the wind-velocity values already derived and computes the values of the pressure gradients at the same points at which he had previously determined the wind velocity. To obtain the pressure gradients in millibars per meter (mb/m), Hoecker used the cyclostrophic wind equation previously introduced.

We must say that in the use of this equation, some assumptions are made; thus, even if they are theoretically justified, the resulting values cannot be considered as close to direct measurements as are the velocity values derived previously. The reader is referred to Long (1960) and Glaser (1959) for the justification for using the cyclostrophic equation.

In Hoecker (1961), the distribution of the pressure with respect to the distance from the tornado axis and the elevation above the ground is presented. Assuming the average forward velocity to be 27 mph for that particular storm, Hoecker also includes distribution of the pressure at ground level. Hoecker's data are very useful

Table 2-7. Pressure drop.

SOURCE	PRESSURE DROP ESTIMATES (IN. HG)	DIFFERENCE BETWEEN ATMOSPHERIC AND REDUCED PRESSURE (PSF)[2]
Abdullah (1955)	–	981 psf
C. F. Brooks (1948)	–	1000–1600 psf
E. M. Brooks (1951)	–	418 psf
Ferrell (1889)	3.35	500 psf
Finley (1882)	13	2000 psf
Humphreys (1940)	–	200 psf
Kessler (1970)	–	418 psf
Logie (1919)		1000 psf
Teesdale (1928)	6.7	1000 psf
	Unofficial Records	
Baier (1896)		–
Flora (1954)		–
Frankenfield (1896)		–
Outram (1904)		401 psf (5.67 in. Hg)
Reynolds (1958)		–
U.S. Weather Bureau (1894)		170.47 psf (2.41 in. Hg)
	Official Records	
Carr (1952)		0.65 in (45.98 psf)
Harkness (1894)		0.38 in. (26.88 psf)
Outram (1904)		0.55 in. (38.9 psf)
	Pressure Values Estimated from Scaling Motion Pictures	
Hoecker (1961)		60 mb (125 psf)

for the design of venting systems because in determining vent sizes, calculations are based on pressure and time factors. The maximum value of the pressure drop computed by Hoecker was 60 mb, which expressed in inches of mercury is 1.8, corresponding to a pressure of 125 psf. Table 2-7 contains some values of pressure drop, including estimated values, unofficial records, official records, and some measurements derived from scaling motion pictures of tornadoes.

Rate of Pressure Change

The rate of pressure change with respect to time, or to the distance from the tornado center, is an essential element for the design of tornado-resistant buildings. If the pressure difference between the inside and the outside of a building must be balanced, a venting system must be designed. Its basic function is to make a certain volume of air escape from the building as the pressure on the outside drops. The size of the outlets is determined according to the rate of pressure drop. Most important also is to establish whether the pressure load could be a dynamic load—which is possible if the time interval in which the pressure drops is equal to the fundamental period of vibration of the structure.

The absence of data for the peak forward speed makes accuracy in terms of rate of pressure change impossible.

Reynolds (1958) suggested an empirical value for the pressure gradient of 1 in. for 120 ft. Assuming a diameter of 0.25 mi and a constant forward motion of 40 mph, he derived that a point at the edge of the limit of destruction would be in the center of the storm after 11 sec. This procedure assumes that the pressure varies linearly from the undisturbed area to a minimum value at the center of the tornado, which is not probable for the entire cross section through the vortex.

Hoecker's analysis (1961) of the Dallas tornado indicated a maximum time rate of pressure change of 26 mb/sec. Notice that such a value was not constant but was observed only for 0.5 sec. Furthermore, this value was derived at a certain distance from the center of the storm (4 sec from the center, which corresponds to approximately 159 ft, for the translational speed of 27 mph used by Hoecker); therefore, it is not necessarily the highest value. Case studies of criteria adopted for the design of some nuclear power plants indicate the following:

1. A pressure differential of 3 psi in 3 sec was adopted for: Peach Bottom Atomic Power Station, Peach Bottom, Pennsylvania; Limerick Generating Station, near Philadelphia, Pennsylvania; Fort St. Vrain Nuclear Generating Station, Platteville, Colo-

rado; Surry Power Station, Surry, Virginia; Forked River Nuclear Generating Station, Forked River, New Jersey; Davis-Besse Nuclear Power Station, Oak Harbor, Ohio; Trojan Nuclear Power Plant, Prescott, Oregon; and others.

2. A pressure differential of 3 psi in 3 sec (1 psi/sec) followed by calm for 2 sec and a repressurization was adopted for: Arkansas Nuclear One, Russellville, Arkansas; Millstone Nuclear Power Station Unit No. 2, Waterford, Connecticut; and others.

3. A pressure differential of 3 psi in 5 sec was adopted for: Oconee Nuclear Station, Seneca, South Carolina; Vermont Yankee Generating Station, Vernon, Vermont; and others.

Horizontal Torque

When the resultant of the wind force and the reaction from the foundation do not lie on the same vertical plane, a horizontal moment is applied to a building. This can occur often because the foundation's center of gravity and the wind resultant do not coincide with the windward face of the building's center of gravity, due to the hyperbolic distribution of the wind force.

For instance, in the tornado in Lubbock, Texas, on May 11, 1970, the 21-story Great Plains Life Building received severe structural damage due to a horizontal torque (Figure 2-39). A service shaft located at one end of the building offset the horizontal resultant of the foundation from the geometrical centroid. A plastic deformation made the top of the steel structure rotate 12 in. to the east with respect to the base, forcing the building to be evacuated, with no prediction for future repairs. This indicates very clearly the suggested design procedure to avoid such torque. A residual horizontal moment due to hyperbolic distribution of the wind force can probably be resisted by a residual strength furnished by redundant members. For small structures such as dwellings, some strengthening methods can be employed. As indicated in Figure 2-40, the rotation of the structure can be prevented by eliminating translations and using tension members at the corners anchored in concrete foundation, or by using steel bracings notched in the studs for wood-framed structures, or by using concrete piers acting as stiffeners.

Figure 2-39. Damage caused by tornado of May 11, 1970, in Lubbock, Texas. Great Plains Life Building, a 22-story steel-frame building, probably the first to be tested by tornado forces. The steel frame deformed permanently, with a 12-in. twist between top of building and its base. Estimated wind velocities were 179 mph at 140-ft elevation, and vertical velocities of 95 mph at 30-ft elevation and 220 mph at 270-ft elevation. (Courtesy of the Institute for Disaster Research, Texas Tech University)

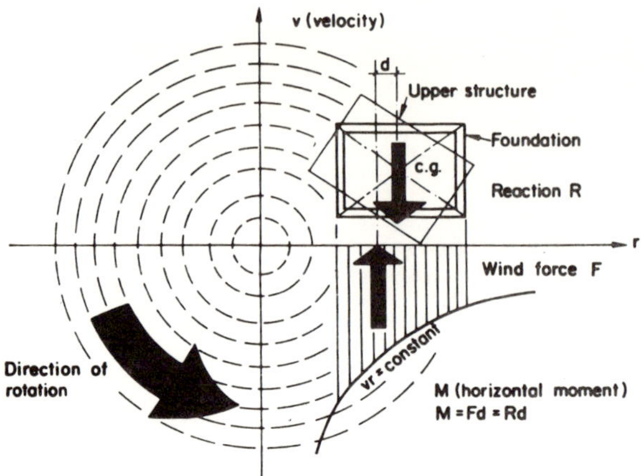

Figure 2-40. Horizontal twist of the structure with respect to the foundation that originates when the resultant of the wind force and the reaction from the foundation do not lie in the same vertical plane.

Figure 2-41. Among the many reports of freak accidents caused by tornadoes, one of the most extraordinary is irrefutably documented by this photograph. The Eads Bridge spanning the Mississippi River in St. Louis, Missouri, which had been struck by a tornado in March 1871 during its construction, was again struck by a tornado on May 17, 1896. Driven by the wind, a 2-by-10-in. wooden plank penetrated the 0.5-in.-thick wrought-iron web of one of the girders *(see arrow)*. (Courtesy United States Weather Bureau)

Foundations

We can see that for small-diameter tornadoes with high forward speed, it is more probable that the rate of loading can be high enough to produce a dynamic magnification with more severe consequences to the structure and foundations. Studies on the resistance of soils under dynamic conditions have indicated that for sandy granular soils, the shear strength and the bearing capacity decrease as the rate of loading increases. This is true, however, only up to a value of 0.002 in./sec for dry or saturated soil conditions. Furthermore, it was found that the wedge of soil under shear stresses was smaller as the rate of loading increased (Casagrande and Shannon 1949, Seed and Lundgren 1954, Whitman 1957, Whitman and Healy 1962). Further studies at Georgia Institute of Technology (Banks 1963, Woodard 1964) indicated that as the rate of loading was increased over 10 in./sec, the bearing capacity of dry sand increased up to the static value and then remained constant. For saturated sand instead, not only did the bearing capacity increase sharply but the ultimate strength became several times larger than the value under static loading. However, at ultimate strength, it was found that the settlements could reach 4% of the diameter of the footing, which cannot be considered acceptable for practical purposes.

Horizontal forces applied to foundations are opposed by reactions generated either by passive soil pressure or by friction between foundation and soil. In the case of large dynamic horizontal loads, researchers studied the response of the system. They concluded that in the case of frictional reactions acting alone, displacements are small and negligible. On the other hand, when reactions from friction and passive pressure act together, it was expected that the lateral displacement could be as high as 4% of the foundation depth. In view of these findings, it is recommended that, to reduce such displacement, the foundation should be so designed to provide a reaction due to pressure that is twice the required minimum.

Missiles

The literature on tornado accidents includes a large number of reports of objects that, driven by tornado winds, have penetrated solid material. For instance, Frankenfield (1896), reported that during the St. Louis tornado of May 17, 1896, a 2-by-10-in. wood plank was driven by the wind into a 0.5-in.-thick web of a wrought-

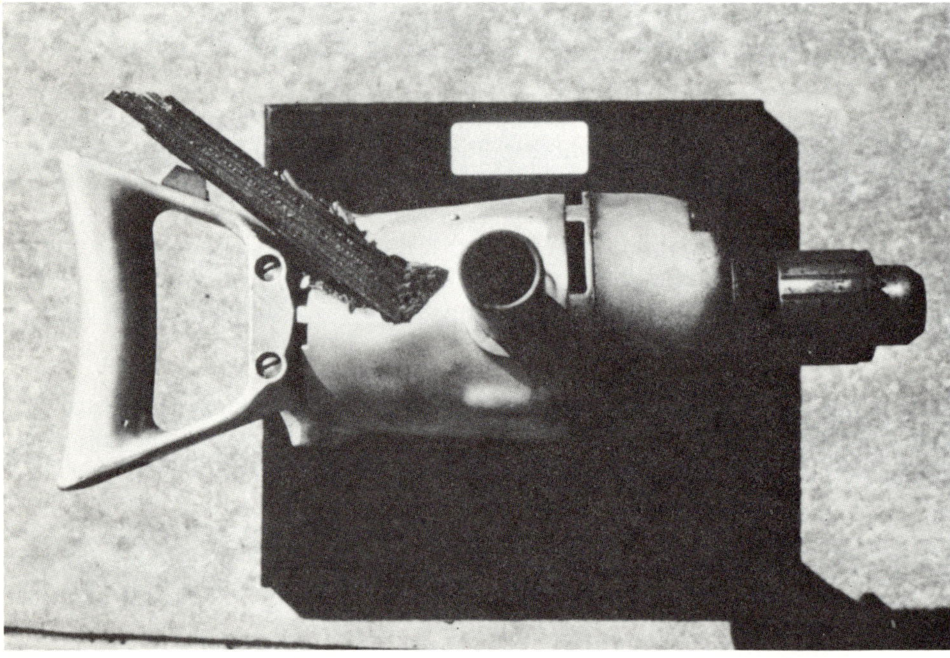

Figure 2-42. Metal casing of a drill penetrated by a piece of wood during tornado of April 30, 1949, in Norman, Oklahoma. (Presently on display in the Aerospace, Mechanical and Nuclear Engineering Department, University of Oklahoma. Photograph by National Severe Storms Laboratory, NOAA)

iron girder on the Eads Bridge. Documented by photographs, the incident provides irrefutable proof of the wind's effect (Figure 2-41).

Splinters of wood penetrating trees, straws penetrating solid wood, and so on are common occurrences during tornadoes. One of the most incredible incidents reported is that of a flimsy piece of wood that perforated the metal brake drum of an automobile during the tornado of May 31, 1947, at Leedy, Oklahoma. A similar accident was reported following the tornado of April 30, 1949, in Norman, Oklahoma. There, a wood splint pierced the metal casing of an electric drill (Figure 2-42).

Experimental work since 1967 has tried to reproduce phenomena of penetration similar to that occurring during tornadoes (Keller and Vonnegut 1976). A pneumatic gun, designed by W. Sykes (1967) of Arthur D. Little, Inc., can accelerate small particles such as straws and splinters to known speeds and reproduce the penetration of similar objects as reported after tornadoes. This gun has been helpful in evaluating wind speeds required to produce the penetration of objects into different materials (Table 2-8, Figures 2-43 and 2-44).

Tornadoes carry aloft a large amount of solids and condensed fluid particles that can move up to the speed of the air in the vortex. Thus, the characteristic dark color of the funnel-shaped column of air is actually the result of the material in suspension, including, of course, the condensation.

The types of debris that can be expected to be flying in the orbits of a tornado vortex are of the most incredible variety. For instance, when the storm moves over the countryside, the most probable materials to be found are loose soil particles, such as dust, sand, gravel, and rocks, plus fragments of vegetation, such as tree limbs that break and come loose. On the other hand, when the storm passes over built-up urban areas, the variety can be much greater since it can include debris from the streets, damaged or destroyed buildings, the interior contents of such buildings, and many other sources. (See Figure 2-44.)

Many objects can be picked up when their weights are small in comparison to the area exposed to the winds. Vehicles, in general, fit this category and can become windborne without reaching the high speeds of air particles. In fact, automobiles roll over and over and are dragged on the ground for short distances until they stop against some obstacle or are ejected from the vortex by centrifugal force (Figure 2-45). The damage they suffer is usually very severe, a fact that makes them very unsafe for their occupants. Likewise, they can inflict severe damage to structures when they collide. Other types of windborne material from urban areas can include items such as street lights, utility poles, signs, fences, manhole covers, electrical transformers, and telephone and electrical poles.

a

b

Figure 2-43. *a,* Pneumatic gun at the Atmospheric Sciences Research Center, State University of New York at Albany. *b,* Gun can fire splinters and straws into blocks of wood at velocities close to the speed of sound. The results of these experiments provide some indication of the wind speeds that must be involved when tornadoes cause straws and splinters to penetrate trees and wooden structures.

Figure 2-44. As the result of a tornado, splintered timbers penetrate the wall of a filling station. (Courtesy of National Weather Service)

Figure 2-45. Damaged car clearly indicates the vulnerability of automobiles during tornadoes. The house in the background sustained substantial damage from the passing tornado. (Courtesy of National Weather Service)

Table 2-8

a. Threshold Speeds Required for Penetration by Toothpicks

TARGET	THRESHOLD SPEED (M/SEC)	THRESHOLD SPEED (MPH)
Ponderosa-pine board (seasoned)	30	67
Fir plywood	70	157
Oak, dark, soft area (seasoned)	200	447
Maple board (seasoned)	215	481
Oak, light, hard area (seasoned)	>230	>515
Glass, 6-mm thick	>300	>671
Cement block	>310	>693

b. Threshold Speed Required for Penetration by Broom Straws

TARGET	THRESHOLD SPEED (M/SEC)	THRESHOLD SPEED (MPH)
Ponderosa pine (seasoned)	65	145
Fir, soft rings (seasoned)	75	168
Oak, dark, soft area (seasoned)	140	313
Dead, barkless elm-tree branch	170	380
Fir, hard rings (seasoned)	>200	>447
Maple (seasoned)	>210	>470
Oak, hard, light area (seasoned)	>215	>481

c. Maximum Observed Projectile Penetration and Associated Speed

PROJECTILE	TARGET[a]	PENETRATION (MM)	PENETRATION (IN.)	SPEED (M/SEC)	SPEED (MPH)
Toothpick	Pine board	Complete, 19	0.75	190+	425
Toothpick	Maple board	3.5	0.14	300	671
Toothpick	Maple board, wet	15	0.59	300	671
Toothpick	Elm branch	6	0.24	220	492
Toothpick	Soft oak	2	0.08	235	526
Toothpick	Plywood	Complete, 10	0.39	200+	447
Toothpick	Glass, 2.5-mm thick	Glass shattered		270	604
Broom straw	Pine	6	0.24	190	425
Broom straw	Pine, wet	6	0.24	190	425
Broom straw	Soft fir	9	0.35	170	380
Broom straw	Elm branch	2	0.08	180	403
Broom straw	Soft oak	2.5	0.1	185	414
Broom straw	Plywood	3.5	0.14	170	380
Bamboo straw, pointed	Pine board	Complete, 19	0.75	180+	403
Bamboo straw, blunt	Pine board	16	0.63	190	425
Bamboo straw, pointed	Maple	7.5	0.30	215	481
Bamboo straw, blunt	Maple	5.5	0.22	180	403

[a] All wood targets were seasoned wood.

Buildings, whether completely destroyed or partially damaged, are probably the highest contributors to the amount of debris. Roofs and walls produce debris such as shingles, boards, nails, plaster, bricks, blocks, and metal sheets. This last item, in particular, is extremely dangerous. In the case of the tornado in Dacca, East Pakistan, on April 14, 1969, where over 600 dead and 5000 injured were counted, the major individual cause of deaths and injuries was, in fact, the corrugated sheet-metal panels popularly used as roofing material. These became deadly missiles during the storm, cutting like razor blades flying in the air. In addition, in the United States, there are similar examples of metal sheets used in buildings for general cladding or roofing, which became windborne when the panels were detached or ripped from their anchors.

Other types of building debris are doors, window frames and glass, chimneys, fans, and air conditioners. Finally, many missiles are also produced by interior furnishings of damaged buildings and all types of stored goods, such as materials from warehouses and outdoor storage areas. (See Figure 2-46.)

Once an object has been set in motion by the wind force, it behaves according to its own aerodynamic characteristics. Whether the object will really be injected in the tornadic vortex, the extent to which it will be accelerated, and the velocity it will finally reach, are all matters that depend on the drag-and-lift coefficients of the object and its orientation with respect to the wind direction. Because of the large variety, it is convenient to simplify the problem by dividing the spectrum of potential missiles into three categories:

1. At one extreme are those objects with low ratios of weight-to-surface area that are assumed to be injected into the vortex and reach the same velocity of the air particles around them. These missiles have good aerodynamic shapes and therefore are expected to remain in orbit for a prolonged time and to be at any height within the vortex.
2. At the other extreme are heavy objects with poor aerodynamic shapes, such as spherical bodies. They never achieve the velocity of the air nor are they really injected into the vortex; rather they roll, tumble, and bounce over the ground.
3. In this last category are grouped those objects with intermediate characteristics that place them between the two previous categories.

Figure 2-46. Pine board driven through a tree by the violent force of a tornado that hit Pryor Creek, Oklahoma, April 27, 1942. (Courtesy of the Oklahoma Historical Society)

Impact Loading

All the various missiles that have been mentioned differ considerably from each other in terms of weight and area of contact when they collide against an obstacle. Some types, whose weight and contact areas are particularly critical, can be selected for design purposes and their effects can be evaluated. At this point, it is convenient to examine the design specifications of some of the nuclear power plants that have been built in the United States. The criteria that have been adopted to satisfy the requirements of the Atomic Energy Commission, in terms of tornado forces, have produced basic models that most of the designers follow. Table 2-9 indicates the missiles that have been considered and divides them into three categories. Missile type 1 includes those missiles that are expected to reach the maximum possible velocity, assumed as the sum of the maximum tangential and forward velocity. The values vary up to 360 mph, which are considered very conservative.

The type 2 missiles are usually longitudinal, slender, and traveling head-on, with a small impact area. In this category, for instance, are heavy and slow missiles such as automobiles and similar objects, that are not really injected into the vortex. Of these, the most commonly used is the 4000-lb automobile traveling at 50 mph.

Finally, type 3 objects are those that have intermediate speed—for example, a 3-in.-diam pipe traveling at 100 mph.

Simplified Analysis of Penetration

When the mass of the missile is small in comparison to that of the struck structure, the energy that the structure absorbs and dissipates is small and, therefore, the verification of structural stability can be neglected. On the other hand, the analysis to be made is the penetration of the missile in the structure, which depends upon the following factors: impact velocity, propagation of stress waves, impact cratering, friction, heating, spalling of the stricken structure, and crushing of the missile when this occurs. In classifying impacts, we can identify two types: high-velocity and low-velocity. The dividing value of the impact speed is 500 ft/sec or 341 mph. Therefore,

Table 2-9. Design tornado load adopted for various nuclear power plants.

STATION and OWNER	MISSILE TYPE 1	MISSILE TYPE 2	MISSILE TYPE 3
St. Lucie, (formerly Hutchinson Island), Nuclear Power Plant Fort Pierce, Fla. Florida Power & Light Co.	2 x 4 in x 10 ft timber velocity: 300 mph	4,000 lbs. automobile velocity: 50 mph.	
Fort St. Vrain Nuclear Generating Station Platteville, Colo. Public Service Co. of Colorado	4 x 12 in x 12 ft timber velocity: 300 mph		
Donald C. Cook Nuclear Power Plant Bridgman, Mich. Indiana & Michigan Electric co.	12 ft x 12 ft x 4 on bolted wood deck (450 lbs) velocity: 200 mph 4 ft x 4 ft corrugated sheet siding (100 lbs) velocity: 225 mph	4,000 lbs. automobile velocity: 50 mph.	2 ½ in dia. x 8 ft. schedule 40 steel pipe velocity: 50 mph.
Arkansas Nuclear One Russeville, Ark. Arkansas Power & Light. Co.	4 x 12 in x 12 ft timber velocity: 300 mph	4,000 lbs. automobile velocity: 50 mph.	
Point Beach Nuclear Power Plant Town of Two Creeks, Wisconsin Wisconsin Electric Power Co.	4 x 12 in x 12 ft timber velocity: 300 mph	4,000 lbs. automobile velocity: 50 mph.	3.4 in. O.D. pipe velocity: 100 mph.
Davis-Basse Nuclear Power Station Oak Harbor, Ohio Toledo Edison Co.	4 in Dia. x 12 ft timber velocity: 250 mph	4,000 lbs. automobile velocity: 50 mph.	
Calvert Cliffs Nuclear Power Plant Lusby, Md. Baltimore Gas and Electric.Co.	4 x 12 in x 12 ft timber velocity: 300 mph	4,000 lbs. automobile velocity: 50 mph.	3 in. Dia. schedule 40 steel pipe velocity: 100 mph.
Oconee Nuclear Station Seneca, S.C. Duke Power Co.	4 x 12 in x 12 ft timber velocity: 300 mph	4,000 lbs. automobile velocity: 50 mph.	

Quad-Cities Station Cordova, Illinois Commonwealth Edison Co. and Iowa-Illinois Ges and Electric Co.	35 ft. telephone pole 13 in. butt dia. (50 lbs. per cu. ft.) velocity: 150 mph.	one-ton mass velocity: 100 mph. contact area: 25 sq. ft.	
Forked River Nuclear Generating Station Forked River, N.J. GPU Service Corporation	4 x 12 in x 12 ft timber (110 lbs.) velocity: 360 mph. impact area: 0.33 sq ft. Street light fixture (25 lbs.) impact area: 0,5 sq ft. velocity: 360 mph. 1-1/2" crushed rock (0.25 lbs.) velocity: 360 mph. impact area: 0.01 sliding panels 1.5 x 50 ft. (400 lbs.) velocity: 360 mph. impact area: 0.4 sq ft.	2,000 lbs. automobile velocity: 100 mph. Impact area: 25 sq. ft. 4,000 lbs. automobile velocity: 100 mph. Impact area: 30 sq. ft.	
	tools (150 lbs.) velocity: 360 mph. impact area: 0.1 sq ft. ductwork (150 lbs.) velocity: 360 mph. impact area: 0.3 sq. ft.		
Forked River Nuclear Generating Station (cont.)			utility pole (1200 lbs) velocity: 200 mph. impact area: 1.0 sq. ft. steel Plate 8' x 8' x 3/8" velocity: 200 mph. impact area: 1.0 sq. ft. crated motor (1,000 lbs.) velocity: 200 mph. contact area: 15 sq. ft.

certain aerodynamic tornado-driven missiles are in the category of high-velocity impact. In this case, we can neglect several factors such as hypervelocity effects, friction, heating, impact crater, nose shape of the missile, and crushing of the missile. On the basis of these assumptions, therefore, there are several empirical formulas among which the most popular is the modified Petry formula that follows:

$$D = K_1 \frac{W}{A}\left[\log_{10}\left(1+\frac{V^2}{215{,}000}\right)\right] \tag{1}$$

$$= K_1 \frac{V}{A} V'$$

where

V' = the velocity factor equal to $\log_{10}\left(1+\frac{V^2}{215{,}000}\right)$

D = depth of penetration into infinite thickness in ft

W = weight of missile in pounds

A = minimum effective cross-sectional area of missile in ft^2

V = impact velocity in ft^2/sec (see Figure 2-47)

K_1 = experimentally obtained coefficient typical of each material (see Table 2-10)

The effective penetration D_1, into a slab of a finite thickness T, varies with the ratio T/D as indicated by:

$$D_1 = D\left[1 + e^{-4\left(\frac{T}{D}-2\right)}\right] \tag{2}$$

D_1 is either equal to D (when the slab thickness is twice D) or larger (when the slab thickness is smaller than D). Thus, the minimum thickness required to avoid a total penetration is: $T = 2D$. In practice, D must be computed first; then a thickness larger than $2D$ is to be selected. (See Figure 2-48.) To illustrate the use of Equation 1, several missiles having different speeds have been selected, and their penetration into various materials has been calculated. Assumptions and results have been included in Table 2-11 (pp. 136–137).

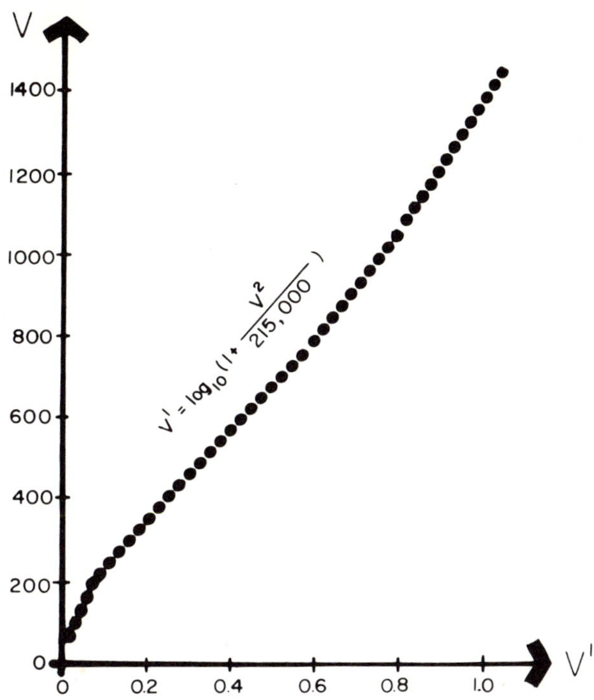

Figure 2-47. Velocity factor in the modified Petry formula expressed as a function of the impact velocity, *V*.

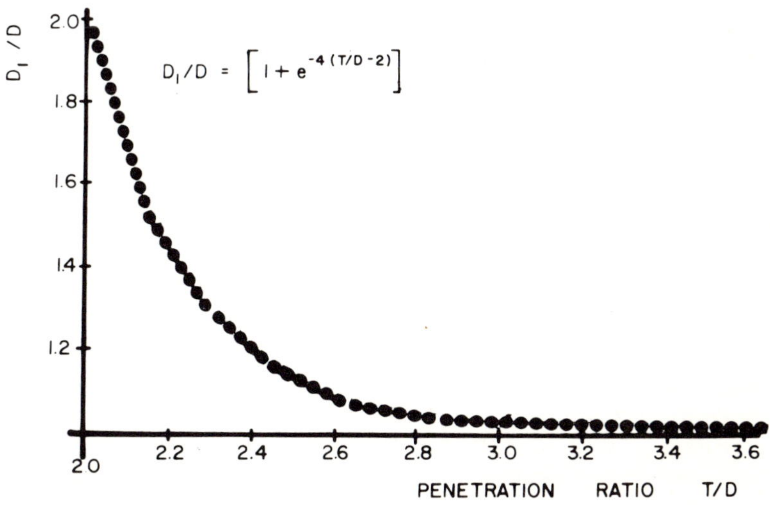

Figure 2-48. D_1/D as a function of T/D in the penetration formula.

Table 2-10. Coefficient K_1 for the modified Petry formula.[a]

MATERIAL	FT³/LB
Limestone	5.38×10^{-3}
Concrete $f'_c = 2200$ psi	7.99×10^{-3}
Reinforced concrete: $f'_c = 3200$ psi and 1.4 percentage of reinforcement	4.76×10^{-3}
Reinforced concrete: $f'_c = 5700$ psi and 1.4 percentage of reinforcement	2.82×10^{-3}
Stone masonry	11.72×10^{-3}
Brickwork	20.78×10^{-3}
Sandy soil	36.7×10^{-3}
Soil with vegetation	48.2×10^{-3}
Soft soil	73.2×10^{-3}

[a] From Samuel and Haman (1939).

Local Effects

Observing the penetration at closer range, we see that a certain amount of material in the impact area is crushed when the stresses generated during the collision exceed the ultimate strength of the material itself; thus, part of the energy is spent in the crushing of the missile, with consequent reduction of penetration. For instance, in the case of a 4-by-12-in. wood plank, 12 ft long, that travels at 200 mph, calculations show that the depth of penetration in a concrete wall is about 4 in. However, in reality, it will be less. In the case of a 4000-lb automobile traveling at 50 mph and striking a reinforced concrete wall, the depth of the concrete that is crushed in the impact is 0.125-in. according to calculations. Formulas used in calculating the penetration of missiles into targets, although they take the characteristics of the target material into consideration, ignore those of the missile material. For instance, if missiles consist of lightweight, low-strength material such as wood, or are hollow objects of low weight in comparison to their volume, their impact resistance is poor and they crush as they collide. The energy spent in the crushing reduces the impact energy, thereby making the effects on the target less severe.

Still using a concrete wall as an example and observing the impact phenomenon further, we see that when a missile strikes the exterior face of the wall, a compressive shock wave originates at the area of impact and travels through the wall toward the interior face (free face). Once there, the shock wave is reflected and becomes a tension pulse that travels back toward the exterior face where the impact occurred.

If the material has poor tensile strength, as in the case of concrete, the tensile force causes spalling of the material starting from the interior face. Such spalling proceeds toward the impact surface, gradually decreasing its intensity as the tensile pulse decreases. Finally, its stops where the tensile strength of the concrete is larger than the tensile stress. Spalling of the material produces a number of small pieces of material that, in turn, become missiles when they move away from the structure with a certain velocity. Such a velocity is generated by the energy that the wall receives from the missile. In reinforced concrete, the depth of spalling is reduced by the reinforcement and is usually found not to exceed 3 in. The velocity that these fragments acquire is fortunately low, i.e., less than 20 ft/sec (13.63 ft/sec as proved by several studies analyzing the spalling effects of high-velocity missiles).

Materials

In very broad terms, it can be said that the impact resistance of a structural member is measured by the amount of energy (strain energy) that the member can absorb during the deformation caused by the load. If we consider a static load, *P*, gradually applied to a member and draw the load-deflection curve, the area under the curve represents the energy absorbed by the member and its impact strength. More specifically, the triangular area represents the elastic-energy-absorbed resilience and the total area represents the elastic plus the plastic energy absorbed (toughness). In other words, the former is the impact strength at the elastic limit, whereas the latter is the impact strength at rupture.

Table 2-11. Missile penetration into various materials.

MISSILE	WEIGHT W (LB)	CROSS-SECTIONAL AREA A (FT^2)	$\frac{W}{A}$	IMPACT VELOCITY V (MPH)	IMPACT VELOCITY V (FT/SEC)	$LOG_{10}\left(1+\frac{V^2}{215{,}000}\right)$	MATERIAL COEFFICIENT K	PENETRATION D (FT)
				a. Brick Masonry				
• 4″x12″x12′ Timber	131	0.274	480	200	294	0.146	20.48×10^{-3}	1.44
• 4″x12″12′ Timber	131	0.274	480	250	367	0.212	20.48×10^{-3}	2.08
• 35-ft Utility pole	1370	0.785	1745	200	294	0.146	20.48×10^{-3}	5.23
• 4000-lb Automobile	4000	20	200	50	73	0.01	20.48×10^{-3}	0.041
• 4000-lb Automobile	4000	30	133	100	147	0.04	20.48×10^{-3}	0.109
• 2000-lb Automobile	2000	25	80	100	147	0.04	20.48×10^{-3}	0.065
• Common brick	4	0.048	82	200	294	0.146	20.48×10^{-3}	0.246
				b. Sandy Soil				
• 4″x12″x12′ Timber	131	0.274	480	200	294	0.146	36.7×10^{-3}	2.57
• 4″x12″x12′ Timber	131	0.274	480	250	367	0.212	36.7×10^{-3}	3.73
• 35-ft Utility pole	1370	0.785	1745	200	294	0.146	36.7×10^{-3}	9.26
• 4000-lb Automobile	4000	20	200	50	73	0.01	36.7×10^{-3}	0.0734
• 4000-lb Automobile	30	133	100	147	0.04	36.7×10^{-3}	0.195	
• 2000-lb Automobile	2000	25	80	100	147	0.04	36.7×10^{-3}	0.117
• Common Brick	4	0.048	82	200	294	0.146	36.7×10^{-3}	0.44
				c. Reinforced Concrete				
• 4″x12″x12′ Timber	131	0.274	480	200	294	0.146	4.76×10^{-3}	0.333
• 4″x12″x12′ Timber	131	0.274	480	250	367	0.212	4.76×10^{-3}	0.484
• 35-ft Utility pole	1370	0.785	1745	200	294	0.146	4.76×10^{-3}	1.21
• 4000-lb Automobile	4000	20	200	50	73	0.01	4.76×10^{-3}	0.0095
• 4000-lb Automobile	4000	30	133	100	147	0.04	4.76×10^{-3}	0.024
• 2000-lb Automobile	2000	25	80	100	147	0.04	4.76×10^{-3}	0.0152
• Common brick	4	0.048	82	200	294	0.146	4.76×10^{-3}	0.0572

Table 2-11. (Continued)

MISSILE	WEIGHT W (LB)	CROSS-SECTIONAL AREA A (FT^2)	$\frac{W}{A}$	IMPACT VELOCITY V (MPH)	IMPACT VELOCITY V (FT/SEC)	$LOG_{10}\left(1+\frac{V^2}{215,000}\right)$	MATERIAL COEFFICIENT K	PENETRATION D (FT)
d. Stone Masonry								
• 4″x12″x12′ Timber	131	0.274	480	200	294	0.146	11.72×10^{-3}	0.82
• 4″x12″x12′ Timber	131	0.274	480	250	367	0.212	11.72×10^{-3}	1.19
• 35-ft Utility pole	1370	0.785	1745	200	294	0.146	11.72×10^{-3}	2.99
• 4000-lb Automobile	4000	20	200	50	73	0.01	11.72×10^{-3}	0.023
• 4000-lb Automobile	4000	30	133	100	147	0.04	11.72×10^{-3}	0.062
• 2000-lb Automobile	2000	25	80	100	147	0.04	11.72×10^{-3}	0.037
• Common brick	4	0.048	82	200	294	0.146	11.72×10^{-3}	0.141
e. Soil with Vegetation								
• 4″x′2′x′2′ Timber	131	0.274	480	200	294	0.146	48.2×10^{-3}	3.36
• 4″x′2′x′2′ Timber	131	0.274	480	250	367	0.212	48.2×10^{-3}	4.9
• 35-ft Utility pole	1370	0.785	1745	200	294	0.146	48.2×10^{-3}	12.25
• 4000-lb Automobile	4000	20	200	50	73	0.01	48.2×10^{-3}	0.096
• 4000-lb Automobile	4000	30	133	100	147	0.04	48.2×10^{-3}	0.256
• 2000-lb Automobile	2000	25	80	100	147	0.04	48.2×10^{-3}	0.154
• Common brick	4	0.048	82	200	294	0.146	48.2×10^{-3}	0.577
f. Soft Soil								
• 4″x12″x12′ Timber	131	0.274	480	200	294	0.146	73.2×10^{-3}	5.13
• 4″x12″x12′ Timber	131	0.274	480	250	367	0.212	73.2×10^{-3}	7.45
• 35-ft Utility pole	1370	0.785	1745	200	294	0.146	73.2×10^{-3}	18.70
• 4000-lb Automobile	4000	20	200	50	73	0.01	73.2×10^{-3}	0.1464
• 4000-lb Automobile	4000	30	133	100	147	0.04	73.2×10^{-3}	0.39
• 2000-lb Automobile	2000	25	80	100	147	0.04	73.2×10^{-3}	0.234
• Common brick	4	0.048	82	200	294	0.146	73.2×10^{-3}	0.88

If we allow only elastic deformations in the structure, we must consider the resilience as the impact capacity. If we can allow some permanent deformations to occur, we can consider the toughness of the material as a characteristic of strength.

The impact resistance of prestressed concrete members has been proved by experimental testing to reach an optimum for a specific percentage of reinforcement; the amount of prestress; and the strength of the concrete. This is so critical that changing the proportions, even by increasing the amount of steel, for instance, will cause a reduction of the impact strength. Is prestressed concrete better than concrete, or vice versa, in resisting impact loads? By testing reinforced versus prestressed concrete beams having equal static strength, it was concluded that there is not a real difference between the two and that the only index of strength is the resilience of the beam. This can be made to vary, although the beams still have the same static strength. It can be accomplished by changing the percentage of reinforcement or prestress and allowing various deflections of the beams under load.

Dwelling Structures

How can we evaluate the effects of tornado forces on different building types? It is particularly helpful in this respect to use the test results collected for structures exposed to nuclear blasts. A certain analogy existing between tornado forces and nuclear explosions makes this possible. In fact, although nuclear blasts and tornadoes are different phenomena, several aspects of the forces involved are comparable, such as the drag force of the wind and the pressure differential on opposite sides of walls, floor slabs, roofs, and so on.

On March 17, 1953, and May 5, 1955, nuclear tests were conducted in Nevada (Figure 2-49).* Buildings of various types were located at different distances from the center of the explosion in order to evaluate damage caused by a variety of loadings. From the analysis of the damage, it is deduced that wood-framed dwellings collapse at a pressure of 720 psf (5 psi) (see Figure 2-49); at 245 psf (1.7 psi), they do not collapse, but their damage is beyond repair. Ordinary masonry construction differs little from wood framing in overall strength. On the other hand, seismic-resistant structures consisting of reinforced concrete panels and reinforced concrete block walls can tolerate pressures of 245 psf (1.2 psi) with damage limited to doors and windows. It is very useful to establish that, for average buildings, broken doors and windows

* Data on these tests are available. See U.S. Dept. of Defense (1962).

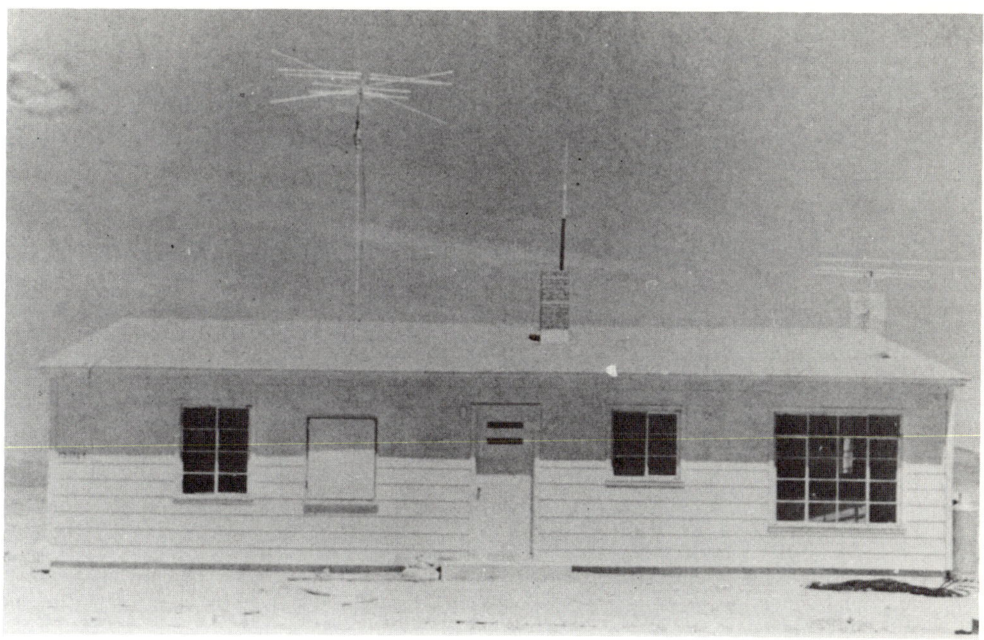

Figure 2-49. Structural testing on houses consisting of different construction types conducted in the Nevada Proving Ground in 1955. (Courtesy of U.S. Atomic Energy Commission) *a,* Rambler-type house before testing. *b,* After testing. Overpressure: 5 psi = 720 psf.

constituted an adequate venting system to balance pressure in most cases. Interestingly, the seismic-resistant structure could resist loads of 720 psf (5 psi) with only negligible additional damage.

The Nevada tests clearly indicate that buildings of ordinary construction are vulnerable to forces similar to those expected in tornadoes. However, the use of special materials such as concrete, and the adoption of special techniques such as those used for earthquake-resistant structures, improve tremendously the strength of the buildings. Typical dwelling structures are usually designed by empirical methods without the benefit of any structural analysis. Building codes are, in general, sufficient to guarantee their stability under conventional loads, and engineering verifications are therefore omitted. Economic factors also do not justify professional fees. Tornado-resistant criteria are unusual conditions for almost any building, excluding nuclear power plants. However, for buildings requiring structural design, the storm-resistant criteria are established by the engineer after the forces involved are defined. Dwelling structures, on the other hand, are left in an unresolved situation.

For small buildings in those areas where tornado frequency is significant, it is proposed to consider the following criteria. The basic structure is assumed to be divided into two zones: a central one referred to as the *primary zone,* and the rest of the building around it, the *secondary zone.* (See Figure 2-50.) The primary zone built in reinforced concrete should satisfy these basic structural requirements:

1. To resist the various forces (conventional loads and tornado forces) applied to the structure itself so that its stability is guaranteed with a high degree of safety.
2. To function as a resisting element for the secondary structure against the action of horizontal forces.
3. To serve as a structural support for conventional vertical loads.

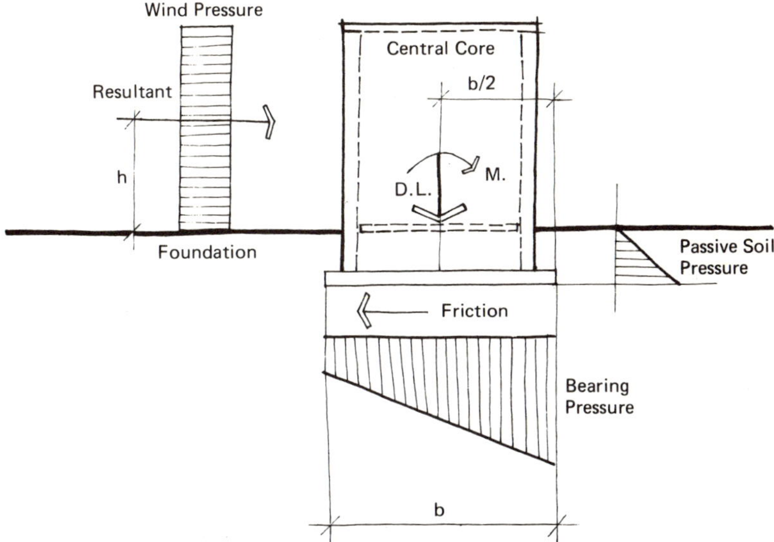

Figure 2-50. Horizontal wind forces applied to the central core produce an increased soil pressure under the foundation and generate a passive pressure against the soil that prevents horizontal translation.

The wind force directly applied to the exterior surface generates a vertical overturning moment that must be balanced by a larger resisting moment. An obvious method to increase the resisting moment is to use grade beams, as indicated in Figure 2-51. A check against the tendency of the structure to slide must be made on the assumption that a frictional force acts between the soil and the foundation and that a passive soil pressure acts on the vertical projection of the underground part of the foundation. Note that these two forces are reactions generated by the horizontal wind force. The maximum soil pressure must be checked against the allowable bearing capacity of the soil. The structure must be analyzed for shear and properly reinforced. Consideration must also be given to the horizontal moment previously mentioned.

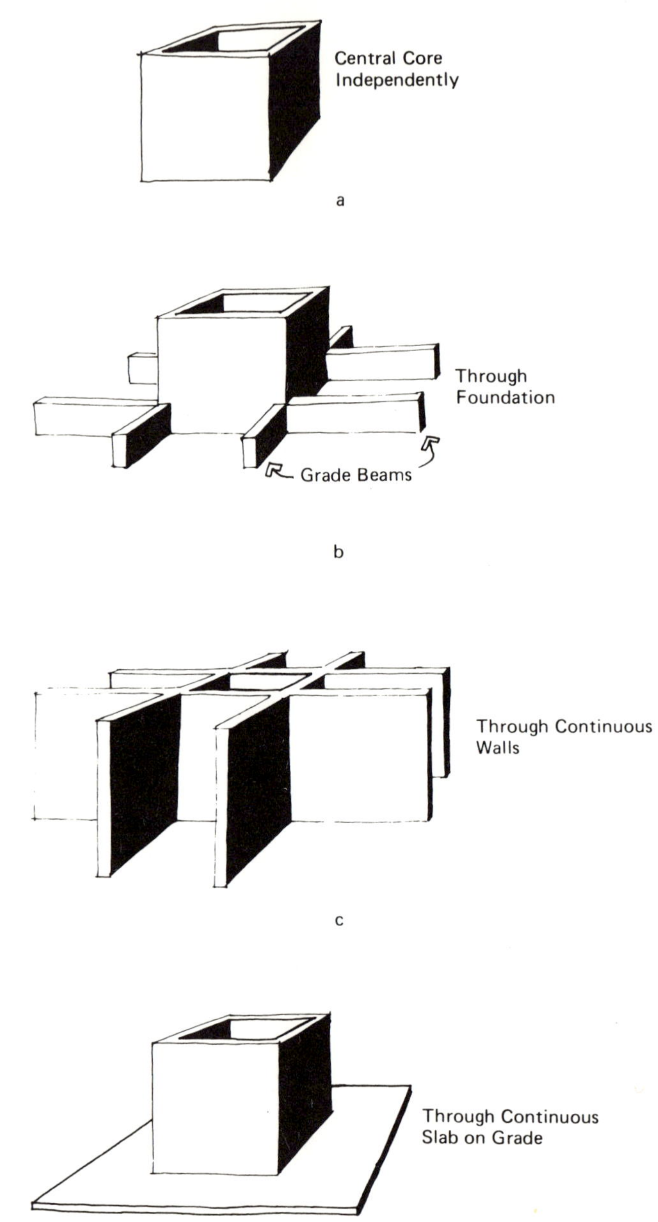

Figure 2-51. The stability of the central core against the wind forces during tornado actions can be provided by the walls themselves (*a*), or can be improved by extending the grade beams, which increase the resisting moment (*b*). Similar results can also be obtained by extending the walls of the core (*c*); enlarging the floor slab outside the enclosed area would also increase the resisting moment (*d*).

A venting system for pressure balance is necessary, but, since the propagation of the pressure wave travels at the local speed of sound, the location of the vents is not critical for small structures. Vents can therefore be located where most convenient. For instance, vents can be installed on the roof slab rather than on the walls, which could then provide better shielding action with fewer openings.

The sizing of the vent area should be based on conservative assumptions. The nuclear tests in Nevada have confirmed that improper size is critical. In some cases, it was found that airflow through first-floor windows was more rapid than through small basement windows, producing a pressure differential between basement and first floor. This caused the floor to collapse (Figure 2-52).

For impact loads from flying debris, it is suggested that reinforced concrete be used for the primary zone. As mentioned above, stiffening of the whole structure should not endanger the stability of the primary zone. To satisfy this condition, it is necessary that supports connecting primary to secondary structure be so designed that they fail when the reactions reach certain preestablished limits. This criterion should be applied to horizontal components of the reactions and to vertical as well.

The floor plan of the primary zone is illustrated by several schemes that compare different arrangements of walls in terms of the area that is shielded (see Figure 2-53). The shielding action is provided by reinforced concrete walls, assuming that doors have no protective capacity. It is recommended, however, that doors be solid, reinforced, and of the sliding type so that they can be retracted behind the walls; however, when closed, they should bear against the walls, acting as slabs supported on three sides.

The secondary zone, which constitutes the larger part of the building, is basically designed on the concept of controlled damage to avoid total collapse, but it is not expected to withstand intact the full fury of a tornado. To keep the additional cost of the building as low as possible, the secondary structure should be built, with few exceptions, using conventional materials and standard construction techniques. It is recommended, for instance, that additional reinforcing elements from the foundations to the upper plate be installed. The most important factor to the stability of the secondary structure, however, is not the stability of the structure itself, but the contribution offered by the primary structure when coupled with it.

Figure 2-52. Pressure differential produced by tornadoes, as well as by atomic blasts, does not occur only between the interior and the exterior of a building, but it can exist between opposite sides of interior partitions, floors, etc. Here, for instance, the difference in window size between the upper floor and the basement allowed more air to move in the upper floor; therefore, the pressure in the upper area was different than that in the basement. This pressure differential was applied to the floor, causing the damage shown. (Courtesy of U.S. Atomic Energy Commission)

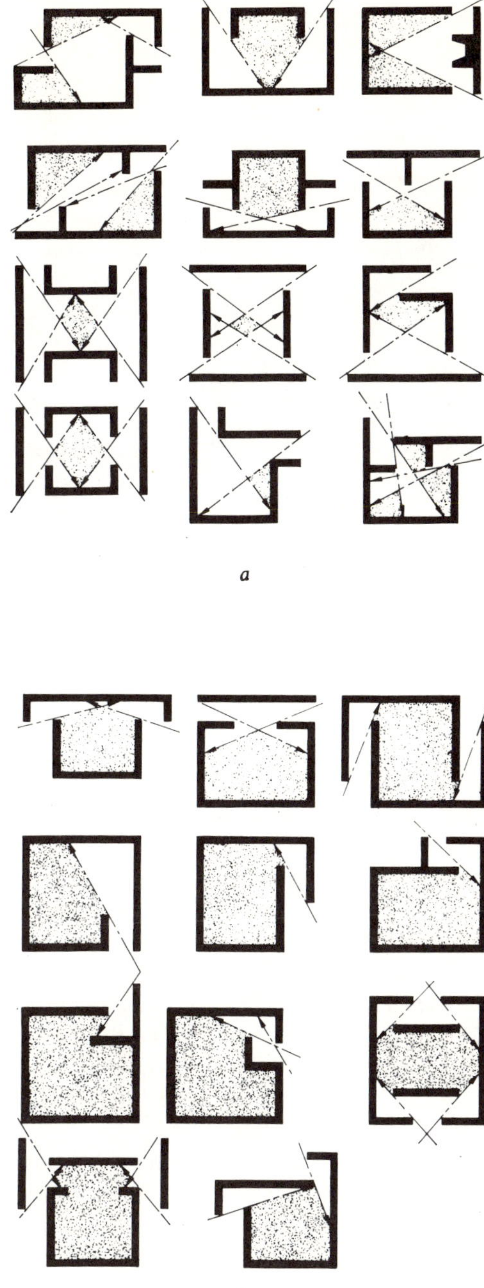

Figure 2-53. Central core: At first glance, there seems to be no problem in designing the arrangement of the core: designating the area; enclosing it with concrete walls; arranging baffles to protect door and window openings from flying debris. However, as we start sketching general schemes, we realize that some work but many do not: i.e., the sheltered area is extremely small for the same amount of wall area. After sketching a number of alternatives, we divide them in two groups: *a,* acceptable; *b,* nonacceptable.

Figure 2-54. Storm cellar is a typical feature on most farms of the midwestern United States. A lifesaver when a tornado is near. (Courtesy of National Oceanic and Atmospheric Administration)

Tornado Shelters

For people living in parts of the world where tornado probability is high or where past tragedy from tornadoes cannot be forgotten, fear is a painful reality. For instance, in some areas of the midwestern United States, tornado watches are a daily occurrence during the tornado season. Tornado warnings (i.e., when tornadoes are actually detected in the vicinity) are frequent. A constant state of insecurity and anxiety is produced. For example, the people of Dacca, Bangladesh, will never forget the tornado that struck their city on April 14, 1969. Decapitating metal sheets flew through the air, 600 died, and 5000 were injured. In view of the danger to human life, the need for some sort of protection cannot be ignored. Even the rugged society that settled the Great Plains in the 1800s recognized the need for tornado shelters and built so-called storm cellars near their houses (Figure 2-54).

Since the 1950s, several plans for tornado shelters in residential areas have been prepared by such agencies as the American Red Cross, the Portland Cement Association, and Kansas State University. However, their architectural qualities would have done more to disrupt than to enhance the appearance of the grounds of conventional city dwellings. Some schematic proposals made by the author and first published in the 1950s are included here (Melaragno 1968). In Figures 2-55 through 2-58, Types A through D illustrate structures made out of reinforced concrete, with pressure-balancing vents and a configuration of the top slab that could be used as a landscape feature. These structures are not completely undergound and thus eliminate feelings of claustrophobia.

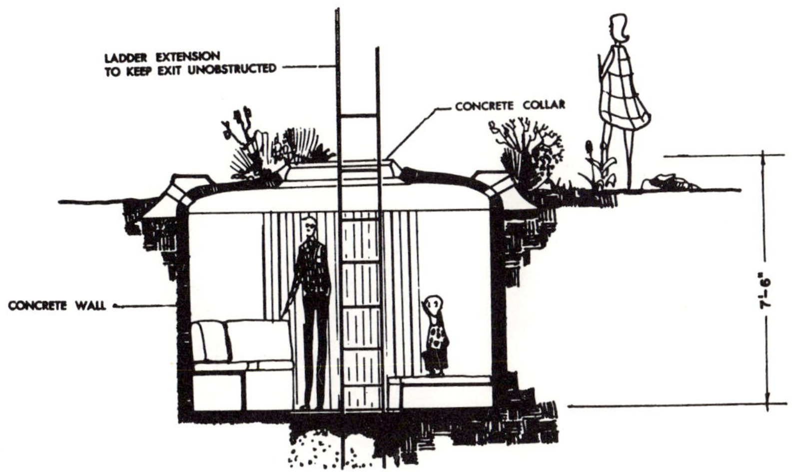

Figure 2-55. Type A outdoor tornado shelter.

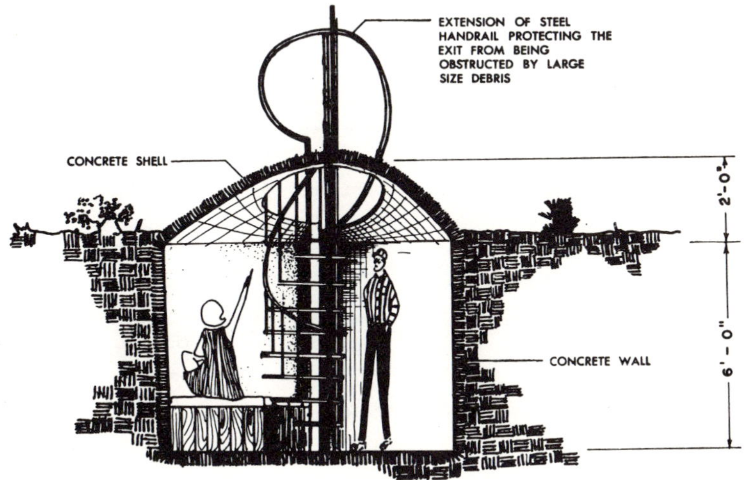

Figure 2-56. Type B outdoor tornado shelter.

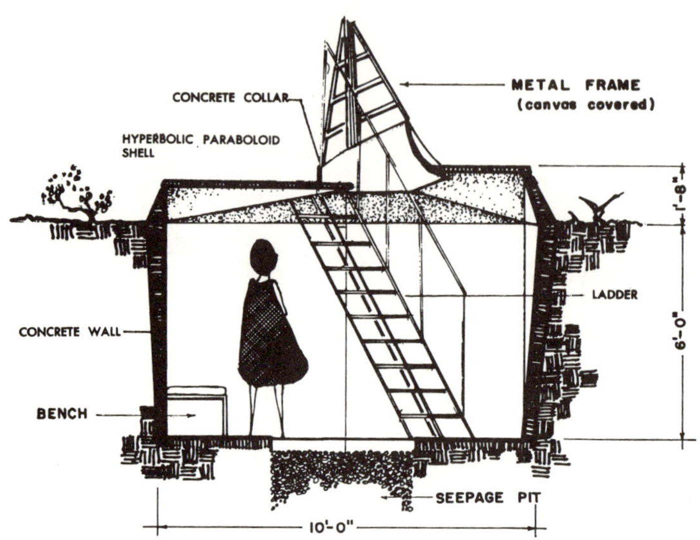

Figure 2-57. Type C outdoor tornado shelter.

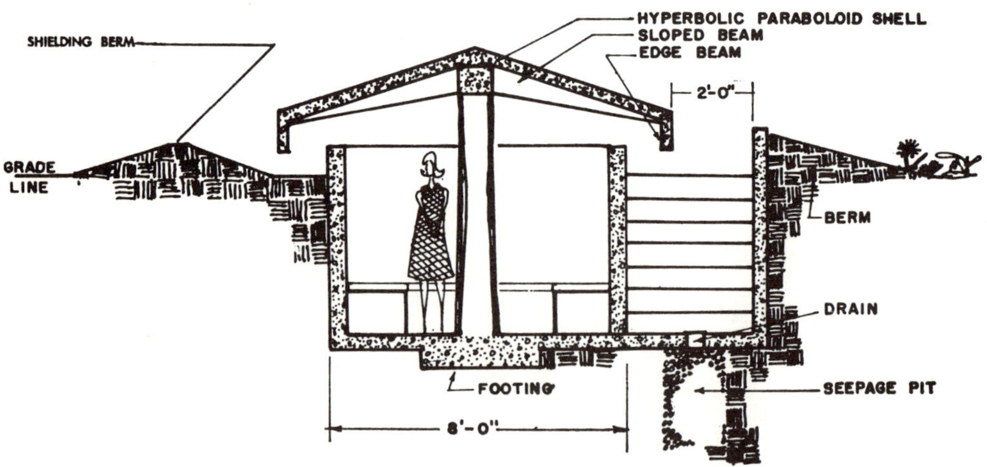

Figure 2-58. Type D outdoor tornado shelter.

Safety Measures

Latest evaluations of tornado-caused casualties and property damage related to many tornadoes indicate the existence of new correlations between safety measures and their effectiveness. The U.S. Department of Health, Education, and Welfare (1979) emphasizes some new points on safety and disproves some previous commonly accepted assumptions. The validity of the study is corroborated by the agencies whose representatives participated in this survey, including the Center for Disease Control, the American Red Cross, and the United States Air Force's School of Health Care Sciences.

The object of the study was the disaster caused by a tornado that struck the southwest corner of Wichita Falls, Texas, on the late afternoon of April 10, 1979. The following data were compiled:

City population: 100,000
Damage area: 7 mi long, 1 mi wide
Deaths: 44
Injuries: 1871 individuals
Damage: over $300 million
Destroyed residential units: over 6000
Warning system: operational and efficient (sounding of sirens 1 hr before and 5 min before tornado touchdown)

Relating the number of deaths to the other parameters, we note the following:

- Deaths are 0.044% of the city population.
- Deaths are 2.3% of the total number (1915) of people physically struck.

When the number of accidents and fatalities among people who stayed indoors during a tornado was compared with that of people who looked for shelter outdoors, it was obvious that dwellings of conventional construction type were much safer. This was true not only for dwellings on the fringe of the tornado path, but also for those in the most severe part of the damage path. In fact, even when the structures were totally demolished, the people inside were protected. Components that were particularly effective in shielding people included basements, closets, hallways, and heavy furniture. Outdoor features such as culverts, underpasses, and ravines were less effective.

Conclusion

The forces with which tornadoes act upon building structures in their path can be summarized as: wind forces per se, air pressure changes between the interior and exterior of buildings, and the impacts by missiles. The wind forces are the result of rotational and translational velocities, and their intensity is inversely proportional to the distance from the tornado eye. Pressure changes are proportional to the wind velocity, and therefore it varies with the distance from the eye. The rate of pressure change that is a function of the diameter is also related to the translational velocity.

The structural response of buildings varies accordingly to the forces involved (wind pressure, pressure change, impact). The lateral resistance of buildings to wind pressure can be based on structural systems such as vertical diaphragms, horizontal diaphragms, shear walls, frame action, and vertical trussing.

The pressure change can be dealt with, either by reducing it or by resisting it. Pressure reduction can be obtained by exterior and interior venting. Pressure resistance implies the design of floors and roofs, as well as walls, to specific pressure loads.

The resistance against the impact of flying debris is most conveniently based on protective structures, which, for conventional buildings, may include baffled walls, special design of windows, and protective shelters.

As mentioned, tornado forces were not considered in the design of buildings until the late 1960s when the construction of nuclear power plants required that tornado forces had to be included in the design of these structures. The United States Atomic Energy Commission (1979) specifies that structures should be designed to resist natural phenomena, including tornadoes. Furthermore, the AEC specifies that nuclear power stations must also be designed to resist the impact loads associated with tornadoes. On this basis, and considering a probability index for tornado occurrence as suggested by Thom (1963), various power plants have been designed following a model that is almost equal for all of them (see Figure 2-59).

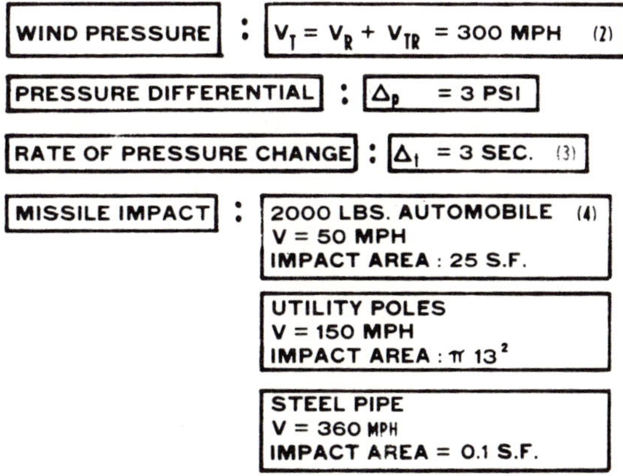

Figure 2-59. Design model for nuclear power plants.

REFERENCES

Abdullah, A. J. 1955. Some aspects of the dynamics of tornadoes. *Monthly Weather Review* 83–94 (April).

American Society of Civil Engineers. 1961. *Wind Forces.* Task Committee on Wind Forces final report. Paper No. 3269 on wind forces. *Trans. Amer. Soc. Civil Eng.* 1124–97.

Asp, M. O. 1950. Tornadoes in Oklahoma, 1857–1949. *Monthly Weather Review* 23–26 (Feb.).

Baier, J. 1896. Low pressure in the St. Louis tornado. *Monthly Weather Review* 332 (Sept.).

Banks, D. C. 1963. A Study of Bearing Capacity in Sands under Dynamic Loadings. M.S. thesis, Georgia Institute of Technology, Atlanta, GA.

Batton, L. J. 1959. *The Nature of Violent Storms.* Garden City, New York: Doubleday & Co.

Bigler, S. G. 1975. *Tornado Damage Surveys.* A & M project 156, scientific report No. 1. Dept. of Oceanography and Meteorology, Texas A & M University, College Station, Texas.

Booker, C. A. 1953. Tower damage provides key to Worcester tornado data. *Electrical World* 170: 20–24.

———. 1954. On transmission towers destroyed by the Worcester, Massachusetts, tornado of June 9, 1953. *Bull. Amer. Meteorol. Soc.* 225, 229 (May).

Boscovich. 1749. Sopra il turbine che la notte tra gli XI, e XII Giugno de MDCCXLIX danneggio una gran parte di Roma [About the tornado that on the night of June 11–12, 1749, damaged a large part of Rome]. Rome.

Brooks, C. F. 1949. *Why the Weather?* New York: Harcourt, Brace & Co.

Brooks, E. M. 1951. Tornadoes and related phenomena. *Compendium of Meteorology.* Boston: American Meteorological Society.

Carr, J. A. 1952. Preliminary reports on the tornadoes of March 21–22, 1952. *Monthly Weather Review* 50–58.

Casagrande, A. and W. L. Shannon. 1949. Strength of soils under dynamic loads. *Trans. Amer. Soc. Civ. Eng.* 114: 755–772.

Chang, C. C. 1966. First real man-made tornado is generated in laboratory "cage" by space scientistst at the Catholic University of America. News item released by Office of Public Information, Catholic University of America.

Clarke, R. H. 1962. *Severe Local Wind Storms in Australia.* Division of Meteorological Physics, Technical Paper No. 13, C.S.I.R.O., Melbourne, Australia.

Dames & Moore. 1972. Tornado Wind Damage Probability and Recurrence for West Valley, New York. U.S. Nuclear Regulatory Commission docket No. 50–201.

Dergarabedian, P. and F. Fendell. 1971. A method for rapid estimation of maximum tangential wind speed in tornadoes. *Monthly Weather Review* 99: 143–145.

Evesson, D. T. 1969. Tornado occurrences in New South Wales. *Australian Meteorological Magazine* 17(3): 143–165 (Sept.).

———. 1970. Tornado Occurrences in New South Wales. Bureau of Meteorology working paper No. 130 August 18, 1965, Sydney, Australia.

Ferrel. 1889. *A Popular Treatise on the Winds.* New York: John Wiley & Sons.

Finley, J. P. 1884. Report on the character of six hundred tornadoes. *Prof. Papers of the Signal Service,* No. VII. Available NOAA library, Silver Spring, MD.

Flora, S. D. 1954. *Tornadoes of the United States.* University of Oklahoma Press.

Frankenfield, H. C. 1896. The tornado of May 27, at St. Louis, Missouri. *Monthly Weather Review* 77–81 (March).

Fujita, T. T. 1971 Proposed Characterization of Tornadoes and Hurricanes by Area and Intensity. SMRP Res. Paper No. 91, University of Chicago, Chicago, Illinois.

Fujita, T. T., D. L. Bradbury, and P. G. Black. 1967. Estimation of Tornado Wind Speed from Characteristic Ground Marks. Satellite and Mesometeorology Research Project Paper 69, Department of Geophysical Sciences, University of Chicago. Presented at Fifth Conference on Severe Local Storms, St. Louis, MO.

———, A. D. Pearson, D. M. Ludlum. 1975. Long-term fluctuation of tornado activities. *Preprints,* Ninth Conference on Severe Local Storms, Norman, OK, Oct. 21–23. Published by American Meteorological Association, Boston, MA.

Gilbert, E. G. and J. M. Walker. 1966. Tornado at the Royal Horticultural Society's Garden Wisley.

Glaser, A. M. 1959. An observational deduction of the structure of a tornado vortex. *Cumulus Dynamics,* Proceedings of the First Conference on Cumulus Convection, May 19–22, Portsmouth, New Hampshire. New York: Pergamon Press.

Gates, D. M. 1972. *Man and His Environment: Climate.* The University of Michigan. New York: Harper & Row.

Hoecker, W. H. 1960. Wind speed and air flow patterns in the Dallas tornado of April 2, 1957. *Monthly Weather Review* 167–80 (May).

———. 1961 Three-dimensional pressure pattern of the Dallas tornado and some resultant implications. *Monthly Weather Review* 533–42.

———, R. G. Beebe, D. T. Williams, J. T. Lee, S. G. Bigler, and E. P. Segner, Jr. 1960. *The Tornadoes at Dallas, Texas, April 2, 1957.* U.S. Department pf Commerce, Weather Bureau Research Paper No. 41.

Howe, G. M. 1975. Tornado path sizes. *J. Applied Meteorol.* 13: 343–347.

Humphreys, W. J. 1940. *Physics of the Air.* New York: McGraw-Hill.

Keller, D. and B. Vonnegut. 1976. Wind speed estimation based on the penetration of straws and splinters into wood. *Weatherwise* 29(5).

Kessler, E. 1970. Tornadoes. *Bull. Amer. Meteorol. Soc.* 926–36.

———. and J. T. Lee. 1976. Normalized Indices of Destruction and Deaths by Tornadoes. NOAA Tech. Memo ERL NSSL-77.

Kert, T. H. and T. J. Dean. 1963. Report on a tornado at Malta.

Lamb, H. H. 1957. Tornadoes in England, May 21, 1950. *Geophys. Memoirs,* No. 99 Air Ministry, Meteorological Office, London.

Logie, J. 1919. Note on tornadoes. *Quart. J. Meteorol. Soc.* 45: 317.

Lamb, H. H. 1965. Tornado destruction of trees at Shalford Park, Guildford.

Long, R. R. 1960. Tornadoes and dust whirls. Johns Hopkins University, Baltimore, MD, Dept of Mechanical Engineering, Technical Report No. 10 (ONR series) or No. 13 (CWB series).

Lamy, F. 1689. *Conjectures phys. sur les collonnes de nue* [theories on tornadoes]. Paris.

Lacy, R. E. 1968. Tornadoes in Britain during 1963–66.

Martin, J. R. 1940. Tornadoes in the United States, 1916–1937, inclusive, Mimeographed summary in NOAA library, Washington, DC.

McDonald, J. R., K. C. Mehta, and J. E. Minor. 1975. Development of Wind-Speed Risk Models for the Savannah River Plant Site. Report for E. I. du Pont de Nemours & Company under purchase order No. AXC 672-W.

Meaden, G. T. 1976. Tornadoes in Britain: their intensities and distribution in space and time. *J. Meteorol.* 1: 242–251.

Melaragno, M. G. 1968. *Outdoor Tornado Shelters for Residential Areas.* Kansas State University, Manhattan, KA.

Montanari, G. 1694. Le forze di Eolo, Dialogo fisico-matematico sopra gli effetti del vortice, o sia Turbine . . . che il gierno 29 luglio 1686 ha scorso e flagellato molte ville, e luoghi del territorio di Mantova, Padova, Verona, etc. [the forces of Aeolus: physico-mathematical dialogue about the effects of the vortex, i.e., tornado, that on July 29, 1686, struck and plagued many villas and locations in the territories of Mantua, Padua, Verona, etc.].

The tornado of Little Rock, Arkansas, October 2, 1894. Editorial in *Monthly Weather Review* 413–414 (Oct.).

Thom, H. C. S. 1963. Tornado probabilities. *Monthly Weather Review* 730–36 (Oct.–Dec.).

U.S. Atomic Energy Commission. 1974. *Design Basis Tornadoes for Nuclear Power Plants.* Regulatory Guide 1.76, U.S. Atomic Energy Commission, Washington, D.C.

U.S. Dept. of Health, Education, and Welfare. 1979. Morbidity and Mortality Report, Vol. 28, No. 17.

U.S. Dept. of Commerce. 1965. Tornado facts. Environmental Services Social Administration, Weather Bureau, Washington, D.C. (L.S. 6403 Rev.).

U.S. Dept. of Defense. 1962. *The Effects of Nuclear Weapons.* U.S. Atomic Energy Commission.

U.S. Dept. of Housing and Urban Development. 1966. HUD recommends ways to beat storm damage. News item released Oct. 7, 1966.

Van Tassel, E. L. 1955. The North Platte Valley tornado outbreak of June 27, 1955. *Monthly Weather Review* 255–64 (Nov.).

Vonnegut, B. 1960. Electrical theory of tornadoes. *J. Geophys. Res.* 65: 203–212.

Outram, R. S. 1904. Storm of August 20, 1904, in Minnesota. *Monthly Weather Review* 32: 365–66 (August).

Pearson, A. D. 1971. Statistics on tornadoes that caused fatalities 1960–1970. *Preprints,* Seventh Conference on Severe Local Storms, Kansas City, MO, Oct. 5–7.

Peltier, 1840. Meteorologie. Observations et recherches experimentales sur les causes qui concourent à la formation des trombes [observations and experimental research on the causes which contribute to the formation of tornadoes]. Paris.

Reynolds, G. W. 1958. Venting and other building practices as practical means of reducing damage from tornado low pressure. *Bull. Amer. Meteorol. Soc.* 14–20 (Jan.).

Samuel, F. J. and C. W. Haman. 1939. *Civil Protection.* London: The Architectural Press. In A. Amirikian. 1950. *Design of Protective Structures.* U.S. Navy NavDocks P-51.

Schafer, J. T., D. L. Kelly, and R. F. Abbey, Jr. 1979. Tornado track characteristics and hazard probabilities. *Preprints,* Fifth International Conference on Wind Engineering, July 8–14, Colorado State University.

Seed, H. B. and R. Lundgren. 1954. Investigation of the effect of transient loading on the strength and deformation characteristics of saturated sands. *Proc. Amer. Soc. Testing materials* 54: 1288–1306.

Seelye, C. J. 1945. Tornadoes in New Zealand. *New Zealand Journal of Science and Technology* 27: 166–174.

Segman, R. 1971. Wet tornado. NOAA, pp. 45–47.

Segner, E. P., Jr. 1960. Estimates of Minimum Wind Forces Causing Structural Damage. U.S. Weather Bureau research paper No. 41.

Skaggs, R. H. 1970. On tornado probabilities. *Proc. Assn. Amer. Geog.* 2: 123–126.

Swanson, A. E., R. E. Stippich, and F. C. Bates. 1967. Paper presented at the Fifth Conference on Severe Storms, St. Louis, MO.

Sykes. 1967. Letter to editor. *Weatherwise* 271 (Dec.).

Teesdale, L. V. 1928. Tornado-resistant construction of buildings possible by venting. *Southern Lumberman* 40–42 (July 7, 1928).

Tepper, M. 1958. Tornadoes. *Scientific American* (May).

——— and Weyer, J.R. 1966. Luminous phenomena in nocturnal tornadoes. *Science* 1213–20 (Sept.).

Wegener, A. 1917. *Wind and Wasserhosen in Europa.* Braunschweig, Germany: F. Vieweg & Sohn.

Wolford, L. V. 1960. Tornado Occurrences in the United States. Technical paper No. 20, revised. U.S. Weather Bureau, Washington, D.C.

World Meteorological Organization. 1966. *International Meteorological Vocabulary.* New York: Unipub.

Woodard, J. M. 1965. An investigation of the Dynamic Bearing Capacity of Footings on Sand. M.S. thesis. Georgia Institute of Technology, Atlanta, GA.

Whitman, R. V. 1957. The behavior of soils under transient loading. *Proceedings of the Fourth International Conference on Soil Mechanics and Foundation Engineering (London)* 1: 207–212.

——— and K. A. Healy. 1962. Shear strength of sands during rapid loadings. *Proc. Amer. Soc. Civil Eng.* 88(SM 2): 99–133.

Wilson, and G. M. Morgan, Jr. 1971. Long-track tornadoes and their significance. *Preprints,* Seventh Conference on Severe Local Storms, Kansas City, MO, Oct. 5–7, 183–186. Published by American Meteorological Society, Boston, MA.

3
Aerodynamic Wind Forces

AWARENESS OF WIND LOADS

Because of the inherent static strength of heavy masonry structures to resist wind forces, these forces were not considered dangerous until major failures began to occur in slender trussed bridges. During the 1880s, bridge collapses, caused mainly by poor lateral resistance against wind loads, reached the astonishing rate of 25 cases per year in the United States alone. Even in Europe, where civil engineering work was more stringently regulated, collapses did occur, although they were far fewer than in America.

With the beginning of the use of iron as a building material, several iron suspension bridges constructed in the early 1800s in Europe and in America collapsed. For instance, the collapse of the Dryburgh Abbey suspension bridge in Scotland in 1818 was caused by aerodynamic instability. The Brighton Chain Pier Bridge, in Brighton, England, crashed in 1836 due to dynamic wind forces similar to those that destroyed the Tacoma Narrows Bridge in the state of Washington in 1940. Other types of bridges also failed in the nineteenth century due to wind action. Of major consequence was the failure of the Tay Bridge in Scotland in 1879, in which 75 people lost their lives. This collapse was investigated by a board of inquiry, which pointed out the major weakness of bridges was their low resistance to wind forces. After this first awareness, however, a century passed before the problem was seriously considered in structural design—except for the work of Alexandre Gustave Eiffel at the end of the century, and the great impact of the aeronautics industry on aerodynamics in general in the early 1900s.

It was not until the collapse of the Tacoma Narrows Bridge in 1940 that full scientific attention was given to the problem (Figure 3-1). The crash, which was caused by the phenomenon of dynamic instability, started a new era of research. The bridge initiated a longitudinal oscillation under a wind speed of 38 mph. This lengthwise deformation (galloping), which produced vertical waves in the structure of up to a maximum of four humps, lasted for 2 hr. Then, a steady speed of 42 mph added a transverse twisting that reached 45° from the horizontal. After a total of 3 hr from the start of the vibrations, the bridge gave way. A Karman vortex, 39 ft long (i.e., equal to the width of the bridge), had set the bridge in motion with a periodical force that coincided with the natural frequency of the bridge. This of course had created a phenomenon of resonance that increased the amplitude of the deformation beyond limits, making the collapse inevitable. Due to the length of time the phenomenon lasted, there was sufficient time for complete documentation of the collapse; in fact, numerous photographs and films were taken.

In reality, the effect of the wind on bridges had been sufficiently proved even before the bridge collapses previously mentioned, but there was a certain inertia to react on the part of the structural engineers. We see the catastrophes caused by tornadoes and the answer has been to accept them as acts of God. Because of their rarity, they could be fatalistically accepted. The same attitude was also taken for the smaller tornadoes of Europe, which were documented starting with the scientific observations of Geminiano Montanari in 1694. Through the centuries, due to its conservatism, the building art has been extremely slow to react to changes. However, the energetic attitude of twentieth-century civil engineering, which stems from the dynamic character of the industrial society of today, has been the moving force behind current wind engineering. (See Figure 3-2.)

Wind engineering originated from aerodynamics, which together with hydrodynamics constitutes the essential part of fluid dynamics. All these disciplines, of course,

Figure 3-1. In 1940, the Tacoma Narrows Bridge, a suspension steel structure in the state of Washington, collapsed after galloping for 2 hours and then twisting for another hour under a steady wind of 42 mph. A spectacular failure, well documented with photographs and films, it inspired a new phase of research in wind forces.

share the same fundamental principles, some of which are discussed in this chapter. More precisely, this chapter will only include the wind-loading aspects of wind engineering. The aspects of wind engineering dealing with the environment are discussed in subsequent chapters.

Figure 3-2. Vulnerability of masonry to high-speed winds. The collapse of the windward wall of a gymnasium building occurred in Joplin, Missouri, in a windstorm with straight winds (not a tornado). The entire unreinforced masonry wall between the two vertical expansion joints collapsed. (Courtesy of the Institute for Disaster Research, Texas Tech University)

VELOCITY PRESSURE

Calling q the *velocity pressure* or *dynamic pressure* or *dynamic head,* as it is usually referred to, its expression is given by:

$$q = \frac{1}{2} \rho V^2$$

Substituting the value of the air density ρ at standard conditions, the previous expression becomes:

$$q = 0.00256\ V^2$$

where

q = pressure in psf
V = wind velocity in mph

The dynamic pressures of wind velocities from 10 to 300 mph are given in Table 3-1.

Table 3-1. Dynamic pressure.

WIND VELOCITY V (MPH)	DYNAMIC PRESSURE $q = 0.00256\ V^2$ (PSF)	WIND VELOCITY V (MPH)	DYNAMIC PRESSURE $q = 0.00256\ V^2$ (PSF)
10	0.26	160	65.5
20	1.0	170	74.0
30	2.3	180	83.0
40	4.1	190	92.4
50	6.4	200	102.4
60	9.2	210	113.
70	12.6	220	124.
80	16.4	230	135.5
90	20.7	240	147.
100	25.6	250	160.
110	31.0	260	173.
120	37.0	270	187.
130	43.3	280	201.
140	50.2	290	215.5
150	57.6	300	230.

What does q really mean? And how is it derived? The value q is the pressure generated by the wind on a building assuming that the moving air comes to a complete stop as it hits the structure. Therefore, q is the total kinetic energy possessed by the air that is transferred from the air to the structure. This, of course, is a theoretical situation; it cannot occur because the air does not lose all its energy but only part of it. The air deviates from its original path as it approaches and passes over the structure, and it continues to move downwind with a residual velocity. If it had stopped, there would have been an accumulation of dead air in front of the structure and the flow would have come to a halt.

The real pressure at any given point on a structure will always be less than the theoretical value q, and it will be expressed for convenience as a percentage of the fixed value q. The dynamic pressure q represents the total kinetic energy of the air that strikes the structure. The expression is derived from the Bernoulli equation, which is valid for any ideal fluid (gases and liquids).

Let us consider a point at a certain distance from the structure where the wind is undisturbed, and let us indicate with p and V, respectively, the undisturbed atmospheric pressure and wind velocity at that point. Let us now consider a point on the structure stricken by the wind, and let us indicate with p_1 and V_1, respectively, the wind pressure and the wind speed at that point. With ρ the air density, the Bernoulli equation for these two points can be written as follows:

$$p+\frac{1}{2}\rho V^2=p_1+\frac{1}{2}\rho V_1{}^2$$

$$p_1-p=\frac{1}{2}(V^2-V_1{}^2)$$

If we assume that the air stops completely as it hits the structure, it follows that $V_1=0$. Thus, substituting:

$$p_1-p=\frac{1}{2}\rho V^2$$

but (p_1-p) is the new wind pressure (the total wind pressure minus the atmospheric pressure), which we called q. Thus:

$$q=\frac{1}{2}\rho V^2$$

WIND PRESSURE p OR STATIC VELOCITY PRESSURE

The real wind pressure p at any given point of a structure is the product of q (dynamic pressure) and a coefficient that varies from point to point for a given structure. Such a coefficient C_P is called the *pressure factor.* Thus:

$$p = C_P q$$

For example, given the structure indicated in Figure 3-3, let us determine the wind pressure at point P_1 on the windward wall and P_2 on the sidewall. The shape coefficients for P_1 and P_2 are, respectively, 0.90 and -0.45. Assume a wind speed $V = 40$ mph.

The solution is:

$$q = 0.00256(40)^2 = 4.1 \text{ psf}$$

Thus:

$$P_1 = C_{P_1} q = 0.90 \times 4.1 = 3.69 \text{ psf}$$
$$P_2 = C_{P_2} q = -0.45 \times 4.1 = -1.84 \text{ psf}$$

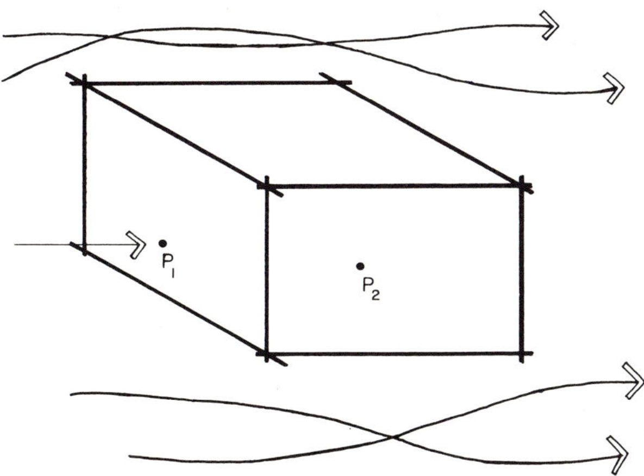

Figure 3-3. Determination of pressure coefficients.

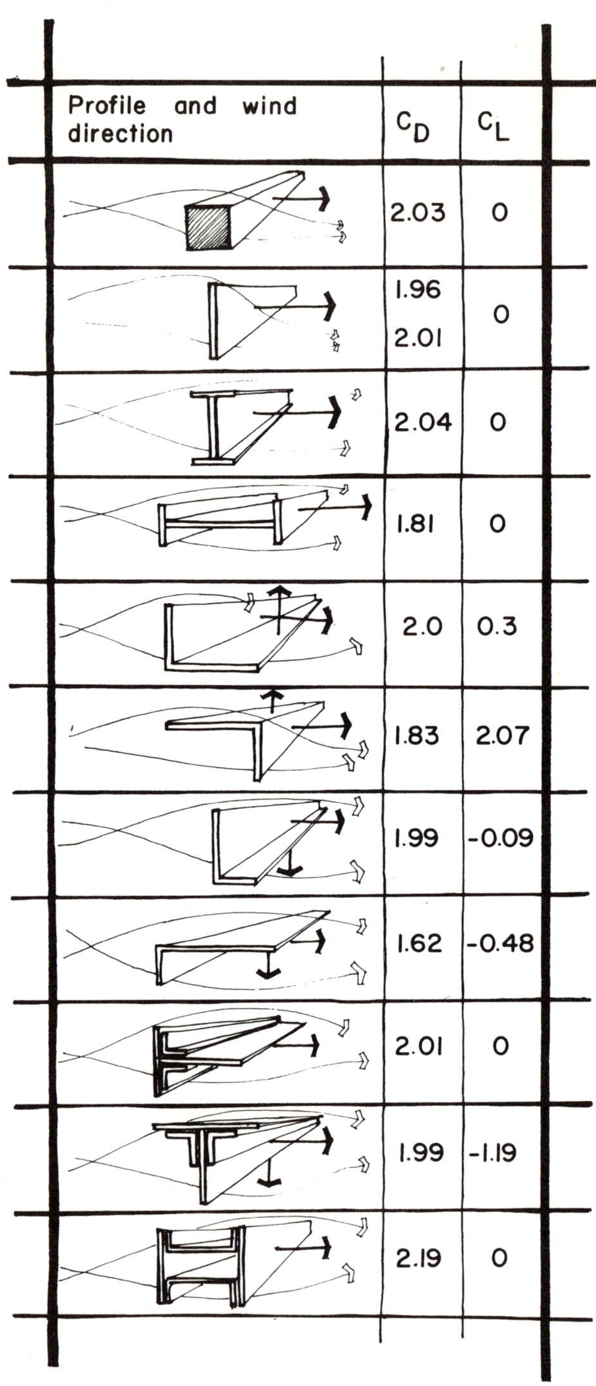

Profile and wind direction	C_D	C_L
	2.03	0
	1.96 2.01	0
	2.04	0
	1.81	0
	2.0	0.3
	1.83	2.07
	1.99	-0.09
	1.62	-0.48
	2.01	0
	1.99	-1.19
	2.19	0

Figure 3-4. Values of the drag coefficient C_D and lift coefficient C_L for various cross-sectional shapes.

WIND FORCE OR DRAG FORCE

If we consider the structure as a whole rather than its individual points, we can determine the total wind force on it—or, as it is also called, the drag force—by two different methods. In theory, such force can be calculated by summing vectorially all the individual pressure forces. In practice, it is computed as the product of the surface area A (measured as the projected area of the structure on a plane perpendicular to the wind direction), the dynamic pressure q, and the "shape factor" C_F. Thus:

$$\text{Wind force} = qAC_F$$

The difference between C_P and C_F is that while the former refers to a single point, the latter refers to the whole structure.

For individual bodies of various configurations, an overall shape factor C_F can be obtained in a wind-tunnel test, and its value can be divided into two basic components: one perpendicular to the wind force C_L (lift coefficient), and the other parallel to the wind force C_D (drag coefficient). From these, we call (qAC_L) the lift force and (qAC_D) the drag force. For various shapes, including square plates, angles, and so on, the coefficients can be found in Figure 3-4. The same kind of terminology is used to describe the lift and drag forces on the wing of an airplane, and actually, we are dealing with the same topic, as in the case of buildings.

PRESSURE COEFFICIENT C_p

Pressure coefficients are determined experimentally. Usually, a model of the structure is placed in a wind tunnel, and gauges are placed at any point on the model where the pressure is to be measured. From the pressure readings and knowing the air speed, the shape coefficients can be derived.

The coefficients are nondimensional, i.e., they are the same regardless of which units are used for p. Positive values are for positive pressure, whereas negative signs indicate suction. The coefficient C_p depends basically on the shape or geometrical characteristics of the structure and at times on the Reynolds number R_e.

Apart from wind-tunnel testing that one could conduct for a specific problem, a large number of pressure coefficients for a likewise large variety of building shapes can be obtained by gathering together the work of many researchers who have worked in this area. Among the most prominent names are Irminger and Nøkkentved (1930, 1936) and Nøkkentved (1932, 1934) from Denmark. The most extensive research on shape coefficients conducted in the United States is that by N. Chien et al. at the Iowa Institute of Hydraulic Research in 1951.

However, the most useful work so far has been conducted by Ackeret (1936) of the Institute for Aerodynamics in Switzerland. His work was incorporated into the Swiss building code and was included by the American Society of Civil Engineers in its report on wind forces on structures (1962).

	α	External Pressure Coefficient, C_{pe}								Internal Pressure Coefficient, C_{pi}			
										Openings uniformly distributed	Openings Mainly on Side		
		A	B	C	D	E	F	G	H		A	B	C
Roofs 0° to 10°, $h{:}b{:}L = 1{:}1{:}1$ 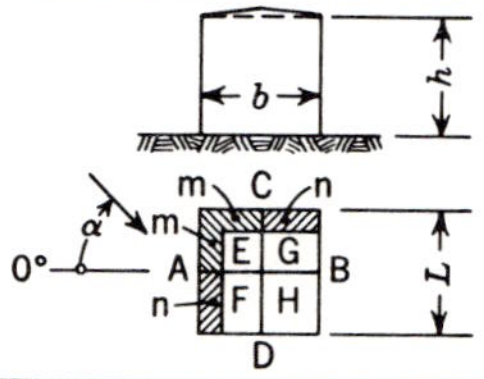	0°	0.9	-0.5	-0.6	-0.6	-0.7	-0.7	-0.5	-0.5	±0.2	0.8	-0.4	-0.5
	15°	0.8	-0.5	-0.7	-0.5	-0.7	-0.6	-0.5	-0.6	±0.2	0.7	-0.4	-0.6
	45°	0.5	-0.5	0.5	-0.5	-0.8	-0.5	-0.5	-0.4	±0.2	0.4	-0.4	0.4
	45°	For m, $C^*_{pe} = -1.2$; for n, $C^*_{pe} = -0.8$											

Table 3-2(1). Small buildings of residential scale. Cubical dimensions. Low-pitch roof. External pressure coefficients for walls and roof, including roof edges. Internal pressure coefficients as a function of location of openings. Various wind directions.

Structure	α	External Pressure Coefficient, C_{pe}								Internal Pressure Coefficient, C_{pi}			
										Openings uniformly distributed	Openings Mainly on Side		
		A	B	C	D	E	F	G	H		A	B	C
0°-10°; $h:b:L = 2.5:2:5$	0°	0.9	-0.5	-0.7	-0.7	-0.6	-0.6	-0.5	-0.5	± 0.2	0.8	-0.4	-0.6
	45°	0.6	-0.5	0.4	-0.5	-0.9	-0.7	-0.6	-0.7	± 0.2	0.5	-0.4	0.3
	90°	-0.5	-0.5	0.9	-0.4	-0.8	-0.2	-0.8	-0.2	± 0.2	-0.4	-0.4	0.8
	45°	For m, $C^{*}_{pe} = -1.5$											

Table 3-2(2). Small buildings of residential scale. Rectangular plan. Height equal to 1.25 the short side and the length equal to 2.5 the short side. Low-pitch roof. External pressure coefficients for walls and roof, including roof edges. Internal pressure coefficients as a function of location of openings. Various wind directions.

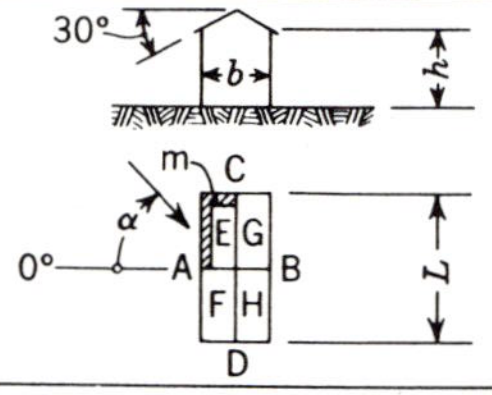

	α	External Pressure Coefficient, C_{pe}								Internal Pressure Coefficient, C_{pi}			
		A	B	C	D	E	F	G	H	Openings uniformly distributed	Openings Mainly on Side A	Openings Mainly on Side B	Openings Mainly on Side C
30°, h, b, $h:b:L = 2.5:2:5$, 0°, α, m, A, B, C, D, E, F, G, H, L	0°	0.9	-0.5	-0.7	-0.7	-0.6	-0.6	-0.5	-0.5	± 0.2	0.8	-0.4	-0.6
	45°	0.6	-0.5	0.4	-0.4	-0.4	-0.5	-0.6	-0.7	± 0.2	0.5	-0.4	0.3
	90°	-0.5	-0.5	0.9	-0.4	-0.7	-0.2	-0.7	-0.2	± 0.2	-0.4	-0.4	0.8
	45°	For m, $C^*_{pe} = -1.2$											

Table 3-2(3). Same as in (2) except the roof has a higher pitch (up to 30°).

	α	External Pressure Coefficient, C_{pe}								Internal Pressure Coefficient, C_{pi}			
		A	B	C	D	E	F	G	H	Openings uniformly distributed	Openings Mainly on Side A	Openings Mainly on Side B	Openings Mainly on Side C
$h:b:L = 2.5:2:5$	0°	0.9	-0.5	-0.8	-0.8	0.3	0.3	-0.6	-0.6	±0.2	0.8	-0.4	-0.7
	45°	0.6	-0.5	0.4	-0.4	0.3	-0.1	-0.5	-0.6	±0.2	0.5	-0.4	0.3
	90°	-0.5	-0.5	0.9	-0.4	-0.8	-0.2	-0.8	-0.2	±0.2	-0.4	-0.4	0.8
	75°	For m, C^*_{pe} = -1.2											

Table 3-2(4). Same as in (2) except the roof has a higher pitch (up to 50°).

	α	External Pressure Coefficient, C_{pe}								Internal Pressure Coefficient, C_{pi}			
										Openings uniformly distributed	Openings Mainly on Side		
		A	B	C	D	E	F	G	H		A	B	C
Closed hall Roofs 0° to 3°, $h{:}b{:}L = 1{:}4{:}4$	0°	0.9	-0.3	-0.4	-0.4	-0.8	-0.8	-0.3	-0.3	±0.2	0.8	-0.2	-0.3
	15°	0.8	-0.3	-0.1	-0.5	-0.7	-0.8	-0.2	-0.3	±0.2	0.7	-0.3	-0.2
	45°	0.5	-0.4	0.5	-0.4	-0.9	-0.6	-0.6	-0.3	±0.2	0.4	-0.4	0.4
	15°	For o, side C, $C^*_{pe} = -0.8$											
	45°	For m, $C^*_{pe} = -2.0$; n, $C^*_{pe} = -1.0$											

Table 3-2(5). Low-height, square building with sides equal to four times the height. Almost flat roof (pitch up to 3° only). External pressure coefficients for walls and roof, including roof edges. Internal pressure coefficients as a function of location of openings. Various wind directions.

	α	External Pressure Coefficient, C_{pe}								Internal Pressure Coefficient, C_{pi}			
										Openings uniformly distributed	Openings Mainly on Side		
		A	B	C	D	E	F	G	H		A	B	C
$h{:}b{:}L = 1{:}8{:}16$	0°	0.8	-0.5	-0.5	-0.5	0.2	0.2	-0.6	-0.6	±0.2	0.7	-0.4	-0.4
	45°	0.5	-0.5	0.4	-0.3	0.1	-0.1	-0.8	-0.5	±0.2	0.4	-0.4	0.3
	90°	-0.3	-0.3	0.9	-0.3	-0.5	-0.1	-0.5	-0.1	±0.2	-0.2	-0.2	0.8
	10° and 90°	For m, $C^*_{pe} = -1.0$											

Table 3-2(6). Low-height, rectangular building. Short and long side are, respectively, 8 and 16 times the height. Medium-pitch roof (up to 30°). External pressure coefficients for walls, including walls and roof corners. Internal pressure coefficients as a function of location of openings. Various wind directions.

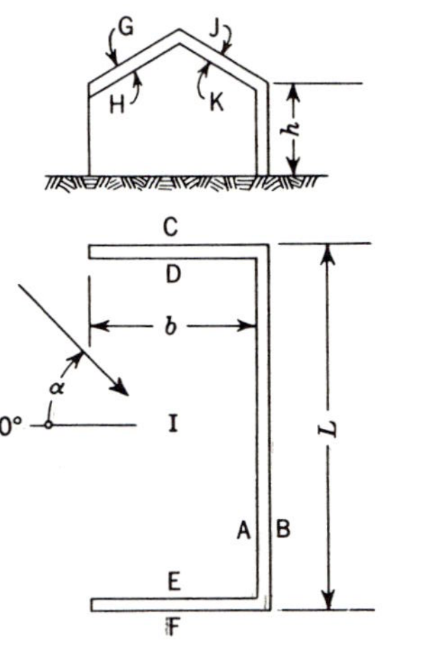

α	A	B	C	D	E	F	G	H	J	K
	External Pressure Coefficient, C_{pe}									
	(I) One long wall open									
0°	0.8	-0.5	-0.7	0.8	0.8	-0.7	-0.3	0.8	-0.4	0.8
45°	0.7	-0.6	0.4	0.6	0.8	-0.4	-0.2	0.6	-0.7	0.7
60°	0.3	-0.7	0.7	0.3	0.4	-0.4	-0.3	0.2	-0.6	0.2
180°	-0.5	0.9	-0.8	-0.5	-0.5	-0.8	-0.4	-0.5	-0.2	-0.5

Table 3-2(7a). One-story rectangular buildings, such as hangars and garages, with one large opening the size of a wall. Roof pitch: 30°. In the plan, the short and long sides are, respectively, two and four times the height at the eave. External pressure coefficients for indoor and outdoor faces of the walls and roof. Various wind directions. (Opening in the long sidewall.)

Roof 30°, $h:b:L = 1:2:4$

α	External Pressure Coefficient, C_{pe}									
	A	B	C	D	E	F	G	H	J	K
	(II) One end wall open									
0°	0.9	-0.7	-0.7	-0.4	-0.7	-0.8	-0.2	-0.7	-0.4	-0.7
45°	0.5	0.7	0.8	-0.5	0.7	-0.4	-0.3	0.7	-0.6	0.8
60°	0.1	0.9	0.9	-0.6	0.9	-0.4	-0.3	0.9	-0.7	0.9
90°	-0.5	0.8	0.8	-0.5	0.8	-0.3	-0.4	0.8	-0.4	0.8

Table 3-2(7b). Opening in the short sidewall.

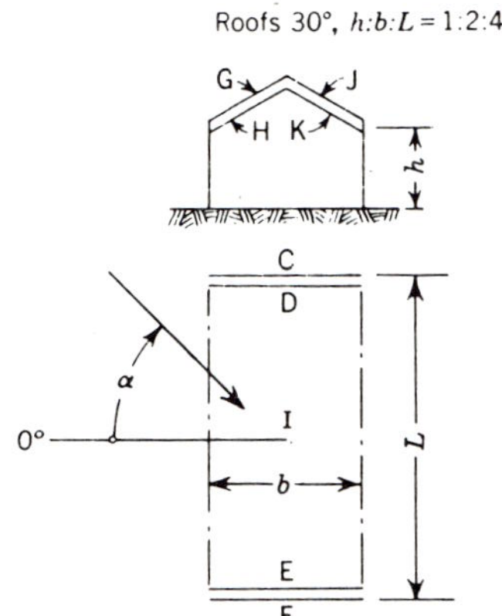

α	A	B	C	D	E	F	G	H	J	K
	External Pressure Coefficient, C_{pe}									
	(I) Two long walls open									
0°			-0.2	-0.7	-0.7	-0.2	0.4	-0.9	-0.5	-0.8
45°			0.5	-0.4	0.5	-0.4	0	-0.3	-0.6	0
60°			0.7	-0.6	0.5	-0.4	-0.3	-0.1	-0.7	-0.3

Table 3-2(8a). Same as in (7), but with openings on the two long sidewalls.

Buildings open on two sides
Roofs 30°, $h:b:L = 1:2:4$

α	External Pressure Coefficient, C_{pe}									
	A	B	C	D	E	F	G	H	J	K
	(II) Two end walls open									
0°	0.9	-0.7	-0.7	-0.4			-0.2	-0.7	-0.4	-0.7
45°	0.5	-0.4	-0.1	-0.8			-0.3	-0.4	-0.8	-0.3
60°	0.3	-0.2	0.1	-0.5			-0.3	-0.1	-0.8	0.1
60°	End Det, c = 0.7, d = -0.6; End Det e = 0.6, f = -0.8									

Table 3-2(8b). Same as in (7), but with openings on the two short sidewalls.

	α	External Pressure Coefficient, C_{pe}										Internal Pressure Coefficient, C_{pi}, with Vents			
		A	B	C	D	E	F	G	H	J	K	F & J open	F & J closed	F only open	J only open
Closed buildings with roof vents Roofs 20° $h{:}b{:}L = 1{:}4{:}8$; $h' = 0.5h$	0°	0.8	-0.5	-0.7	-0.7	-0.2	0.6	-1.0	-0.6	-0.5	-0.6	-0.2	±0.2	0.5	-0.4
	45°	0.4	-0.5	0.4	-0.5	-0.3	0.2	-1.3	-1.4	-1.0	-0.7	-0.5	±0.2	0.1	-0.9
	90°	-0.4	-0.4	0.8	-0.3	-0.4	-0.2	-0.3	-0.3	-0.2	-0.4	-0.3	±0.2	-0.2	-0.2
	0° and 45°	For m, $C^*_{pe} = -1.2$; n, $C^*_{pe} = -2.4$													

Table 3-2(9). Buildings with a central clerestory with vents, typical of rural religious buildings. Upper and lower roofs pitched. Rectangular plan with short and long sides, respectively, four and eight times the height. External pressure coefficients for walls and roofs, including roof edges. Internal pressure coefficients as a function of location of openings. Various wind directions.

	α	External Pressure Coefficient, C_{pe}								Internal Pressure Coefficient, C_{pi}				
										Openings uniformly distributed	Openings Mainly in			
		A	B	C	D	E	F	G	H		A	B	C	Roof EF
Roofs 0° to 15° $h{:}b{:}L = 2.5{:}1{:}1$	0°	0.9	-0.6	-0.7	-0.7	-0.8	-0.8	-0.8	-0.8	±0.2	0.8	-0.5	-0.6	
	15°	0.8	-0.5	-0.9	-0.6	-0.8	-0.8	-0.7	-0.7	±0.2	0.7	-0.5	-0.8	
	45°	0.5	-0.5	0.5	-0.5	-0.8	-0.7	-0.7	-0.5	±0.2	0.4	-0.4	0.4	
	45°	For m, C^{*}_{pe} = -1.0; n, C^{*}_{pe} = -0.8												

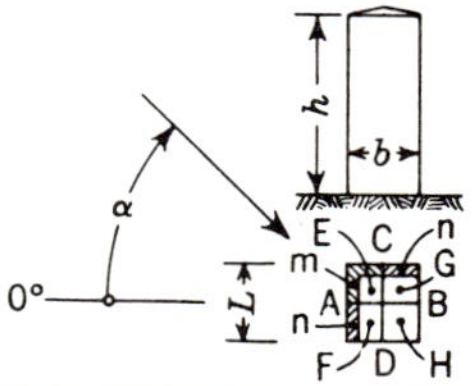

Table 3-2(10). Slender, square plane buildings with a height equal to 2.5 times the plan side. Low-pitch roof (up to 150°). External pressure coefficients for walls and roof, including roof edges. Internal pressure coefficients as a function of location of openings. Various wind directions.

		External Pressure Coefficient, C_{pe}								Internal Pressure Coefficient, C_{pi}				
										Openings uniformly distributed	Openings Mainly in			
	α	A	B	C	D	E	F	G	H		A	B	C	Roof EF
Tall buildings closed $h{:}b{:}L = 2{:}1{:}2$	0°	0.9	-0.5	-0.8	-0.8	-1.0	-1.0	-0.5	-0.5	± 0.2	0.8	-0.4	-0.7	
	45°	0.6	-0.5	0.4	-0.4	-0.3	-0.4	-0.5	-0.6	± 0.2	0.5	-0.4	0.3	
	90°	-0.6	-0.6	0.9	-0.4	-0.7	-0.5	-0.7	-0.5	± 0.2	-0.5	-0.5	0.8	
	0°	For m, $C^*_{pe} = -1.2$												

Table 3-2(11). Slender, rectangular plan buildings with long side equal to twice the short side and the height equal also to twice the short side. Roof pitch: 30°. External pressure coefficients for walls and roof, including roof edges. Internal pressure coefficients as a function of location of openings. Various wind directions.

Shed roof
$h:b:L = 1:2.4:12$

α	External Pressure Coefficient, C_{pe}								Internal Pressure Coefficient, C_{pi}				
									Openings uniformly distributed	Openings Mainly in			
	A	B	C	D	E	F	G	H		A	B	C	Roof EF
0°	0.9	-0.5	-0.6	-0.6	-0.5	-0.5	-0.5	-0.5	±0.2	0.8	-0.4	-0.5	-0.4
45°	0.5	-0.6	0.4	-0.4	-1.2	-0.7	-1.1	-0.7	±0.2	0.4	-0.5	0.3	-0.8
90°	-0.4	-0.3	0.9	-0.2	-0.3	0	-0.3	0	±0.2	-0.2	-0.1	0.8	0
180°	-0.4	0.8	-0.7	-0.7	0.1	0.1	0.2	0.2	±0.2	-0.3	0.7	-0.6	0
45°	For m, $C^{*}_{pe} = -1.4$												

Table 3-2(12). Low, rectangular buildings with shed roof. Short and long side are, respectively, 2.4 and 12 times the building height. Roof pitch: 30°. External pressure coefficients for walls and roof, including roof edges. Internal pressure coefficients as a function of location of openings. Various wind directions.

$h{:}b{:}L = 1{:}1{:}5$

α	External Pressure Coefficient, C_{pe}								Internal Pressure Coefficient, C_{pi}				
									Openings uniformly distributed	Openings Mainly in			
	A	B	C	D	E	F	G	H		A	B	C	Roof EF
0°	0.9	-0.5	-0.6	-0.6	0.6	0.6	-0.5	-0.5	±0.2	0.8	-0.4	-0.5	0.5
45°	0.5	-0.8	0.4	-0.5	0.2	-0.1	-1.0	-0.8	±0.2	0.4	-0.7	0.3	0
90°	-0.4	-0.4	0.9	-0.3	-0.4	0	-0.4	0	±0.2	-0.1	-0.1	0.8	-0.1
180°	-0.5	0.9	-0.6	-0.6	-0.5	-0.5	-0.1	-0.1	±0.2	-0.4	0.8	-0.5	-0.4
45°	For m, C^*_{pe} = -1.3												

Table 3-2(13). Low, rectangular buildings whose short and long sides are, respectively, one and five times the height of the building at the eave. Double-pitched roof with different slopes. External pressure coefficients for walls and roof, including roof edges. Internal pressure coefficients as a function of location of openings. Various wind directions.

Closed connecting passage between large walls
$h:b:L = 1:1:10$

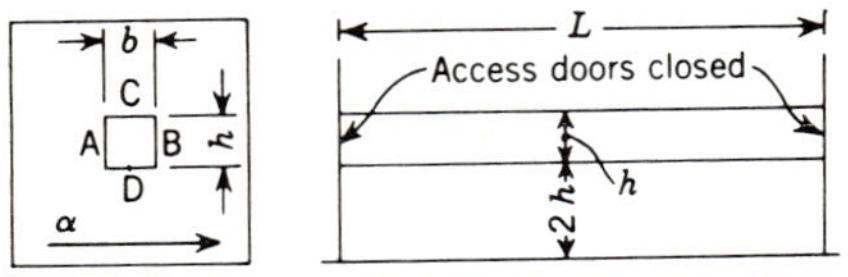

α	External Pressure Coefficient, C_{pe}								Internal Pressure Coefficient, C_{pi}				
									Openings uniformly distributed	Openings Mainly in			
	A	B	C	D	E	F	G	H		A	B	C	Roof EF
0°	0.8	-1.2	-1.4	-1.5					-0.5	0.7	-1.1	-1.3	

Table 3-2(14). Enclosed tubular pedestrian bridge above street level connecting two buildings. Square cross-sectional area, while the length is 10 times the height or the width of the tube. Bottom of the bridge is twice the bridge height above the street. External pressure coefficients for top, bottom, and sidewalls. Wind direction perpendicular to long side.

	α	External Pressure Coefficient, C_{pe}							
						End Surfaces			
	α	A	B	C	D	J	K	L	M
Roof +30° α = 0°-45°, A-D full length α = 90°, A-D part length L'	0°	0.6	-1.0	-0.5	-0.9				
	45°	0.1	-0.3	-0.6	-0.3				
	90°	-0.3	-0.4	-0.3	-0.4	0.8	-0.4	0.3	-0.3
	45°	For m, $C^*_{pe,top}$ = -1.0; $C_{pe,bottom}$ = -0.2							
	90°	Tangential acting friction, R = 0.05 q : L : b							

Table 3-2(15). Double-pitched sheltering canopy with no walls below, typical of train stations. Roof pitch: 30°. The rectangular, horizontal projection with a length five times the width. Height of the canopy eaves above the ground is equal to one-half the canopy width. External pressure coefficients for the roof surfaces over and below, including the two triangular surfaces at the ends. Various wind directions.

	α	External Pressure Coefficient, C_{pe}							
						End Surfaces			
		A	B	C	D	J	K	L	M
Roof +30° $\alpha = 0°$-45°-180°, A-D full length $\alpha = 90°$, A-D part length L' $h = 0.5\,b$, $h' = 0.8\,h$, $L = 5b$, $L' = b$, $R_{90°}$	0°	0.1	0.8	-0.7	0.9				
	45°	-0.1	0.5	-0.8	0.5				
	90°	-0.4	-0.5	-0.4	-0.5				
	180°	-0.3	-0.6	0.4	-0.6				
	45°	For m, $C^{*}_{pe,top} = -1.5$; $C_{pe,bottom} = 0.5$							
	90°	End surface friction load, see above							

Table 3-2(16). Same as in (15), but with one of the long sides protected by a standing train or stored material.

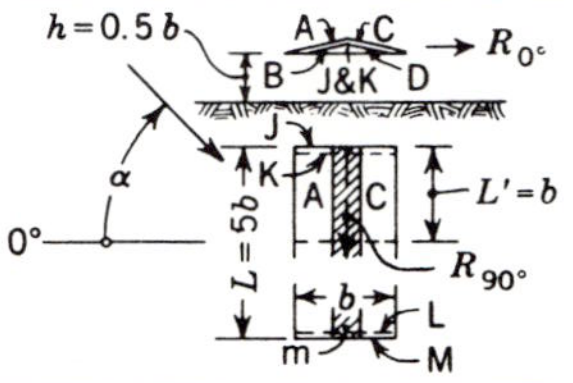

	α	External Pressure Coefficient, C_{pe}							
						End Surfaces			
		A	B	C	D	J	K	L	M
Roof +10° α = 0°-45°, A-D full length α = 90°, A-D part length L'	0°	-1.0	0.3	-0.5	0.2				
	45°	-0.3	0.1	-0.3	0.1				
	90°	-0.3	0	-0.3	0	0.8	-0.6	0.3	-0.4
	0°	For m, $C^*_{pe,top}$ = -1.0; $C^*_{pe,bottom}$ = 0.4							
	0° and 90°	Tangential acting friction, R = 0.1 q : L : b							

Table 3-2(17). Same as in (15), but with roof pitch of only 10°.

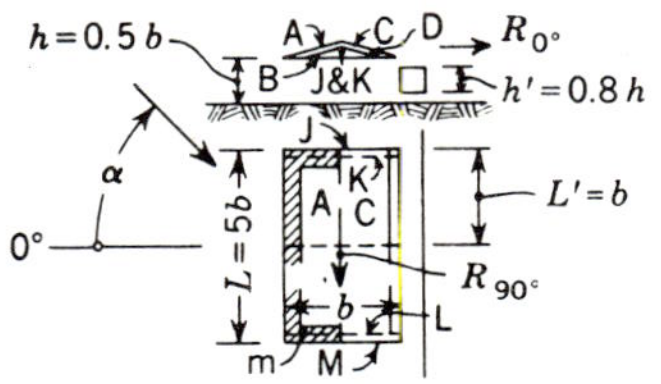

	α	External Pressure Coefficient, C_{pe}			
		A	B	C	D
Roof +10° α = 0°-45°-180°, A-D full length α = 90°, A-D part length L'	0°	-1.3	0.8	-0.6	0.7
	45°	-0.5	0.4	-0.3	0.3
	90°	-0.3	0	-0.3	0
	180°	-0.4	-0.3	-0.6	-0.3
	0°	For m, $C^*_{pe,top}$ = -1.6; $C^*_{pe,bottom}$ = 0.9			
	0° and 180°	End surface friction load, see above			

Table 3-2(18). Same as in (15), but with roof pitch of only 10°.

	α	External Pressure Coefficient, C_{pe}							
						End Surfaces			
		A	B	C	D	J	K	L	M
Roof −10° $\alpha = 0°$-$45°$, A-D full length $\alpha = 90°$, A-D part length L' $h = 0.5b$, $L = 5b$, $L' = b$, $R_{0°}$, $R_{90°}$, m	0°	0.3	-0.7	0.2	-0.9				
	45°	0	-0.2	0.1	-0.3				
	90°	-0.1	0.1	-0.1	0.1				
	0°	For m, $C^*_{pe,top} = 0.4$; $C^*_{pe,bottom} = -1.5$							
	0° and 90°	Tangential acting friction, R = 0.1 q : L : b							

Table 3-2(19). Same type of structure as described in (15), except that the pitch of the canopy is reversed; i.e., the roof has a valley at the centerline in place of a ridge line.

	α	External Pressure Coefficient, C_{pe}			
		A	B	C	D
Roof – 10° $\alpha = 0°$-45°-180°, A-D full length $\alpha = 90°$, A-D part length L' 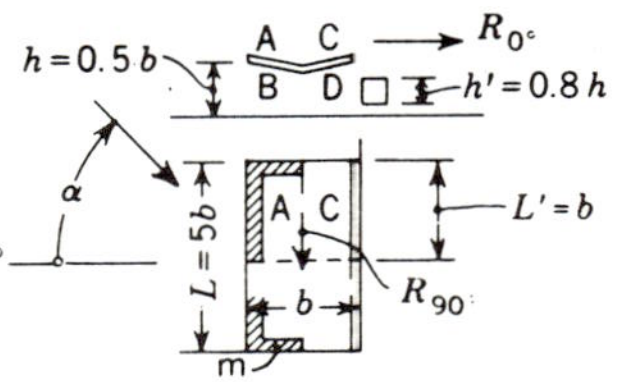	0°	-0.7	0.8	-0.6	0.6
	45°	-0.4	0.3	-0.2	0.2
	90°	-0.1	0.1	-0.1	0.1
	180°	-0.4	-0.2	-0.6	-0.3
	0°	For m, $C^*_{pe,top} = -1.1$; $C^*_{pe,bottom} = 0.9$			
	0° and 180°	Tangential acting friction, R = 0.1 q : L : b			

Table 3-2(20). Same as in (19), but with one of the long sides protected by a standing train or stored material.

	α	External Pressure Coefficient, C_{pe}																Internal Pressure Coefficient, C_{pi}			
		A	B	C	D	E	F	G	H	J	K	L	M	N	O	P	Q	Openings uniformly distributed	Openings mainly in A	Openings mainly in B	Openings mainly in C
$R_\alpha = 0.1\ q{:}b{:}L$ $h{:}b{:}L = 1{:}4{:}5$	0°	0.9	-0.3	-0.4	-0.4	0.6	-0.6	-0.6	-0.5	-0.5	-0.4	-0.3	-0.3					±0.2	0.8	-0.2	-0.3
	45°	0.5	-0.4	0.5	-0.3	0.2	-0.8	-0.5	-0.4	-0.2	-0.4	-0.2	-0.5					±0.2	0.4	-0.3	0.4
	90°	-0.4	-0.4	0.9	-0.3	-0.3	-0.4	-0.4	-0.4	-0.4	-0.4	-0.4	-0.3					±0.2	-0.3	-0.3	0.8
	180°	-0.3	0.9	-0.3	-0.3	-0.2	-0.3	-0.3	-0.4	-0.4	-0.6	-0.6	-0.1					±0.2	-0.2	0.8	-0.2
	0° and 180°	For m, $C^*_{pe} = -1.3$; n, $C^*_{p3} = -2.0$																			

Table 3-2(21). North-light or sawtooth roofed buildings with rectangular floor plan whose short and long sides are, respectively, four and five times the height. Roof pitches are 60° and 30° for the two directions of the slope. External pressure coefficients for walls and roof, including roof edges. Internal pressure coefficients as a function of location of openings. Various wind directions.

	α	External Pressure Coefficient, C_{pe}																Internal Pressure Coefficient, C_{pi}			
		A	B	C	D	E	F	G	H	J	K	L	M	N	O	P	Q	Openings uniformly distributed	Openings mainly in A	Openings mainly in B	Openings mainly in C
$h:b:L = 1:3:4$	0°	0.9	-0.5	-0.6	-0.6	-0.8	-0.8	-0.4	-0.4	-1.0	-0.4	-0.5	-0.5					± 0.2	0.8	-0.4	-0.5
	45°	0.5	-0.5	0.5	-0.4	-0.6	-0.5	-0.5	-0.5	-0.5	-0.5	-0.5	-0.5					±0.2	0.4	-0.4	0.4
	90°	-0.5	-0.5	0.9	-0.4	-0.8	-0.4	-0.8	-0.4	-0.4	-0.4	-1.0	-0.4					±0.2	-0.4	-0.4	0.8
	0° and 90°	For m, C^*_{pe} = -1.1; n, C^*_{pe} = -1.5																			

Table 3-2(22). Low building with flat roof but inclined roof edges. Rectangular floor plan whose short and long sides are, respectively, three and four times the height. External pressure coefficients for walls, roof, and roof edges. Internal pressure coefficients as a function of location of openings. Various wind directions.

h:b:L = 1:12:12
r = 5/6b
Y = 0.1b
X = 0.1b

α	External Pressure Coefficient, C_{pe}																Openings uniformly distributed with windows and doors closed	Openings mainly in		
	A	B	C	D	E	F	G	H	J	K	L	M	N	O	P	Q		A with window Y open	C with all gates open	C with only gate X open
0°	0.7	-0.2	-0.3	-0.3	-0.1	-0.5	-0.8	-0.8	-0.4	-0.1							±0.2	0.4	-0.1	-1.5
30°	0.6	-0.3	0.2	-0.4	-0.1	-0.4	-0.7	-0.9	-0.7	-0.4							±0.2	0.7	0.6	0.7
90°	-0.3	-0.3	0.9	-0.3							-0.8	-0.7	-0.5	-0.3	-0.1	-0.1	±0.2	-1.0	0.8	0.4
30°	For m, C^*_{pe} = -1.8 with $C^*_{pe,min}$ = -2.5																			

Table 3-2(23). Buildings with cylindrical roof with smooth surface. Typical of hangars or industrial buildings. The sides of the square plan are 12 times the height at the eave. The rise at the crown is one-sixth of the building length or width. External pressure coefficients for walls, roof, and edges. Internal pressure coefficients as a function of location of openings. Various wind directions.

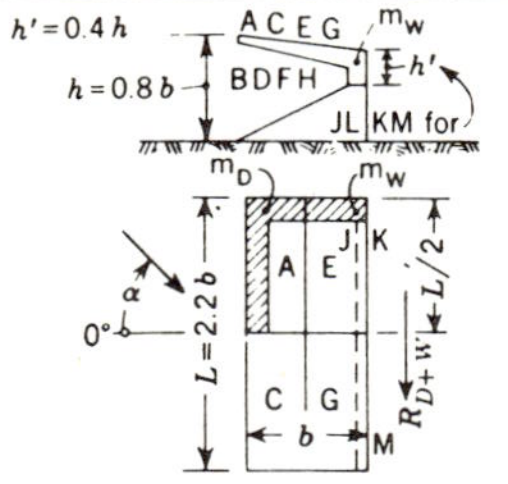

α	A	B	C	D	E	F	G	H	J	K	L	M
	External Pressure Coefficient, C_{pe}											
	C_{pe} at top and bottom of roof								C_{pe} at front and back of wall			
0°	-1.0	0.9	-1.0	0.9	-0.7	0.9	-0.7	0.9	0.9	-0.5	0.9	-0.5
45°	-1.0	0.7	-0.7	0.4	-0.5	0.8	-0.5	0.3	0.8	-0.6	0.4	-0.4
135°	-0.4	-1.1	-0.7	-1.0	-0.9	-1.1	-0.9	-1.0	-1.1	0.6	-1.0	0.4
180°	-0.6	-0.3	-0.6	-0.3	-0.6	-0.3	-0.6	-0.3	-0.3	0.9	-0.3	0.9
45°	For m_D, $C^*_{pe,top} = -2.0$; $C^*_{pe,bottom} = 1.0$											
60°	For m_W, $C^*_{pe,k} = -1.0$; $C^*_{pe,j} = 1.0$											
90°	Tangential acting friction, $R_D = 0.1\ q : b : L$; $R_W = 0.1\ q : h : L$											

Table 3-2(24). Grandstand structure, open on three sides but with closed back wall supporting the canopy. External pressure coefficients for front and back wall and for top and bottom of roof, including roof edges and the tangential friction. Various wind directions.

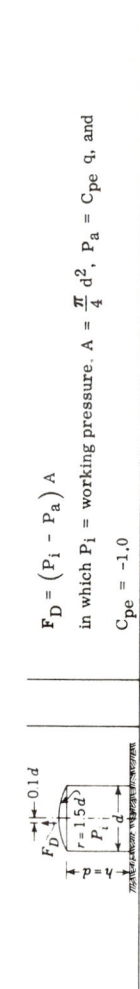

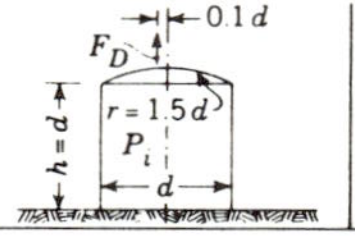

$$F_D = \left(P_i - P_a\right) A$$

in which P_i = working pressure. $A = \frac{\pi}{4} d^2$, $P_a = C_{pe}\ q$, and

$C_{pe} = -1.0$

Table 3-2(25). Cylindrical building typical of reservoirs and low-rise domical roof. Height at the edge is equal to the diameter of the building. External pressure coefficient for the lifting force on the roof.

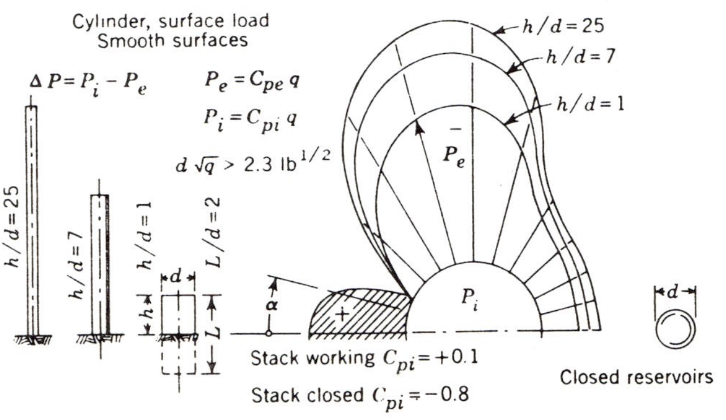

α	Value of h/d		
	1	7	25
	Value of L/d		
	2	14	50
	Value of C_{pe}		
0°	1.0	1.0	1.0
15°	0.8	0.8	0.8
30°	0.1	0.1	0.1
45°	-0.7	-0.8	-0.9
60°	-1.2	-1.7	-1.9
75°	-1.6	-2.2	-2.5
90°	-1.7	-2.2	-2.6
105°	-1.2	-1.7	-1.9
120°	-0.7	-0.8	-0.9
135°	-0.5	-0.6	-0.7
150°	-0.4	-0.5	-0.6
165°	-0.4	-0.5	-0.6
180°	-0.4	-0.5	-0.6

Table 3-2(26). Vertical cylindrical structures with smooth surfaces, such as smokestacks. External pressure coefficients for various zones of the cylinder expressed as a function of the angle α. The coefficients are given for three types of cylinders that differ from each other in terms of h/d ratio (height/diameter) and L/d ratio (total height including underground portion/diameter). Internal pressure coefficients C_{pi} for stack open to the atmosphere on top (working condition) and for closed stack.

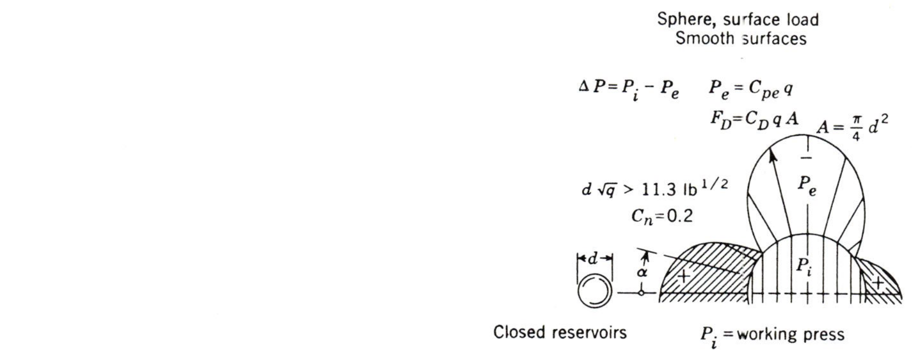

Sphere, surface load
Smooth surfaces

$\Delta P = P_i - P_e$ $P_e = C_{pe} q$

$F_D = C_D q A$ $A = \frac{\pi}{4} d^2$

$d \sqrt{q} > 11.3$ lb$^{1/2}$

$C_n = 0.2$

Closed reservoirs

P_i = working press

α	Value of C_{pe}
0°	1.0
15°	0.9
30°	0.5
45°	-0.1
60°	-0.7
75°	-1.1
90°	-1.2
105°	-1.0
120°	-0.6
135°	-0.2
150°	0.1
165°	0.3
180°	0.4

Table 3-2(27). Spherical buildings with smooth surface. External pressure coefficients for various zones on the sphere expressed as a function of the angle α.

Upright cylinder
$F_D = C_D q A$
$d\sqrt{q} > 2.3\ \text{lb}^{1/2}$
$A = dh$

$h/d =$	25	7	1

Cross Section	Description	Value of C_D		
	Smooth surface; metal, timber, concrete	0.45	0.5	0.55
	Rough surface; round ribs, h = 2% d	0.7	0.8	0.9
d	Very rough surface; sharp ribs, h = 8% d	0.8	1.0	1.2
	Smooth, rough surface; sharp edges	1.0	1.2	1.4

Table 3-2(28). Vertical cylindrical buildings or structures having different surface conditions (smoothness, geometry, materials) and an h/d ratio (height/diameter). Drag coefficients C_D are given for various surface conditions and three h/d ratios.

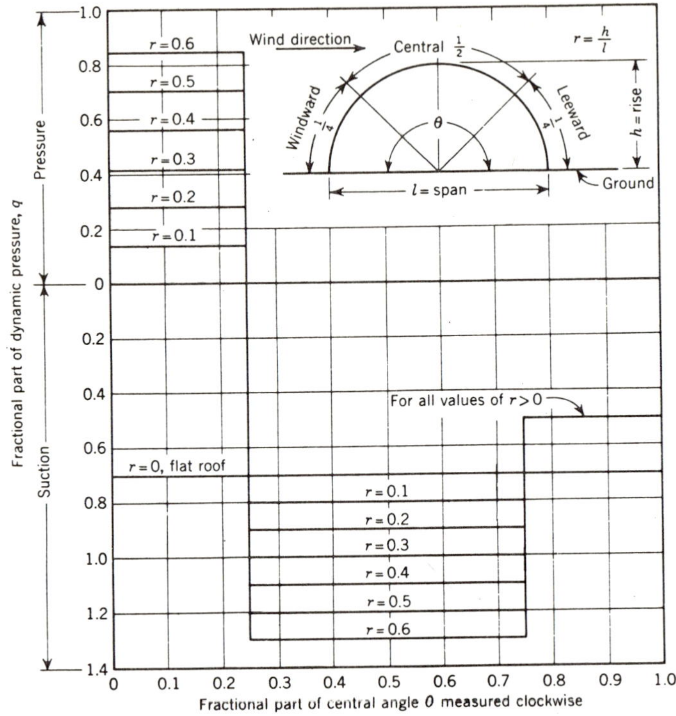

Figure 3-5. Exterior shape coefficient for cylindrical roof configurations having the longitudinal axis horizontal and perpendicular to the wind. Note that the roof is assumed to spring from the ground (to be distinguished from similar roofs on top of walls or columns).

The buildings included in Table 3-2 (Ackeret 1936) have a variety of geometrical configurations different from each other. Such differences can be based on floor plan, roof shapes, layout of walls, and so on, or on different ratios between height, width, and depth of similar building shapes. The wind direction is also not always the same, and its angle α varies. The pressure coefficients are divided into C_{pe} (external pressure coefficient) and C_{pi} (internal pressure coefficient). Notice that such coefficients have been found for various surfaces of the building walls and roof. Such areas are indicated by the letters *A, B, C, D,* etc., and at times with *m* and *n* for areas near the roof edge where pressures are particularly critical. For the internal pressure coefficient, a distinction is made as to where most of the windows and door openings are applied, such as on the windward wall or leeward wall, or on the roof. (See Figures 3-5 and 3-6.)

Notice that these coefficients are applied only to specific parts of the building and that their effect has to be combined to have an overall wind force on the building as a whole.

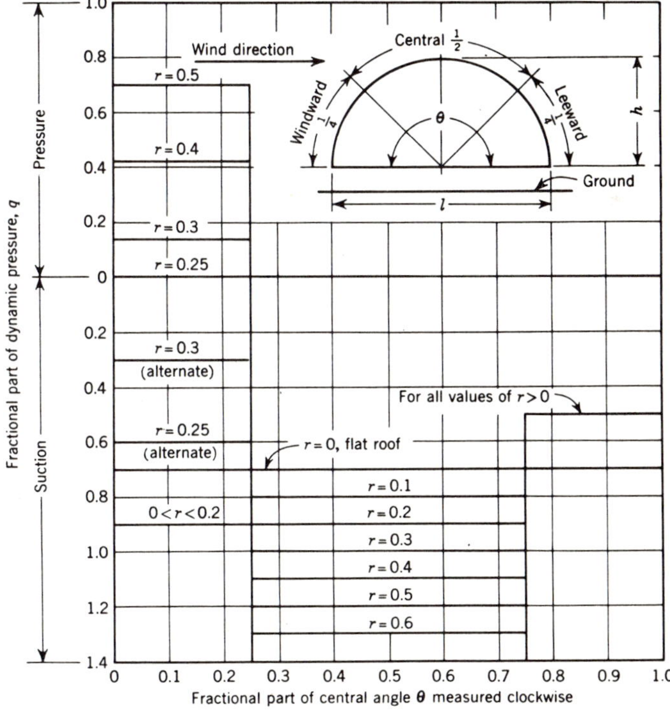

Figure 3-6. Exterior shape coefficient for cylindrical roof configurations having the longitudinal axis horizontal and perpendicular to the wind. (Note that the roof is assumed to be on top of a structure and not springing from the ground.)

THE REYNOLDS NUMBER (Re)

The Reynolds number (Re), named after the Irish physicist Sir Osborne Reynolds (1842–1912), is a dimensionless quantity that does not depend on the units being used. It is a fundamental parameter in aerodynamics, especially for determining the shape factors of structures and consequently the wind pressure. Its importance is also in the law of similitude for aerodynamic modeling.

The Reynolds number is defined as the ratio of the inertial force ρV^2h^2 (i.e., qA) to the viscous drag force μVh. The former is the obstructing force generated by the frontal impact as the air collides against an obstacle. It is high for blunt bodies and minimal for airfoils and other aerodynamic shapes. The latter is instead the force developed by the viscosity existing between air strata just like friction between solid laminae. This force is maximal for airfoils and streamlined shapes, and minimal for blunt bodies. In all cases, however, both inertial and viscous forces are always present at the same time, although in different proportions. (See Figures 3-7 through 3-9.)

Accordingly, Re is expressed as follows:

$$\text{Re} = \frac{\text{Inertial forces}}{\text{Viscous forces}} = \frac{\rho V^2 h^2}{\mu V h} = \frac{\rho V h}{\mu} = \frac{Vh}{\nu}$$

where

ρ = air density
V = wind velocity
h = diameter or width of structure
μ = air viscosity

$$\nu = \frac{\mu}{\rho} = \text{kinematic viscosity} = 1.82 \times 10^{-5}\ \text{ft}^2/\text{sec}$$

Obviously, the larger the inertial force, or the wind speed or the diameter of the structure, the larger is the Re. Vice versa, as the previous parameter decreases, so does Re. The kinematic viscosity of the air ν remains constant and does not affect a variation of Re.

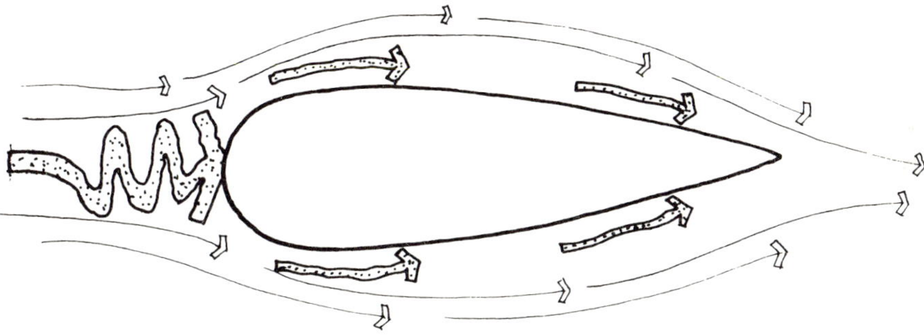

Figure 3-7. Inertial and viscous forces on a body engulfed by wind. The frontal resistance to the wind is, of course, the inertial force that slows down the velocity and diverts the paths of the air. The frictional force between the sides of the body and the air as well as the forces parallel to the wind that slow down the layers of air, are the viscous forces in question. Note that the ratio between the two is Re (Reynolds number).

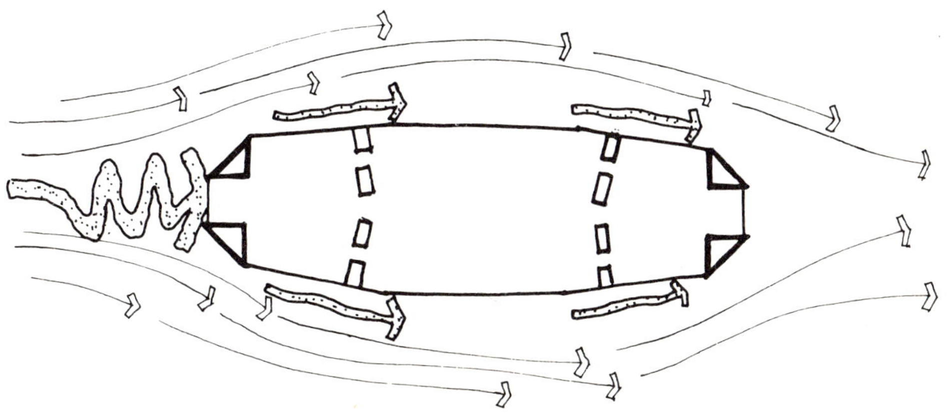

Figure 3-8. Inertial and viscous forces applied to a building by the wind.

Figure 3-9a. Small inertial force on the narrow frontal area of the building and large viscous forces on the wide lateral area when the wind blows as shown (small Re).

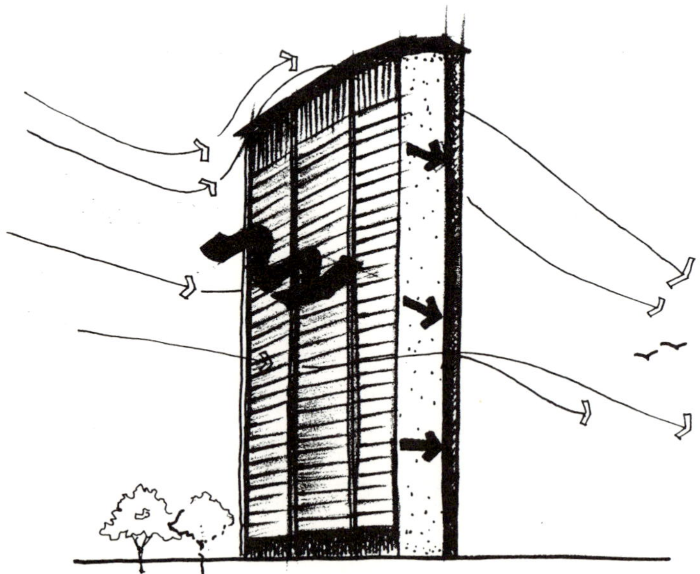

Figure 3-9b. Large inertial forces on the wide frontal area of the building and low viscous forces on the narrow sides when the wind blows as shown (large Re).

Let us look at an example. Given that 50 mph wind speed = 73.33 ft/sec, then, for a circular tank, 10 ft in diameter, the Reynolds number is calculated as follows:

$$Re = \frac{Vh}{\nu} = \frac{73.33 \times 10}{1.82 \times 10^{-5}} \cong 40 \times 10^{6}$$

The Re varies considerably. There is a certain value called the *critical Reynolds number* that varies between 10^5 and 3.6×10^6, according to the roughness of the surface. When Re is less than the critical value, the airflow in the boundary layer is laminar, and the shape factor C_F or drag coefficient C_D is large. When Re is higher than the critical value, the flow in the boundary layer is turbulent and C_F or C_D decreases.

AERODYNAMICS

Aerodynamics is the science that studies the phenomena associated with the relative motion of air and a body immersed in it. Naturally, as in the case of a flying airplane or a building engulfed by the wind, there is no substantial difference whether the air is stationary and the body moves, or vice versa. Of course, this is a simplification. The presence of the ground and the variability of the wind velocity make quite a difference between the two cases; however, the general principles apply to both. Only very basic principles of aerodynamics are required to understand the problems concerning building design; therefore, the following discussion will be limited to a few essential points.

AIRFLOW

Airflow can be either *laminar* or *turbulent.* Laminar flow consists of air particles moving along smooth linear paths, or, as the word implies, air moves in stratified laminae or layers that indicate smoothness of flow. Turbulent flow is the opposite; i.e., the air particles move in an erratic and random fashion in all directions, breaking the smoothness of the flow.

STREAMLINES

In the graphical representation of airflow around an obstacle, the paths of air particles are indicated by lines called *streamlines.* Streamlines denote a laminar flow; they do not exist in turbulent flow. It is interesting to notice that as the streamlines from a certain spacing get closer together, the air speed increases at that point; conversely, as they separate, the speed decreases.

BOUNDARY LAYER

That portion of the air where the speed of the air particles is modified by the presence of a body (for instance, a moving plane in still air or a still building in flowing wind) is called a *boundary layer.* We could say, therefore, that the boundary layer is that portion of the air bounded inside by the body itself and outside by the edge of undisturbed air.

Ludwig Prandlt (1875–1953), originator of the German school of aerodynamics and the founder of the boundary layer theory, discovered that attached to a body that is in relative motion with air, is a thin film of air that stays with the body. Each superimposed layer of air is retarded by friction, but less and less as the distance between laminae and body increases. Note that the flow within the boundary layer can be laminar or contain areas of turbulence.

CONTINUITY

Continuity or "principle of conservation of mass" states that the mass of air entering a stream tube (a bundle of streamlines) is the same at the end and therefore at any cross section in between. Thus, from Figure 3-10:

$$\rho V_1 A_1 = \rho V_2 A_2$$

Thus:

$$V_1 A_1 = V_2 A_2$$

This principle also applies in hydraulics to water flowing in a pipe.

MODEL LAWS

Wind-tunnel models can be either rigid or aeroelastic. Rigid models can be subdivided into: static rigid, rotating rigid, and spring-mounted rigid. Aeroelastic models can be categorized

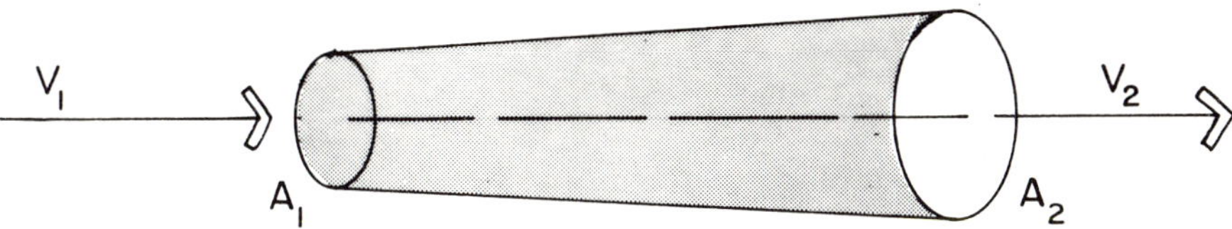

Figure 3-10. Because of the principle of continuity, the mass of the air does not change as the air flows through section 1, or 2, or any other section.

as static aeroelastic and dynamic aeroelastic. For wind analysis of buildings, elastic models are mostly used.

$$m_1 D_{model} = D_{prototype}$$
$$m_1^2 \Delta_{model} = \Delta_{prototype}$$
$$m_2 E_{model} = E_{prototype}$$
$$e_{model} = e_{prototype}$$
$$\sqrt{m_2} \cdot V_{model} = V_{prototype}$$
$$\frac{\sqrt{m_2}}{m_1} n_{model} = n_{prototype}$$
$$\delta_{model} = \delta_{prototype}$$

where

Δ = area
D = typical major length
V = air velocity
m_1 and m_2 = factors of proportionality
n = natural frequency of vibration
e = damping
δ = density

Stresses and moments can be derived from the following relationship:

$$m_2 \sigma_{model} = \sigma_{prototype}$$
$$m_1 d_{model} = d_{prototype}$$

Wind-tunnel tests on models can be related to the actual condition of structures exposed to the wind, if the relationships between the model and the prototype of structure are respected:

$$Re_{model} = Re_{prototype}$$

TESTING APPARATUS

Wind tunnels, originally built for aerodynamic testing of aircraft, can also be used for building design for wind forces. Basically divided into two categories—open jet and closed jet—wind tunnels are used for quantitative analysis of several aerodynamic parameters.

On the other hand, when the objective of the study is just of quantitative nature, the sophisticated wind-tunnel types just mentioned are unnecessary. Much simpler types can be used instead. For instance, the visualization of the windflow through and around buildings can be shown with low-speed air currents that use smoke plumes, bubbles, filaments, membranes, and so on.

Teaching aids in architectural schools sometimes use practical devices that can be quickly improvised for approximate demonstration of windflow. For instance, a fan and some cardboard can quickly be converted into a rudimental wind tunnel. Small models of buildings can also be quickly built using a frame of balsa wood or cardboard and applying paper or plastic over it to simulate roofs and walls. By placing models into the improvised wind tunnels, the concavity or convexity of the membrane surfaces of the model show, respectively, areas of positive pressure or negative pressure (suction).

Another quick method is to attach light threads or filaments to cardboard models inserted in the tunnel. In addition, another simple apparatus for the visualization of flow patterns is the "water table." It consists of a horizontal plane over which a thin layer of water (about 0.5 in.) flows in a laminar flow. Water flowing from a faucet over the plane and discharging in a sink can be quickly improvised. The laminar flow is obtained with a vertical screen upstream with respect to the model. A series of syringe needles discharging small droplets of dye in the water generates a series of colored streamlines that flow in, through, and around a model placed on the plane and demonstrate the pattern flow, which can be photographed.

SELF-EXCITED OSCILLATIONS

Self-excited oscillations are the type most commonly observed when buildings vibrate due to the effect of the wind. Den Hartog (1947) described this type of motion. "In a self-excited oscillation the alternative force that sustains the motion is created or controlled by the motion itself; if the motion stops, the alternating force disappears."

A common cause of self-excited oscillation is the vortex-shedding phenomenon usually known as the Karman vortex, named after Theodore von Kármán (1881–1963), who was a student of Prandtl at Göttingen University. Each vortex that separates from the building on the leeward side produces a force on the building itself that is perpendicular to the direction of the wind. In this case, the vortex is frequently determined only by the wind alone, if the structure is at rest. When the wind velocity reaches a certain value, referred to as the "critical velocity," the structure starts to oscillate under the effect of the vortices and the frequency of the vortex shedding is no longer controlled by the wind but by the structure itself. As the wind velocity increases, the structure remains in motion up to a certain value, after which it comes to rest and stays at rest even if the wind velocity increases. Should the wind velocity continue to rise, it will reach another critical velocity, and the structure will again start to oscillate up to a certain value of the velocity, after which it will come to rest once more. The phenomenon, therefore, is periodical, and there are many critical wind velocities for any given structure. As mentioned, the structure starts to oscillate at the critical velocity and stops oscillating at a certain value, which is not very well defined. Between such values and the next value of the critical velocity, the structure is at rest. When the vortex frequency is equal to the natural frequency of the structure, the excitation is harmonic. If, instead, the vortex frequency is equal to a multiple of the natural frequency of the structure, the excitation is then subharmonic. (See Appendix A.)

Prismatic structures in the wind can be excited by the Karman vortex; however, the oscillation varies if the building is circular in cross section; or if it has edges and, if so, whether these edges are sharp or rounded, etc. In prismatic buildings with round cross sections, the Karman vortex phenomenon theoretically consists of the formation of individual vortices that appear in the leeward side of the building, on each side of it, where the airstream separates from the edges of the building. (See Figure 3-11a.) If D is the diamteer of the building's cross section, the vortices on one of the two sides are spaced 4.3 D apart. The building vibrations excited by the Karman vortex have a frequency of $n = 0.207V$ cycles per second, where V is the wind velocity. Notice that the two lines of vortex shedding parallel to the wind are spaced 1.2 D apart, and the rotation of the vortices in one line is opposite the

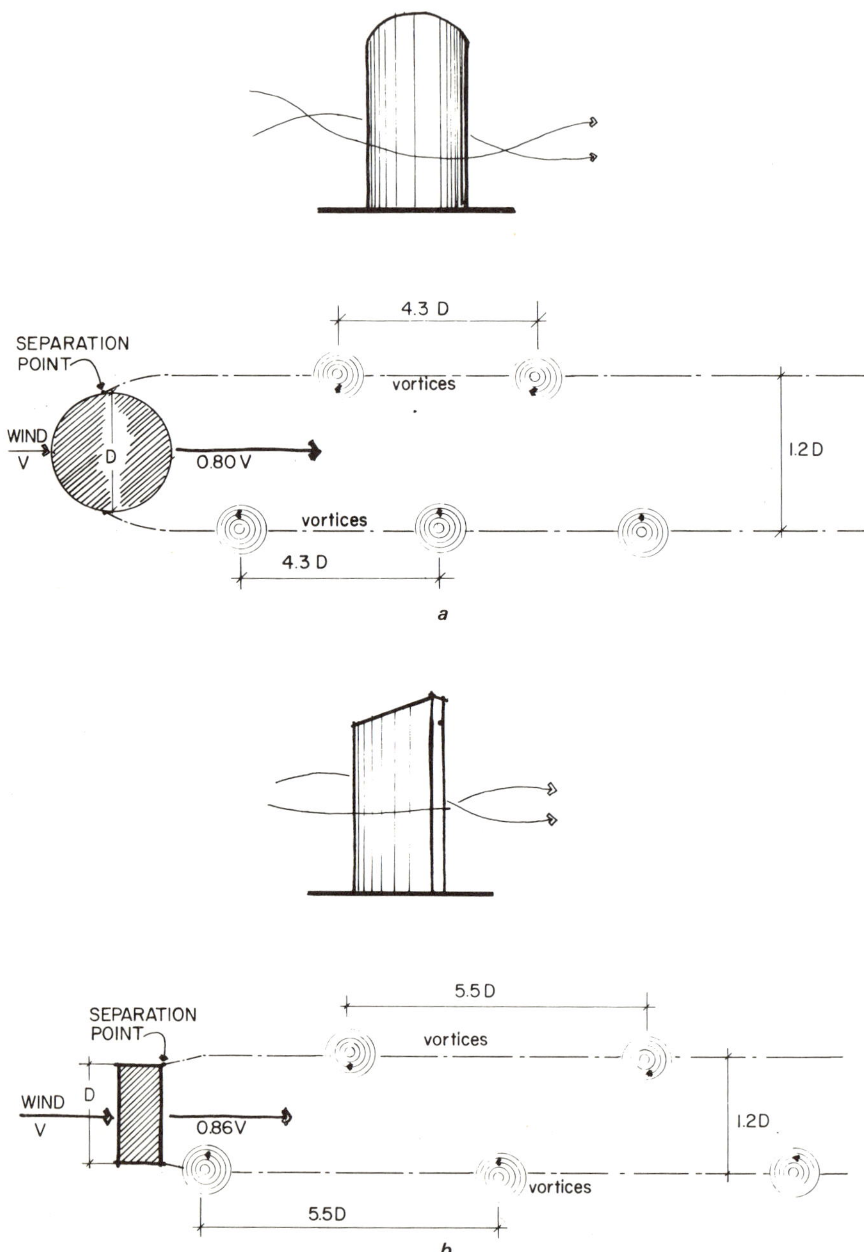

Figure 3-11. *a,* The Karman vortex phenomenon for circular cross-sections. *b,* The Karman vortex phenomenon for sharp corners or flat plates.

vortices of the other line. Furthermore, the vortices in one of the lines are staggered with respect to those in the other line. This alternative pattern, which is similar to that of streetlights commonly observed in cities, was the reason for the name given to the phenomenon: "Karman vortex street" (or "Karmansche Wirbelstrasse"). Notice also that this mechanism is sufficiently dependable for a distance of approximately $7D$ on the leeward side of the building and that the vortex shedding is an accurate periodic phenomenon only up to an Re value equal to 3×10^5, which is referred to as the "critical Reynolds number." Above the value, the Karman vortex phenomenon is aperiodic. However, most of the damage suffered by tall, flexible structures, such as has been reported for slender smokestacks, occurred when Re was above the critical value. Consequently, even if the phenomenon is aperiodical and erratic, its consequences are nevertheless destructive as if it were periodical. The range below and above the critical Re are called *subcritical* and *supercritical,* respectively. The oscillation of the above mentioned structures was in the low, natural frequency, it should be noted. In conclusion, apart from the periodicity and aperiodicity, the difference in the consequences between the subcritical and the supercritical conditions is minimal.

Experiments on flat plates exposed with their face perpendicular to the wind, a condition similar to that of a rectangular cross-sectional building facing the wind, show that the frequency of the vortex shedding is such that two consecutive eddies of the same sign on one side of the building are $5.5D$ apart, where D is the width of the face of the building facing the wind (see Figure 3-11b). This is also valid for any wind angle, in addition to the 90° with the windward face of the building. The distance between the two parallel lines of eddy shedding is $1.2D$, just as in the case of the circular cross-sectional building. Furthermore, for buildings with sharp edges, it was found that the vortex-shedding phenomenon is independent of the Reynolds number, when Re is larger than 1000.

WIND-RESISTANT FRAMING SYSTEMS

The resistance of high-rise buildings to wind, as well as to earthquakes, is the main determinant in the formulation of new structural systems that evolve during the continuous effort of structural engineers to increase building height while containing the deflection within acceptable limits and minimizing the amount of materials. With the abandonment of masonry in favor of steel at the end of the nineteenth century, the steel skeleton has been continuously revised, giving birth to several improvements that culminated in the 1970s in the Sears Tower in Chicago. Concrete systems have also progressed since the beginning of the century, generating a variety of alternatives and improvements that have pushed the upper limits of building heights to values that were considered utopian a few decades ago. Considering steel and reinforced concrete structures, high-rise systems capable of resisting efficiently the large wind forces encountered at high altitudes offer a great number of structural typologies to choose from. In a general classification of the available alternatives, steel frames are distinguished from those of concrete.

Among the major wind-resistant steel systems, we can include the following:

- Frame with rigid connection (lateral braces excluded) (Fig. 3-12)
- Bracket bracing (Fig. 3-13)
- Knee bracing (Fig. 3-14)
- The narrow K-bracing (Fig. 3-15)
- Portal bracing (Fig.3-16)
- Full diagonal X-bracing (Fig. 3-17)
- Full diagnoal K-bracing (Fig. 3-18)
- Tapered frame (Fig. 3-19)
- Circular tube frame (Fig. 3-20)
- Rectangular tube frame (Fig. 3-21)
- Trussed core with cap-trusses and suspended system (Fig. 3-22)
- Trussed core with cap-trusses, suspended system, and tie-downs (Fig. 3-23)
- Trussed core with multiple trusses, suspended system, and tie-downs (Fig. 3-24)
- Trussed tube (Fig. 3-25)
- Tapered trussed tube (Fig. 3-26)
- Tube bundle (Fig. 3-27)

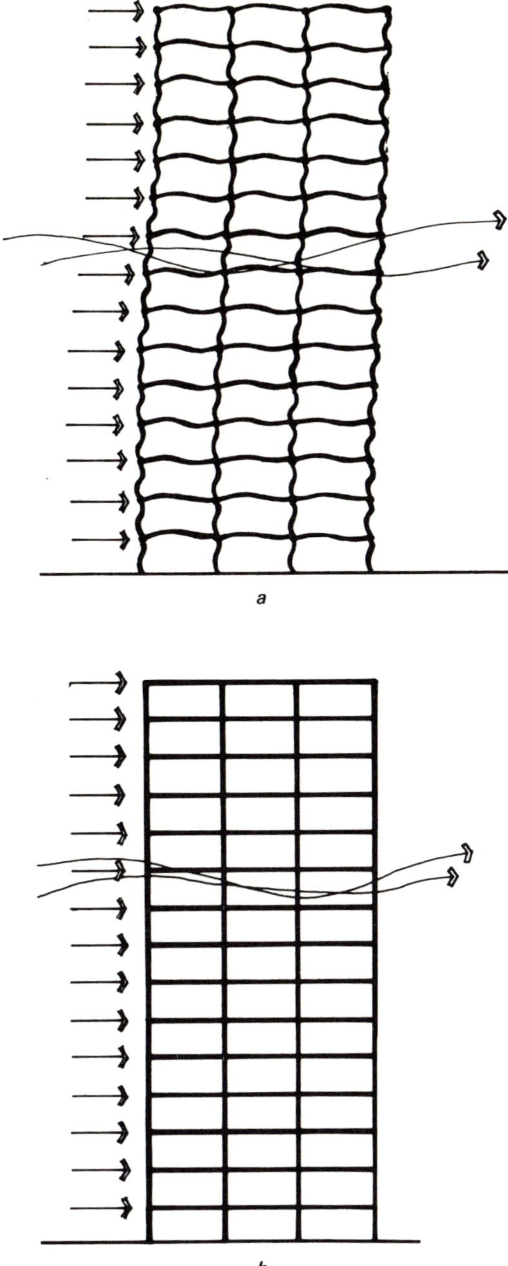

Figure 3-12. *a,* Typical rigid frame with rigid moment-resistant connections. *b,* Typical distortion of rigid frames under wind loads only.

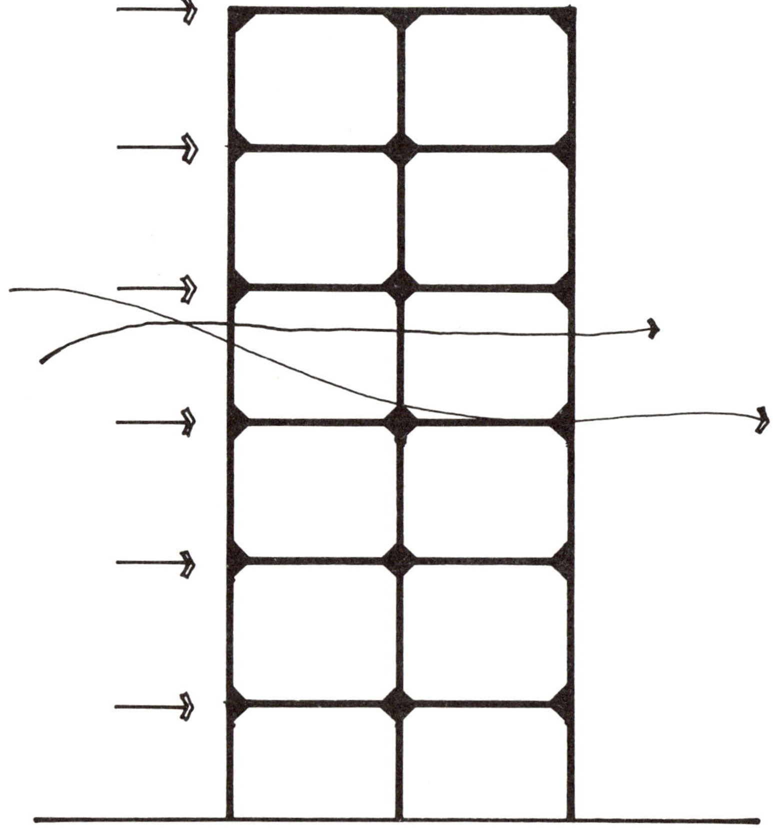

Figure 3-13. Bracket-type bracing involving only small plates produces an efficient moment-resistant connection.

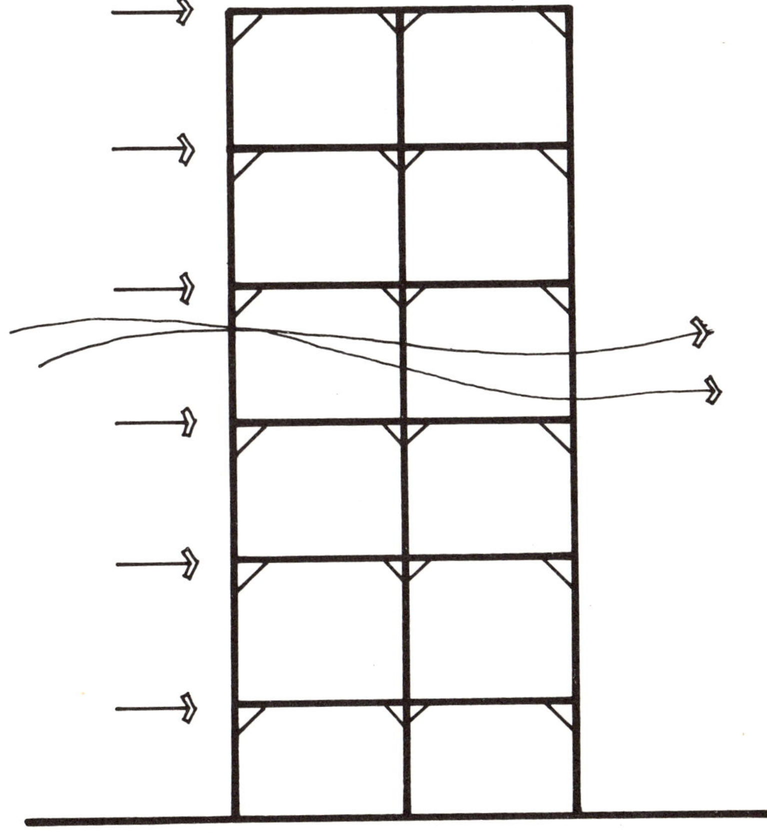

Figure 3-14. Knee bracing.

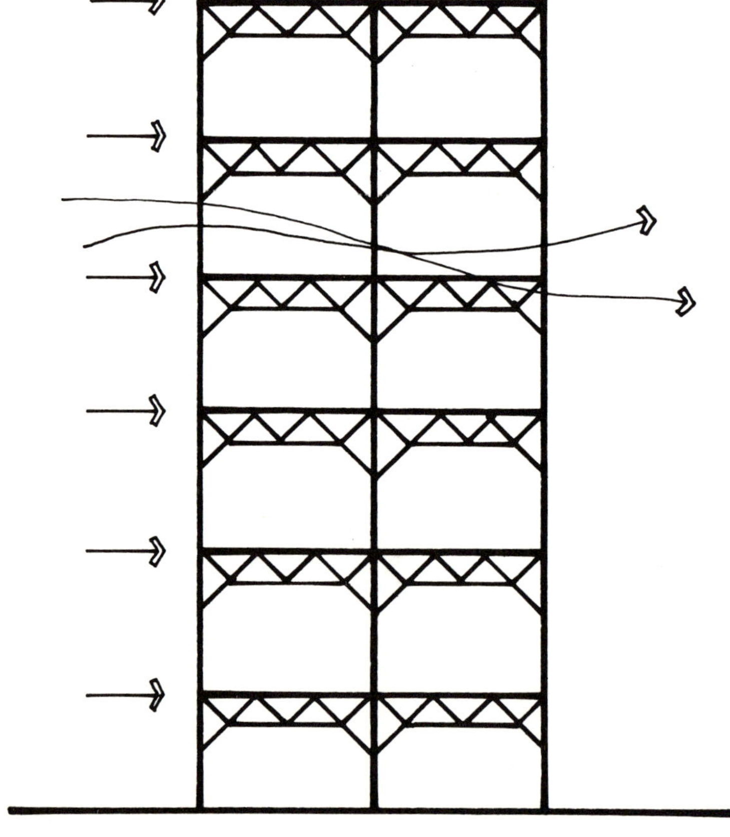

Figure 3-15. Portal bracing.

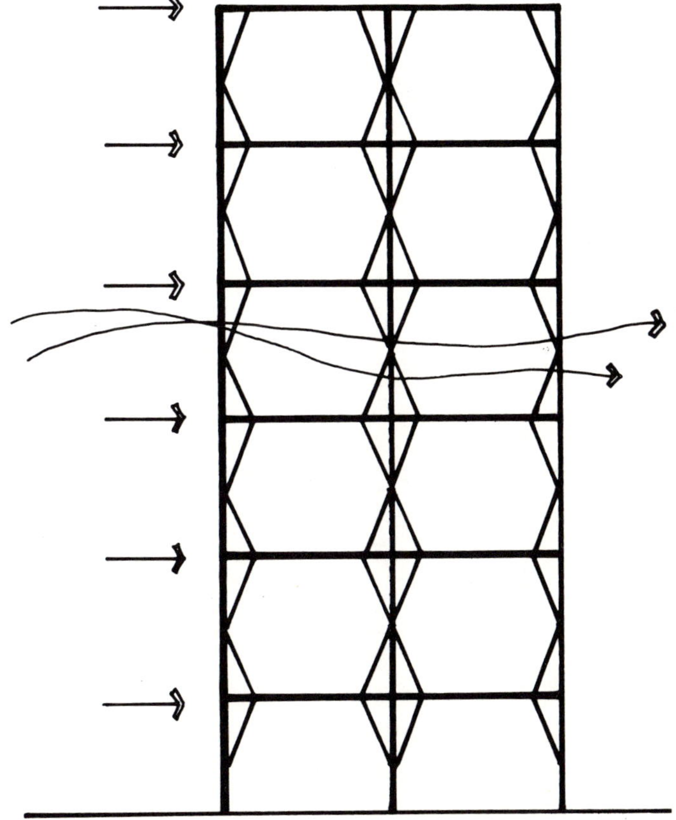

Figure 3-16. Narrow K-bracing.

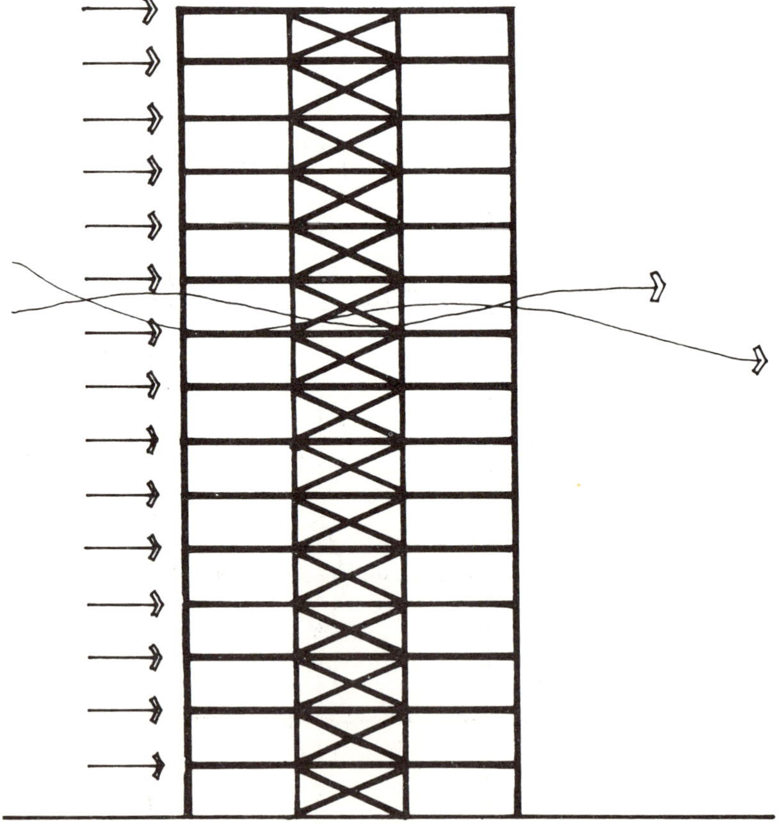

Figure 3-17. Full-length diagonal X-bracing.

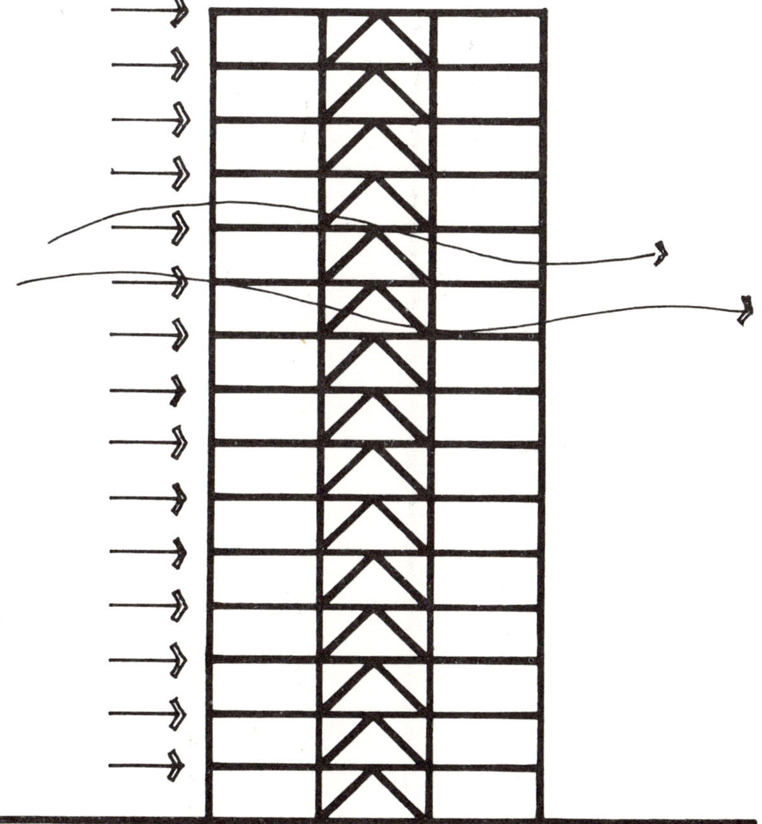

Figure 3-18. Full-length diagonal K-bracing; more efficient than the X-bracing because the shorter length of the braces elongates less, thus reducing the lateral drift of the structure.

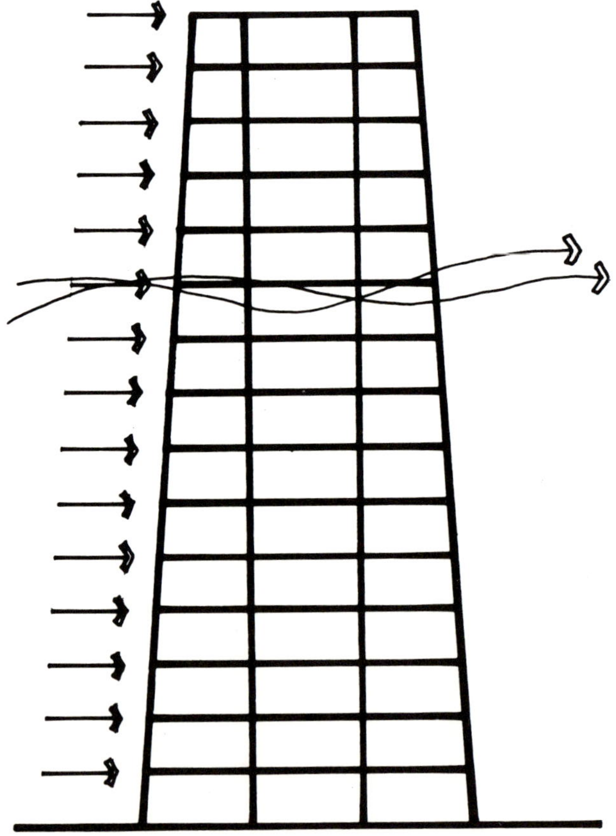

Figure 3-19. Tapered frame system whose efficiency in wind resistance is much higher than the conventional frame with vertical columns.

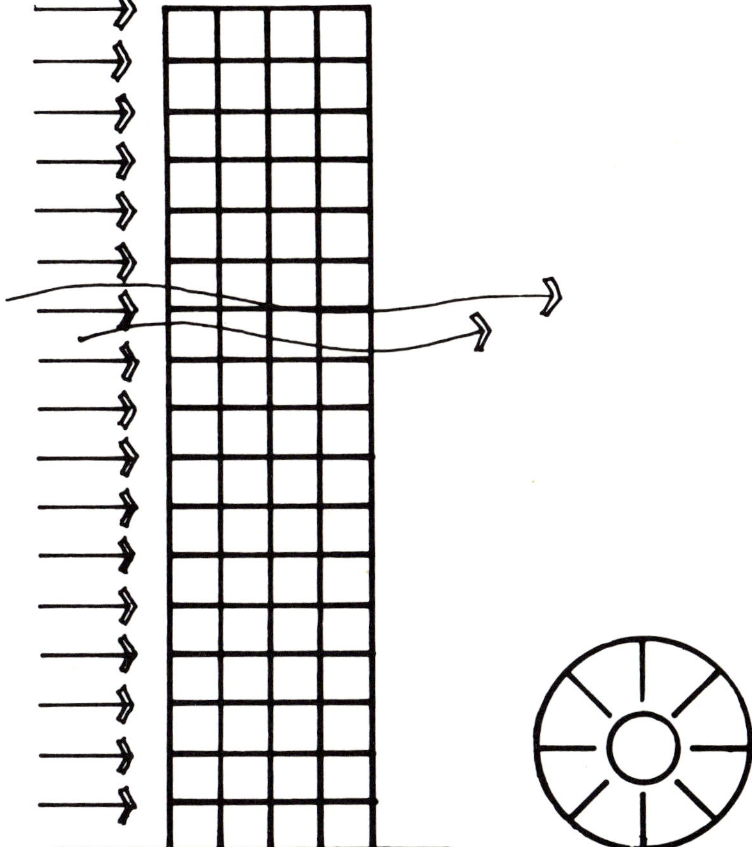

Figure 3-20. The circular tube-frame consisting of a rigid frame of closely spaced columns and beams that acts as a hollow tube cantilevered from the ground in resisting wind forces. Notice that the central core that will share the vertical load is not active in resisting wind forces.

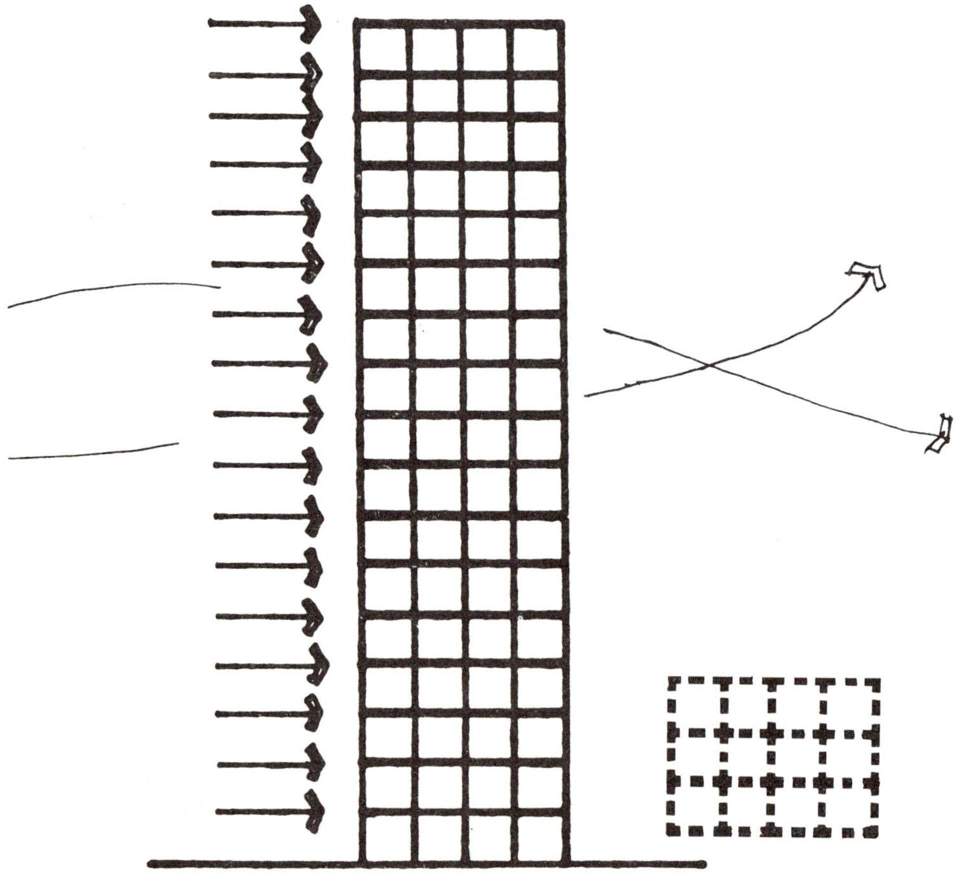

Figure 3-21. The prismatic tube-frame, structurally similar to the circular tube-frame except for the rectangular shape of the cross section (see Figure 3-20).

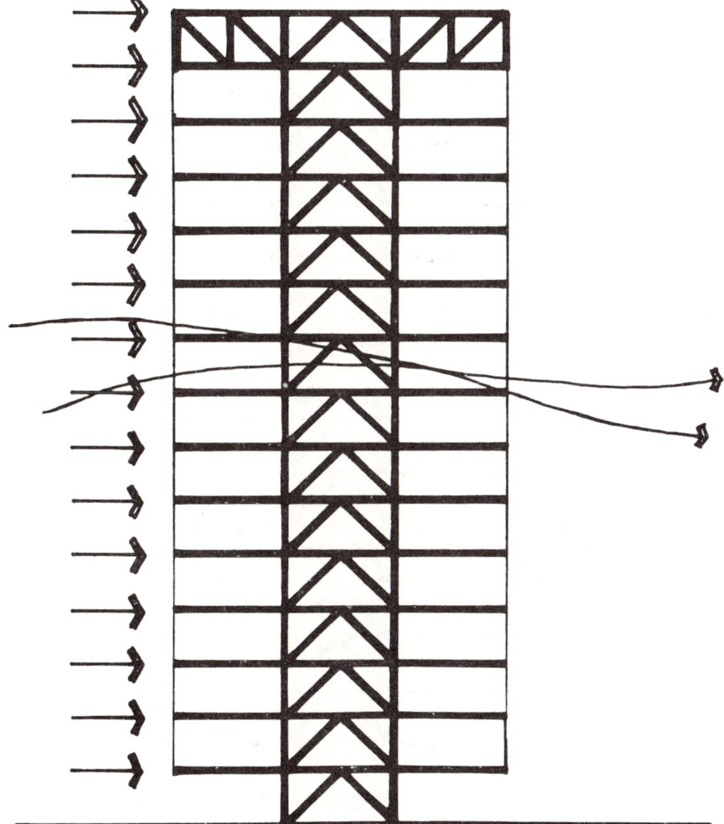

Figure 3-22. The trussed core, with cap trusses and suspended floors without tie-downs, depends on the vertical truss for lateral stability against the wind. Although not very efficient, it does allow the free space under the building.

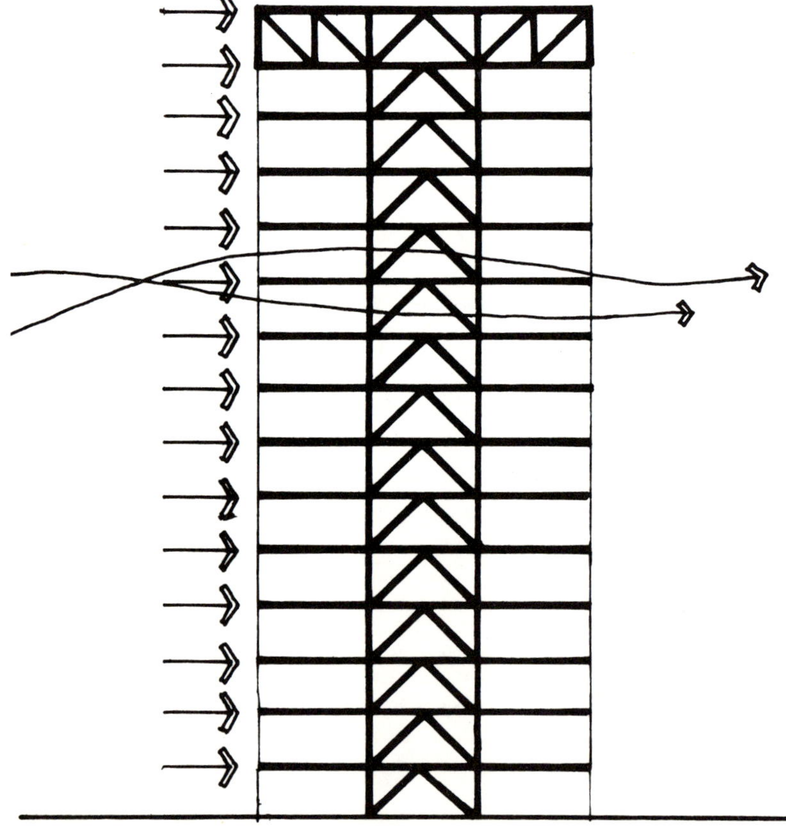

Figure 3-23. Similar to the system of Figure 3-22, the trussed core with cap trusses, suspended floors, and tie-downs has a substantially higher resistance to wind forces, although the tie-downs eliminate the free space at the ground level.

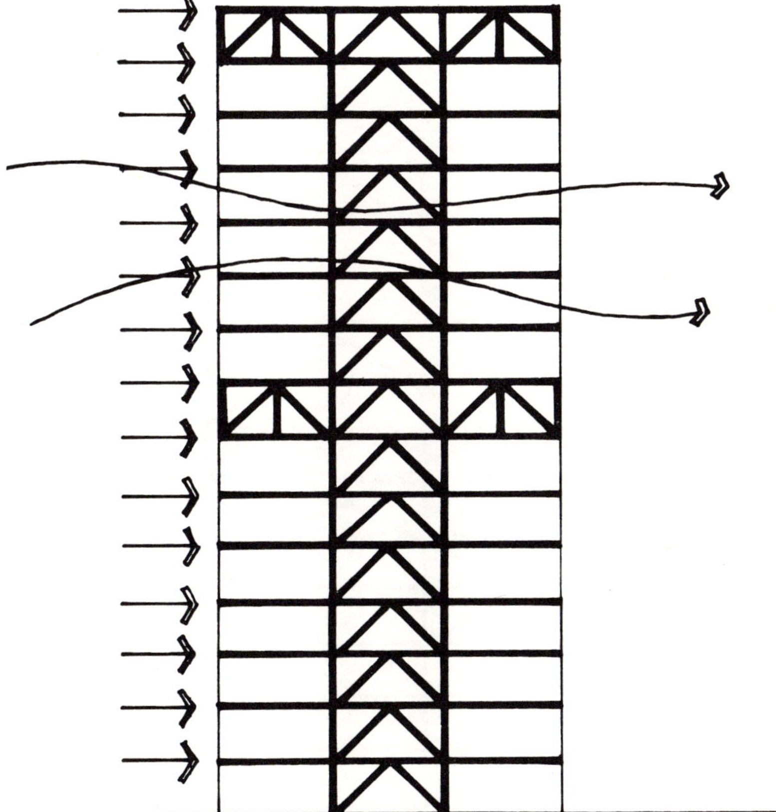

Figure 3-24. The trussed core with multiple trusses is similar to the previous system of Figure 3-23, but the additional trusses reduce considerably the side sways due to wind.

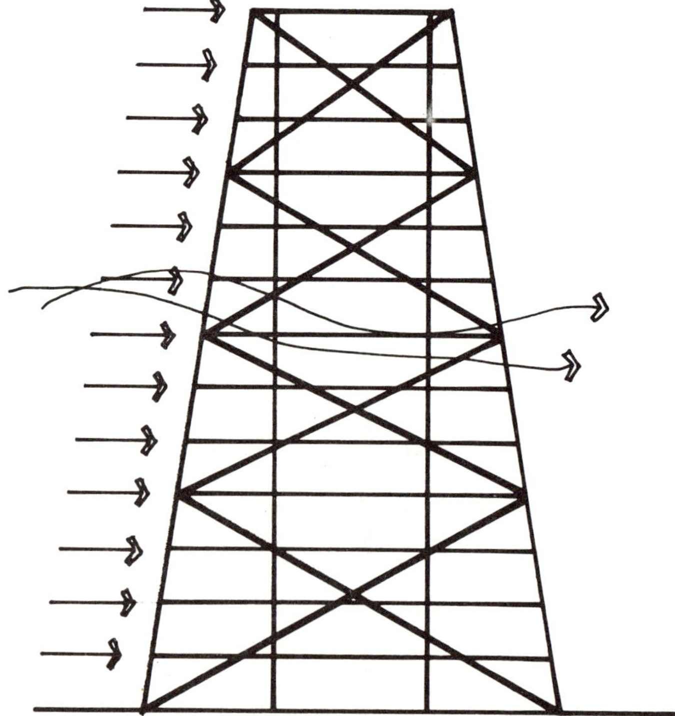

Figure 3-25. Sloped trussed frame, typical example of which is the John Hancock Center in Chicago. Similar to the previous system (Figure 3-24), except for the sloped facade, which increases the efficiency.

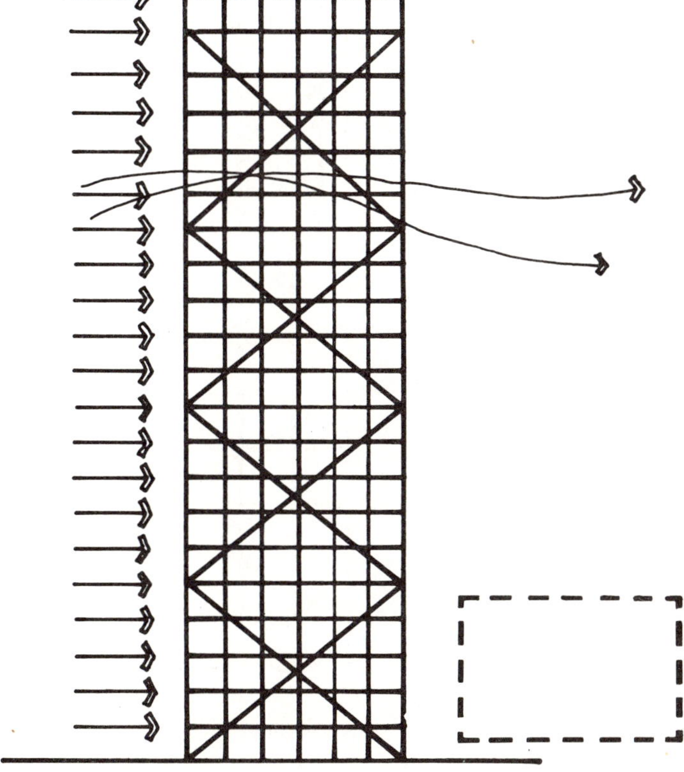

Figure 3-26. The trussed frame is a combination of a rigid frame and exterior diagonal braces.

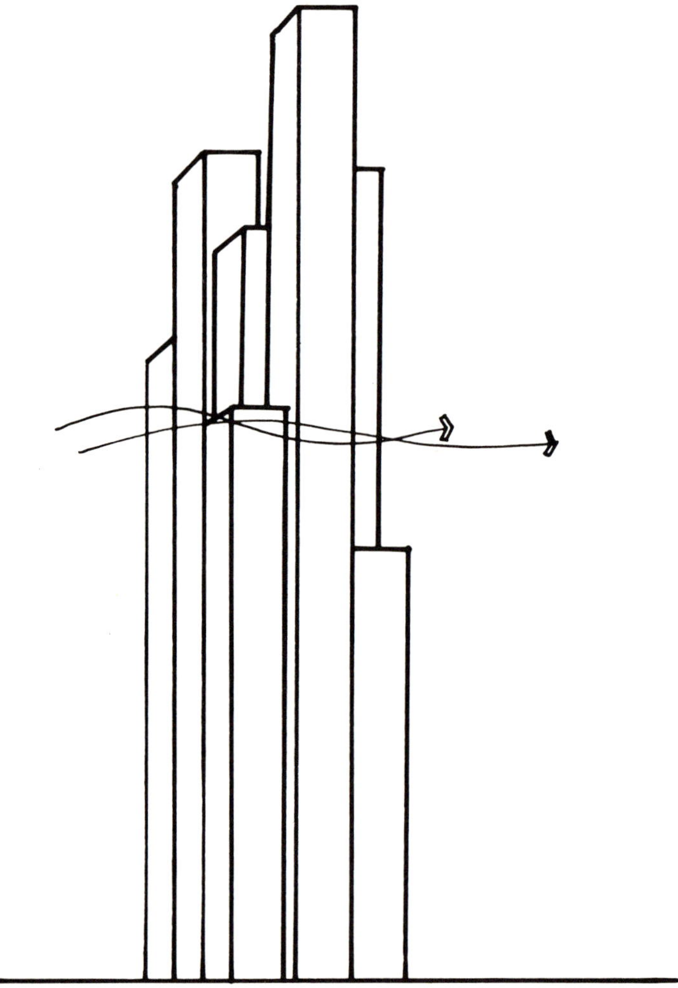

Figure 3-27. The bundle-tube concept used for the Sears Tower in Chicago. A combination of individual structures that form an integral unit in resisting wind forces. Its high structural efficiency can be better understood by comparing the sum of the moments of inertia of the individual tubes with respect to their own neutral axes with the moment of inertia of all the tubes with respect to one common neutral axis.

Steel frames with rigid connections represent the minimal effort in structural wind resistance that can only be applied to small buildings with low wind forces and low slenderness. Rigid connections between steel members prior to the acceptance of welding have been realized with rivets and bolts with a certain degree of difficulty. In general, we can distinguish the following types of column-to-beam connections.

The *angle bracket* connection consists of two angles, one at the bottom and one at the top of the beam, that bolt the lower and upper flanges of the beam to the flange of the column. However, the flexibility of these two angles reduces the moment resistance of the joint, making this connection the least dependable. It is used mostly for light wind loads. (See Figure 3-28.)

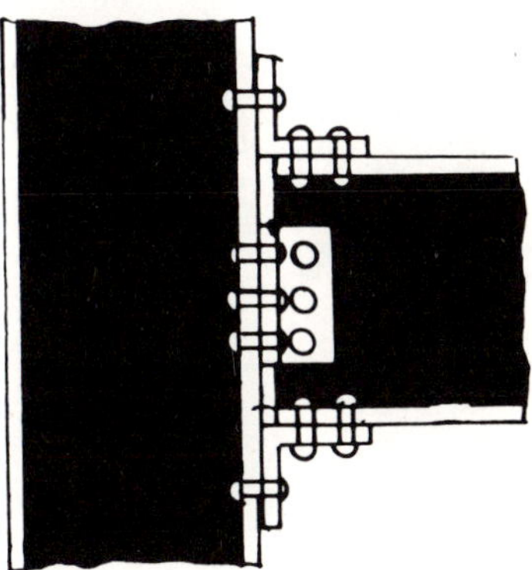

Figure 3-28. Angle bracket connection. A moment-resistant connection of minimal stiffness due to the flexibility of the angles.

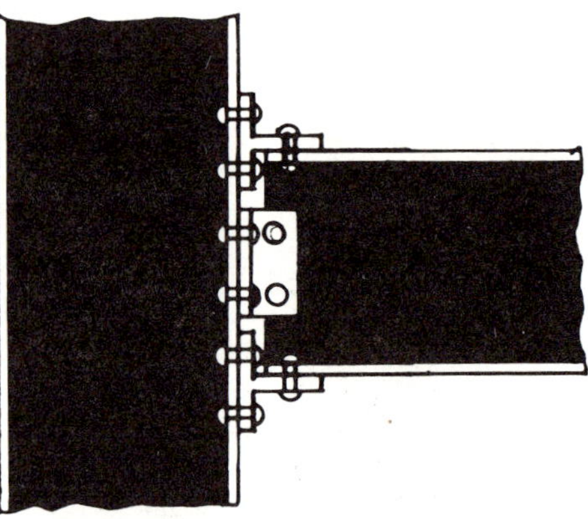

Figure 3-29. Split-beam or structural tee connection. Very popular and efficient moment-resistant connection because of the stiffness of the tees.

The *split-beam* or *structural tee* connection is a very efficient moment-resistant joint of great popularity (see Figure 3-29). While the vertical loads in this connection are carried by two web angles, the moments are transferred from the top and bottom flanges of the beam to the flange of the column by two stiff Ts. It is the rigidity of the Ts in comparison to the flexibility of the angles of the previously mentioned connection that makes the split beam a much better joint for wind loads.

The *heavy bracket* bolted connection is a complex assembly that provides an effective moment transfer between columns and beams and a very stiff joint (see Figure 3-30). However, its high construction cost makes its efficiency questionable from the economic point of view.

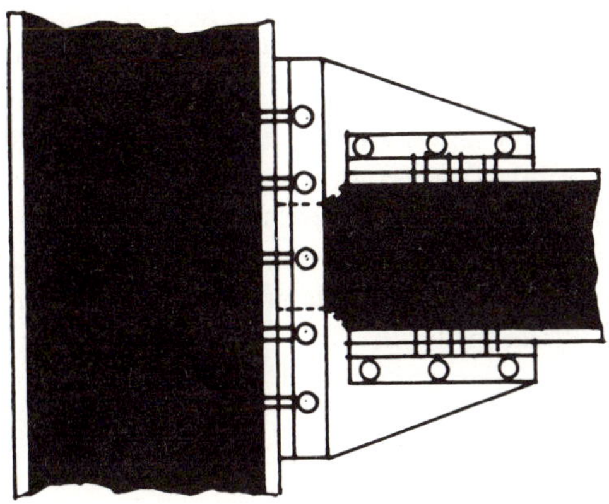

Figure 3-30. Heavy-bracket bolted connection. A complex and expensive moment-resistant connection; not very popular.

The *welded* moment-resistant connection is the most efficient type because of the large moment-carrying capacity it possesses and also because of the low quantity of parts involved. The moment-induced tension in the upper flange of the beam is transferred to the column flange through the top plate that is in turn fillet-welded to the beam and butt-welded to the column (see Figure 3-31a). Note that this plate is usually omitted at the connection made at the other end of the same beam (see Figure 3-31b). In the second connection, in fact, the flange of the beam and that of the column are directly welded to each other without the interposition of the top plate. The welded connection carries the vertical loads by means of a stiffened seat on which the beam rests. The compression induced by the moment is transferred directly from the lower flange of the beam to the flange of the column. Notice that stiffening plates are usually added perpendicularly to the column flange to add rigidity to it under the action of the moment.

The rigid frame previously discussed has a limited practical application up to about 10 stories. For higher buildings, the rigid-frame concept is still valid, but its realization is not by means of a rigid joint at the intersection of columns and beams. Instead, the connection includes a simple joint carrying only shear and short diagonal braces that carry the moment by triangulation. The resultant is an economical yet efficient moment-resistant connection. The short braces are usually small enough not to obstruct the space around the joint. Bracket bracing, knee bracing, narrow K-bracing, and portal bracing are among those most popularly used. With these structural systems, advantage is taken of the small cost of the simple standard connections and of the great benefit of diagonal bracings, which provide lateral wind resistance as well as continuity at the end of the girders.

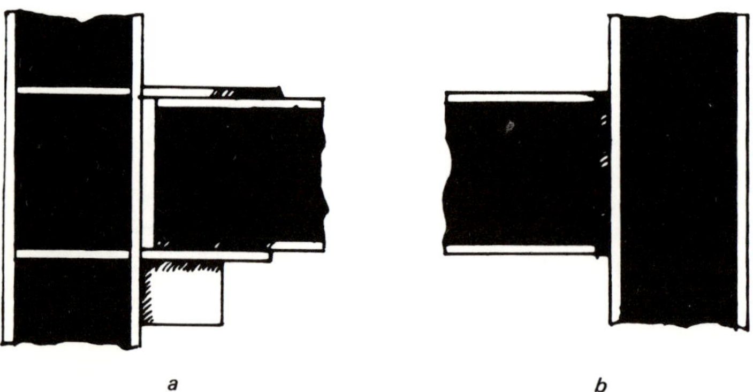

Figure 3-31. Typical moment-resistant welded connection. *a,* Connection at one end of the beam using intermediate plates; *b,* Connection at the other end of the beam where web and flanges of the beam are welded directly to the flange of the column.

The bracket bracing that involves small plates and other profiles does not take much room and is easily concealed in the exterior walls or above the suspended ceiling for the interior columns (see Figure 3-13). The knee-bracing system, which can also be concealed by a suspended ceiling, consists of small diagonal members connecting the beams with the columns below them (see Figure 3-14). The narrow K-bracing system consisting of diagonal members connecting the beams with the columns can also be concealed within narrow walls or column casings (see Figure 3-16). The portal bracing system, on the other hand, is employed when trussed girders are used for the floor systems and some sort of a knee brace that is part of the truss, engages the column and provides a rigid connection between column and truss (see Figure 3-15).

The full-length diagonal X-bracing (Figure 3-17) and the full-length diagonal K-bracing (Figure 3-18) provide two structural systems of great efficiency that are totally different from the braced rigid frames previously discussed. However, because both use bracing and simple connections between columns and beams, they are sometimes confused with rigid-frame bracing. The full-length diagonal bracing (whether X or K) consists of vertical trusses anchored at the base, loaded horizontally by wind forces, and acting as vertical cantilevered trusses. These trusses are the braced portals consisting of columns and beams simply connected and of X- or K-shaped diagonal members. These large trusses are very efficient in resisting lateral forces and their horizontal drift is greatly reduced in comparison to the previous systems. The vertical truss action eliminates any moment in the columns and beams due to wind forces, but it induces only axial tension and compression, which are more efficiently resisted than bending forces. A small drawback, easily offset by all the other advantages, is that the beams simply connected at each end will have larger positive moments due to vertical loads.

The full-length diagonal X-bracing systems are very popular. They are usually designed with the assumption that only one leg of the X is active—that is, acting in tension—while the other is neglected in the computation. This has the double advantage of reducing the truss from hyperstatic to isostatic, which reduces the calculations, and allows the use of slender, tensile members rather than heavy compressive ones. Obviously, the neglected members will work in tension when the wind direction is reversed.

The full-length diagonal K-bracing equivalent to the previous one differs only because the length of the legs of the K are shorter than those of the X. Since the elongation of a tensile member is proportional to the length of the member, the elongation of the legs of the K is shorter than that in the legs of the X. The conclusion is a shorter, horizontal drift for the K system under equal wind forces.

The tapered-frame system consists of a rigid frame whose exterior columns are inclined toward the interior of the building (see Figure 3-19). The resulting structure, whose difference from conventional rigid frames is just the geometric configuration, produces a much more rigid behavior that limits the lateral drift. Current literature shows a 50% reduction in the lateral sway when the exterior columns slope 8% from the vertical, in comparison to a structure with vertical exterior columns.

The circular tube frame consists basically of a rigid frame of closely spaced columns and beams rigidly connected on the exterior cylindrical surface (see Figure 3-20). Although internal columns may share the vertical loads, the exterior circular frame acts like a hollow tube fixed at the foundation and cantilevered vertically when loaded with wind forces. The continuity of columns and beams gives the image and the structural strength of a tube pierced by the spaces left empty between columns and beams.

The prismatic tube frame is completely similar to the circular tube in its structural behavior against wind forces (see Figure 3-21). The only variation is the rectangular configuration of the building cross section that could accentuate the intensity of the wind forces in comparison to the wind forces on the circular tube.

The trussed core with cap trusses and suspended floors departs considerably from the other systems (see Figure 3-22). The central core usually trussed with K braces (more rigid than the X types) carries a horizontal truss on the top from which the rest of the stories are hanging. The central core is therefore acting as a vertical cantilevered truss in resisting the horizontal wind forces, but being centrally located, its efficiency is not very high. Its basic feature is to leave a column-free space at the ground level under the building.

The trussed core, with its cap trusses, suspended floors, and tie-downs, is a substantial improvement over the previous systems (see Figure 3-23). The tie-downs, in fact, acting in tension along the exterior of the building, increase tremendously the lever arm of the resisting structure against wind forces. The vertical central core, therefore, acts in compression, while the tie-downs provide the tension that creates the resisting moment.

The trussed core, with its multiple trusses, suspended floors, and tie-downs, stems from the core just mentioned because, in addition to the cap truss, other horizontal trusses are provided at different heights (see Figure 3-24). The horizontal trusses that stiffen the structure at critical points have the effect of reducing considerably the lateral drift.

The trussed frame is a combination of a rigid frame and diagonal braces placed on the exterior walls (see Figure 3-25). Such braces, usually X-shaped, span across several stories to form an exterior vertical truss in cooperation with the columns and the horizontal girders. The rigid frame, including all columns and beams with rigid connections, share, together with the exterior trusses, the lateral resistance against wind loads.

The trussed frame can be improved in its strength for wind loads if the exterior columns are sloped with their tops toward the interior of the building rather than being vertical (see Figure 3-26). The criterion of the increased stiffness for sloped exterior columns has already been discussed previously. A typical application is, of course, the John Hancock Center in Chicago.

The bundle tube was used by Skidmore Owings & Merrill for the Sears Tower in Chicago (Figure 3-27), which is the tallest building in the world. In reinforced concrete, the rigid frame is a system that in recent years has increased the upper limit from 20 to 60 stories with improved structural schemes. For instance, the combination of a rigid frame and shear walls creates an assembly in which the two systems share the functions: the frame supporting mostly vertical loads, and the shear walls carrying the horizontal wind loads (see Figure 3-32).

A variation on these systems consists in the simultaneous action of the frame and the shear walls in resisting horizontal forces together (see Figure 3-33). The

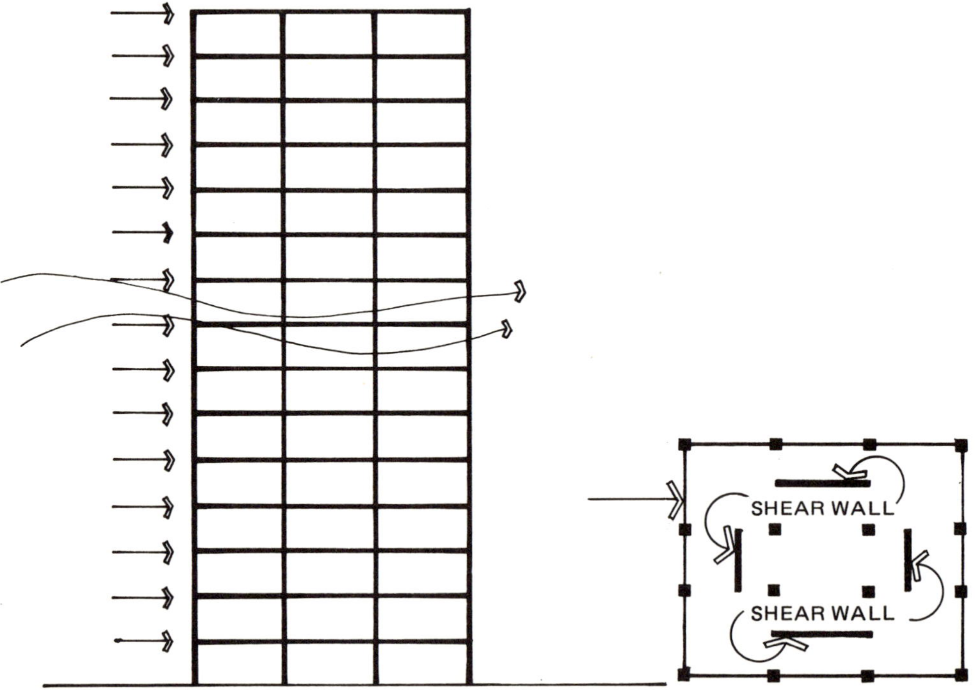

Figure 3-32. Combination of rigid frame and shear walls for concrete structures, in which the two elements act independently in resisting the wind load, most of which is carried by the walls, which are generally more rigid than the frame.

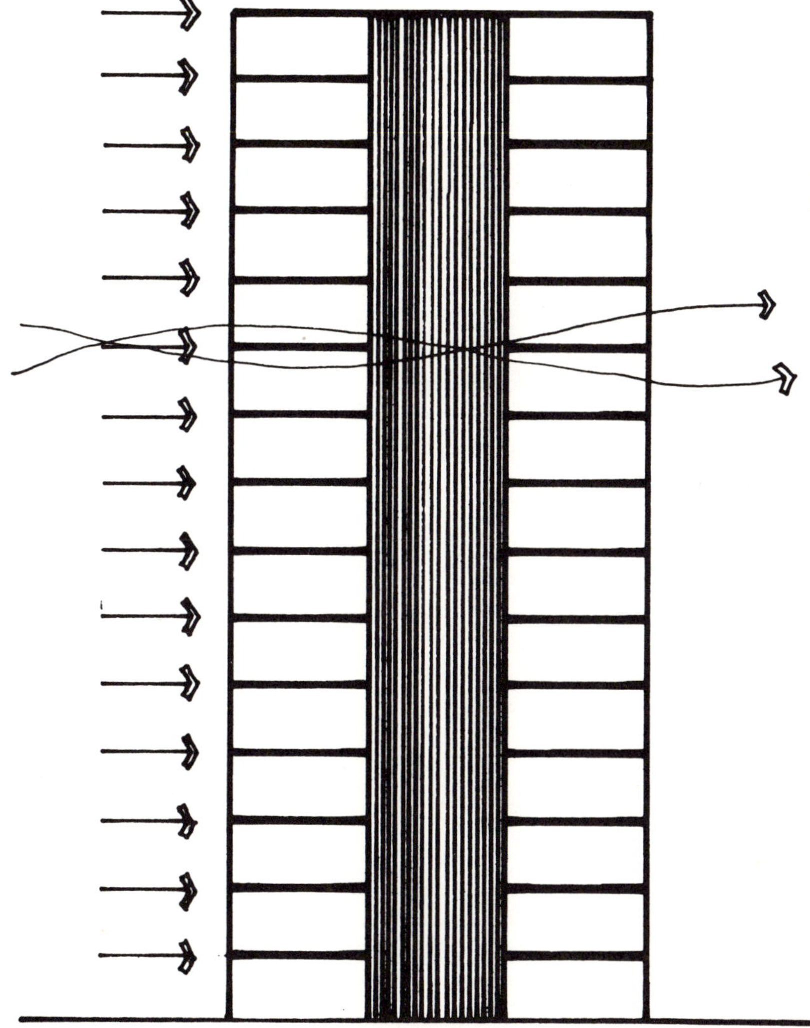

Figure 3-33. Frame and shear walls, acting simultaneously to resist wind forces, are a variation of the previous system.

wall could be continuous or pierced for door and window openings, and its relative stiffness must be evaluated to assess the frame-wall interaction. Current literature reports that the reciprocal action of walls and frame is such that, above a certain height, the contribution of walls in resisting lateral sways may be negative and therefore detrimental with respect to the sway of the frame alone. These observations have suggested separating the wall from the frame above a certain critical height to be determined in each case.

The framed tube, like the circular tube frame in steel or the prismatic tube frame, is a rigid frame structure of columns and beams closely spaced and arranged on the outside walls of a building (see Figure 3-34). The cross section of the building, whether circular or rectangular, may be seen as the cross section of a tubular element subjected to bending. The building acts like a vertical tube cantilevered from the ground. The exterior walls of the building can be considered as continuous surfaces of concrete pierced by the spaces left open between columns and beams. The structure concentrated along the exterior walls carries simultaneously the wind forces and the vertical loads.

A further improvement of the previous structure is the so-called tube in tube (see Figure 3-35). It consists of an exterior structure similar to the one previously mentioned and an interior one rigidly connected to the exterior tube by the stiff slabs at each floor. Both carrying vertical and horizontal forces, the interior and exterior tubes are structurally combined with a consequential increase in efficiency.

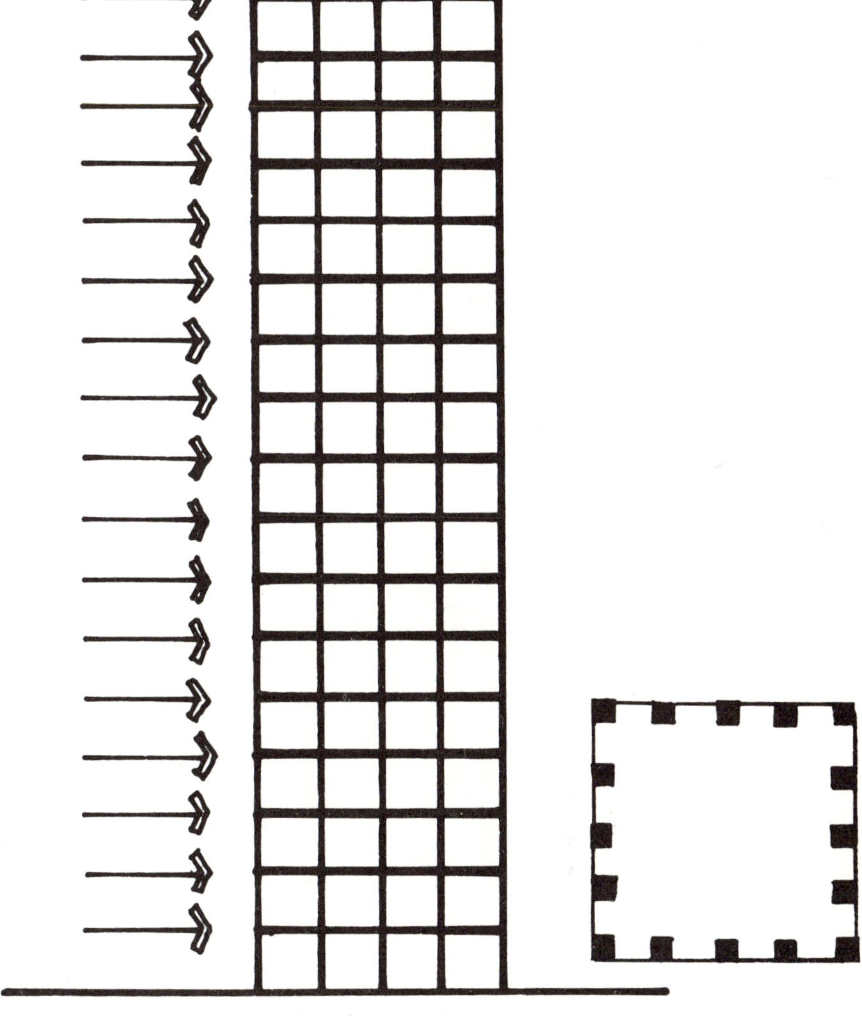

Figure 3-34. A framed tube in concrete behaves like the circular or prismatic tube-frame mentioned for steel structures (Figures 3-20 and 3-21).

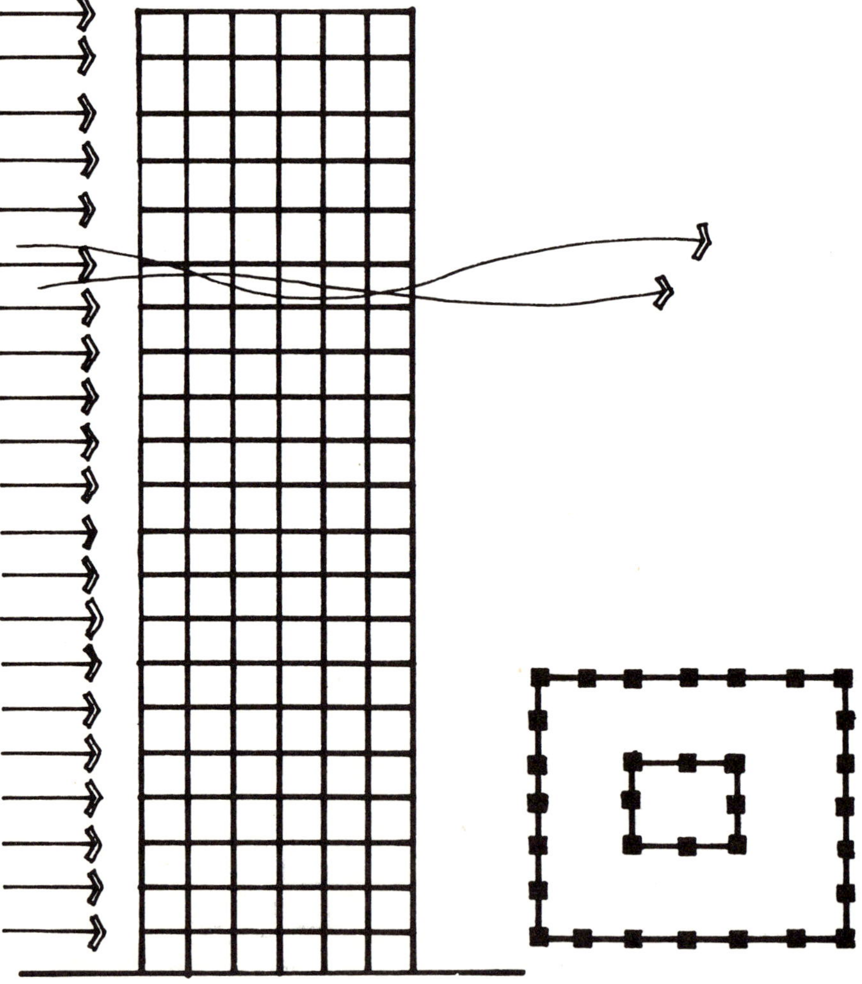

Figure 3-35. The "tube-in-tube," made up of an exterior rigid frame and an interior rigid frame, consists of closely spaced columns and is similar to the system described in Figure 3-34. It is the most efficient system for a reinforced concrete structure and has increased the height to its actual maximum. (Water Tower Place, Chicago, Illinois, 859 ft high.)

ENERGY FLOW AND DISSIPATION

Energy, defined as the potential for performing work, manifests itself as the product of a force times the distance through which it acts. In the interaction of wind and building, a certain amount of energy flows from the wind to the building and is then dissipated to the atmosphere as heat. How much of the wind's kinetic energy is transferred to the building depends upon the capacity of the building to absorb and dissipate it. For a constant wind force, the larger the amplitude of the building's sway, the greater is the energy absorbed and the amount to be dissipated. High-amplitude long-period sways and low-amplitude short-period vibrations, generated by various types of wind-exciting causes, depend on the characteristics of the wind, on one hand, and the dynamic characteristics of the structure, on the other. For instance, static wind pressure will deflect the building and maintain the deformation steady without additional energy flow. If the wind ceases, the building will recover its original posture after a series of free vibrations that use the energy that was stored in the structure during the first deformation. In practical terms, however, the constant change of wind velocity in direction and magnitude generates various building motions since the exciting force varies. This action can be visualized as a continuous buffeting of the winds against the building, which induces deflection in the direction of the wind and pumps energy from the wind to the building. (Note that at times the deformation can be torsional due to the geometrical configuration of the structure.) The building, however, can also deform in a direction perpendicular to that of the wind under the excitation of a series of rhythmical eddies (that is, the Karman vortex).

Under all of these conditions, energy flows from the wind to the building, and is then dissipated. Wider sways will dissipate more energy so that the annoyance to the building's occupants is minimized. This requires absorbing the least amount of energy from the wind as possible and dissipating the amount absorbed through some mechanism other than the vibration of the building. For instance, a mass acting as a shock absorber can take the energy away from the building and dissipate it in the form of heat. Sometimes, the damping mechanism is built into the structure—that is, into the building connections capable of dissipating energy through friction and ultimately through heat. The examples presented in the following pages will illustrate more specifically various types of damping devices that have been applied recently to major structures.

The World Trade Center in New York City (Figure 3-36) employs a unique damping mechanism, never used before, that should be widely accepted once time has proved its durability. The floor trusses framing into the exterior columns are connected to the columns by means of an interposed device on the lower cord. Such a device consists of two layers of neoprene (by the 3M Company) sandwiched between three steel plates. The viscoelastic properties of the neoprene absorb and dissipate the vibrations transferred from the wind to the steel structure. A total of 10,000 of these devices are mounted on each of the two towers, constituting a huge shock absorber, which dampens the vibrations.

In contrast to the World Trade Center, for which energy-dissipation concepts were included in the original design, the John Hancock Tower in Boston had to be retrofitted with energy dissipators. After suffering constant wind damage, the building was reinforced at an additional cost of $15 million, which brought the total cost to $150 million at the dollar value of 1968, when construction was begun. Out of 10,344 windows in the glass facade, over 5000 popped out during windstorms. The temporary replacement of these windows with plywood earned the building the nickname "Plywood Palace."

Final remedy consisted of adding two spring-mounted masses (each consisting of 600,000 lb of lead) and reinforcing the steel frame with 300 L-shaped beams. This retrofitting, of course, points out the critical need for an exhaustive wind analysis during the original design of wind-sensitive structures.

Figure 3-36. The World Trade Center, a giant shock absorber that includes 10,000 viscoelastic dampers. Each damper consists of two neoprene pads within three steel plates and connects the floor trusses to the exterior columns.

At times, the amplitude of building sway exceeds the values that were predicted in the design stage. Exceptional windstorms, naturally of short duration, can reveal the wind sensitivity of certain buildings. For example, the Gulf & Western Building in New York City (Figure 3-37) had to be evacuated during a storm of particular intensity, which occurred in January 1973. The building sway, according to the press, terrified the occupants of the building. No structural damage was reported because this slender and efficient structure has a coefficient of safety of substantial magnitude. In general, the popping out of windowpanes and damage to partitions and other nonstructural elements are often unavoidable in certain windy locations unless considerable expense is incurred to strengthen the weak parts and dampen the vibrations.

For the Citicorp Center, a 914-ft steel tower in New York, the damping mechanism for the wind-induced vibrations consists of a tuned mass damper mounted on the top of the building. Built by MTS Systems Corporation, the mechanism includes an 800,000-lb mass of concrete whose motion counteracts the motion of the building to a certain extent, while the residual energy is dissipated through a dashpot mechanism. The mass is allowed to slide horizontally over a steel surface due to the interposition of a film of oil under pressure at 12 points of contact (oil bearings). At these points, the oil lifts the mass an infinitesimal amount and reduces the friction. A control device is activated by the vibration of the building and starts the oil pump, which makes the mass free to slide. If the oil pressure is reduced, the mass can be made to stop.

The period of motion of the mass must be tuned up to the period of vibration of the building. This is done by adjusting the two horizontal pneumatic springs that limit the horizontal traveling distance of the mass. Each of these two springs includes a piston that is controlled by pressurized nitrogen; thus, the tuning is done by bleeding the nitrogen out of the cylinders. This, of course, is done once initially and then periodically if required. The dashpot, however, has to be tuned constantly due to the different period of vibrations set up by winds having different characteristics.

Figure 3-37. The Gulf & Western Building, near New York City's Columbus Circle, is an example of a modern slender, steel structure whose lack of structural damping resulted in its oscillating excessively at high wind speeds. In January 1973, the building had to be evacuated temporarily during high winds, as reported in the *Wall Street Journal,* yet its structural integrity was not impaired. (Courtesy of Gulf & Western Industries, Inc.)

BUILDING INTERFERENCE

We have previously seen that in the leeward part of a building in a windstream, a strong suction zone is created, with a formation of eddies in the wake, and at times a rhythmical generation of vortices (the Karman vortices). Buildings within this critical zone can be affected, with consequences of unexpected magnitude as indicated by many case studies. For instance, an impressive case involves the huge John Hancock Building in Chicago. In its wake, in fact, at a distance of approximately 120 ft, is another tall building, the Water Tower Place, with a record height for reinforced concrete buildings of 859 ft (Figure 3-38). Under certain types of wind conditions, the aerodynamic interference between the two buildings produces a negative wind pressure (suction) of approximately 50 psf.

Figure 3-38. Under certain conditions, aerodynamic interference can be a severe problem when one structure is in the wake of the other. This is true for Chicago's 859-ft-tall Water Tower Place, the world's tallest reinforced-concrete building, built 120 ft from the John Hancock Building. The suction produced by the latter on the concrete structure of the former at times causes a pressure load of 50 psf, a considerable value even for such a windy city as Chicago.

FLUTTER IN SUSPENDED ROOFS

Lightweight, suspended-roof structures are derivatives of the century-old suspension bridge. They are critically affected by flutter caused by the wind blowing over the roof. Flutter of light-roof decks, more precisely referred to as *aerodynamic instability,* is a dynamic load that affects not just the deck but the whole cable structure down to the supports. Wind-induced vibrations can be critical to the integrity of the structure if their frequency approaches the natural frequencies that are inherent characteristics of the particular structure. In fact, when the two frequencies coincide, resonance occurs. At resonance, the amplitude of the vibrations increases hyperbolically until failure takes place. We are all familiar with such a phenomenon. It is the same one that causes the shattering of glasses when the natural frequency of the glass is matched by that of sound. Resonance was also the factor that made the Tacoma Narrows Bridge collapse when a 40-mph wind of certain characteristics made the bridge vibrate for hours with increasing humps and sags.

Naturally, the problem would be completely eliminated if the frequency of the induced vibrations could be calculated and the structure designed to have natural frequencies quite different from that of the induced vibrations. However, natural frequencies are difficult to calculate exactly, and the frequency of the forced vibrations is usually unknown because of the large spectrum of potential sources.

How can we eliminate or reduce flutter? One method could be to increase the rigidity of the roof by using heavy concrete decks; however, it has been proved that the approach is unsatisfactory. Also, the use of stabilizing cables to tie the main suspension cables to the ground has proved inefficient. The concept of increasing the deck weight, especially, is self-defeating because the major advantage of any suspension system is the minimization of mass and weight. What is left then?

The suggested method consists of combining the main cable system with a secondary one so that the two exchange the energy of vibration from the wind in such a manner that resonance is avoided. How is this done? The wind sets the secondary cables in motion because they are directly connected to the deck; the main cables carry the load. If the two cables have different frequencies, resonance is avoided. (See Figure 3-39.)

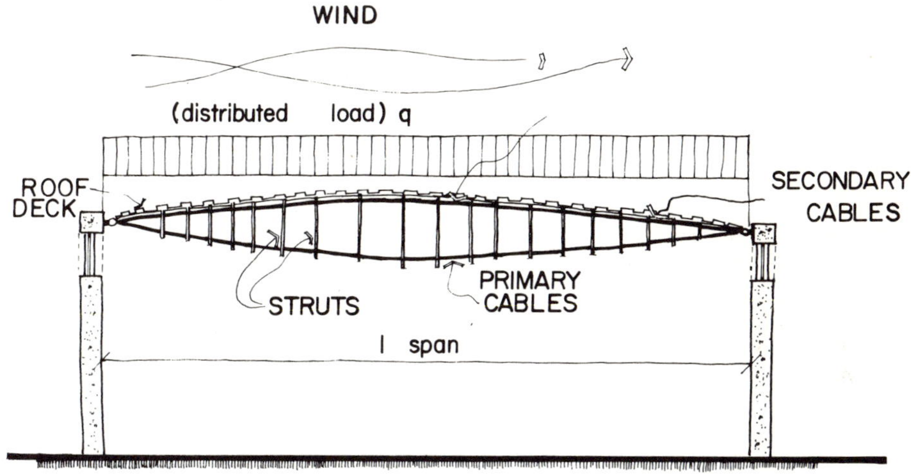

Figure 3-39. Suggested methods for damping flutter in light suspended roofs.

In fact, it is suggested that a secondary system of cables be added to the main cable system that carries the load. The secondary system connected to the primary one is designed so that its natural frequency of vibration is always different from that of the primary system. Therefore, the two systems vibrate with different geometrical shapes that will never coincide. For instance, when at a given time and point one cable tends to go up, the other tends to go down. A dampening effect is the result. Even if one cable should be excited to vibrate in resonance with the forced vibrations induced by the wind, the other cable will still act as a dampener and avoid resonance. The main advantage of the system is that the structure can still be light, as originally conceived, and not lose the inherent advantages of economy of suspended structures.

Example: Let us consider a practical example of a self-dampening system. The main concept is to have a primary and secondary system of cables combined together, with each system having a different period of vibration.

Consider the two cable systems of Figure 3-39. The top cables are the secondary system, and the bottom cables are the primary system. The two systems are separated by vertical struts in compression. The roof connected to the top cables transfers the load to the bottom cables.

Prestressing tension forces, T_b and T_u, are applied, respectively, to the bottom and upper cables by spreading the two cables apart by means of struts. When a vertical downward load is applied, the upper tension is reduced to $(T_u - \Delta T_u)$, and the bottom tension is increased to $(T_b + \Delta T_b)$. Therefore, when the tension increases in one system, it automatically decreases in the other. The natural frequency of the primary bottom cables is:

$$f_b = n\frac{\pi}{l}\sqrt{\frac{T_b + \Delta T_b}{q/g}}$$

The natural frequency of the secondary upper cables is:

$$f_u = n\frac{\pi}{l}\sqrt{\frac{T_u - \Delta T_u}{q/g}}$$

where

f_b = natural frequency of bottom cables
f_u = natural frequency of upper cables
n = any integer indicating the different modes of vibrations—for $n = 1$, the frequency is the lowest and is called the *fundamental*
q = distributed loads on the cables
l = span of the cables
g = acceleration of gravity
T_b = prestressing tensile force in the bottom cables
T_u = prestressing tensile force in the upper cables
ΔT_b = increase or decrease in tensile force due to q in the bottom cables
ΔT_u = increase or decrease in tensile force due to q in the upper cables

Obviously, the frequencies in the top and bottom cables will always differ, as indicated by the formulas.

WIND LOADS ON BUILDINGS

Tall slender buildings, towers, and stacks with low natural damping are particularly sensitive to wind loads, especially wind gusts, and to the forces produced by vortex shedding, and the conditions of dynamic instability generated by galloping and flutter. The design of such structures cannot be based only on a pressure calculated as the product of the dynamic pressure q and the shape coefficient C_p. The effects of gustiness and the response of the building due to its own construction characteristics have to be included. Sensitive structures therefore must be designed according to special wind pressures p_z that are calculated as follows:

Example (Figure 3-40):
Given:

type of frame: steel
location: urban site, type A, in Charlotte, North Carolina
width: 50 ft
$c=$ depth: 50 ft
$h=$ height: 500 ft above grade
$N=$ number of stories: 42

Find: wind pressure on the building.
Solution:

For the given location, assume 80 mph as the design wind speed from the chart in Figure 2-9b. The expected dynamic head of wind at 30 ft above ground is:

$$q_{30} = 0.00256(80)^2 = 16.38 \text{ psf} \tag{3-1}$$

The wind pressure p_z (in psf) at any elevation z (in ft) above ground is

$$p_z = K_z G q_{30} \tag{3-2}$$

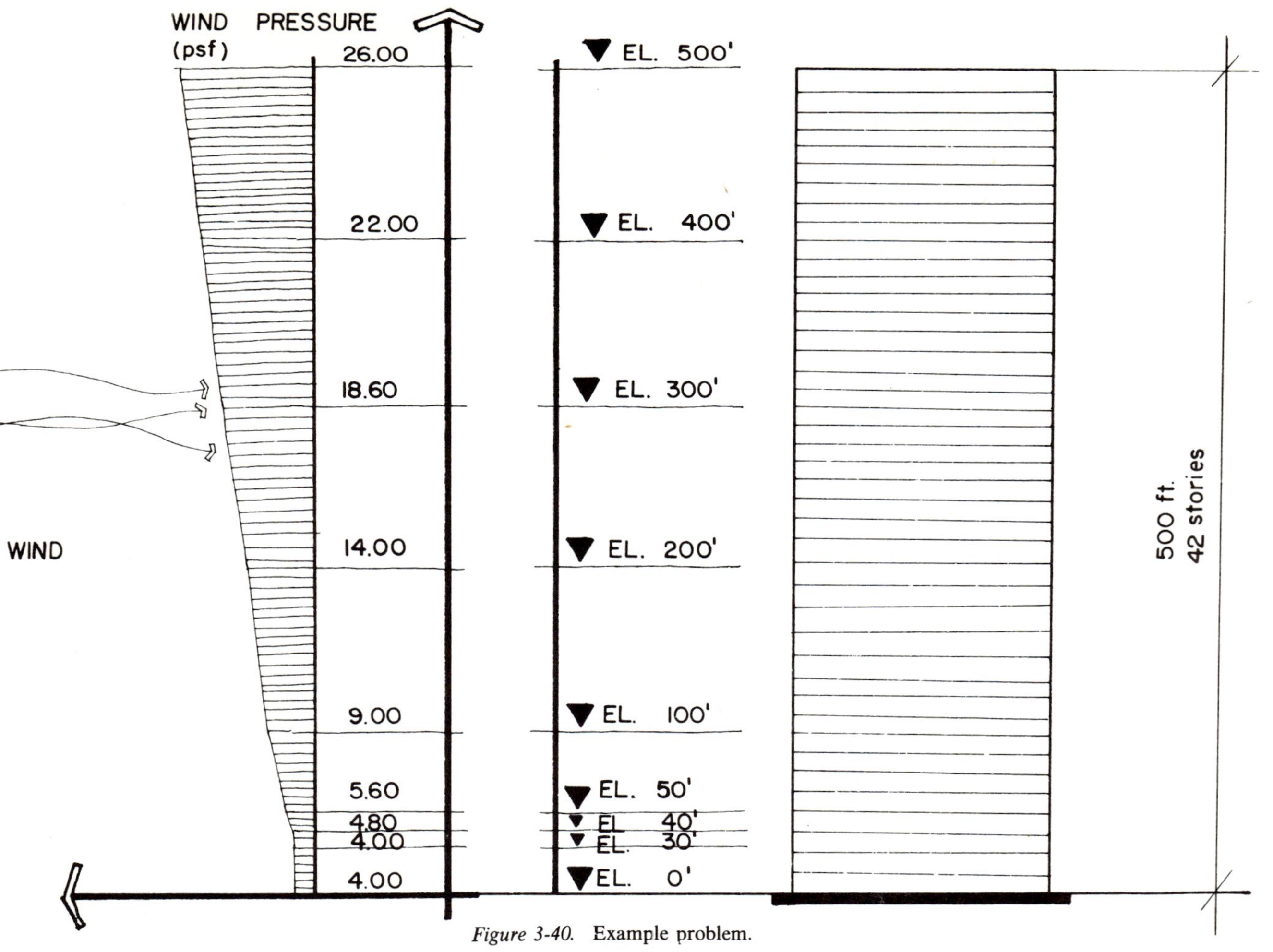

Figure 3-40. Example problem.

Table 3-3. Effect of terrain roughness on wind velocity.

TERRAIN	z_G (FT)	$1/\alpha$
Urban (A)	1500	1/3
Suburban (B)	1200	1/4.5
Open country (C)	900	1/7

where K_z is the velocity pressure coefficient and G is the gust response factor.

K_z is computed for each elevation z above ground from:

$$K_z = \frac{(V_z)^2}{V_{30}} \tag{3-3}$$

where V_z and V_{30} are, respectively, the wind velocities at z (ft) and 30 ft above ground. V_z is given by the following:

$$V_z = V_{30}\left(\frac{900}{30}\right)^{1/7}\left(\frac{z}{z_G}\right)^{1/\alpha} \tag{3-4}$$

Substituting Equation 3–4 into 3–3, we find:

$$K_z = \left(\frac{900}{30}\right)^{2/7}\left(\frac{z}{z_G}\right)^{2/\alpha} \tag{3-5}$$

where z_G, the gradient height, is that elevation in ft above ground at which the ground friction is no longer affecting the wind velocity. At the airport where the wind speed is usually measured, $z_G = 900$ ft. z_G can be found according to the type of exposure (urban, suburban, open country) in Table 3-3. The coefficient $1/a$ is also a function of the exposure and can be found from the same table. Thus, in this case:

$$z_G = 1500 \text{ ft}$$

and

$$1/\alpha = 1/3 \tag{3-6}$$

Substituting the values for z_G and $1/a$ into Equation 3–5, this becomes:

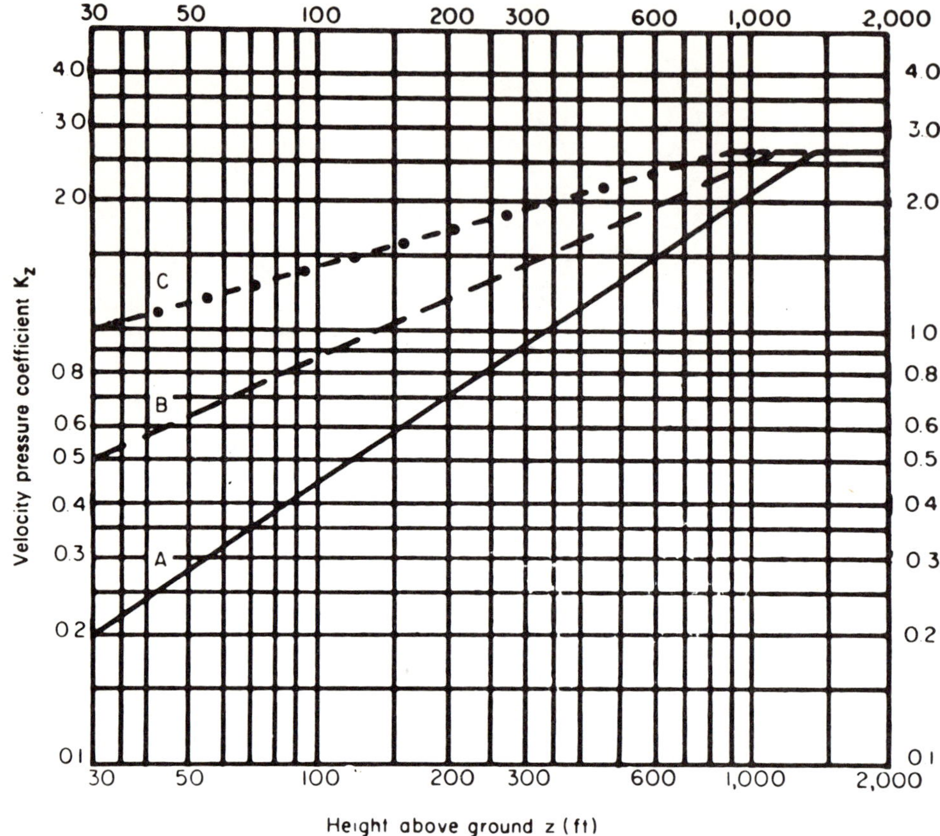

Figure 3-41. Velocity pressure coefficient K_z as a function of height z.

$$K_z = \left(\frac{900}{30}\right)^{2/7}\left(\frac{z}{1500}\right)^{2/3}$$

K_z for the different elevation z can be calculated from Equation 3–3 or found directly from the chart in Figure 3-41. Thus, the values of K_z for various height z are:

$z = 30$ ft, $K_{30} = 0.2$
$z = 40$ ft, $K_{40} = 0.24$
$z = 50$ ft, $K_{50} = 0.28$
$z = 100$ ft, $K_{100} = 0.45$
$z = 200$ ft, $K_{200} = 0.70$
$z = 300$ ft, $K_{300} = 0.93$
$z = 400$ ft, $K_{400} = 1.10$
$z = 500$ ft, $K_{500} = 1.30$

The gust response factor G is given by:

$$G = 0.65 + 3.31\, T_{z'}\, R \tag{3–7}$$

$T_{z'}$ is the turbulence intensity factor, which is computed only at elevation, $z' = \frac{2}{3}h$, where h is the height of the building. Thus:

$$z' = \frac{2}{3}(500) = 333 \text{ ft} \tag{3–8}$$

and

$$T_{z'} = \frac{2.35\,(C)^{1/2}}{(z'/30)^{1/\alpha}} \tag{3–9}$$

where C is the surface drag coefficient that varies according to the type of terrain (urban, suburban, or open country). In this case, C equals 0.025 (urban). Substituting the value for C and $1/\alpha = 1/3$, $T_{z'}$ is:

$$T_{z'} = T_{333} = \frac{2.35(0.025)^{1/2}}{(333/30)^{1/3}} = 0.16 \tag{3–10}$$

$T_{z'}$ can also be read directly from the chart on Figure 3-42.

R in Equation 3–7 is the response factor. It consists of two terms: the first term in the expression, $(0.785\ PF/\beta)^{1/2}$, includes the dynamic properties of the structure; the second term, $[S/(1 + 0.002c)]^{1/2}$, includes the size properties of the building and the characteristics of the surrounding area. If the building is insensitive to wind (usually when the building height h is less than 5 times the building width c), the calculation is simplified by assuming $R = 1.2$ (Defense Civil Preparedness Agency

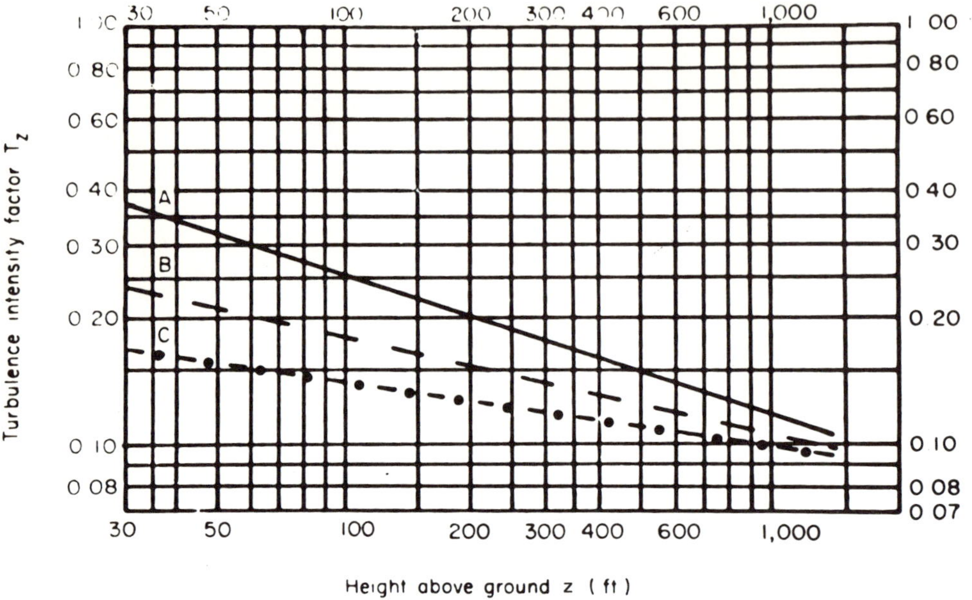

Figure 3-42. Turbulence intensity factor T_z as a function of height z.

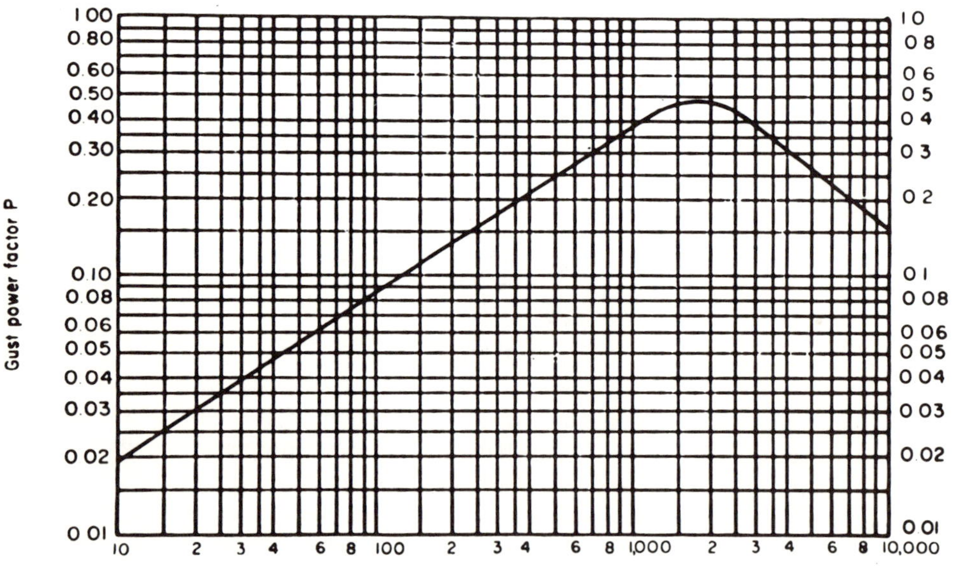

Figure 3-43. Gust power factor P as a function of the expression: $\frac{1.12\sqrt{K_{30}}\,V_{30}}{f}$.

1973). Such value for R is conservative, and its assumption greatly simplifies the calculations for the gust response factor G_F. In fact, building dimension, natural frequency, and the damping ratio of the structure—in other words, the dynamic characteristics of the building, are no longer required. For wind-sensitive buildings, R must be calculated from

$$R = \left[0.785\left(\frac{PF}{\beta}\right) + \frac{S}{1 + 0.002c}\right]^{1/2} \tag{3–11}$$

where β is the damping coefficient that depends on the mechanical and aerodynamic characteristics of the structure. If such characteristics are not known, it is suggested that the following values be used:

$\beta = 0.01$ (for steel structures)
$\beta = 0.02$ (for concrete structures)

Therefore, in this problem $\beta = 0.01$.

P is the gust power factor that can be determined from the graph in Figure 3-43 as a function of the factor

$$\frac{1.12\sqrt{K_{30}}\ V_{30}}{f} \tag{3–12}$$

Table 3-4. Fundamental period of vibration of buildings.

SOURCE	FORMULAS	SYMBOLS
Structural Engineers Association of California	$T = 0.06\sqrt{\frac{H^2}{b}}$	T: period (sec) H: height of bldg (ft) b: width of bldg (ft) (in the direction of vibrations)
Modified Structural Engineers Association of California	$T = 0.05\frac{H}{\sqrt{b}}$	Same
Joint Committee on Lateral Forces	$T = 0.06\frac{H}{\sqrt{b}}$	Same
Taniguchi (Tokyo Institute of Technology)	$T = KN$	K: coefficient that varies from 0.07 to 0.09 N: number of stories in the building
Franklin and Carder (based on 500 tests conducted by the U.S. Geological Survey)	$T = 0.006H$	Same. Foundation on average compacted soil.
Franklin and Carder	$T = 0.0058H$	Same. Foundation on compact soil.
Franklin and Carder	$T = 0.0059H$	Same. Foundation on soft soil.

where f is the fundamental period of vibration (see Table 3-4) of the structure (cycles per second) in the direction of the wind. In absence of precise data, f is assumed to be $10/N$, where N is the number of stories of the structure. Thus, $f = 10/42 = 0.24$ c/s, and the factor in Equation 3–12 becomes $\frac{1.12\sqrt{0.2}(80)}{0.24} = 170$. From Figure 3-43, P is found to be 0.14.

F is the gust correlation factor that can be determined from Figure 3-44 as a function of h/c and the factor

$$\frac{0.88\,fh}{V_{30}\sqrt{K_h}} \tag{3–13}$$

where h and c are, respectively, the height and the width of the structure (in ft) in the direction of the wind, and K_h is the velocity pressure coefficient at elevation h.

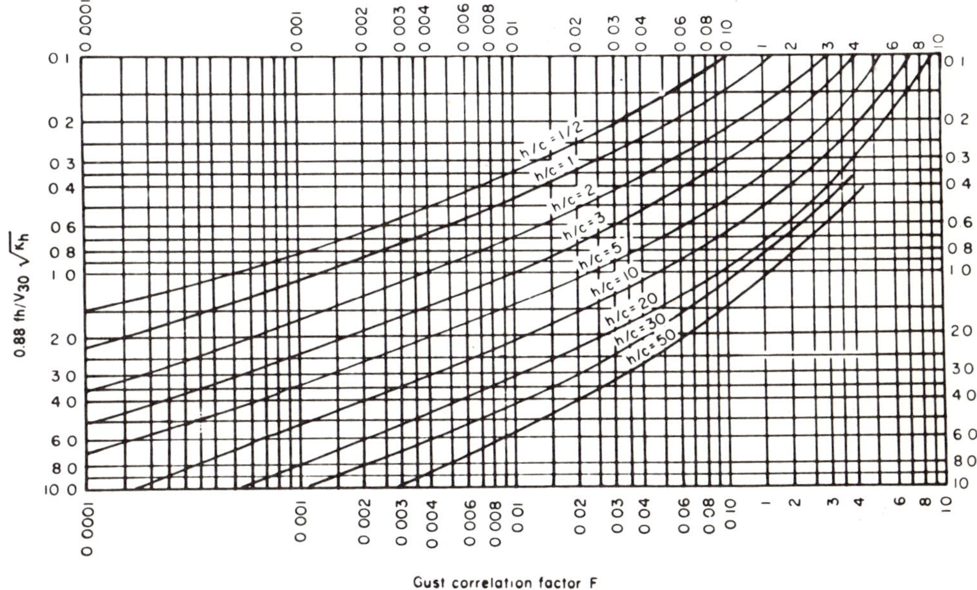

Figure 3-44. Gust correlation factor F as a function of the expression: $\dfrac{0.88fh}{V_{30}\sqrt{K_h}}$.

Thus, for $\frac{h}{c} = \frac{500}{50} = 10$, and $K_h = K_{500} = 1.30$ (obtained from Figure 3-41 for an elevation $h = 500$ ft, in an urban location), the factor (Equation 3–13) is equal to

$$\frac{(0.88)(0.24)(500)}{(80)\sqrt{1.3}} = 1.16$$

Therefore, from Figure 3-44, we find:

$$F = 0.045$$

S is the structure size factor and can be determined directly from the graph in Figure 3-45 as a function of the height of the structure h and the type of terrain. Thus, for $h = 500$ ft and urban area (A), S equals 0.85.

Now, it is possible to compute R; thus from Eq. 3-11:

$$R = \left[\frac{(0.785)(0.14)(0.045)}{0.01} + \frac{0.85}{1+(0.002)(50)}\right]^{1/2} = 1.1$$

Substituting R and $T_{z'}$ in Equation 3–7, we find:

$$G = 0.65 + (3.31)(0.16)(1.1) = 1.23$$

Having calculated G for the building and q_{30}, we compute K_z at various elevations z, and we obtain P_z (see Table 3-5).

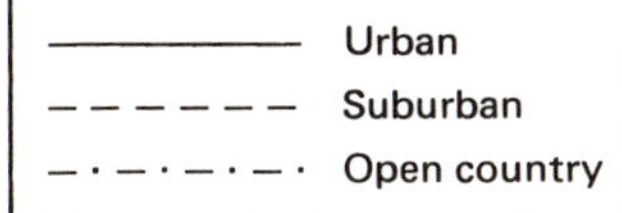

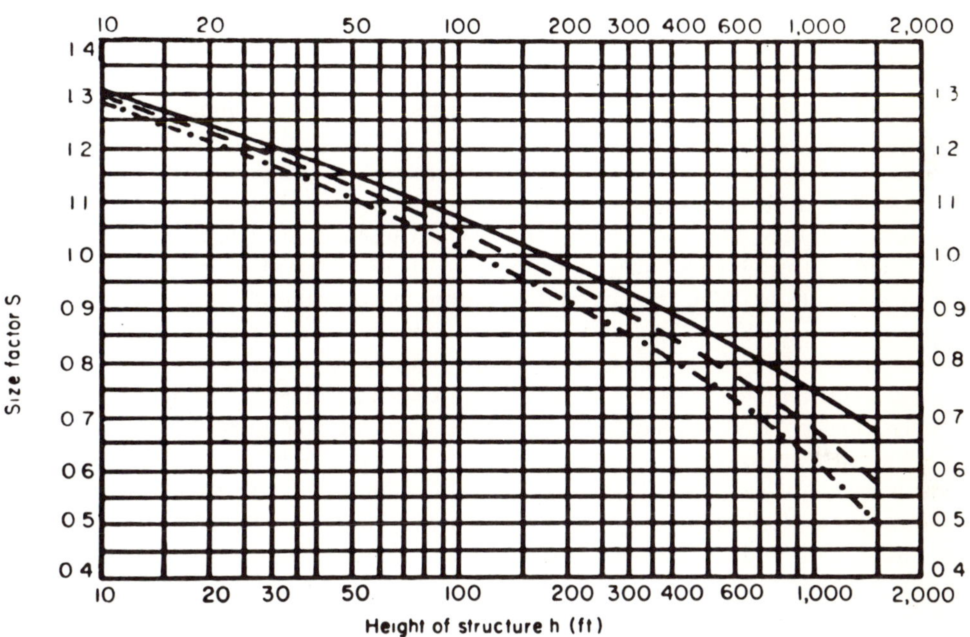

Figure 3-45. Size factor S as a function of the height of the structure h.

Table 3-5

Z	K_z	$G \times q_{30}$	$P_z = K_z G q_{30}$
30	0·2	20	4.0
40	0·24	20	4.8
50	0·28	20	5.60
100	0·45	20	9.00
200	0·70	20	14.00
300	0·93	20	18.60
400	1·10	20	22.00
500	1·30	20	26.00

WIND LOADS ON RIGID FRAMES

The wind load applied to a building is first experienced by the building itself as a pressure force distributed over the exterior walls and roof of the building. Such pressure, then, is transferred to the resisting structure and in turn reaches the ground as it passes through the foundation. If the structure is a rigid frame, the wind loading is transferred from walls to the connections between columns and beams whose entirety constitutes the structure. The wind pressure is thus changed into a series of concentrated horizontal forces applied at the joints of the frame. Forces on the windward walls will push the frame, while forces in the areas of suction will pull it.

As we analyze a planar frame, we take into account only the components of the forces in that plane, and the remaining components are considered when we analyze the perpendicular frames. It is interesting to note that when we talk about rigid frames, we are usually referring to either a single portal or to a multistory or a multibay portal frame. The behavior of each portal, as well as the behavior of the frame consisting of several portals, is practically the same under horizontal wind forces. Figures 3-46 and 3-47 show distorted shape and the moment diagram for a single portal and a multiple portal, respectively, for horizontal wind forces. These illustrations should help the reader to visualize the general behavior of rigid frames to resist lateral forces. The rigid frame, however, loses its efficiency as the building height increases above 20 or 30 stories, and, in recent years, several other structural concepts have been successfully adopted for higher buildings. The addition of shear walls, for one, or the concept of "tube in tube" are recent realizations of new structural concepts in skyscraper engineering to resist wind forces.

A simplification of structural design for lateral forces included the so-called portal method and the cantilever method, which have been satisfactorily applied for many years for buildings up to 20 and 30 stories high. However, with the practical use of computer programs, an exact structural analysis can now be economically conducted to check several design options and optimize a final solution. It is convenient, therefore, for the reader to observe in practical terms how to conduct such an analysis on an average building using an available program. Developed during the 1960s at MIT, STRUDL II is a program with wide applications that can be used all around the world wherever an IBM computer is found. This program, used in the following example, is recommended to the reader.

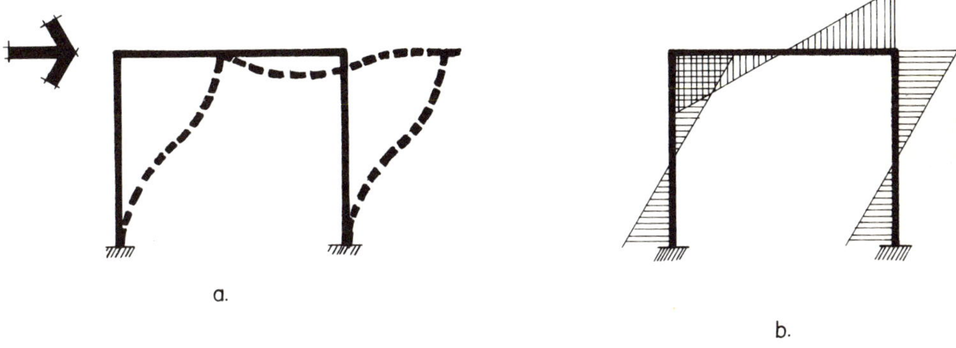

Figure 3-46. *a,* Single portal distortion under wind loads. *b,* Bending moments diagram for single portals under wind loads.

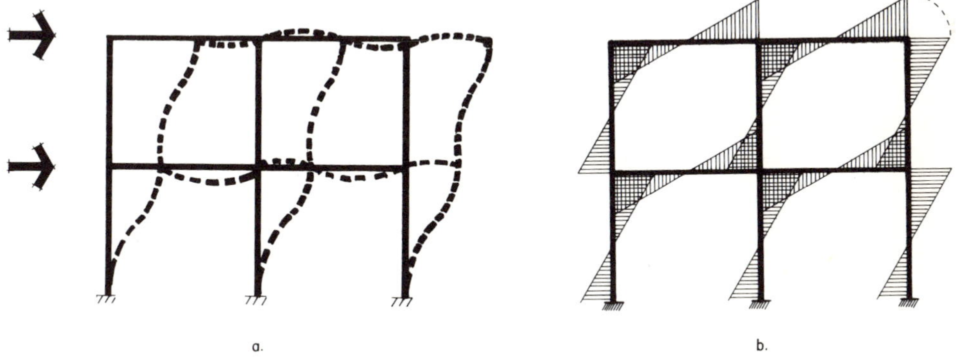

Figure 3-47. *a,* Distortion of two-story, two-bay frame under wind loads. *b,* Bending moments diagram for two-story, two-bay frame under wind loads.

the analysis performed with the STRUDL II program. Note that the remarks in parentheses have been added to the printout for the convenience of the reader. They are *not* part of the commands; therefore, readers should not expect such remarks on the printout of a similar problem.

Example: A 20-story building, 240 ft high, was designed only to indicate the structural behavior of a multistory rigid frame under lateral wind forces, with no consideration given to other design parameters and loads. The computer printout on pp. 276–320 includes the input that has been formulated for the problem, and the printout of the analysis performed with the STRUDL II program. Note that the remarks in parentheses have been added to the printout for the convenience of the reader. They are *not* part of the commands; therefore, readers should not expect such remarks on the printout of a similar problem.

WIND LOAD FACTORS

The design of structures to resist wind loads is taken into consideration by building codes with various criteria. In conventional elastic design in steel, the American Institute of Steel Construction (1970) suggests that wind loads combined with live and dead loads can be assumed to be only three-fourths of the total value; or, the allowable stresses can be increased by one-third, which produces the same effect. The basic concept for this assumption is that it is improbable that full live loads and full wind loads can coincide at the same time. This condition, of course, has to be checked against full dead and live loads combined to determine which of the two is more stringent. Similarly, in ultimate-strength design methods in reinforced concrete, the American Concrete Institute (1976) specifies that the required ultimate load-carrying ability of a member (U) should be:

$$U = 0.75(1.4D + 1.7L + 1.7W)$$

In this formula, D, L, and W, are, respectively, the dead load, the live load, and the wind load on a structure. Note that the 0.75-factor is the same that is used for the elastic design method in steel and the working-stress method in concrete. The formula above has to be checked against: $U = 1.4D + 1.7L$.

The same ultimate strength design method in concrete considers the following formula:

$$U = 0.9D + 1.3W$$

This formula includes conditions in which overturning moments control, such as in the case of tall buildings.

HUMAN COMFORT AND BUILDING MOTION

The theory of elasticity tells us that all materials are deformed when stressed and all structures therefore deflect under loads. Tall slender buildings under wind loads deflect considerably so that at times the comfort of the people living in them can be seriously disturbed.

When a building deflects because of wind forces, we distinguish a steady deflection due to the constant component of the wind velocity, and an additional deflection produced by vibrations around the deflected position that are caused by the gustiness of the wind. The steady deflection is a matter of structural concern, but it does not affect people's comfort. The vibrations instead cause the discomfort. The following parameters are the most significant in the way people react to vibrations: building's acceleration, change of acceleration, frequency or period of vibration, and amplitude of deflection. To establish criteria of human reaction to vibrations, the relationships among such parameters must be based on actual tests. Experimental data in this field are scarce and incomplete. Some design firms try to establish their own design criteria. A relationship of general validity that relates the building acceleration to people's reaction is indicated in Table 3-6.

Table 3-6. Reaction to building acceleration.

	BUILDING ACCELERATION	
PEOPLE TOLERANCE	(FT/SEC^2)	% OF GRAVITY
Imperceptible	0 to 16.1	0 to 0.5
Perceptible	16.1 to 48.3	0.5 to 1.5
Disturbing	48.3 to 161	1.5 to 5.0
Very disturbing	161 to 403	5.0 to 15.0
Unbearable	403 to over 403	15.0 to over 15.0

A correlation between the acceleration of the building, the amplitude, and the period of vibration is expressed by the following:

$$T = 329\ A/a$$

where

T = period (sec)
A = amplitude (in.)
a = acceleration (ft/sec^2)

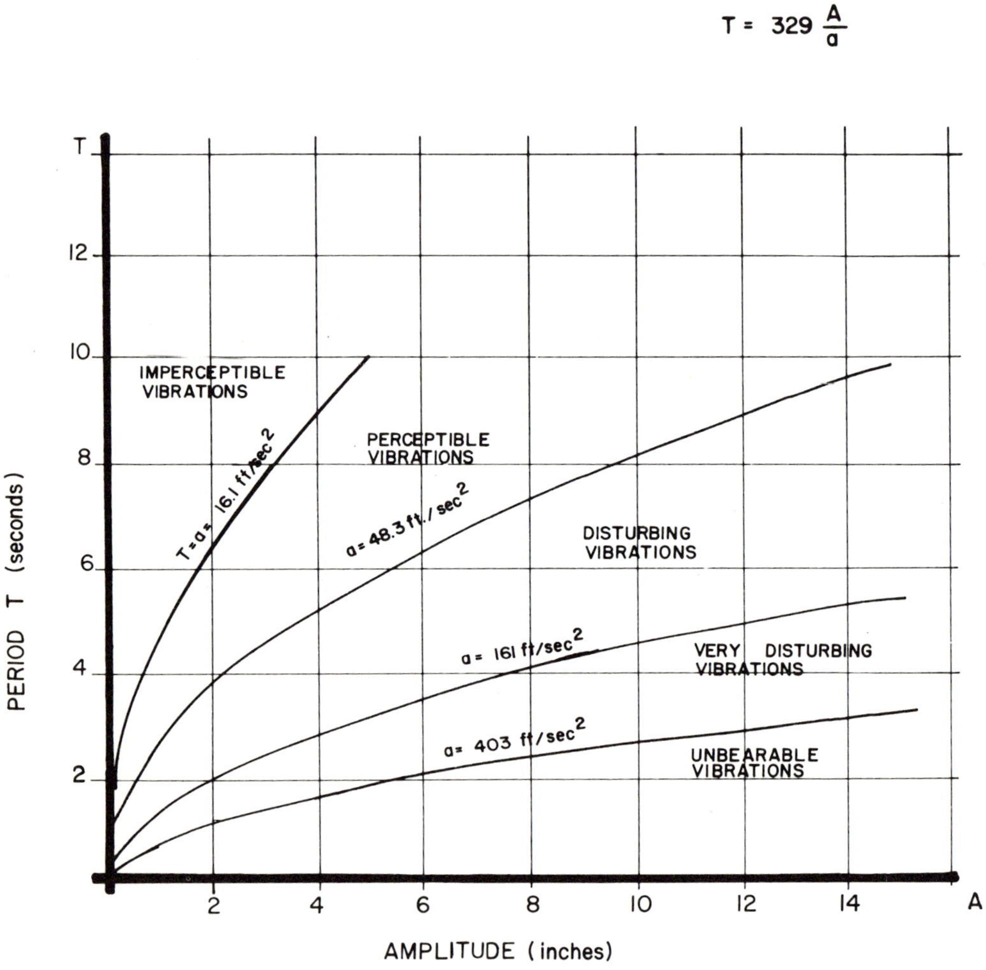

Figure 3-48. Human response to building motion.

If A and T are the variables, and if we give a the four values (0.5, 1.5, 5.0, and 15.0) seen in Table 3-4, we can plot the preceding equation and obtain four diagrams (see Figure 3-48). Knowing the amplitude and period of vibration of a given building, we can see to what extent it satisfies the comfort of its occupants. All points above and to the left of the curve are governed by the curve immediately below or to the right.

For instance, for a building swaying with a period of 8 sec and an amplitude of deflection of 2 in., the motion is not perceived by the occupants. For the Empire State Building, if the vibrations had a period of 6 sec and an amplitude of deflection of 3.5 in. (actually measured), the motion would barely be perceptible.

Recent experimental work by Yamada et al. (1977) on people's reaction to the motion of tall buildings includes new levels of tolerance expressed in terms of period of vibration and acceleration of buildings. The work conducted on various human subjects takes into account age, sex, and body posture of the individuals being tested during the forced vibration of a controlled moving structure. The ability of the subjects to walk, to climb or descend stairs, and to perform a task such as tracing a drawing under different accelerations and periods of vibration were tested. It is interesting that the sensitivity to building motions was higher for females in comparison to males, for younger in comparison to older people, for standing versus sitting body postures, and for orientation of the body parallel to the building motion.

FORM RESPONSE TO WIND

Form response of buildings to the forces of the prevailing winds can be at times a strong design concept that successfully dominates other parameters. The structural benefits of reducing wind pressure for slender, sensitive high-rise buildings is a major economic factor in structural considerations. Furthermore, the implication in the optimization of natural ventilation adds significance to the use of form as a response to wind. Tall, slender buildings exposed to high pressures, gustiness, eddy-shedding phenomena, and so on, can attain successful configurations that minimize inertial forces versus viscous forces and can produce turbulence with their skin so that Karman vortices are not formed.

Wind factors in site analysis included at an early stage of design are strong points to be emphasized in architectural education.

SKYSCRAPERS

The unit cost of a high-rise structure in dollars per square foot of floor area increases with building height. This is due in part to the vertical dead and live loads that accumulate at the lower levels and in part to the horizontal loads of wind and, in some cases, of earthquakes. While the vertical loads make the cost increase linearly (constant rate of increase) with height, the horizontal forces make the cost increase with an exponential curve. This is particularly evident for steel frames because steel cost is proportional to its weight. The weight per square foot of floor area for various building heights is shown in Figure 3-49. Of course, earthquake forces are not considered in zones where their probability is low; however, wind forces are always included in the design of tall slender structures. It is the wind, therefore, that is the major factor that limits the height of skyscrapers. (See Figure 3-50.)

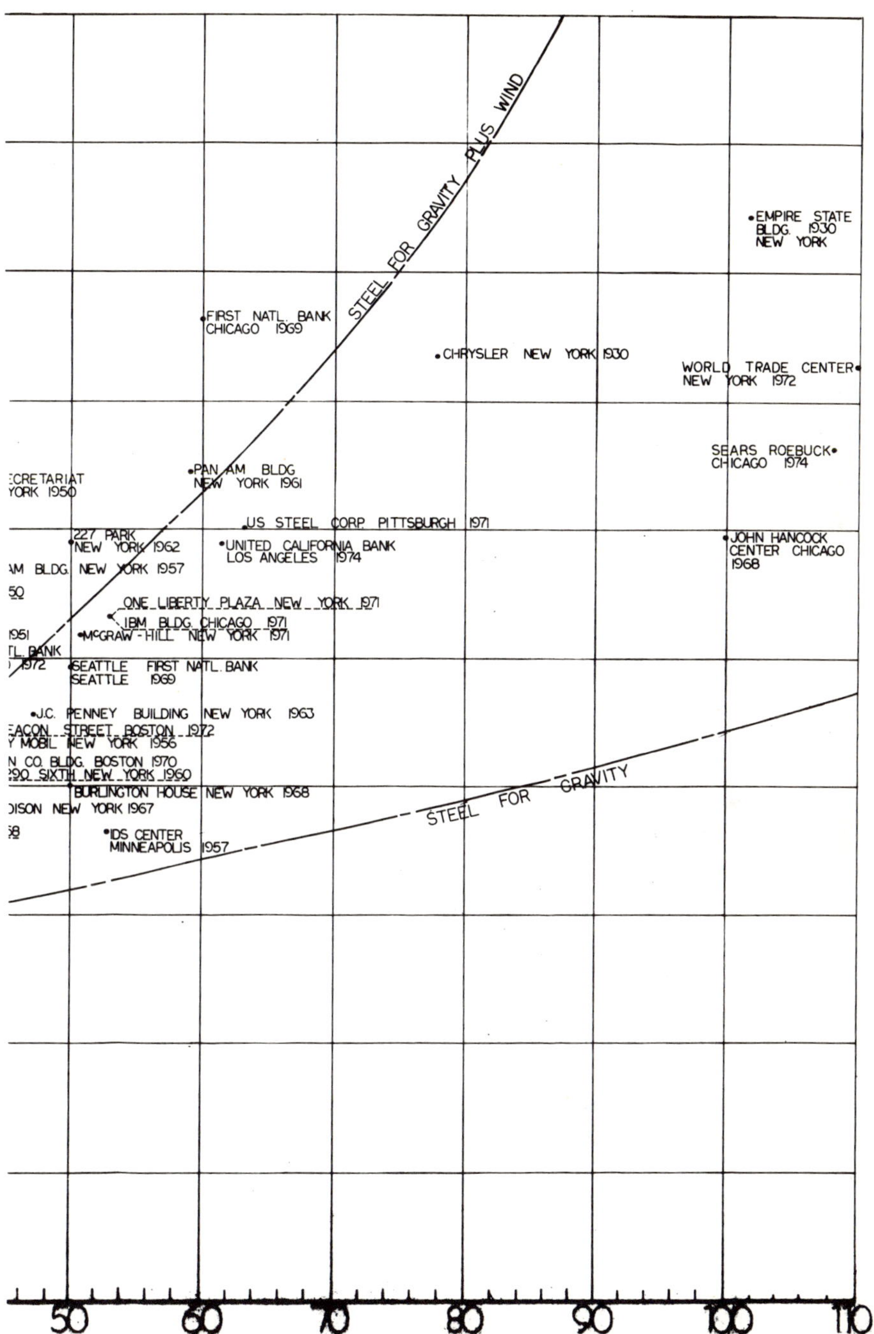

Figure 3-49. Unit weight of steel frame vs. building slenderness.

In fact, in the 1950s, when Frank Lloyd Wright proposed the construction of the "Illinois," a "mile-high" skyscraper for Chicago, the exciting concept of a structure of such size was frustrated by the enormous wind force that would have required the lower part of the building to be a solid mass of steel according to structural engineers from Skidmore Owings & Merrill. The author remembers vividly the comments of the famous structural engineer Pier Luigi Nervi on Wright's proposal: "paper architecture." Such projects are so numerous, he claimed, that they could fill up large museums. But, on the other hand, he believed that we need dreamers who can anticipate the future in architecture. In fact, only a few decades ago, nobody in the structural engineering profession anticipated the present heights that reinforced concrete structures have reached. We have seen in this chapter the evolution of the various steel and concrete structural systems, examples of which are listed in Tables 3-7 and 3-8. According to the present state of the art, a drastic increase in the height of man-built structures is purely a matter of speculation. More precisely, in an atmosphereless environment with no wind, such as on the lunar surface, buildings could reach dizzying heights, but on earth, the omnipresent wind can allow maximum heights only in regions where predictable wind speeds are lowest.

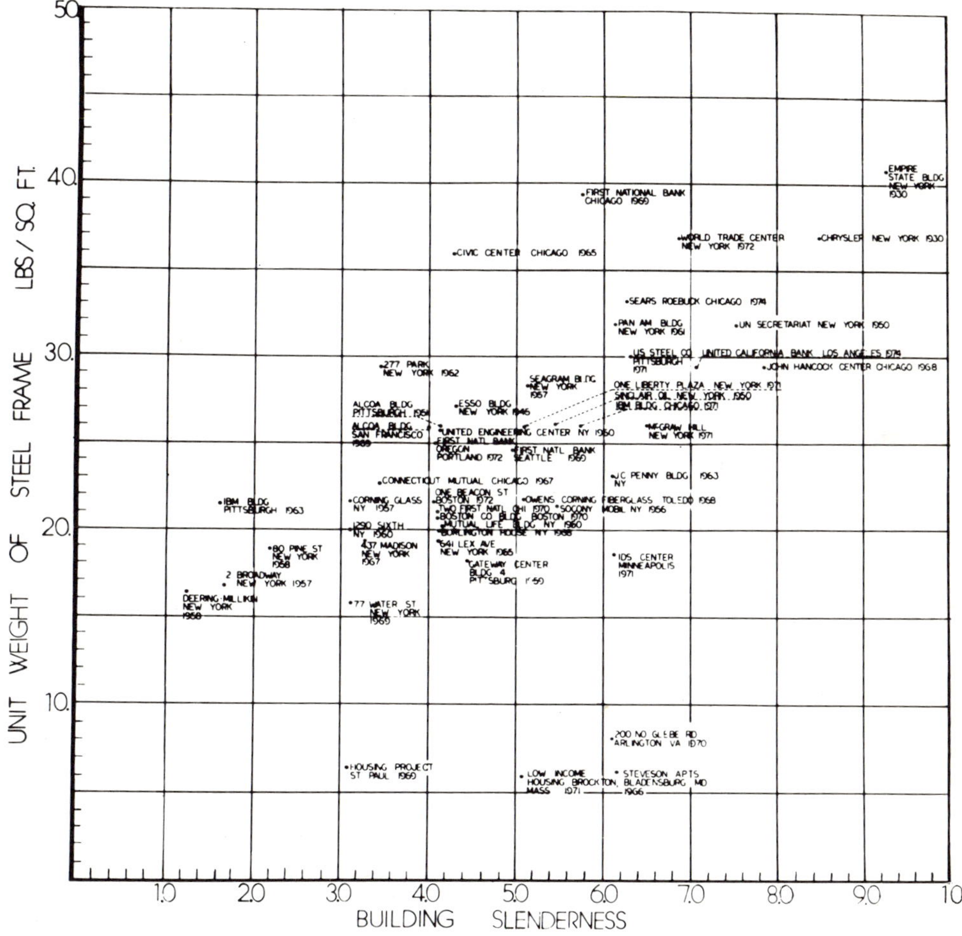

Figure 3-50. The response of high-rise buildings to wind forces results in severe structural penalties expressed as an increase of the weight of the structure per square foot of building area. Shown here is the unit weight of steel frames of known buildings as a function of the building slenderness. As the slenderness increases, the unit weight increases also due to the wind load.

Table 3-7. Wind-resistant steel structures.

BUILDING	ARCHITECT	CITY	YEAR	NO. STORIES	WEIGHT lbs/ft²	WEIGHT kg/mg²
Empire State	Shreve, Lamb & Harmon	New York	1930	102	42.2	98
Chrysler	Wm. Van Allen	New York	1930	77	37	181
Esso Bldg., Rapid City	Carson & Lundin	New York	1945	32	27.2	133
UN Secretariat		New York	1950	42	32	156
Sinclair Oil	Carson & Lundin	New York	1950	27	26.6	130
Mutual Life Insurance Bldg.		New York	1950			
Alcoa Bldg.	Harrison & Abramovitz	Pittsburgh	1951	30	25.7	125
641 Lexington Avenue		New York	1955	33	19	93
Socony Mobile	Harrison, Abramovitz	New York	1956	42	22	108
Corning Glass	Harrison, Abramovitz & Abbe	New York	1957	26	22	108
Seagram Bldg.	Miss. Vauder Roche & Philip Johnson	New York	1957	42	28	137
2 Broadway	Emery Roth & Sons	New York	1957	30	16.8	82
80 Pine St		New York	1958	38	18.7	91
Gateway Center Bldg. #4		Pittsburgh	1959	22	17.9	87
United Engineering Center		New York	1960	20	26.6	125
Sperry Rand	Emery Roth & Sons	New York	1960	43	20	98
Pan Am Bldg.	Emery Roth & Sons, Walter Gropius & Pieto Belluschi	New York	1961	59	32	157
Chemical Bank NY Trust Co Bldg.	Emery Roth & Sons	New York	1962	50	29	144
J. C. Penney Bldg.	Shreve, Lamb & Harmon	New York	1963	46	23	112
Civic Center	SOM, C. F. Murphy Associates Loebl Schlossman, Bennet and Dart	Chicago	1965	31	37	181

Connecticut Mutual		Chicago	1967	25	22.5	110
437 Madison Ave		New York	1967	41	18.9	92
John Hancock Center	SOM	Chicago	1968	100	29.5	144
Owens	Corning Fiberglas	Toledo	1968	30	22	108
Burlington House		New York	1968	50	50	98
First National Bank	C. F. Murphy Associates & The Parkins & Will Partnership	Chicago	1969	60	30	147
Alcoa Bldg.	SOM	San Francisco	1969	26	26	127
Seattle	First National Bank	Seattle	1969	50	24.5	120
77 Water St.	Kaufman	New York	1969	29	16	78
Two 1st National Plaza		Chicago	1970	30	21.3	104
Boston Co. Bldg.		Boston	1970	42	21	103
US Steel Corp.	Harrison, Abramovitz & Abbe	Pittsburgh	1971	30	6.3	31
One Liberty Plaza	I. M. Pei	New York	1971	54	26.5	129
IBM Bldg.	Minoru Yamasaki	Chicago	1971	54	26.5	129
McGraw-Hill	Harrison, Abramovitz & Harris	New York	1971	51	26	127
Investors Diversified Service Center	Philip Johnson	Minneapolis	1971	54	17.9	87
World Trade Center	Minoru Yamasaki w/ Emery Roth & Sons, Eng. Skilling, Helle, Christiansen, Robertson	New York	1972	110	37	181
1st National Bank, Oregon		Portland	1972	40	25	122
One Beacon St		Boston	1972	40	22	108
Sears Tower	SOM	Chicago	1974	109	33	161

Table 3-8. Wind-resistant concrete framing.

BUILDING	OCCUPANCY TYPE	LOCATION	YEAR	NO. STORIES	HEIGHT (ft)	WIND-RESISTANT FRAMING
36 Central Pk. So.	Apts.–hotel	New York City	1970	43	461	Shear wall-frame interaction
Standard Bank	Office	Johannesburg	1971	40	459	Suspended Structure
[not named]	Office	Miami, Florida	–	40	456	Shear wall-frame interaction
Park Regis	Apts.	Sydney, Australia	1968	51	450	Shear wall-frame interaction
State Government Office Block	Office	Sydney, Australia	1967	38	450	Shear wall core, steel external columns
Main Place	Office	Dallas, Texas	1968	34	450	Shear wall-frame interaction
Italia Building	Office	Sao Paulo, Brazil	1956	45	443	Frame
Harlem River	Apts.	New York City	–	46	440	Shear wall-frame interaction
LaSalle Plaza	Office	Chicago	1972	40	440	Exter. frame & core
St. Martin's Tower	Office	Sydney, Australia	–	35	437	Shear wall-frame interaction
Gulf Life Tower	Office	Jacksonville, Florida	–	27	430	Shear wall-frame interaction
Jerome Ave.	Apts.	New York City	1971	42	430	Shear wall-frame interaction
Madison & 89th St	Apts.	New York City	1971	38	430	Shear wall-frame interaction
2 Illinois Center	Office	Chicago	–	37	430	Shear walls at service core
North Point	Office	Sydney, Australia	–	38	430	–
Xerox Building	Office	Rochester, New York	–	30	430	Framed tube
Canoco Building	Office	Houston, Texas	1973		429	Shear wall-frame interaction
National Life Bldg.	Office	Nashville, Tennessee	1970	30	428	Shear wall-frame interaction
East End Av.	Apts.	New York City	1971	42	426	Shear wall-frame interaction
Colony North	Gar'g–apts.	New Jersey	–	45	425	Shear wall-frame interaction
L.T.V. Tower	Office	Dallas, Texas	1963	31	422	Shear wall-frame interaction
860 UN Plaza	Office–apt.	New York City	1965	41	420	Shear wall-frame interaction
111 Wacker Drive	Office	Chicago	1970	36	420	Shear wall-frame interaction
Laclede Gas Bldg.	Office	St. Louis, Missouri	1969	31	420	–
2626 Lakeview Av.	Apts.	Chicago	1967	42	418	Shear wall-frame interaction
110 William St.	Office	New York City	1958	31	418	–
Nonalco (Arrowhead)	Office	Mexico City	1963	24	417	–
Pirelli Building	Office	Milan, Italy	1955	32	414	Frame
3rd & 92nd	Apts.	New York City	–	42	411	Shear wall-frame interaction
Atlas Building	Apt–office	Buenos Aires	1957	42	410	–
Carlyle Apts.	Apts.	Chicago	1962	40	410	Shear wall core
Demetrio Eliades	Apts.	Buenos Aires	–	39	410	–

Lanray	Hotel	Sydney, Australia	–	36	410	Shear wall-frame interaction
Institute of Technology	Educational	Sydney, Australia	–		410	Shear wall
Barbican Redevelopment	Housing	London	–	43	409	–
1st Ave. & 55th St.	Apts.	New York City	1970	38	409	Shear wall-frame interaction
Gateway III, Union Sta.	Office	Chicago	1971	35	408	Composite (conc. tube-int. steel framing)
Intern'l Trade Center	Office	New Orleans	1965	33	407	–
Outer Drive East	Apts.	Chicago	1964	40	405	Shear wall-frame interaction
Broadway & 44th	Office–theatre	New York City	1972	33	405	Shear wall-frame interaction
Bentall Centre, Tower III	Office	Vancouver, British Columbia	1973	27	404	R/c core-peripheral ductile space frame
1300 N. LakeShore Dr.	Apts.	Chicago	1962	40	402	Slab and shear walls
FOCSA Bldg., Vedada	Apts.	Cuba	1955	39	402	Shear wall
1130 S. Michigan	Apts.	Chicago	1965	44	400	Shear wall-flat plate
Water Tower Building	Apts.–hotel commercial	Chicago	1975	76	859	Multicell-tube
MLC Center	Office	Sydney, Australia	u.c.	62	760	Shear wall-frame interaction
Peachtree Center Plaza	Hotel	Atlanta, Georgia	1976	70	722	–
One Shell Plaza	Office	Houston, Texas	1970	52	714	Tube-in-tube
Edificio Puegeot	Office	Buenos Aires, Argentina	1962	63	660	Coupled shear walls
Carlton Centre	Office	Johannesburg	1972	30	660	Shear wall-frame interaction
Lake Point Tower	Apts.	Chicago	1965	70	645	Reinforced concrete core & flat plate
1000 Lake Shore Plaza	Apts.	Chicago	1964	55	640	Shear wall-frame interaction
Place Victoria	Office	Montreal	1964	47	624	Shear wall-frame interaction
A.M.P. Society II	Office	Sydney	1973	50	612	Shear wall-frame interaction
Tour Fiat	Office	Paris	1973	49	610	Tubular ex. bearing wall
Quantas	Office	Sydney, Australia	u.c.	50	605	Shear wall-frame interaction
Collins Tower	Office–hotel	Melbourne, Australia	u.c.	45	600	Shear wall-frame interaction
ANZ Tower	Office	Melbourne, Australia	u.c.	45	600	Shear wall-frame interaction
One Shell Square	Office	New Orleans, Louisiana	1971	50	590	Composite (concrete tube steel interior)
Marina City	Apts.–parking	Chicago	1962	60	588	Shear wall-frame interaction
Mid-Continental Plaza	Office	Chicago	1972	50	582	Shear wall-frame interaction

Table 3-8 (Continued)

BUILDING	OCCUPANCY TYPE	LOCATION	YEAR	NO. STORIES	HEIGHT (ft)	WIND-RESISTANT FRAMING
Nauru House	Office	Melbourne, Australia	u.c.	51	580	Shear wall-frame interaction
Australia Square Tower	Office	Sydney	1968	45	562	Shear wall-frame interaction
Newberry House	Apts.–commercial	Chicago	1972	58	560	Shear wall-frame interaction
No 3 Park Avenue	Mixed use	New York City	1975	40	556	Shear wall-frame interaction
Galleria	Office–school	New York City	1975	57	552	Shear wall-frame interaction
Central Manhattan	Office	New York	1973	42	542	Slipformed conc. core, steel frame exterior
CB 21	Residential-commercial	Paris	1971	51	540	Coupled shear walls
Manufacturer's Life Center	Apts.	Toronto, Canada	1973	55	533	Shear wall-frame interaction
Frontier Towers	Office	Chicago	u.c.	42	525	Shear wall-frame interaction
Center Square Complex	Apts.	Philadelphia	1972	57	520	Shear wall-frame interaction
West Tower	Office	Chicago	1968	40	518	Shear wall-frame interaction
111 E. Chestnut	Office	New York City	1972	42	515	Shear wall-frame interaction
Royal Centre Office Bldg.	Office	Vancouver, British Columbia	1946	34	507	Frame
Branco de Estado,	Hotel	Sao Paulo, Brazil	1962	50	500	Coupled shear walls
Americana Hotel	Hotel	New York City	1972	45	496	Shear wall-frame interaction
Landmark Hotel	Office	Vancouver, British Columbia	1964	38	491	Tube-in-tube
CBS	Apts.	New York City	1966	47	483	Shear wall-frame interaction
Excelsior House	Office	New York City	1964	38	478	Tube-in-tube
Brunswick Bldg.	Apts.	Chicago	1969	42	471	Shear wall-frame interaction
Broadway & 64th St.	Office	New York City	1968	32	469	Shear wall-frame interaction
FRD PO & Tower	Office	New York City	1970	40	467	Shear wall-frame interaction
1250 Broadway	Office	New York City	u.c.	38	465	Shear walls at service core
East Point Apt.	Apts.	Chicago	1966	44	400	Shear wall-flat plate
DeWitt Chestnut Apt.	Apts.	Chicago	1965	43	400	Framed tube
Brooks Tower	Apts.–office	Denver, Colorado	1966	43	400	Shear wall
1020 Rush St.	Apts.	Chicago	1969	40	400	Shear wall
Edgewater Twin Towers	Apts.	Chicago	1966	40	400	Shear wall-frame interaction
1010 Commons	Office	New Orleans	1970	33	400	Shear wall-frame interaction
Time-Life Bldg.	Office	Chicago	1969	30	400	Shear wall-frame interaction
Carlton Centre Hotel	Hotel	Johannesburg	1972	30	400	Shear wall-frame interaction

MOBILE-HOME PROTECTION

Mobile homes, a yet-to-be-perfected product of the industrial era and a radical departure from the conservative methodologies of the building art, are particularly vulnerable to wind forces. Their particular light weight and lack of anchorage to the ground are the major causes for their poor lateral stability at high-speed wind loads. In fact, it is estimated that in the United States, over 5000 mobile homes yearly are severely damaged or totally destroyed. Such high figures suggest some positive measures be taken. The Defense Civil Preparedness Agency (1972) suggests (1) proper orientation to the prevailing wind to minimize the exposed area, (2) use of natural or artificial windbreaks, and (3) anchoring with "tie-downs." (See Figure 3-51.)

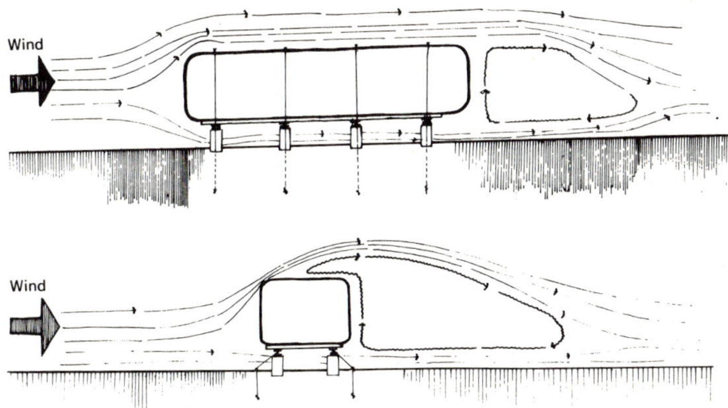

Figure 3-51. As the wind flows over, around, and underneath a typical mobile home, to minimize the wind overturning moment, it is advisable to orient the longitudinal axis in the direction of the prevailing wind.

Tie-downs include two types: frame tie and "over-the-top." The frame tie connects the frame to the ground, but it is not sufficient because the wind could detach the unit from the frame and blow it off. The over-the-top tie, on the other hand, connects the unit to the ground and completes the necessary protection. (See Figure 3-52.)

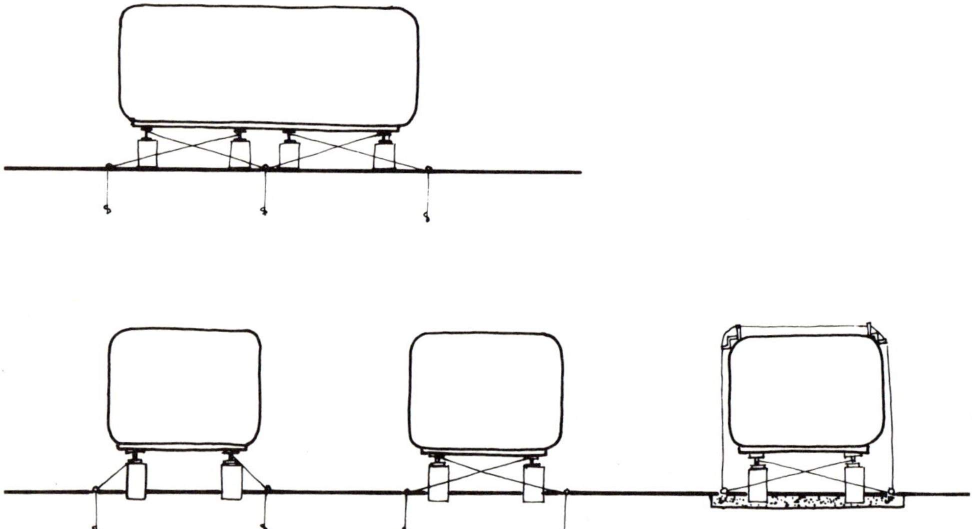

Figure 3-52. Various arrangements of tiedowns for single and double mobile-home units. Note the two-frame types and the overhead type.

Tie-downs consist of galvanized steel straps, or cables or wire ropes, connected to ground anchors through drop-forged turnbuckles with closed eyes. Galvanized steel anchors are screwed into the ground to a depth of not less than 4 ft to provide sufficient reaction. Table 3-9, based on a tensile capacity of 4800 lb per tie, indicates the number and type of tie required for mobile homes of different length at various wind speeds.

Table 3-9. Tie-down anchorage requirements for mobile homes.

WIND VELOCITY (mph)	10 AND 12 FT WIDE, 30 TO 50 FT LONG		10 AND 12 FT WIDE, 50 TO 60 FT LONG		12 AND 14 FT WIDE, 60 TO 70 FT LONG	
	NO. OF FRAME TIES	NO. OF OVER-THE-TOP TIES	NO. OF FRAME TIES	NO. OF OVER-THE-TOP TIES	NO. OF FRAME TIES	NO. OF OVER-THE-TOP TIES
70	3	2	4	2	4	2
80	4	3	5	3	5	3
90	5	4	6	4	7	4
100	6	5	7	5	8	6
110	7	6	9	6	10	7

COMPUTER ANALYSIS OF WIND LOAD ON A HIGH-RISE BUILDING

INPUT

STRUDL 'MGM' 'WIND ON 20-STORY FRAME'

TYPE PLANE FRAME XY (All joints are assumed fixed except when released.)
UNITS KIPS FEET
JOINT COORDINATES

```
1  X 0  Y 0 S (Support)
2  X 0  Y 12
3  X 0  Y 24
4  X 0  Y 36
5  X 0  Y 48
6  X 0  Y 60
7  X 0  Y 72
8  X 0  Y 84
9  X 0  Y 96
10 X 0  Y 108
11 X 0  Y 120
12 X 0  Y 132
13 X 0  Y 144
14 X 0  Y 156
15 X 0  Y 168
16 X 0  Y 180
17 X 0  Y 192
18 X 0  Y 204
19 X 0  Y 216
20 X 0  Y 228
21 X 0  Y 240
22 X 30 Y 240
23 X 30 Y 228
24 X 30 Y 216
25 X 30 Y 204
26 X 30 Y 192
27 X 30 Y 180
28 X 30 Y 168
29 X 30 Y 156
```

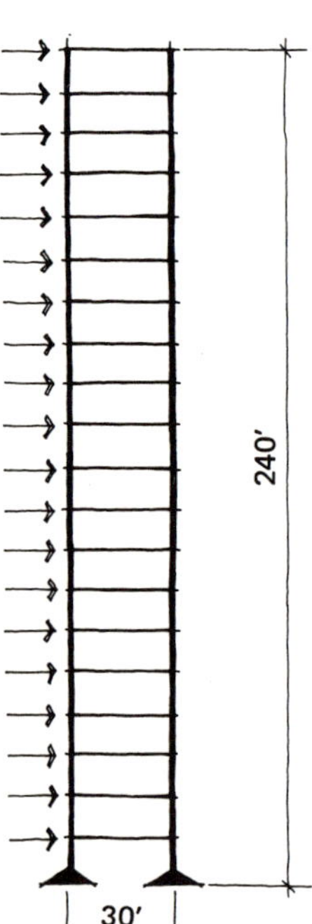

30 X 30 Y 144
31 X 30 Y 132
32 X 30 Y 120
33 X 30 Y 108
34 X 30 Y 96
35 X 30 Y 84
36 X 30 Y 72
37 X 30 Y 60
38 X 30 Y 48
39 X 30 Y 36
40 X 30 Y 24
41 X 30 Y 12
42 X 30 Y 0 S (Support)

JOINT RELEASE MOMENT Z
1 42 (These two joints are changed to fixed hinges.)

MEMBER INCIDENCE (First number identifies the member, second number identifies the "start" joint, third number identifies the "end" joint.)
1 1 2 (Column 1st floor)
2 2 3 (Column 2nd floor)
3 3 4 (Column 3rd floor)
4 4 5 (Column 4th floor)
5 5 6 (Column 5th floor)
6 6 7 (Column 6th floor)
7 7 8 (Column 7th floor)
8 8 9 (Column 8th floor)
9 9 10 (Column 9th floor)
10 10 11 (Column 10th floor)
11 11 12 (Column 11th floor)
12 12 13 (Column 12th floor)
13 13 14 (Column 13th floor)
14 14 15 (Column 14th floor)
15 15 16 (Column 15th floor)
16 16 17 (Column 16th floor)
17 17 18 (Column 17th floor)

18 18 19 (Column 18th floor)
19 19 20 (Column 19th floor)
20 20 21 (Column 20th floor)
21 42 41 (Column 1st floor)
22 41 40 (Column 2nd floor)
23 40 39 (Column 3rd floor)
24 39 38 (Column 4th floor)
25 38 37 (Column 5th floor)
26 37 36 (Column 6th floor)
27 36 35 (Column 7th floor)
28 35 34 (Column 8th floor)
29 34 33 (Column 9th floor)
30 33 32 (Column 10th floor)
31 32 31 (Column 11th floor)
32 31 30 (Column 12th floor)
33 30 29 (Column 13th floor)
34 29 28 (Column 14th floor)
35 28 27 (Column 15th floor)
36 27 26 (Column 16th floor)
37 26 25 (Column 17th floor)
38 25 24 (Column 18th floor)
39 24 23 (Column 19th floor)
40 23 22 (Column 20th floor)
41 2 41 (Girder 2nd floor)
42 3 40 (Girder 3rd floor)
43 4 39 (Girder 4th floor)
44 5 38 (Girder 5th floor)
45 6 37 (Girder 6th floor)
46 7 36 (Girder 7th floor)
47 8 35 (Girder 8th floor)

```
48  9   34  (Girder 9th floor)
49  10  33  (Girder 10th floor)
50  11  32  (Girder 11th floor)
51  12  31  (Girder 12th floor)
52  13  30  (Girder 13th floor)
53  14  29  (Girder 14th floor)
54  15  28  (Girder 15th floor)
55  16  27  (Girder 16th floor)
56  17  26  (Girder 17th floor)
57  18  25  (Girder 18th floor)
58  19  24  (Girder 19th floor)
59  20  23  (Girder 20th floor)
60  21  22  (Girder roof)

UNITS INCHES
MEMBER 1 TO 10 PROP PRISMATIC AX 40 IZ 1590 (Columns from 1st to 10th floor: W 14 × 136)
MEMBER 21 TO 30 PROP PRISMATIC AX 40 IZ 1590 (Columns from 1st to 10th floor: W 14 × 136)
MEMBER 11 TO 20 PROP PRISMATIC AX 14. 1 IZ 485 (Columns from 11th to 20th floor: W 14 × 48)
MEMBER 31 TO 40 PROP PRISMATIC AX 14. 1 IZ 485 (Columns from 11th to 20th floor: W 14 × 48)
MEMBER 41 TO 60 PROP PRISMATIC AX 22. 4 IZ 2100 (All girders: W 24 × 76)
CONSTANT E 29000 ALL (Modulus of elasticity for steel)
LOADING 1 (Wind load from left)
JOINT 2 TO 20 LOAD FORCE X 8. 3 (Positive wind pressure)
JOINT 1 21 LOAD FORCE X 4. 15 (Positive wind pressure)
JOINT 23 TO 41 LOAD FORCE X 7. 2 (Negative wind pressure: suction)
JOINT 22 42 LOAD FORCE X 3. 6 (Negative wind pressure: suction)
STIFFNESS ANALYSIS
UNIT INCH KIPS
LIST FORCES DISTORSIONS LOADS REACTIONS DISPLACEMENTS ALL
```

FIRST OUTPUT

```
*****************************************
*  RESULTS OF LATEST ANALYSES  *
*****************************************
```

PROBLEM—MGM   TITLE—WIND ON 20-STORY FRAME

ACTIVE UNITS INCH KIP RAD DEGF SEC

ACTIVE STRUCTURE TYPE PLANE FRAME

ACTIVE COORDINATE AXES X Y

LOADING—1

MEMBER FORCES

MEMBER	JOINT	——— FORCE ———		——— MOMENT —
		AXIAL	SHEAR Y	BENDING Z
1	1	-1239. 9987793	151. 1289825	0. 0000000
1	2	1239. 9987793	-151. 1289825	21762. 5742187
2	2	-1074. 4538574	143. 3710480	8035. 6406250
2	3	1074. 4538574	-143. 3710480	12609. 7890625
3	3	-953. 3605957	135. 6256104	9187. 0664062
3	4	953. 3605957	-135. 6256104	10343. 0234375
4	4	-846. 1528320	127. 8748016	8954. 3359375
4	5	846. 1528320	-127. 8748016	9459. 6367187
5	5	-746. 6850586	120. 1249237	8444. 5937500
5	6	746. 6850586	-120. 1249237	8853. 4023437
6	6	-653. 7202148	112. 3749237	7880. 2343750
6	7	653. 7202148	-112. 3749237	8301. 7539062
7	7	-567. 0095215	104. 6249237	7306. 2148437
7	8	567. 0095215	-104. 6249237	7759. 7773437
8	8	-486. 4970703	96. 8749237	6732. 4375000

8	9	486. 4970703	-96. 8749237	7217. 5507812
9	9	-412. 1481934	89. 1249237	6165. 2617187
9	10	412. 1481934	-89. 1249237	6668. 7304687
10	10	-343. 8349609	81. 3749237	5627. 6445312
10	11	343. 8349609	-81. 3749237	6090. 3515625
11	11	-280. 9331055	73. 6249237	5232. 0000000
11	12	280. 9331055	-73. 6249237	5369. 9921875
12	12	-225. 1137848	65. 8749237	4677. 4843750
12	13	225. 1137848	-65. 8749237	4808. 5078125
13	13	-175. 5600433	58. 1249695	4111. 1640625
13	14	175. 5600433	-58. 1249695	4258. 8281250
14	14	-132. 2055511	50. 3749695	3544. 9746094
14	15	132. 2055511	-50. 3749695	3709. 0217285
15	15	-95. 0449829	42. 6249847	2979. 8815918
15	16	95. 0449829	-42. 6249847	3158. 1157227
16	16	-64. 0779114	34. 8749695	2415. 9658203
16	17	64. 0779114	-34. 8749695	2606. 0307617
17	17	-39. 3041382	27. 1250000	1853. 2478027
17	18	39. 3041382	-27. 1250000	2052. 7517090
18	18	-20. 7228241	19. 3748627	1291. 8828125
18	19	20. 7228241	-19. 3748627	1498. 0988770
19	19	-8. 3230143	11. 6256819	733. 8803711
19	20	8. 3230143	-11. 6256819	940. 2182617
20	20	-1. 9697351	3. 8731174	203. 2981110
20	21	1. 9697351	-3. 8731174	354. 4306641
21	42	1239. 9987793	151. 1207275	0. 0
21	41	-1239. 9987793	-151. 1207275	21761. 3867187
22	41	1074. 4538574	143. 3787384	8036. 5234375
22	40	-1074. 4538574	-143. 3787384	12610. 0117187
23	40	953. 3605957	135. 6241150	9186. 8945312
23	39	-953. 3605957	-135. 6241150	10342. 9804687
24	39	846. 1528320	127. 8749847	8954. 3593750
24	38	-846. 1528320	-127. 8749847	9459. 6367187
25	38	746. 6850586	120. 1248627	8444. 5820312
25	37	-746. 6850586	-120. 1248627	8853. 3945312
26	37	653. 7202148	112. 3748627	7880. 2304687

26	36	-653. 7202148	-112. 3748627	8301. 7500000
27	36	567. 0095215	104. 6248627	7306. 2070312
27	35	-567. 0095215	-104. 6248627	7759. 7734375
28	35	486. 4970703	96. 8748627	6732. 4335937
28	34	-486. 4970703	-96. 8748627	7217. 5468750
29	34	412. 1481934	89. 1248627	6165. 2578125
29	33	-412. 1481934	-89. 1248627	6668. 7265625
30	33	343. 8349609	81. 3748627	5627. 6406250
30	32	-343. 8349609	-81. 3748627	6090. 3437500
31	32	280. 9331055	73. 6249237	5231. 9960937
31	31	-280. 9331055	-73. 6249237	5369. 9882812
32	31	225. 1137848	65. 8749237	4677. 4843750
32	30	-225. 1137848	-65. 8749237	4808. 5039062
33	30	175. 5600433	58. 1249390	4111. 1601562
33	29	-175. 5600433	-58. 1249390	4258. 8242187
34	29	132. 2055511	50. 3749542	3544. 9726562
34	28	-132. 2055511	-50. 3749542	3709. 0195312
35	28	95. 0449829	42. 6249542	2979. 8796387
35	27	-95. 0449829	-42. 6249542	3158. 1137695
36	27	64. 0779114	34. 8749695	2415. 9645996
36	26	-64. 0779114	-34. 8749695	2606. 0307617
37	26	39. 3041382	27. 1249542	1853. 2458496
37	25	-39. 3041382	-27. 1249542	2052. 7468262
38	25	20. 7228241	19. 3750916	1291. 8898926
38	24	-20. 7228241	-19. 3750916	1498. 1228027
39	24	8. 3230143	11. 6242952	733. 8305664
39	23	-8. 3230143	-11. 6242952	940. 0683594
40	23	1. 9697351	3. 8768759	203. 5960388
40	22	-1. 9697351	-3. 8768759	354. 6740723
41	2	0. 5420492	-165. 5447998	-29798. 2187500
41	41	-0. 5420492	165. 5447998	-29797. 9140625
42	3	0. 5545785	-121. 0937958	-21796. 8437500
42	40	-0. 5545785	121. 0937958	-21796. 8906250
43	4	0. 5491731	-107. 2074890	-19297. 3554687
43	39	-0. 5491731	107. 2074890	-19297. 3398437

44	5	0. 5501274	-99. 4679260	-17904. 2226562
44	38	-0. 5501274	99. 4679260	-17904. 2226562
45	6	0. 5499775	-92. 9646149	-16733. 6367187
45	37	-0. 5499775	92. 9646149	-16733. 6289062
46	7	0. 5500007	-86. 7109222	-15607. 9726562
46	36	-0. 5500007	86. 7109222	-15607. 9648437
47	8	0. 5499973	-80. 5122986	-14492. 2187500
47	35	-0. 5499973	80. 5122986	-14492. 2109375
48	9	0. 5499970	-74. 3489227	-13382. 8164062
48	34	-0. 5499970	74. 3489227	-13382. 8085937
49	10	0. 5500024	-68. 3131714	-12296. 3750000
49	33	-0. 5500024	68. 3131714	-12296. 3671875
50	11	0. 5499873	-62. 9019318	11322. 3515625
50	32	-0. 5499873	62. 9019318	-11322. 3437500
51	12	0. 5500016	-55. 8193207	-10047. 4804687
51	31	-0. 5500016	55. 8193207	-10047. 4726562
52	13	0. 5499975	-49. 5537415	-8919. 6718750
52	30	-0. 5499975	49. 5537415	-8919. 6679687
53	14	0. 5499980	-43. 3544617	-7803. 8046875
53	29	-0. 5499980	43. 3544617	-7803. 8007812
54	15	0. 5499985	-37. 1605530	-6688. 9023437
54	28	-0. 5499985	37. 1605530	-6688. 8984375
55	16	0. 5499948	-30. 9671021	-5574. 0781250
55	27	-0. 5499948	30. 9671021	-5574. 0781250
56	17	0. 5500183	-24. 7737579	-4459. 2773437
56	26	-0. 5500183	24. 7737579	-4459. 2734375
57	18	0. 5498686	-18. 5813141	-3344. 6357422
57	25	-0. 5498686	18. 5813141	-3344. 6376953
58	19	0. 5508004	-12. 3998146	-2231. 9797363
58	24	-0. 5508004	12. 3998146	-2231. 9536133
59	20	0. 5474274	-6. 3532801	-1143. 5158691
59	23	-0. 5474274	6. 3532801	-1143. 6638184
60	21	0. 2768772	-1. 9697351	-354. 4306641
60	22	-0. 2768772	1. 9697351	-354. 6740723

MEMBER DISTORTIONS

MEMBER	DISTORTION AXIAL (in.)	DISTORTION SHEAR Y (in.)	ROTATION - BENDING Z
1 (Column—Left)	0. 1539309 (Elong.)	1. 6311331	0. 0339819
2 (Column—Left)	0. 1333805 (Elong.)	-0. 2594418	0. 0071424
3 (Column—Left)	0. 1183482 (Elong.)	-0. 6019400	0. 0018050
4 (Column—Left)	0. 1050397 (Elong.)	-0. 6332644	0. 0007890
5 (Column—Left)	0. 0926919 (Elong.)	-0. 6022905	0. 0006383
6 (Column—Left)	0. 0811515 (Elong.)	-0. 5590384	0. 0006582
7 (Column—Left)	0. 0703874 (Elong.)	-0. 5136133	0. 0007082
8 (Column—Left)	0. 0603927 (Elong.)	-0. 4682435	0. 0007575
9 (Column—Left)	0. 0511632 (Elong.)	-0. 4243575	0. 0007861
10 (Column—Left)	0. 0426830 (Elong.)	-0. 3871174	0. 0007225
11 (Column—Left)	0. 0989346 (Elong.)	-1. 2516804	0. 0007064
12 (Column—Left)	0. 0792770 (Elong.)	-1. 1171408	0. 0006707
13 (Column—Left)	0. 0618260 (Elong.)	-0. 9738975	0. 0007559
14 (Column—Left)	0. 0465581 (Elong.)	-0. 8307494	0. 0008398
15 (Column—Left)	0. 0334715 (Elong.)	-0. 6884108	0. 0009124
16 (Column—Left)	0. 0225660 (Elong.)	-0. 5469403	0. 0009730
17 (Column—Left)	0. 0138415 (Elong.)	-0. 4063522	0. 0010213
18 (Column—Left)	0. 0072978 (Elong.)	-0. 2667661	0. 0010556
19 (Column—Left)	0. 0029311 (Elong.)	-0. 1296258	0. 0010563
20 (Column—Left)	0. 0006937 (Elong.)	-0. 0128179	0. 0007737
21 (Column—Right)	-0. 1539309 (Short.)	1. 6310434	0. 0339800
22 (Column—Right)	-0. 1333805 (Short.)	-0. 2595572	0. 0071414
23 (Column—Right)	-0. 1183482 (Short.)	-0. 6019171	0. 0018052
24 (Column—Right)	-0. 1050397 (Short.)	-0. 6332676	0. 0007890
25 (Column—Right)	-0. 0926919 (Short.)	-0. 6022895	0. 0006383
26 (Column—Right)	-0. 0811515 (Short.)	-0. 5590381	0. 0006582
27 (Column—Right)	-0. 0703874 (Short.)	-0. 5136129	0. 0007082
28 (Column—Right)	-0. 0603927 (Short.)	-0. 4682432	0. 0007575

29 (Column—Right)	-0. 0511632 (Short.)	-0. 4243572	0. 0007861
30 (Column—Right)	-0. 0426830 (Short.)	-0. 3871172	0. 0007225
31 (Column—Right)	-0. 0989346 (Short.)	-1. 2516804	0. 0007064
32 (Column—Right)	-0. 0792770 (Short.)	-1. 1171398	0. 0006787
33 (Column—Right)	-0. 0618260 (Short.)	-0. 9738970	0. 0007559
34 (Column—Right)	-0. 0465581 (Short.)	-0. 8307490	0. 0008398
35 (Column—Right)	-0. 0334715 (Short.)	-0. 6884104	0. 0009124
36 (Column—Right)	-0. 0225660 (Short.)	-0. 5469400	0. 0009730
37 (Column—Right)	-0. 0138415 (Short.)	-0. 4063522	0. 0010213
38 (Column—Right)	-0. 0072978 (Short.)	-0. 2667638	0. 0010557
39 (Column—Right)	-0. 0029311 (Short.)	-0. 1296381	0. 0010557
40 (Column—Right)	-0. 0006937 (Short.)	-0. 0129045	0. 0007734
41 (Girder)	-0. 0003004 (Short.)	10. 5689001	0. 0000009
42 (Girder)	-0. 0003073 (Short.)	7. 7308626	-0. 0000002
43 (Girder)	-0. 0003043 (Short.)	6. 8443689	0. 0000000
44 (Girder)	-0. 0003049 (Short.)	6. 3502502	-0. 0000000
45 (Girder)	-0. 0003048 (Short.)	5. 9350662	-0. 0000000
46 (Girder)	-0. 0003048 (Short.)	5. 5358162	-0. 0000000
47 (Girder)	-0. 0003048 (Short.)	5. 1400824	-0. 0000000
48 (Girder)	-0. 0003048 (Short.)	4. 7466011	-0. 0000000
49 (Girder)	-0. 0003048 (Short.)	4. 3612633	-0. 0000000
50 (Girder)	-0. 0003048 (Short.)	4. 0157976	-0. 0000000
51 (Girder)	-0. 0003048 (Short.)	3. 5636272	-0. 0000000
52 (Girder)	-0. 0003048 (Short.)	3. 1636190	-0. 0000000
53 (Girder)	-0. 0003048 (Short.)	2. 7678432	-0. 0000000
54 (Girder)	-0. 0003048 (Short.)	2. 3724117	-0. 0000000
55 (Girder)	-0. 0003048 (Short.)	1. 9770079	-0. 0000000
56 (Girder)	-0. 0003048 (Short.)	1. 5816116	0. 0000000
57 (Girder)	-0. 0003047 (Short.)	1. 1862698	-0. 0000000
58 (Girder)	-0. 0003052 (Short.)	0. 7916446	0. 0000001
59 (Girder)	-0. 0003034 (Short.)	0. 4055283	-0. 0000004
60 (Girder)	-0. 0001534 (Short.)	0. 1256227	-0. 0000007

RESULTANT JOINT SUPPORTS LOADS—

JOINT		——— FORCE ———		——— MOMENT ———
		X FORCE	Y FORCE	Z MOMENT
1	GLOBAL	-155. 2789154	-1239. 9987793	0. 0000000
42	GLOBAL	-154. 7206726	1239. 9987793	0. 0

RESULTANT JOINT FREE JOINTS LOADS—

JOINT		——— FORCE ———		——— MOMENT ———
		X FORCE	Y FORCE	Z MOMENT
2	GLOBAL	8. 2999916	-0. 0000000	0. 0000000
3	GLOBAL	8. 2999916	0. 0	-0. 0000000
4	GLOBAL	8. 2999916	-0. 0000000	0. 0000000
5	GLOBAL	8. 2999916	0. 0	-0. 0000000
6	GLOBAL	8. 2999916	-0. 0000000	-0. 0000000
7	GLOBAL	8. 2999916	0. 0000000	0. 0000000
8	GLOBAL	8. 2999916	-0. 0000000	0. 0000000
9	GLOBAL	8. 2999916	-0. 0000000	-0. 0000000
10	GLOBAL	8. 2999916	0. 0000000	0. 0000000
11	GLOBAL	8. 2999916	0. 0000000	0. 0000000
12	GLOBAL	8. 2999954	-0. 0000000	-0. 0000000
13	GLOBAL	8. 2999916	-0. 0000000	-0. 0000000

14	GLOBAL	8. 2999916	0. 0000000	-0. 0000000
15	GLOBAL	8. 2999916	0. 0000000	-0. 0000000
16	GLOBAL	8. 2999954	-0. 0000000	0. 0000000
17	GLOBAL	8. 2999916	-0. 0000000	-0. 0000000
18	GLOBAL	8. 2999916	0. 0000000	-0. 0000000
19	GLOBAL	8. 2999954	-0. 0000000	-0. 0000000
20	GLOBAL	8. 2999954	-0. 0000000	-0. 0000000
21	GLOBAL	4. 1499910	0. 0000000	-0. 0000000
22	GLOBAL	3. 5999985	-0. 0000000	-0. 0000000
23	GLOBAL	7. 1999950	-0. 0000000	-0. 0000000
24	GLOBAL	7. 1999950	0. 0000000	0. 0000000
25	GLOBAL	7. 1999950	-0. 0000000	-0. 0000000
26	GLOBAL	7. 1999950	-0. 0000000	-0. 0000000
27	GLOBAL	7. 1999950	0. 0000000	-0. 0000000
28	GLOBAL	7. 1999950	0. 0000000	0. 0000000
29	GLOBAL	7. 1999950	-0. 0000000	-0. 0000000
30	GLOBAL	7. 1999950	0. 0000000	-0. 0000000
31	GLOBAL	7. 1999950	0. 0000000	-0. 0000000
32	GLOBAL	7. 1999950	0. 0000000	-0. 0000000
33	GLOBAL	7. 1999950	-0. 0000000	-0. 0000000
34	GLOBAL	7. 1999950	0. 0000000	-0. 0000000
35	GLOBAL	7. 1999950	0. 0000000	-0. 0000000
36	GLOBAL	7. 1999950	0. 0000000	-0. 0000000
37	GLOBAL	7. 1999950	0. 0000000	-0. 0000000
38	GLOBAL	7. 1999950	0. 0000000	-0. 0000000
39	GLOBAL	7. 1999950	0. 0000000	-0. 0000000
40	GLOBAL	7. 1999950	0. 0000000	-0. 0000000
41	GLOBAL	7. 1999950	0. 0000000	0. 0000000

RESULTANT JOINT DISPLACEMENTS— SUPPORTS

JOINT		DISPLACEMENT		ROTATION
		X DISP.	Y DISP.	Z ROT.
1	GLOBAL	0. 0	0. 0	-0. 0641951
42	GLOBAL	0. 0	0. 0	-0. 0641924

RESULTANT JOINT DISPLACEMENTS— FREE JOINTS

JOINT		DISPLACEMENT		ROTATION
		X DISP. (in.)	Y DISP. (in.)	Z ROT.
2	GLOBAL	7. 6129608	0. 1539309 (at 1st floor)	-0. 0302132
3	GLOBAL	12. 2231073	0. 2873114 (at 2nd floor)	-0. 0230708
4	GLOBAL	16. 1472321	0. 4056596 (at 3rd floor)	-0. 0212658
5	GLOBAL	19. 8427734	0. 5106993 (at 4th floor)	-0. 0204768
6	GLOBAL	23. 3937225	0. 6033912 (at 5th floor)	-0. 0198385
7	GLOBAL	26. 8095093	0. 6845427 (at 6th floor)	-0. 0191803
8	GLOBAL	30. 0850830	0. 7549301 (at 7th floor)	-0. 0184721
9	GLOBAL	33. 2133026	0. 8153228 (at 8th floor)	-0. 0177146
10	GLOBAL	36. 1885529	0. 8664861 (at 9th floor)	-0. 0169284
11	GLOBAL	39. 0133667	0. 9091690 (at 10th floor)	-0. 0162059
12	GLOBAL	42. 5987091	1. 0081034 (at 11th floor)	-0. 0154995
13	GLOBAL	45. 9477844	1. 0873804 (at 12th floor)	-0. 0148288
14	GLOBAL	49. 0570221	1. 1492062 (at 13th floor)	-0. 0140729

15	GLOBAL	51. 9142761	1. 1957645 (at 14th floor)	-0. 0132332
16	GLOBAL	54. 5082703	1. 2292356 (at 15th floor)	-0. 0123208
17	GLOBAL	56. 8294067	1. 2518015 (at 16th floor)	-0. 0113478
18	GLOBAL	58. 8698425	1. 2656431 (at 17th floor)	-0. 0103265
19	GLOBAL	60. 6236267	1. 2729416 (at 18th floor)	-0. 0092709
20	GLOBAL	62. 0882721 (max.)	1. 2758722 (at 19th floor)	-0. 0082146
21	GLOBAL	63. 2839966 (max.)	1. 2765656 (at roof level)	-0. 0074410
22	GLOBAL	63. 2838440	-1. 2765656 (at roof level)	-0. 0074417
23	GLOBAL	62. 0879669	-1. 2758722 (at 19th floor)	-0. 0082151
24	GLOBAL	60. 6233215	-1. 2729416 (at 18th floor)	-0. 0092708
25	GLOBAL	58. 8695374	-1. 2656431 (at 17th floor)	-0. 0103266
26	GLOBAL	56. 8291016	-1. 2518015 (at 16th floor)	-0. 0113478
27	GLOBAL	54. 5079651	-1. 2292356 (at 15th floor)	-0. 0123208
28	GLOBAL	51. 9139709	-1. 1957645 (at 14th floor)	-0. 0132332
29	GLOBAL	49. 0567169	-1. 1492062 (at 13th floor)	-0. 0140729
30	GLOBAL	45. 9474792	-1. 0873804 (at 12th floor)	-0. 0148288
31	GLOBAL	42. 5984039	-1. 0081034 (at 11th floor)	-0. 0154995
32	GLOBAL	39. 0130615	-0. 9091690 (at 10th floor)	-0. 0162059
33	GLOBAL	36. 1882477	-0. 8664861 (at 9th floor)	-0. 0169284
34	GLOBAL	33. 2129974	-0. 8153228 (at 8th floor)	-0. 0177146
35	GLOBAL	30. 0847778	-0. 7549301 (at 7th floor)	-0. 0184721
36	GLOBAL	26. 8092041	-0. 6845427 (at 6th floor)	-0. 0191803
37	GLOBAL	23. 3934174	-0. 6033912 (at 5th floor)	-0. 0198385
38	GLOBAL	19. 8424683	-0. 5106993 (at 4th floor)	-0. 0204768
39	GLOBAL	16. 1469269	-0. 4056596 (at 3rd floor)	-0. 0212658
40	GLOBAL	12. 2228003	-0. 2873114 (at 2nd floor)	-0. 0230710
41	GLOBAL	7. 6126604	-0. 1539309 (at 1st floor)	-0. 0302124

OUTPUT DECIMALS 5 (Input command)
SECTION FRACTIONAL DS 0 0. 1 (Input command)
LIST SECTION FORCE MEMBER 1 TO 60 (Input command)

SECOND OUTPUT

```
****************************************
*  RESULTS OF LATEST ANALYSES  *
****************************************
```

PROBLEM—MGM   TITLE—WIND ON 20-STORY FRAME

ACTIVE UNITS INCH KIP RAD DEGF SEC

ACTIVE STRUCTURE TYPE PLANE FRAME

ACTIVE COORDINATE AXES X Y

INTERNAL MEMBER RESULTS

MEMBER SECTION FORCES

MEMBER 1 (1st floor column—Left)

LOADING 1

DISTANCE	——— FORCE ———		——— MOMENT ———
FROM START	AXIAL	Y SHEAR	Z BENDING
0. 0 FR	1239. 99878	-151. 12898	0. 0
0. 100	1239. 99878	-151. 12898	2176. 25781
0. 200	1239. 99878	-151. 12898	4352. 51172
0. 300	1239. 99878	-151. 12898	6528. 76953
0. 400	1239. 99878	-151. 12898	8705. 02344
0. 500	1239. 99878	-151. 12898	10881. 28125
0. 600	1239. 99878	-151. 12898	13057. 53906
0. 700	1239. 99878	-151. 12898	15233. 79297
0. 800	1239. 99878	-151. 12898	17410. 04297
0. 900	1239. 99878	-151. 12898	19586. 30078
1. 000	1239. 99878	-151. 12898	21762. 55469

MEMBER 2 (2nd floor column)

LOADING 1

DISTANCE	FORCE		MOMENT
FROM START	AXIAL	Y SHEAR	Z BENDING
0. 0 FR	1074. 45386	-143. 37105	-8035. 63672
0. 100	1074. 45386	-143. 37105	-5971. 09375
0. 200	1074. 45386	-143. 37105	-3906. 55469
0. 300	1074. 45386	-143. 37105	-1842. 01172
0. 400	1074. 45386	-143. 37105	222. 52966
0. 500	1074. 45386	-143. 37105	2287. 07178
0. 600	1074. 45386	-143. 37105	4351. 61328
0. 700	1074. 45386	-143. 37105	6416. 15234
0. 800	1074. 45386	-143. 37105	8480. 69531
0. 900	1074. 45386	-143. 37105	10545. 23828
1. 000	1074. 45386	-143. 37105	12609. 78125

MEMBER 3 (3rd floor column)

LOADING 1

DISTANCE	FORCE		MOMENT
FROM START	AXIAL	Y SHEAR	Z BENDING
0. 0 FR	953. 36060	-135. 62561	-9187. 05859
0. 100	953. 36060	-135. 62561	-7234. 05078
0. 200	953. 36060	-135. 62561	-5281. 04297
0. 300	953. 36060	-135. 62561	-3328. 03760
0. 400	953. 36060	-135. 62561	-1375. 02979
0. 500	953. 36060	-135. 62561	577. 97754
0. 600	953. 36060	-135. 62561	2530. 98560
0. 700	953. 36060	-135. 62561	4483. 99219
0. 800	953. 36060	-135. 62561	6437. 00000
0. 900	953. 36060	-135. 62561	8390. 00781
1. 000	953. 36060	-135. 62561	10343. 01562

MEMBER 4 (4th floor column)

LOADING 1

DISTANCE	FORCE		MOMENT
FROM START	AXIAL	Y SHEAR	Z BENDING
0. 0 FR	846. 15283	-127. 87480	-8954. 33203
0. 100	846. 15283	-127. 87480	-7112. 93359
0. 200	846. 15283	-127. 87480	-5271. 53906
0. 300	846. 15283	-127. 87480	-3430. 14453
0. 400	846. 15283	-127. 87480	-1588. 74780
0. 500	846. 15283	-127. 87480	252. 64754
0. 600	846. 15283	-127. 87480	2094. 04370
0. 700	846. 15283	-127. 87480	3935. 43970
0. 800	846. 15283	-127. 87480	5776. 83594
0. 900	846. 15283	-127. 87480	7618. 23047
1. 000	846. 15283	-127. 87480	9459. 62891

MEMBER 5 (5th floor column)

LOADING 1

DISTANCE	FORCE		MOMENT
FROM START	AXIAL	Y SHEAR	Z BENDING
0. 0 FR	746. 68506	-120. 12492	-8444. 58594
0. 100	746. 68506	-120. 12492	-6714. 78516
0. 200	746. 68506	-120. 12492	-4984. 98828
0. 300	746. 68506	-120. 12492	-3255. 19165
0. 400	746. 68506	-120. 12492	-1525. 39380
0. 500	746. 68506	-120. 12492	204. 40379
0. 600	746. 68506	-120. 12492	1934. 20166
0. 700	746. 68506	-120. 12492	3663. 99951
0. 800	746. 68506	-120. 12492	5393. 79687
0. 900	746. 68506	-120. 12492	7123. 59375
1. 000	746. 68506	-120. 12492	8853. 39062

MEMBER 6 (6th floor column)

LOADING 1

DISTANCE	FORCE		MOMENT
FROM START	AXIAL	Y SHEAR	Z BENDING
0. 0 FR	653. 72021	-112. 37492	-7880. 23047
0. 100	653. 72021	-112. 37492	-6262. 03125
0. 200	653. 72021	-112. 37492	-4643. 83203
0. 300	653. 72021	-112. 37492	-3025. 63672
0. 400	653. 72021	-112. 37492	-1407. 43872
0. 500	653. 72021	-112. 37492	210. 75903
0. 600	653. 72021	-112. 37492	1828. 95679
0. 700	653. 72021	-112. 37492	3447. 15454
0. 800	653. 72021	-112. 37492	5065. 35156
0. 900	653. 72021	-112. 37492	6683. 55078
1. 000	653. 72021	-112. 37492	8301. 75000

MEMBER 7 (7th floor column)

LOADING 1

DISTANCE	FORCE		MOMENT
FROM START	AXIAL	Y SHEAR	Z BENDING
0. 0 FR	567. 00952	-104. 62492	-7306. 20703
0. 100	567. 00952	-104. 62492	-5799. 60937
0. 200	567. 00952	-104. 62492	-4293. 01172
0. 300	567. 00952	-104. 62492	-2786. 41479
0. 400	567. 00952	-104. 62492	-1279. 81689
0. 500	567. 00952	-104. 62492	226. 78122
0. 600	567. 00952	-104. 62492	1733. 37866
0. 700	567. 00952	-104. 62492	3239. 97754
0. 800	567. 00952	-104. 62492	4746. 57422
0. 900	567. 00952	-104. 62492	6253. 17187
1. 000	567. 00952	-104. 62492	7759. 76953

MEMBER 8 (8th floor column)

LOADING 1

DISTANCE	FORCE		MOMENT
FROM START	AXIAL	Y SHEAR	Z BENDING
0.0 FR	486.49707	-96.87492	-6732.43359
0.100	486.49707	-96.87492	-5337.43359
0.200	486.49707	-96.87492	-3942.43750
0.300	486.49707	-96.87492	-2547.43970
0.400	486.49707	-96.87492	-1152.44092
0.500	486.49707	-96.87492	242.55647
0.600	486.49707	-96.87492	1637.55469
0.700	486.49707	-96.87492	3032.55273
0.800	486.49707	-96.87492	4427.55078
0.900	486.49707	-96.87492	5822.54687
1.000	486.49707	-96.87492	7217.54687

MEMBER 9 (9th floor column)

LOADING 1

DISTANCE	FORCE		MOMENT
FROM START	AXIAL	Y SHEAR	Z BENDING
0.0 FR	412.14819	-89.12492	-6165.25781
0.100	412.14819	-89.12492	-4881.85937
0.200	412.14819	-89.12492	-3598.46167
0.300	412.14819	-89.12492	-2315.06372
0.400	412.14819	-89.12492	-1031.66553
0.500	412.14819	-89.12492	251.73273
0.600	412.14819	-89.12492	1535.13086
0.700	412.14819	-89.12492	2818.52856
0.800	412.14819	-89.12492	4101.92578
0.900	412.14819	-89.12492	5385.32422
1.000	412.14819	-89.12492	6668.72266

MEMBER 10 (10th floor column)

LOADING 1

DISTANCE	FORCE		MOMENT
FROM START	AXIAL	Y SHEAR	Z BENDING
0. 0 FR	343. 83496	-81. 37492	-5627. 63672
0. 100	343. 83496	-81. 37492	-4455. 83594
0. 200	343. 83496	-81. 37492	-3284. 04077
0. 300	343. 83496	-81. 37492	-2112. 24268
0. 400	343. 83496	-81. 37492	-940. 44434
0. 500	343. 83496	-81. 37492	231. 35397
0. 600	343. 83496	-81. 37492	1403. 15186
0. 700	343. 83496	-81. 37492	2574. 95068
0. 800	343. 83496	-81. 37492	3746. 74854
0. 900	343. 83496	-81. 37492	4918. 54687
1. 000	343. 83496	-81. 37492	6090. 34375

MEMBER 11 (11th floor column)

LOADING 1

DISTANCE	FORCE		MOMENT
FROM START	AXIAL	Y SHEAR	Z BENDING
0. 0 FR	280. 93311	-73. 62492	-5231. 99609
0. 100	280. 93311	-73. 62492	-4171. 79297
0. 200	280. 93311	-73. 62492	-3111. 59863
0. 300	280. 93311	-73. 62492	-2051. 40063
0. 400	280. 93311	-73. 62492	-991. 20215
0. 500	280. 93311	-73. 62492	68. 99623
0. 600	280. 93311	-73. 62492	1129. 19385
0. 700	280. 93311	-73. 62492	2189. 39282
0. 800	280. 93311	-73. 62492	3249. 59155
0. 900	280. 93311	-73. 62492	4309. 78906
1. 000	280. 93311	-73. 62492	5369. 98828

MEMBER 12 (12th floor column)

LOADING 1

DISTANCE	FORCE		MOMENT
FROM START	AXIAL	Y SHEAR	Z BENDING
0. 0 FR	225. 11378	-65. 87492	-4677. 48047
0. 100	225. 11378	-65. 87492	-3728. 88159
0. 200	225. 11378	-65. 87492	-2780. 28369
0. 300	225. 11378	-65. 87492	-1831. 68481
0. 400	225. 11378	-65. 87492	-883. 08691
0. 500	225. 11378	-65. 87492	65. 51147
0. 600	225. 11378	-65. 87492	1014. 11011
0. 700	225. 11378	-65. 87492	1962. 70776
0. 800	225. 11378	-65. 87492	2911. 30664
0. 900	225. 11378	-65. 87492	3859. 90552
1. 000	225. 11378	-65. 87492	4808. 5000

MEMBER 13 (13th floor column)

LOADING 1

DISTANCE	FORCE		MOMENT
FROM START	AXIAL	Y SHEAR	Z BENDING
0. 0 FR	175. 56004	-58. 12497	-4111. 16406
0. 100	175. 56004	-58. 12497	-3274. 16577
0. 200	175. 56004	-58. 12497	-2437. 16772
0. 300	175. 56004	-58. 12497	-1600. 16870
0. 400	175. 56004	-58. 12497	0763. 16943
0. 500	175. 56004	-58. 12497	73. 82961
0. 600	175. 56004	-58. 12497	910. 82886
0. 700	175. 56004	-58. 12497	1747. 82764
0. 800	175. 56004	-58. 12497	2584. 82666
0. 900	175. 56004	-58. 12497	3421. 82666
1. 000	175. 56004	-58. 12497	4258. 82422

MEMBER 14 (14th floor column)

LOADING 1

DISTANCE	FORCE		MOMENT
FROM START	AXIAL	Y SHEAR	Z BENDING
0. 0 FR	132. 20555	-50. 37497	-3544. 97461
0. 100	132. 20555	-50. 37497	-2819. 57471
0. 200	132. 20555	-50. 37497	-2094. 17578
0. 300	132. 20555	-50. 37497	-1368. 77686
0. 400	132. 20555	-50. 37497	-643. 37769
0. 500	132. 20555	-50. 37497	82. 02155
0. 600	132. 20555	-50. 37497	807. 42090
0. 700	132. 20555	-50. 37497	1532. 81982
0. 800	132. 20555	-50. 37497	2258. 21875
0. 900	132. 20555	-50. 37497	2983. 61865
1. 000	132. 20555	-50. 37497	3709. 01758

MEMBER 15 (15th floor column)

LOADING 1

DISTANCE	FORCE		MOMENT
FROM START	AXIAL	Y SHEAR	Z BENDING
0. 0 FR	95. 04498	-42. 62498	-2979. 88159
0. 100	95. 04498	-42. 62498	-2366. 08179
0. 200	95. 04498	-42. 62498	-1752. 28271
0. 300	95. 04498	-42. 62498	-1138. 48291
0. 400	95. 04498	-42. 62498	-524. 68384
0. 500	95. 04498	-42. 62498	89. 11549
0. 600	95. 04498	-42. 62498	702. 91479
0. 700	95. 04498	-42. 62498	1316. 71387
0. 800	95. 04498	-42. 62498	1930. 51367
0. 900	95. 04498	-42. 62498	2544. 31274
1. 000	95. 04498	-42. 62498	3158. 11255

MEMBER 16 (16th floor column)

LOADING 1

DISTANCE	FORCE		MOMENT
FROM START	AXIAL	Y SHEAR	Z BENDING
0. 0 FR	64. 07791	-34. 87497	-2415. 96582
0. 100	64. 07791	-34. 87497	-1913. 76562
0. 200	64. 07791	-34. 87497	-1411. 56689
0. 300	64. 07791	-34. 87497	-909. 36792
0. 400	64. 07791	-34. 87497	-407. 16846
0. 500	64. 07791	-34. 87497	95. 03099
0. 600	64. 07791	-34. 87497	597. 23047
0. 700	64. 07791	-34. 87497	1099. 42969
0. 800	64. 07791	-34. 87497	1601. 62866
0. 900	64. 07791	-34. 87497	2103. 82861
1. 000	64. 07791	-34. 87497	2606. 02759

MEMBER 17 (17th floor column)

LOADING 1

DISTANCE	FORCE		MOMENT
FROM START	AXIAL	Y SHEAR	Z BENDING
0. 0 FR	39. 30414	-27. 12500	-1853. 24878
0. 100	39. 30414	-27. 12500	-1462. 64771
0. 200	39. 30414	-27. 12500	-1072. 04883
0. 300	39. 30414	-27. 12500	-681. 44897
0. 400	39. 30414	-27. 12500	-290. 84912
0. 500	39. 30414	-27. 12500	99. 75055
0. 600	39. 30414	-27. 12500	490. 35034
0. 700	39. 30414	-27. 12500	880. 95020
0. 800	39. 30414	-27. 12500	1271. 54980
0. 900	39. 30414	-27. 12500	1662. 14966
1. 000	39. 30414	-27. 12500	2052. 74976

MEMBER 18 (18th floor column)

DISTANCE	——— FORCE ———		——— MOMENT ———
LOADING 1			
0. 0 FR	20. 72282	-19. 37486	-1291. 88281
0. 100	20. 72282	-19. 37486	-1012. 88501
0. 200	20. 72282	-19. 37486	-733. 88696
0. 300	20. 72282	-19. 37486	-454. 88892
0. 400	20. 72282	-19. 37486	-175. 89098
0. 500	20. 72282	-19. 37486	103. 10712
0. 600	20. 72282	-19. 37486	382. 10498
0. 700	20. 72282	-19. 37486	661. 10327
0. 800	20. 72282	-19. 37486	940. 10132
0. 900	20. 72282	-19. 37486	1219. 09888
1. 000	20. 72282	-19. 37486	1498. 09668

MEMBER 19 (19th floor column)

LOADING 1

DISTANCE	——— FORCE ———		——— MOMENT ———
FROM START	AXIAL	Y SHEAR	Z BENDING
0. 0 FR	8. 32301	-11. 62568	-733. 87988
0. 100	8. 32301	-11. 62568	-566. 46973
0. 200	8. 32301	-11. 62568	-399. 06030
0. 300	8. 32301	-11. 62568	-231. 65060
0. 400	8. 32301	-11. 62568	-64. 24089
0. 500	8. 32301	-11. 62568	103. 16879
0. 600	8. 32301	-11. 62568	270. 57837
0. 700	8. 32301	-11. 62568	437. 98828
0. 800	8. 32301	-11. 62568	605. 39795
0. 900	8. 32301	-11. 62568	772. 80786
1. 000	8. 32301	-11. 62568	940. 21753

MEMBER 20 (20th floor column)

LOADING 1

DISTANCE	FORCE		MOMENT
FROM START	AXIAL	Y SHEAR	Z BENDING
0. 0 FR	1. 96974	-3. 87312	-203. 29817
0. 100	1. 96974	-3. 87312	-147. 52524
0. 200	1. 96974	-3. 87312	-91. 75243
0. 300	1. 96974	-3. 87312	-35. 97954
0. 400	1. 96974	-3. 87312	19. 79332
0. 500	1. 96974	-3. 87312	75. 56618
0. 600	1. 96974	-3. 87312	131. 33905
0. 700	1. 96974	-3. 87312	187. 11192
0. 800	1. 96974	-3. 87312	242. 88478
0. 900	1. 96974	-3. 87312	298. 65771
1. 000	1. 96974	-3. 87312	354. 43042

MEMBER 21 (1st floor column—Right)

LOADING 1

DISTANCE	FORCE		MOMENT
FROM START	AXIAL	Y SHEAR	Z BENDING
0. 0 FR	-1239. 99878	-151. 12073	0. 00400
0. 100	-1239. 99878	-151. 12073	2176. 14380
0. 200	-1239. 99878	-151. 12073	4352. 27734
0. 300	-1239. 99878	-151. 12073	6528. 41406
0. 400	-1239. 99878	-151. 12073	8704. 55078
0. 500	-1239. 99878	-151. 12073	10880. 69141
0. 600	-1239. 99878	-151. 12073	13056. 82812
0. 700	-1239. 99878	-151. 12073	15232. 96484
0. 800	-1239. 99878	-151. 12073	17409. 10156
0. 900	-1239. 99878	-151. 12073	19585. 23047
1. 000	-1239. 99878	-151. 12073	21761. 37109

MEMBER 22 (2nd floor column)

LOADING 1

DISTANCE	——— FORCE ———		——— MOMENT ———
FROM START	AXIAL	Y SHEAR	Z BENDING
0. 0 FR	-1074. 45386	-143. 37874	-8036. 52344
0. 100	-1074. 45386	-143. 37874	-5971. 86719
0. 200	-1074. 45386	-143. 37874	-3907. 21851
0. 300	-1074. 45386	-143. 37874	-1842. 56567
0. 400	-1074. 45386	-143. 37874	222. 08641
0. 500	-1074. 45386	-143. 37874	2286. 73877
0. 600	-1074. 45386	-143. 37874	4351. 39062
0. 700	-1074. 45386	-143. 37874	6416. 04297
0. 800	-1074. 45386	-143. 37874	8480. 69531
0. 900	-1074. 45386	-143. 37874	10545. 34766
1. 000	-1074. 45386	-143. 37874	12610. 00391

MEMBER 23 (3rd floor column)

LOADING 1

DISTANCE	——— FORCE ———		——— MOMENT ———
FROM START	AXIAL	Y SHEAR	Z BENDING
0. 0 FR	-953. 36060	-135. 62411	-9186. 88672
0. 100	-953. 36060	-135. 62411	-7233. 89844
0. 200	-953. 36060	-135. 62411	-5280. 91406
0. 300	-953. 36060	-135. 62411	-3327. 92871
0. 400	-953. 36060	-135. 62411	-1374. 94189
0. 500	-953. 36060	-135. 62411	578. 04370
0. 600	-953. 36060	-135. 62411	2531. 02979
0. 700	-953. 36060	-135. 62411	4484. 01562
0. 800	-953. 36060	-135. 62411	6437. 00000
0. 900	-953. 36060	-135. 62411	8389. 98828
1. 000	-953. 36060	-135. 62411	10342. 97266

MEMBER 24 (4th floor column)

LOADING 1

DISTANCE	FORCE		MOMENT
FROM START	AXIAL	Y SHEAR	Z BENDING
0.0 FR	-846.15283	-127.87498	-8954.35937
0.100	-846.15283	-127.87498	-7112.95703
0.200	-846.15283	-127.87498	-5271.55859
0.300	-846.15283	-127.87498	-3430.16260
0.400	-846.15283	-127.87498	-1588.76367
0.500	-846.15283	-127.87498	252.63504
0.600	-846.15283	-127.87498	2094.03369
0.700	-846.15283	-127.87498	3935.43262
0.800	-846.15283	-127.87498	5776.82812
0.900	-846.15283	-127.87498	7618.23047
1.000	-846.15283	-127.87498	9459.62891

MEMBER 25 (5th floor column)

LOADING 1

DISTANCE	FORCE		MOMENT
FROM START	AXIAL	Y SHEAR	Z BENDING
0.0 FR	-746.68506	-120.12486	-8444.58203
0.100	-746.68506	-120.12486	-6714.78125
0.200	-746.68506	-120.12486	-4984.98828
0.300	-746.68506	-120.12486	-3255.19165
0.400	-746.68506	-120.12486	-1525.39478
0.500	-746.68506	-120.12486	204.40230
0.600	-746.68506	-120.12486	1934.19873
0.700	-746.68506	-120.12486	3663.99658
0.800	-746.68506	-120.12486	5393.79297
0.900	-746.68506	-120.12486	7123.58984
1.000	-746.68506	-120.12486	8853.38672

MEMBER 26 (6th floor column)

LOADING 1

DISTANCE	—— FORCE ——		—— MOMENT ——
FROM START	AXIAL	Y SHEAR	Z BENDING
0. 0 FR	-653. 72021	-112. 37486	-7880. 22656
0. 100	-653. 72021	-112. 37486	-6262. 02734
0. 200	-653. 72021	-112. 37486	-4643. 82812
0. 300	-653. 72021	-112. 37486	-3025. 63477
0. 400	-653. 72021	-112. 37486	-1407. 43677
0. 500	-653. 72021	-112. 37486	210. 75954
0. 600	-653. 72021	-112. 37486	1828. 95679
0. 700	-653. 72021	-112. 37486	3447. 15356
0. 800	-653. 72021	-112. 37486	5065. 35156
0. 900	-653. 72021	-112. 37486	6683. 54687
1. 000	-653. 72021	-112. 37486	8301. 74609

MEMBER 27 (7th floor column)

LOADING 1

DISTANCE	—— FORCE ——		—— MOMENT ——
FROM START	AXIAL	Y SHEAR	Z BENDING
0. 0 FR	-567. 00952	-104. 62486	-7306. 20312
0. 100	-567. 00952	-104. 62486	-5799. 60547
0. 200	-567. 00952	-104. 62486	-4293. 00781
0. 300	-567. 00952	-104. 62486	-2786. 41260
0. 400	-567. 00952	-104. 62486	-1279. 81470
0. 500	-567. 00952	-104. 62486	226. 78172
0. 600	-567. 00952	-104. 62486	1733. 37866
0. 700	-567. 00952	-104. 62486	3239. 97559
0. 800	-567. 00952	-104. 62486	4746. 57031
0. 900	-567. 00952	-104. 62486	6253. 16797
1. 000	-567. 00952	-104. 62486	7759. 76562

MEMBER 28 (8th floor column)

LOADING 1

DISTANCE	FORCE		MOMENT
FROM START	AXIAL	Y SHEAR	Z BENDING
0. 0 FR	-486. 49707	-96. 87486	-6732. 42969
0. 100	-486. 49707	-96. 87486	-5337. 42969
0. 200	-486. 49707	-96. 87486	-3942. 43457
0. 300	-486. 49707	-96. 87486	-2547. 43774
0. 400	-486. 49707	-96. 87486	-1152. 43970
0. 500	-486. 49707	-96. 87486	242. 55698
0. 600	-486. 49707	-96. 87486	1637. 55371
0. 700	-486. 49707	-96. 87486	3032. 55176
0. 800	-486. 49707	-96. 87486	4427. 54687
0. 900	-486. 49707	-96. 87486	5822. 54297
1. 000	-486. 49707	-96. 87486	7217. 54297

MEMBER 29 (9th floor column)

LOADING 1

DISTANCE	FORCE		MOMENT
FROM START	AXIAL	Y SHEAR	Z BENDING
0. 0 FR	-412. 14819	-89. 12486	-6165. 25000
0. 100	-412. 14819	-89. 12486	-4881. 85156
0. 200	-412. 14819	-89. 12486	-3598. 45752
0. 300	-412. 14819	-89. 12486	-2315. 06079
0. 400	-412. 14819	-89. 12486	-1031. 66309
0. 500	-412. 14819	-89. 12486	251. 73422
0. 600	-412. 14819	-89. 12486	1535. 13086
0. 700	-412. 14819	-89. 12486	2818. 52856
0. 800	-412. 14819	-89. 12486	4101. 92578
0. 900	-412. 14819	-89. 12486	5385. 32031
1. 000	-412. 14819	-89. 12486	6668. 71875

MEMBER 30 (10th floor column)

LOADING 1

DISTANCE	FORCE		MOMENT
FROM START	AXIAL	Y SHEAR	Z BENDING
0. 0 FR	-343. 83496	-81. 37486	-5627. 63281
0. 100	-343. 83496	-81. 37486	-4455. 83203
0. 200	-343. 83496	-81. 37486	-3284. 03760
0. 300	-343. 83496	-81. 37486	-2112. 23975
0. 400	-343. 83496	-81. 37486	-940. 44287
0. 500	-343. 83496	-81. 37486	231. 35448
0. 600	-343. 83496	-81. 37486	1403. 15186
0. 700	-343. 83496	-81. 37486	2574. 94873
0. 800	-343. 83496	-81. 37486	3746. 74658
0. 900	-343. 83496	-81. 37486	4918. 54297
1. 000	-343. 83496	-81. 37486	6090. 33984

MEMBER 31 (11th floor column)

LOADING 1

DISTANCE	FORCE		MOMENT
FROM START	AXIAL	Y SHEAR	Z BENDING
0. 0 FR	-280. 93311	-73. 62492	-5231. 99609
0. 100	-280. 93311	-73. 62492	-4171. 79687
0. 200	-280. 93311	-73. 62492	-3111. 60059
0. 300	-280. 93311	-73. 62492	-2051. 40283
0. 400	-280. 93311	-73. 62492	-991. 20410
0. 500	-280. 93311	-73. 62492	68. 99423
0. 600	-280. 93311	-73. 62492	1129. 19189
0. 700	-280. 93311	-73. 62492	2189. 39062
0. 800	-280. 93311	-73. 62492	3249. 58960
0. 900	-280. 93311	-73. 62492	4309. 78516
1. 000	-280. 93311	-73. 62492	5369. 98437

MEMBER 32 (12th floor column)

LOADING 1

DISTANCE	FORCE		MOMENT
FROM START	AXIAL	Y SHEAR	Z BENDING
0. 0 FR	-225. 11378	-65. 87492	-4677. 48047
0. 100	-225. 11378	-65. 87492	-3728. 88354
0. 200	-225. 11378	-65. 87492	-2780. 28564
0. 300	-225. 11378	-65. 87492	-1831. 68677
0. 400	-225. 11378	-65. 87492	-883. 08911
0. 500	-225. 11378	-65. 87492	65. 50948
0. 600	-225. 11378	-65. 87492	1014. 10791
0. 700	-225. 11378	-65. 87492	1962. 70581
0. 800	-225. 11378	-65. 87492	2911. 30469
0. 900	-225. 11378	-65. 87492	3859. 90356
1. 000	-225. 11378	-65. 87492	4808. 50000

MEMBER 33 (13th floor column)

LOADING 1

DISTANCE	FORCE		MOMENT
FROM START	AXIAL	Y SHEAR	Z BENDING
0. 0 FR	-175. 56004	-58. 12494	-4111. 16406
0. 100	-175. 56004	-58. 12494	-3274. 16455
0. 200	-175. 56004	-58. 12494	-2437. 16675
0. 300	-175. 56004	-58. 12494	-1600. 16772
0. 400	-175. 56004	-58. 12494	-763. 16919
0. 500	-175. 56004	-58. 12494	73. 82954
0. 600	-175. 56004	-58. 12494	910. 82837
0. 700	-175. 56004	-58. 12494	1747. 82666
0. 800	-175. 56004	-58. 12494	2584. 82568
0. 900	-175. 56004	-58. 12494	3421. 82471
1. 000	-175. 56004	-58. 12494	4258. 82031

MEMBER 34 (14th floor column)

LOADING 1

DISTANCE	FORCE		MOMENT
FROM START	AXIAL	Y SHEAR	Z BENDING
0. 0 FR	-132. 20555	-50. 37495	-3544. 97363
0. 100	-132. 20555	-50. 37495	-2819. 57373
0. 200	-132. 20555	-50. 37495	-2094. 17578
0. 300	-132. 20555	-50. 37495	-1368. 77588
0. 400	-132. 20555	-50. 37495	-643. 37769
0. 500	-132. 20555	-50. 37495	82. 02124
0. 600	-132. 20555	-50. 37495	807. 42017
0. 700	-132. 20555	-50. 37495	1532. 81885
0. 800	-132. 20555	-50. 37495	2258. 21777
0. 900	-132. 20555	-50. 37495	2983. 61670
1. 000	-132. 20555	-50. 37495	3709. 01562

MEMBER 35 (15th floor column)

LOADING 1

DISTANCE	FORCE		MOMENT
FROM START	AXIAL	Y SHEAR	Z BENDING
0. 0 FR	-95. 04498	-42. 62495	-2979. 87964
0. 100	-95. 04498	-42. 62495	-2366. 08081
0. 200	-95. 04498	-42. 62495	-1752. 28174
0. 300	-95. 04498	-42. 62495	-1138. 48291
0. 400	-95. 04498	-42. 62495	-524. 68384
0. 500	-95. 04498	-42. 62495	89. 11517
0. 600	-95. 04498	-42. 62495	702. 91431
0. 700	-95. 04498	-42. 62495	1316. 71289
0. 800	-95. 04498	-42. 62495	1930. 51172
0. 900	-95. 04498	-42. 62495	2544. 31177
1. 000	-95. 04498	-42. 62495	3158. 11060

MEMBER 36 (16th floor column)

LOADING 1

DISTANCE	FORCE		MOMENT
FROM START	AXIAL	Y SHEAR	Z BENDING
0. 0 FR	-64. 07791	-34. 87497	-2415. 96460
0. 100	-64. 07791	-34. 87497	-1913. 76465
0. 200	-64. 07791	-34. 87497	-1411. 56567
0. 300	-64. 07791	-34. 87497	-909. 36670
0. 400	-64. 07791	-34. 87497	-407. 16724
0. 500	-64. 07791	-34. 87497	95. 03186
0. 600	-64. 07791	-34. 87497	597. 23120
0. 700	-64. 07791	-34. 87497	1099. 42969
0. 800	-64. 07791	-34. 87497	1601. 62988
0. 900	-64. 07791	-34. 87497	2103. 82861
1. 000	-64. 07791	-34. 87497	2606. 02759

MEMBER 37 (17th floor column)

LOADING 1

DISTANCE	FORCE		MOMENT
FROM START	AXIAL	Y SHEAR	Z BENDING
0. 0 FR	-39. 30414	-27. 12495	-1853. 24683
0. 100	-39. 30414	-27. 12495	-1462. 64673
0. 200	-39. 30414	-27. 12495	-1072. 04785
0. 300	-39. 30414	-27. 12495	-681. 44922
0. 400	-39. 30414	-27. 12495	-290. 85010
0. 500	-39. 30414	-27. 12495	99. 74892
0. 600	-39. 30414	-27. 12495	490. 34814
0. 700	-39. 30414	-27. 12495	880. 94727
0. 800	-39. 30414	-27. 12495	1271. 54590
0. 900	-39. 30414	-27. 12495	1662. 14575
1. 000	-39. 30414	-27. 12495	2052. 74463

MEMBER 38 (18th floor column)

LOADING 1

DISTANCE	FORCE		MOMENT
FROM START	AXIAL	Y SHEAR	Z BENDING
0. 0 FR	-20. 72282	-19. 37509	-1291. 88989
0. 100	-20. 72282	-19. 37509	-1012. 88867
0. 200	-20. 72282	-19. 37509	-733. 88794
0. 300	-20. 72282	-19. 37509	-454. 88647
0. 400	-20. 72282	-19. 37509	-175. 88554
0. 500	-20. 72282	-19. 37509	103. 11562
0. 600	-20. 72282	-19. 37509	382. 11670
0. 700	-20. 72282	-19. 37509	661. 11792
0. 800	-20. 72282	-19. 37509	940. 11914
0. 900	-20. 72282	-19. 37509	1219. 11987
1. 000	-20. 72282	-19. 37509	1498. 12085

MEMBER 39 (19th floor column)

LOADING 1

DISTANCE	FORCE		MOMENT
FROM START	AXIAL	Y SHEAR	Z BENDING
0. 0 FR	-8. 32301	-11. 62430	-733. 83008
0. 100	-8. 32301	-11. 62430	-566. 43994
0. 200	-8. 32301	-11. 62430	-399. 05029
0. 300	-8. 32301	-11. 62430	-231. 66072
0. 400	-8. 32301	-11. 62430	-64. 27095
0. 500	-8. 32301	-11. 62430	103. 11880
0. 600	-8. 32301	-11. 62430	270. 50854
0. 700	-8. 32301	-11. 62430	437. 89819
0. 800	-8. 32301	-11. 62430	605. 28809
0. 900	-8. 32301	-11. 62430	772. 67773
1. 000	-8. 32301	-11. 62430	940. 06763

MEMBER 40 (20th floor column)

LOADING 1

DISTANCE	FORCE		MOMENT
FROM START	AXIAL	Y SHEAR	Z BENDING
0. 0 FR	-1. 96974	-3. 87688	-203. 59604
0. 100	-1. 96974	-3. 87688	-147. 76898
0. 200	-1. 96974	-3. 87688	-91. 94205
0. 300	-1. 96974	-3. 87688	-36. 11508
0. 400	-1. 96974	-3. 87688	19. 71190
0. 500	-1. 96974	-3. 87688	75. 53886
0. 600	-1. 96974	-3. 87688	131. 36586
0. 700	-1. 96974	-3. 87688	187. 19286
0. 800	-1. 96974	-3. 87688	243. 01985
0. 900	-1. 96974	-3. 87688	298. 84668
1. 000	-1. 96974	-3. 87688	354. 67383

MEMBER 41 (2nd floor girder)

LOADING 1

DISTANCE	FORCE		MOMENT
FROM START	AXIAL	Y SHEAR	Z BENDING
0. 0 FR	-0. 54205	165. 54480	29798. 20312
0. 100	-0. 54205	165. 54480	23838. 58984
0. 200	-0. 54205	165. 54480	17878. 97266
0. 300	-0. 54205	165. 54480	11919. 37891
0. 400	-0. 54205	165. 54480	5959. 76562
0. 500	-0. 54205	165. 54480	0. 15888
0. 600	-0. 54205	165. 54480	-5959. 44922
0. 700	-0. 54205	165. 54480	-11919. 05859
0. 800	-0. 54205	165. 54480	-17878. 63672
0. 900	-0. 54205	165. 54480	-23838. 21875
1. 000	-0. 54205	165. 54480	-29797. 78906

MEMBER 42 (3rd floor girder)

LOADING 1

DISTANCE	—— FORCE ——		—— MOMENT ——
FROM START	AXIAL	Y SHEAR	Z BENDING
0. 0 FR	-0. 55458	121. 09380	21796. 85937
0. 100	-0. 55458	121. 09380	17437. 48437
0. 200	-0. 55458	121. 09380	13078. 12109
0. 300	-0. 55458	121. 09380	8718. 74609
0. 400	-0. 55458	121. 09380	4359. 37109
0. 500	-0. 55458	121. 09380	-0. 00051
0. 600	-0. 55458	121. 09380	-4359. 37109
0. 700	-0. 55458	121. 09380	-8718. 75000
0. 800	-0. 55458	121. 09380	-13078. 10937
0. 900	-0. 55458	121. 09380	-17437. 45312
1. 000	-0. 55458	121. 09380	-21796. 79687

MEMBER 43 (4th floor girder)

LOADING 1

DISTANCE	—— FORCE ——		—— MOMENT ——
FROM START	AXIAL	Y SHEAR	Z BENDING
0. 0 FR	-0. 54917	107. 20749	19297. 33984
0. 100	-0. 54917	107. 20749	15437. 88281
0. 200	-0. 54917	107. 20749	11578. 41406
0. 300	-0. 54917	107. 20749	7718. 94922
0. 400	-0. 54917	107. 20749	3859. 45361
0. 500	-0. 54917	107. 20749	0. 01418
0. 600	-0. 54917	107. 20749	-3859. 45361
0. 700	-0. 54917	107. 20749	-7718. 91797
0. 800	-0. 54917	107. 20749	-11578. 37500
0. 900	-0. 54917	107. 20749	-15437. 81641
1. 000	-0. 54917	107. 20749	-19297. 26172

MEMBER 44 (5th floor girder)

LOADING 1

DISTANCE	FORCE		MOMENT
FROM START	AXIAL	Y SHEAR	Z BENDING
0.0 FR	-0.55013	99.46793	17904.22266
0.100	-0.55013	99.46793	14323.38281
0.200	-0.55013	99.46793	10742.53906
0.300	-0.55013	99.46793	7161.69922
0.400	-0.55013	99.46793	3580.85571
0.500	-0.55013	99.46793	0.01234
0.600	-0.55013	99.46793	-3580.83057
0.700	-0.55013	99.46793	-7161.67187
0.800	-0.55013	99.46793	-10742.50391
0.900	-0.55013	99.46793	-14323.32422
1.000	-0.55013	99.46793	-17904.14062

MEMBER 45 (6th floor girder)

LOADING 1

DISTANCE	FORCE		MOMENT
FROM START	AXIAL	Y SHEAR	Z BENDING
0.0 FR	-0.54998	92.96461	16733.63281
0.100	-0.54998	92.96461	13386.90234
0.200	-0.54998	92.96461	10040.17969
0.300	-0.54998	92.96461	6693.45703
0.400	-0.54998	92.96461	3346.73267
0.500	-0.54998	92.96461	0.00859
0.600	-0.54998	92.96461	-3346.71558
0.700	-0.54998	92.96461	-6693.43750
0.800	-0.54998	92.96461	-10040.15234
0.900	-0.54998	92.96461	-13386.85547
1.000	-0.54998	92.96461	-16733.55859

MEMBER 46 (7th floor girder)

LOADING 1

DISTANCE	FORCE		MOMENT
FROM START	AXIAL	Y SHEAR	Z BENDING
0. 0 FR	-0. 55000	86. 71092	15607. 96875
0. 100	-0. 55000	86. 71092	12486. 37500
0. 200	-0. 55000	86. 71092	9364. 78125
0. 300	-0. 55000	86. 71092	6243. 19141
0. 400	-0. 55000	86. 71092	3121. 60059
0. 500	-0. 55000	86. 71092	0. 00937
0. 600	-0. 55000	86. 71092	-3121. 58276
0. 700	-0. 55000	86. 71092	-6243. 17187
0. 800	-0. 55000	86. 71092	-9364. 75391
0. 900	-0. 55000	86. 71092	-12486. 32812
1. 000	-0. 55000	86. 71092	-15607. 89844

MEMBER 47 (8th floor girder)

LOADING 1

DISTANCE	FORCE		MOMENT
FROM START	AXIAL	Y SHEAR	Z BENDING
0. 0 FR	-0. 55000	80. 51230	14492. 21484
0. 100	-0. 55000	80. 51230	11593. 77344
0. 200	-0. 55000	80. 51230	8695. 33203
0. 300	-0. 55000	80. 51230	5796. 89062
0. 400	-0. 55000	80. 51230	2898. 45068
0. 500	-0. 55000	80. 51230	0. 00939
0. 600	-0. 55000	80. 51230	-2898. 43164
0. 700	-0. 55000	80. 51230	-5796. 87109
0. 800	-0. 55000	80. 51230	-8695. 30469
0. 900	-0. 55000	80. 51230	-11593. 72656
1. 000	-0. 55000	80. 51230	-14492. 15234

MEMBER 48 (9th floor girder)

LOADING 1

DISTANCE	FORCE		MOMENT
FROM START	AXIAL	Y SHEAR	Z BENDING
0. 0 FR	-0. 55000	74. 34892	13382. 80078
0. 100	-0. 55000	74. 34892	10706. 23828
0. 200	-0. 55000	74. 34892	8029. 67969
0. 300	-0. 55000	74. 34892	5353. 12109
0. 400	-0. 55000	74. 34892	2676. 56177
0. 500	-0. 55000	74. 34892	0. 00142
0. 600	-0. 55000	74. 34892	-2676. 55859
0. 700	-0. 55000	74. 34892	-5353. 11719
0. 800	-0. 55000	74. 34892	-8029. 66797
0. 900	-0. 55000	74. 34892	-10706. 21094
1. 000	-0. 55000	74. 34892	-13382. 75391

MEMBER 49 (10th floor girder)

LOADING 1

DISTANCE	FORCE		MOMENT
FROM START	AXIAL	Y SHEAR	Z BENDING
0. 0 FR	-0. 55000	68. 31317	12296. 37500
0. 100	-0. 55000	68. 31317	9837. 09766
0. 200	-0. 55000	68. 31317	7377. 82422
0. 300	-0. 55000	68. 31317	4918. 55469
0. 400	-0. 55000	68. 31317	2459. 28174
0. 500	-0. 55000	68. 31317	0. 00896
0. 600	-0. 55000	68. 31317	-2459. 26367
0. 700	-0. 55000	68. 31317	-4918. 53516
0. 800	-0. 55000	68. 31317	-7377. 80078
0. 900	-0. 55000	68. 31317	-9837. 05859
1. 000	-0. 55000	68. 31317	-12296. 31641

MEMBER 50 (11th floor girder)

LOADING 1

DISTANCE	——— FORCE ———		——— MOMENT ———
FROM START	AXIAL	Y SHEAR	Z BENDING
0. 0 FR	-0. 54999	62. 90193	11322. 35156
0. 100	-0. 54999	62. 90193	9057. 87891
0. 200	-0. 54999	62. 90193	6793. 41016
0. 300	-0. 54999	62. 90193	4528. 94531
0. 400	-0. 54999	62. 90193	2264. 47583
0. 500	-0. 54999	62. 90193	0. 00766
0. 600	-0. 54999	62. 90193	-2264. 46069
0. 700	-0. 54999	62. 90193	-4528. 92969
0. 800	-0. 54999	62. 90193	-6793. 39062
0. 900	-0. 54999	62. 90193	-9057. 84375
1. 000	-0. 54999	62. 90193	-11322. 29687

MEMBER 51 (12th floor girder)

LOADING 1

DISTANCE	——— FORCE ———		——— MOMENT ———
FROM START	AXIAL	Y SHEAR	Z BENDING
0. 0 FR	-0. 55000	55. 81932	10047. 48047
0. 100	-0. 55000	55. 81932	8037. 98437
0. 200	-0. 55000	55. 81932	6028. 48828
0. 300	-0. 55000	55. 81932	4018. 99658
0. 400	-0. 55000	55. 81932	2009. 50171
0. 500	-0. 55000	55. 81932	0. 00702
0. 600	-0. 55000	55. 81932	-2009. 48779
0. 700	-0. 55000	55. 81932	-4018. 98267
0. 800	-0. 55000	55. 81932	-6028. 46875
0. 900	-0. 55000	55. 81932	-8037. 94922
1. 000	-0. 55000	55. 81932	-10047. 43359

MEMBER 52 (13th floor girder)

LOADING 1

DISTANCE	FORCE		MOMENT
FROM START	AXIAL	Y SHEAR	Z BENDING
0. 0 FR	-0. 55000	49. 55374	8919. 67187
0. 100	-0. 55000	49. 55374	7135. 73828
0. 200	-0. 55000	49. 55374	5351. 80469
0. 300	-0. 55000	49. 55374	3567. 87354
0. 400	-0. 55000	49. 55374	1783. 93970
0. 500	-0. 55000	49. 55374	0. 00608
0. 600	-0. 55000	49. 55374	-1783. 92773
0. 700	-0. 55000	49. 55374	-3567. 86157
0. 800	-0. 55000	49. 55374	-5351. 78906
0. 900	-0. 55000	49. 55374	-7135. 71094
1. 000	-0. 55000	49. 55374	-8919. 63281

MEMBER 53 (14th floor girder)

LOADING 1

DISTANCE	FORCE		MOMENT
FROM START	AXIAL	Y SHEAR	Z BENDING
0. 0 FR	-0. 55000	43. 35446	7803. 80469
0. 100	-0. 55000	43. 35446	6243. 04297
0. 200	-0. 55000	43. 35446	4682. 28125
0. 300	-0. 55000	43. 35446	3121. 52563
0. 400	-0. 55000	43. 35446	1560. 76465
0. 500	-0. 55000	43. 35446	0. 00568
0. 600	-0. 55000	43. 35446	-1560. 75366
0. 700	-0. 55000	43. 35446	-3121. 51367
0. 800	-0. 55000	43. 35446	-4682. 26562
0. 900	-0. 55000	43. 35446	-6243. 01562
1. 000	-0. 55000	43. 35446	-7803. 76562

MEMBER 54 (15th floor girder)

LOADING 1

DISTANCE FROM START	FORCE AXIAL	FORCE Y SHEAR	MOMENT Z BENDING
0. 0 FR	-0. 55000	37. 16055	6688. 90234
0. 100	-0. 55000	37. 16055	5351. 12109
0. 200	-0. 55000	37. 16055	4013. 34253
0. 300	-0. 55000	37. 16055	2675. 56372
0. 400	-0. 55000	37. 16055	1337. 78369
0. 500	-0. 55000	37. 16055	0. 00449
0. 600	-0. 55000	37. 16055	-1337. 77490
0. 700	-0. 55000	37. 16055	-2675. 55469
0. 800	-0. 55000	37. 16055	-4013. 32959
0. 900	-0. 55000	37. 16055	-5351. 10156
1. 000	-0. 55000	37. 16055	-6688. 87109

MEMBER 55 (16th floor girder)

LOADING 1

DISTANCE FROM START	FORCE AXIAL	FORCE Y SHEAR	MOMENT Z BENDING
0. 0 FR	-0. 54999	30. 96710	5574. 07812
0. 100	-0. 54999	30. 96710	4459. 26172
0. 200	-0. 54999	30. 96710	3344. 44873
0. 300	-0. 54999	30. 96710	2229. 63379
0. 400	-0. 54999	30. 96710	1114. 81885
0. 500	-0. 54999	30. 96710	0. 00346
0. 600	-0. 54999	30. 96710	-1114. 81177
0. 700	-0. 54999	30. 96710	-2229. 62769
0. 800	-0. 54999	30. 96710	-3344. 43970
0. 900	-0. 54999	30. 96710	-4459. 24609
1. 000	-0. 54999	30. 96710	-5574. 05469

MEMBER 56 (17th floor girder)

LOADING 1

DISTANCE	FORCE		MOMENT
FROM START	AXIAL	Y SHEAR	Z BENDING
0. 0 FR	-0. 55002	24. 77376	4459. 27734
0. 100	-0. 55002	24. 77376	3567. 42358
0. 200	-0. 55002	24. 77376	2675. 56763
0. 300	-0. 55002	24. 77376	1783. 71387
0. 400	-0. 55002	24. 77376	891. 85840
0. 500	-0. 55002	24. 77376	0. 00341
0. 600	-0. 55002	24. 77376	-891. 85181
0. 700	-0. 55002	24. 77376	-1783. 70679
0. 800	-0. 55002	24. 77376	-2675. 55859
0. 900	-0. 55002	24. 77376	-3567. 40869
1. 000	-0. 55002	24. 77376	-4459. 25391

MEMBER 57 (18th floor girder)

LOADING 1

DISTANCE	FORCE		MOMENT
FROM START	AXIAL	Y SHEAR	Z BENDING
0. 0 FR	-0. 54987	18. 58131	3344. 63477
0. 100	-0. 54987	18. 58131	2675. 70776
0. 200	-0. 54987	18. 58131	2006. 78076
0. 300	-0. 54987	18. 58131	1337. 85376
0. 400	-0. 54987	18. 58131	668. 92725
0. 500	-0. 54987	18. 58131	0. 00037
0. 600	-0. 54987	18. 58131	-668. 92651
0. 700	-0. 54987	18. 58131	-1337. 85278
0. 800	-0. 54987	18. 58131	-2006. 77783
1. 000	-0. 54987	18. 58131	-3344. 62354

MEMBER 58 (19th floor girder)

LOADING 1

DISTANCE	FORCE		MOMENT
FROM START	AXIAL	Y SHEAR	Z BENDING
0. 0 FR	-0. 55080	12. 39981	2231. 97876
0. 100	-0. 55080	12. 39981	1785. 58569
0. 200	-0. 55080	12. 39981	1339. 19287
0. 300	-0. 55080	12. 39981	892. 80005
0. 400	-0. 55080	12. 39981	446. 40698
0. 500	-0. 55080	12. 39981	0. 01390
0. 600	-0. 55080	12. 39981	-446. 37915
0. 700	-0. 55080	12. 39981	-892. 77222
0. 800	-0. 55080	12. 39981	-1339. 16382
0. 900	-0. 55080	12. 39981	-1785. 55371
1. 000	-0. 55080	12. 39981	-2231. 94360

MEMBER 59 (20th floor girder)

LOADING 1

DISTANCE	FORCE		MOMENT
FROM START	AXIAL	Y SHEAR	Z BENDING
0. 0 FR	-0. 54743	6. 35328	1143. 51685
0. 100	-0. 54743	6. 35328	914. 79883
0. 200	-0. 54743	6. 35328	686. 08081
0. 300	-0. 54743	6. 35328	457. 36304
0. 400	-0. 54743	6. 35328	228. 64510
0. 500	-0. 54743	6. 35328	-0. 07289
0. 600	-0. 54743	6. 35328	-228. 79085
0. 700	-0. 54743	6. 35328	-457. 50879
0. 800	-0. 54743	6. 35328	-686. 22607
0. 900	-0. 54743	6. 35328	-914. 94263
1. 000	-0. 54743	6. 35328	-1143. 65869

MEMBER 60 (Roof girder)

LOADING 1

DISTANCE	FORCE		MOMENT
FROM START	AXIAL	Y SHEAR	Z BENDING
0. 0 FR	-0. 27688	1. 96974	354. 43066
0. 100	-0. 27688	1. 96974	283. 52026
0. 200	-0. 27688	1. 96974	212. 60979
0. 300	-0. 27688	1. 96974	141. 69942
0. 400	-0. 27688	1. 96974	70. 78893
0. 500	-0. 27688	1. 96974	-0. 12148
0. 600	-0. 27688	1. 96974	-71. 03192
0. 700	-0. 27688	1. 96974	-141. 94235
0. 800	-0. 27688	1. 96974	-212. 85260
0. 900	-0. 27688	1. 96974	-283. 76245
1. 000	-0. 27688	1. 96974	-354. 67236

REFERENCES

Ackeret, J. 1936. Der windruck auf schornsteine mit kreisquerschnitt. *Schweiz. Bauzeitung* 108: 25.

American Institute of Steel Construction. 1970. *Manual of Steel Construction.* American Concrete Institute.

ASCE. 1962. Wind Forces on Structures. Transactions of the American Society of Civil Engineers. Paper No. 3269. 126(II): 1124–1198.

Defense Civil Preparedness Agency. 1972. *Protecting Mobile Homes from High Winds.* Department of Defense, Washington, D.C.

———. 1973. *Multi-Protection Design.* TR-20 (Vol. 6), Preliminary Draft, Department of Defense, Washington, D.C., p. 41.

Den Hartog, J. P. 1947. *Mechanical Vibrations.* New York: McGraw-Hill Book Co.

Irminger, J. O. V. and Nøkkentved, C. 1930. *Wind Pressures on Buildings.* Copenhagen.

Nøkkentved, C. 1932, 1934. Wind pressure on buildings. Publ. Int. Assoc. Bridge and Structural Engineering, Zurich, Switzerland, Vol. I, p. 365, 1932; Vol. II, p. 257.

Yamada, M. and T. Goto. 1977. *Human Response to Tall Buildings.* Stroudsburg, Pennsylvania: Dowden, Hutchinson & Ross, pp. 58–71

4
Ventilation

From the Latin *ventus* ("wind"), *ventilation* is an air current, either through open spaces or through the interior of buildings. With regard to buildings, natural ventilation is the process that exchanges indoor air for fresh outdoor air without mechanical power. The only natural forces involved are the wind pressure and/or the temperature differential between the indoor and outdoor air.

A general understanding of the ventilation process includes a knowledge of:

- wind energy
- wind velocity with respect to direction and magnitude
- pressure coefficients
- inlet size in itself, with respect to the outlet
- number of inlets
- inlet shape
- inlet orientation with respect to the wind direction
- outlet size in itself, with respect to the inlet
- pattern of airflow within the interior of the building (interior layout)
- distribution of the air velocity through the building
- ventilation rate (ft^3/hr)

The energy required to sustain the air motion through a building is the kinetic energy of the wind: $\frac{1}{2}\rho V^2$, where ρ and V are the density and the velocity of the air, respectively. This is in reality a pressure, which can be expressed as: $q = 0.002558\ V^2$, where q is in psf and V in mph.

The magnitude of the wind velocity (wind speed) is proportional to the attainable energy; the direction of the wind with respect to the building is also essential in order to establish the wind pressure at the inlet and outlet, which in turn controls the energy that can be used for ventilation.

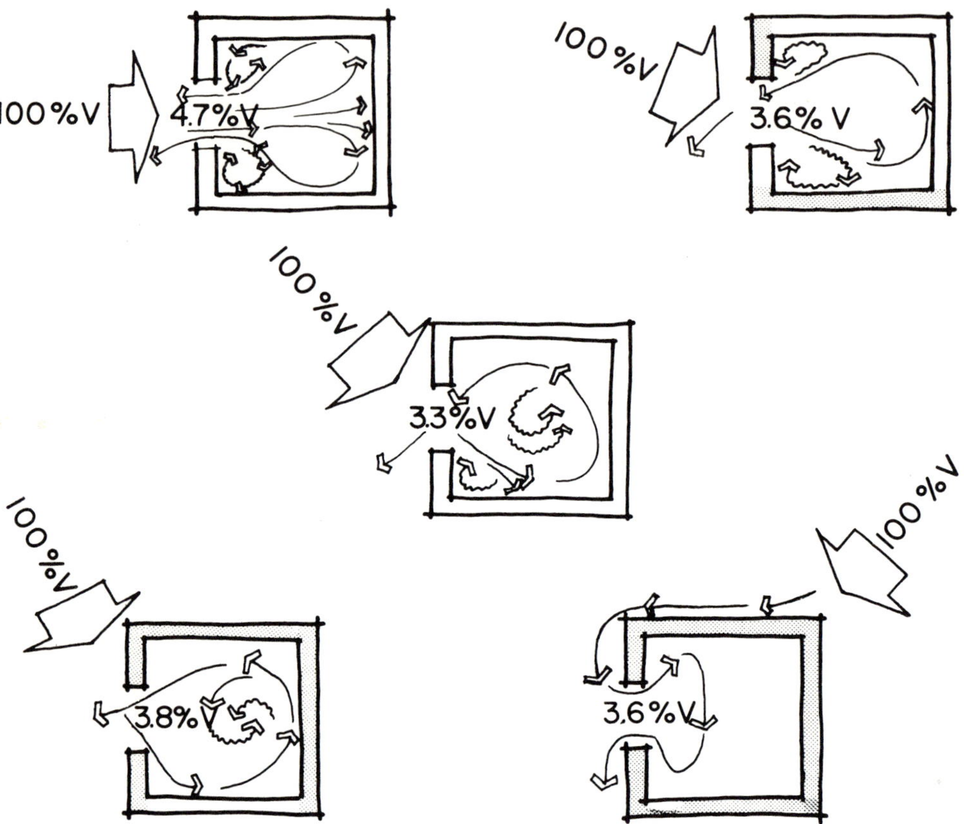

Figure 4-1. The average air velocity as a function of the free wind speed for various wind directions. Notice that the maximum speed value is attained when the wind is perpendicular to the aperture. Ventilation through a simple space having only one aperture that functions simultaneously an inlet and outlet is poor.

The ventilation process consists of air entering a building through one or more inlets (vents, windows, doors, etc.) and escaping through outlets. The driving force is the difference in pressure, Δp, between the wind pressure over the inlet and that at the outlet. The larger the difference, the greater the driving force. But the wind pressure varies from point to point over the building surface and is given by:

$$p = C_{pe}q$$

where C_{pe} is the pressure coefficient found experimentally through wind-tunnel tests. It is necessary therefore to establish the pressure coefficients either through tables for simple shapes or by direct wind-tunnel tests. Knowing the pressure coefficients, it is possible to locate the inlets and outlets so that their pressure difference is the largest possible.

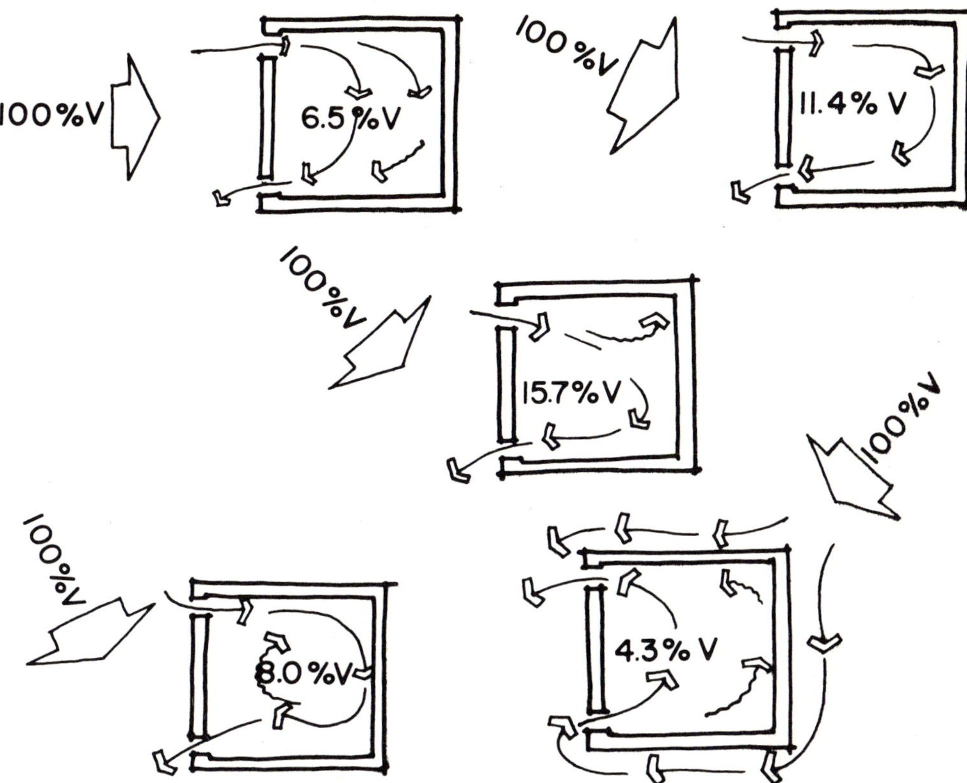

Figure 4-2. Ventilation through a simple space having two apertures on the same wall. Since one acts as an inlet while the other acts as an outlet, the ventilation is greatly improved in comparison to that for the single aperture. The average air speeds, expressed as a percentage of the free wind speed, show that as the wind direction changes from perpendicular to the aperture to a more inclined direction, the ventilation increases up to a point and then decreases. Note that the circulation is more effective when the apertures are on the windward rather than the leeward side.

The size of the inlet per se, independent of other factors, has particular effects in cases where there is only one aperture in the space (absence of cross-ventilation) that functions also as outlet. That is, the air enters at one part and exits at another part of the same aperture. When the wind is perpendicular to the inlet, tests have shown that there are little differences as the size varies; however, if the wind is oblique to the inlet, an increase in size raises considerably the average wind speed within the space. (See Figure 4-1.)

The size of the inlet with regard to that of the outlet has a significant influence on the air velocity inside the building, especially when inlet and outlet areas increase at the same rate. As their sizes expand, the interior air velocity rises sharply. When one increases but the other does not, the velocity does not rise. This, of course, is intuitive because an impediment, regardless of where it occurs, at the entrance or at the exit, slows down the flow.

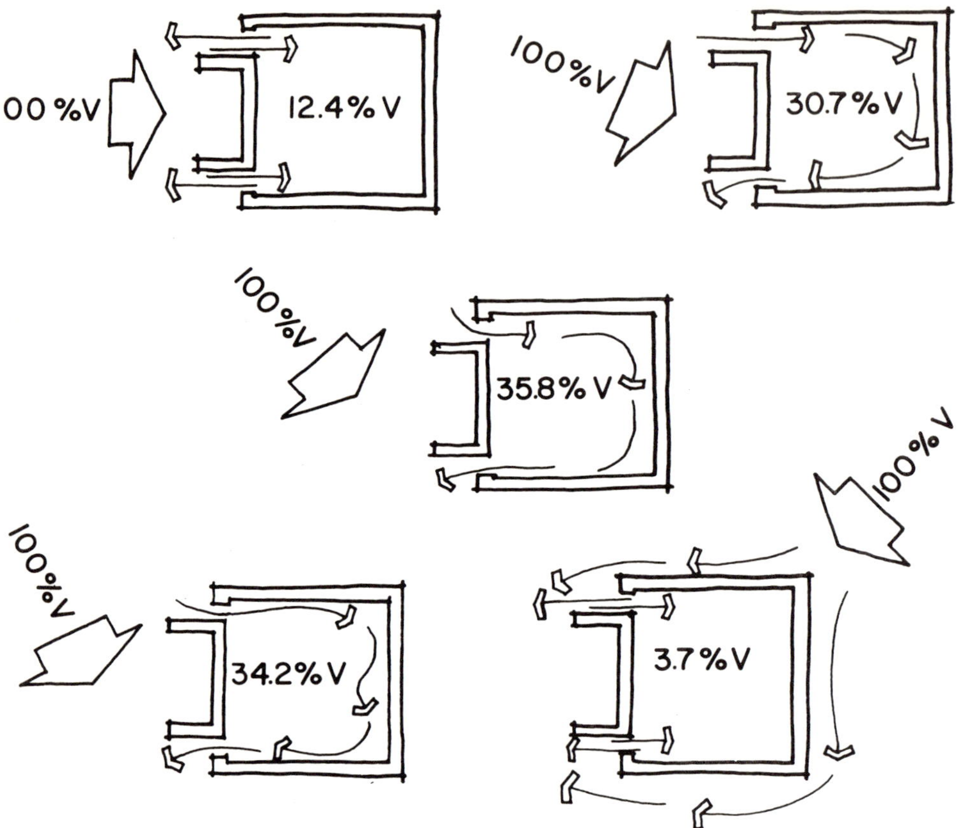

Figure 4-3. Ventilation through a simple space having two apertures with baffles. Notice the dramatic improvement in comparison to similar conditions but with no baffles. Note, however, that when the apertures are in the leeward side, the efficiency of the baffles drops.

When inlet and outlet have different sizes, there is not much significance to identifying which is the larger. In fact, the average air velocity is approximately the same in either case (a little higher when the outlet is larger than the inlet).

The number of apertures within a building space without cross-ventilation is very important because they can be subdivided into inlets and outlets or be partially both an inlet and outlet with a considerable effect on the interior velocity of the air. For instance, if we compare a given building space with one window to the same space with two windows (Figure 4-2), then, assuming equal window area in both cases, we see that the average interior velocity is higher in the latter case. Furthermore, this is accentuated when the wind is at an angle rather than perpendicular to the window.

The shape of the inlet can produce large differences in the interior velocity, especially when cross-ventilation is missing. Baffles installed on the side of windows, for instance, can increase the ventilation considerably (Figure 4-3). In the case of a space with two windows on the same wall and the wind blowing at an angle, the interior air velocity can rise up to three times the original value when baffles are installed.

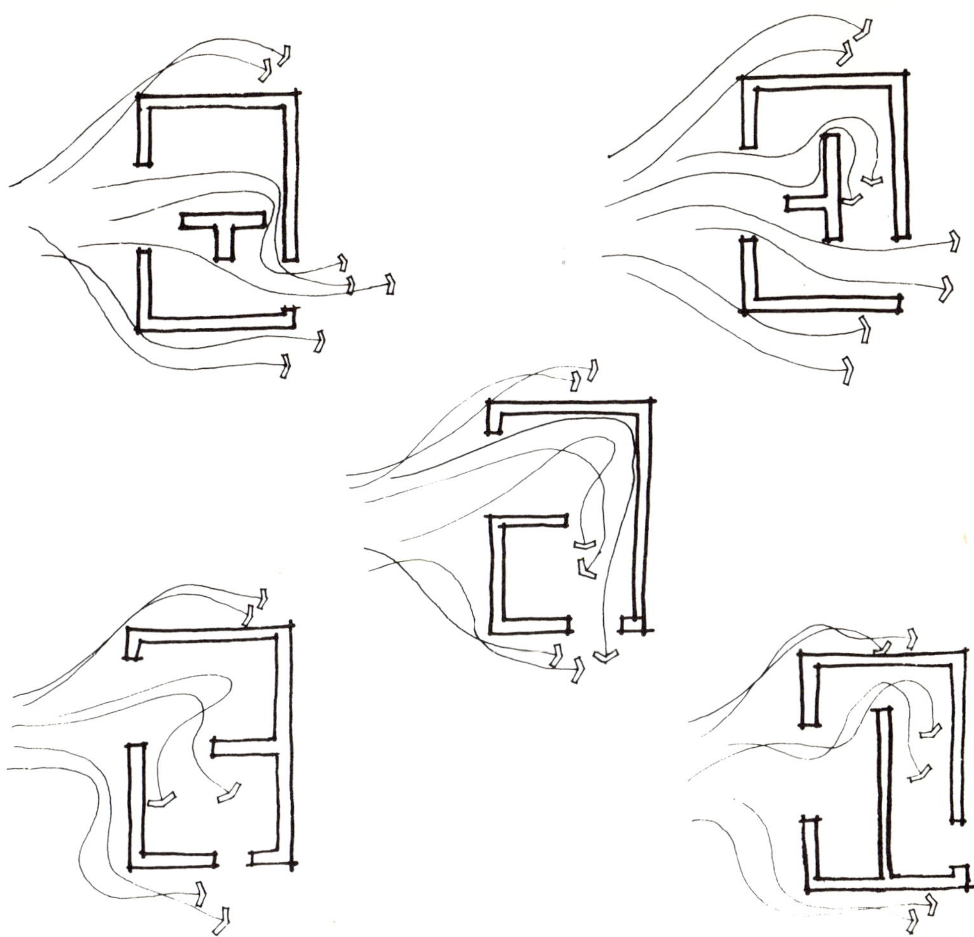

Figure 4-4. Effects of partitions and location of apertures on the patterns of interior air circulation. Higher velocities are attained when air particles traveling from inlet to outlet must follow longer paths than others.

Contrary to first impressions, when the wind is at an angle different than 90° with the surface of an inlet, the interior velocity is higher than it would be if the wind were perpendicular to the inlet. The increase can be up to 300% in some cases.

The size of the outlet has particular significance only when examined as a function of that of the inlet, as previously seen. Outlets, of course, should be located in zones of suction to obtain pressure differentials that maximize ventilation.

As the air enters the building, it circulates all the way through it to the outlet, following a pattern that is determined by the geometry of the interior space. The influence of the obstructions and the tortuosity of the pattern of the airflow are such that systematical generalizations are not meaningful. Each case should be analyzed experimentally. Examples of air-circulation patterns and velocities within various space configurations are illustrated in Figures 4-4 through 4-6. In the latter two, the findings, determined through model tests, are based on a square space of constant size where the relative location of inlets and outlets and the location of an intermediate partition, when used, are arranged in different ways.

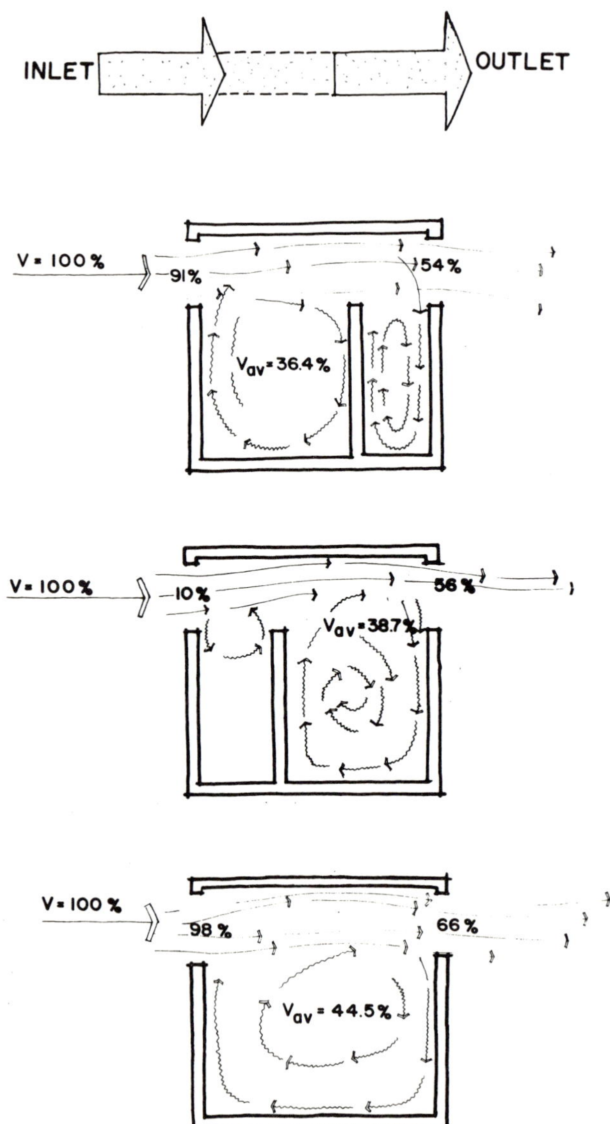

Figure 4-5. Patterns of airflow and average velocity values for the natural ventilation of a simple space in which various arrangements of an interior partition (when present) and different locations of inlet and outlet have been tested experimentally. Note that the pattern from inlet to outlet is an unobstructed line.

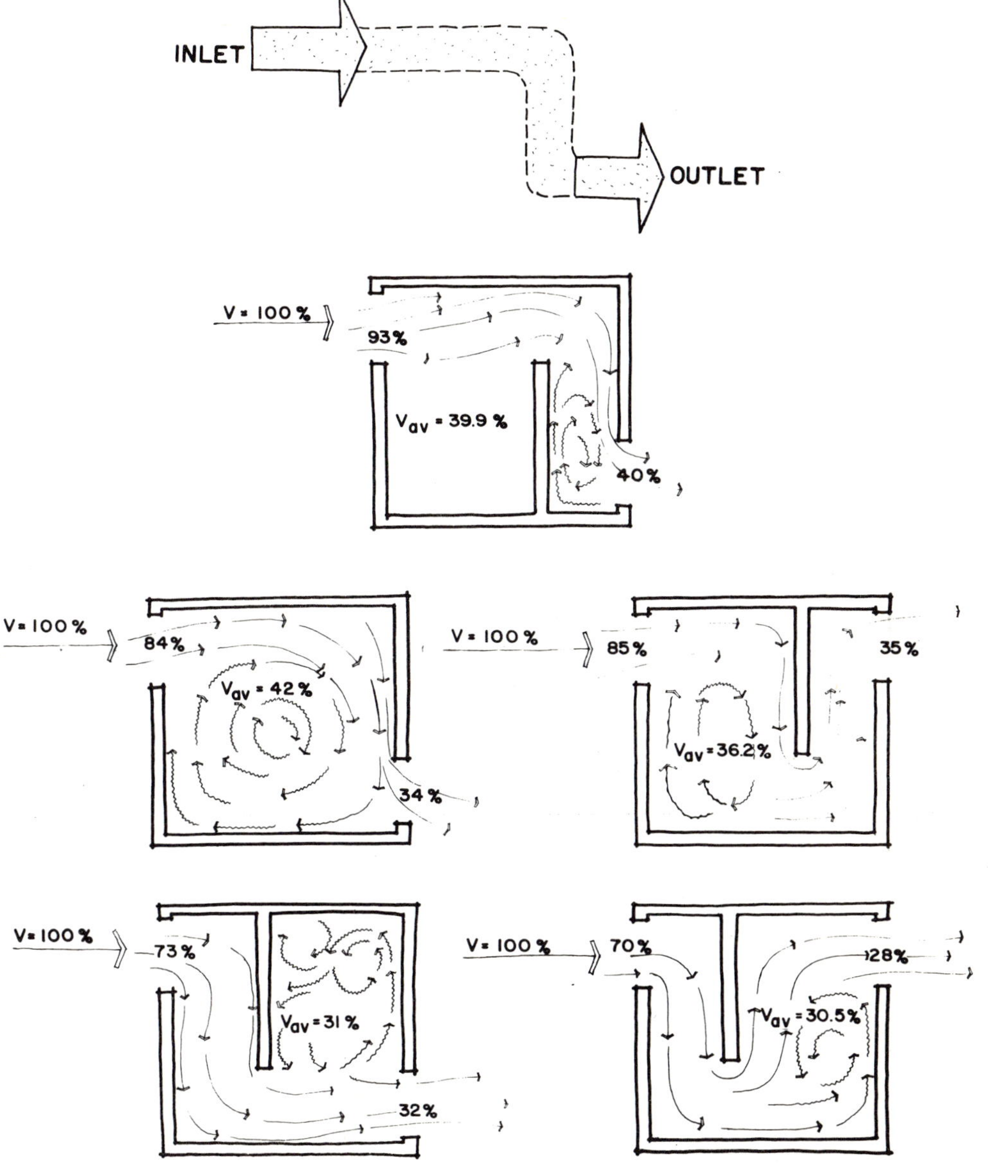

Figure 4-6. Patterns of airflow and average velocity values for the natural ventilation of a simple space in which various arrangements of an interior partition (when present) and different locations of inlet and outlet have been tested experimentally. Notice the bent pattern from inlet to outlet.

The distribution of the air velocity throughout a building varies considerably. It is convenient to show the variation in magnitude by expressing the interior velocity at any point as a function of the free velocity of the wind outside.

We can distinguish between individual local velocities and average overall speed. Both values are equally important. Local values indicate zones of poor circulation where, in some cases, air is still and is not changed. This, of course, is important to know. It is also necessary to find where the velocity is too high, since it can create annoying drafts or more severe problems. The average overall velocity, on the other hand, gives the required datum to calculate the rate of airflow for a given building.

The ventilation rate (volume of air/unit of time) is a quantitative value that depends on all the parameters previously discussed. Desirable values can be established for a specific use of the space, and, on this basis, all the design parameters, such as inlet and outlet sizes and locations, can be established. For an existing building under given conditions (wind direction, speed, sizes, and location of apertures, etc.), the ventilation rate is the parameter to be computed.

In countries such as the United States, where energy had been particularly inexpensive until recent times, mechanical ventilation was the accepted trend, which less-developed countries looked upon as an example to be followed. In the rest of the world, however, passive systems using wind were either the only alternative available or, as in several European countries, a choice very clearly established. For instance, the sealed fenestration for air-conditioning efficiency had been violently rejected in some European countries where centuries of experience in civilized living habits could not allow people to renounce the pleasure of fresh air ventilation through windows.

Of course, the energy crisis has halted the trend toward mechanical ventilation. A refinement in architectural design, which capitalizes upon the free force of wind and combines it with forced ventilation, is the model for the present. However, the advocacy of passive systems does not deny the validity of the sophisticated methodology of the industrialized societies.

FUNCTIONS OF NATURAL VENTILATION

A certain amount of natural ventilation takes place spontaneously by infiltration through doors, windows, and cracks. Ventilation, however, should be controlled to satisfy specific needs in a calculated amount. Air exchanges, measured in cubic feet per hour, cubic meters per hour, or in number of air exchanges per hour, are required for the proper comfort of the occupants in various amounts. The basic reasons for exchanging indoor air with fresh air are oxygen requirements, removal of odors and humidity, and cooling.

EFFICIENCY

The equation of continuity: $Q = AV$, indicates that the rate of airflow Q (ft^3/sec) is equal to the area of the inlet $A(ft^2)$ times the air velocity at the inlet V (ft/sec). Thus, for a given inlet area, the efficiency increases as the velocity increases.

The equation of continuity can be applied not only at the inlet but also at the outlet and at any other point in the building. It shows that the air velocity within the building is not uniform by any means, as mentioned previously. Annoying drafts and areas of low velocity can be observed in model studies (see Table 4-1). In judging the overall efficiency of ventilation in a building, it is more convenient to consider an average value of the air velocity indoors, V_i. The higher V_i is, the higher is the efficiency of ventilation. However, note that natural ventilation may not be available at all times when it is desired, particularly in the summer when ventilation would help lower the temperature and wind is not available. Thus, in building design, the natural ventilation should be complemented with a standby auxiliary mechanical system.

Table 4-1. Comparative interior air velocities for naturally ventilated spaces under different wind directions and with different number of apertures and locations.

CONDITIONS	WIDTH OF APERTURE / WIDTH OF WALL = 0.66		WIDTH OF APERTURE / WIDTH OF WALL = 1.0	
	V_{avg}	V_{max}	V_{avg}	V_{max}
Single aperture in windward wall, wind direction perpendicular	13%	18%	16%	20%
Single aperture in windward wall, wind direction at an angle	15%	33%	23%	36%
Single aperture in leeward wall, wind direction at an angle	17%	44%	17%	39%
Two apertures in leeward wall, wind direction at an angle	22%	56%	23%	50%
One aperture in windward wall, another in adjacent wall, wind direction perpendicular to inlets	45%	68%	51%	103%
One aperture in windward wall, another in adjacent wall, wind direction at an angle	37%	118%	40%	110%
One aperture in windward wall, another in leeward wall, wind direction perpendicular to inlet	35%	65%	37%	102%
One aperture in windward wall, another in leeward wall, wind direction at an angle	42%	83%	42%	94%

Note: Interior air velocities expressed as a percentage of the velocity of the free wind.

VERTICAL POSITIONING OF APERTURES

The height of inlets and outlets above the floor of the space to be ventilated greatly influences the effectiveness of ventilation. Also very important is the position of one with respect to the other.

When inlet and outlet are both high above the floor, the air current is always near the ceiling and totally above the body level without interaction with the lower

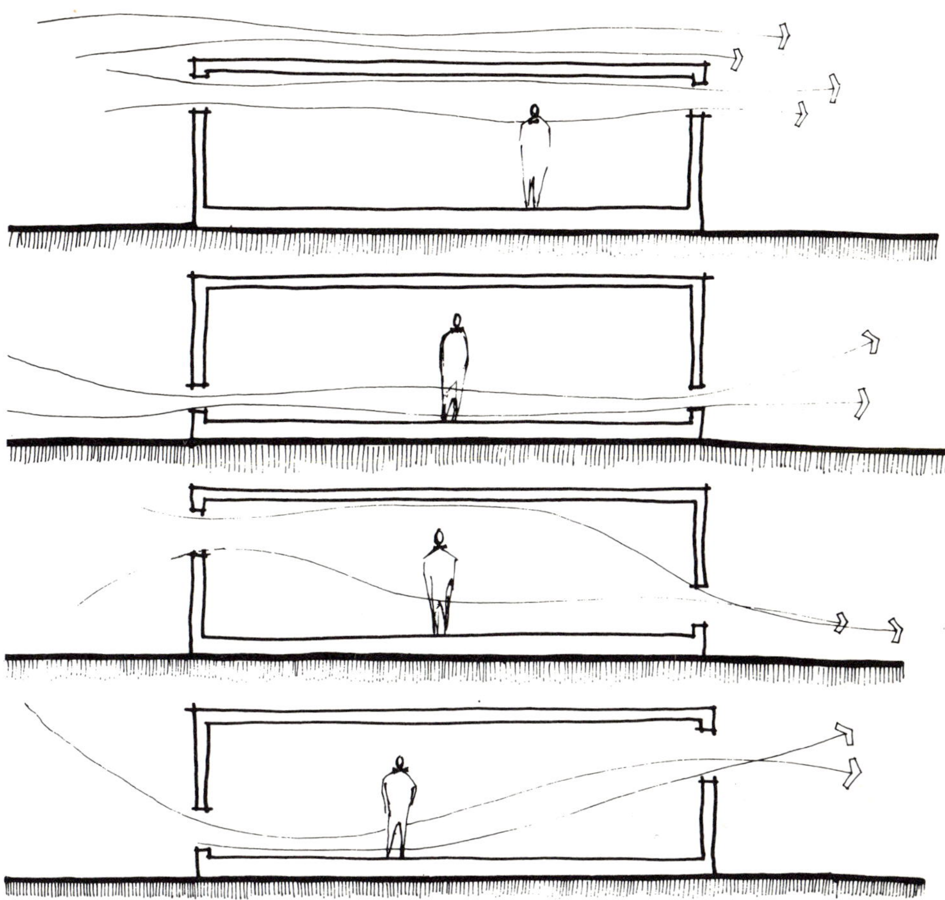

Figure 4-7. The effect of positioning the apertures at various heights above the floor influences the efficiency of the natural ventilation in a given space. The relative height of inlet and outlet with respect to each other is also relevant. *a,* Inlet and outlet both high. Airflow only near ceiling. No air current at body level. Good for removing hot air in warm season. Layers of still air at low levels. *b,* Inlet and outlet both low. Airflow only near floor. Layers of still air near ceiling. Not recommended for hot climates. *c,* Inlet higher than outlet. Good interaction of air layers. Currents at body level. Pocket of warm still air over the outlet. *d,* Inlet lower than outlet. Airflow upward pushing and pulling warm air out from ceiling.

zone. Warm air on the top would be removed, which is a positive aspect in the summer, but the ventilation does not include the whole space nor does it give direct relief to the occupants. When both apertures are low, all the air current is at body level, thus allowing direct contact with the occupants of the space, but leaving inactive the upper layers where heat accumulates. (See Figure 4-7.)

Locating the inlet at a higher level than the outlet expands the motion of air through a better vertical distribution of air velocities. An upper pocket of stationary air, however, is left in the space over the outlet.

The most efficient ventilation occurs when the inlet is low and the outlet is high above the floor. The air current crosses all the air layers, envelops the occupants, and rises, gradually pulling out the warm air near the ceiling.

These generalizations, however, can be easily offset by other conditions, and therefore have little influence in the solution of practical cases. On the other hand, they point out the factors that are involved in the design of architectural spaces for natural ventilation.

CROSS-VENTILATION

Cross-ventilation is a term that distinguishes a particular case in the natural ventilation process of buildings. We speak of cross-ventilation in a given space when the space is connected to the outside by inlets and outlets placed, respectively, in zones of positive pressure and zones of suction. For instance, one or more openings on the windward side (wall or roof side facing the wind) and one or more openings on the leeward side constitute a typical case of cross-ventilation. This implies, of course, a maximum pressure differential between inlets and outlets that guarantees an efficient circulation. Sometimes, however, the term is wrongly applied to conditions that seem to be as those just described, but which in reality are not. For instance, the presence of several openings in a given building may give the impression that some of them are in positive zones and others are in negative zones. In reality, however, all openings may be within zones of pressure of equal sign as the wind shifts. The value of the pressure and its respective sign can vary and lead the observer to erroneous conclusions.

Let us consider the effect of the size of the inlet and that of the outlet on the efficiency of ventilation for cross-ventilated spaces. If the inlet and outlet are aligned and both are perpendicular to the wind, increasing the outlet size while keeping the inlet size constant produces very little improvement in the circulation. If we let L be the width of the wall, as the inlet width is kept at $1/3L$ and the outlet is increased from $1/3L$ to $2/3L$ and to full width L, the average internal velocity is, respectively, $0.35V$, $0.39V$, and $0.44V$. Such changes are obviously minimal and negligible (see Figure 4-8a–c).

Similarly, if the inlet size is kept at $2/3L$ and the outlet size varies from $1/3L$ to $2/3L$ and to L, the interior air velocity varies, respectively, from $0.34V$, to $0.37V$, and $0.35V$ (see Figure 4-8d–f).

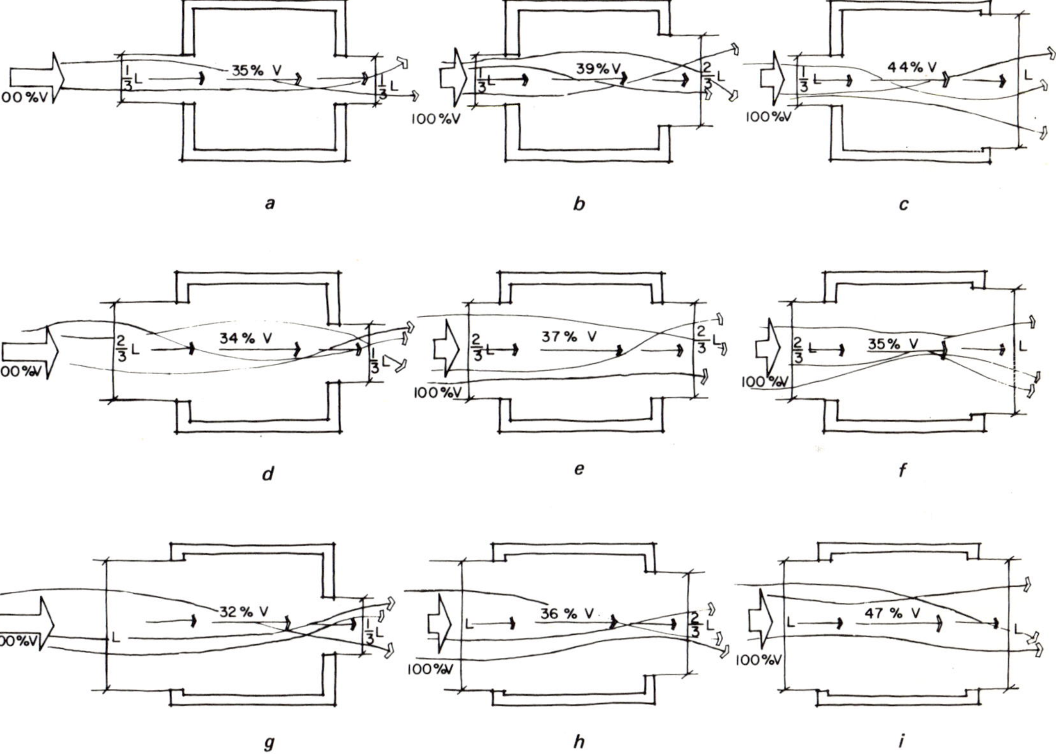

Figure 4-8. Effects of inlet and outlet sizes in cross-ventilated spaces: openings on opposite walls; wind perpendicular to inlet.

By the same token, if the inlet size is kept equal to the wall width L, and the outlet is increased from $1/3L$ to $2/3L$ and full width L, the internal air velocity increases, respectively, from $0.32V$ to $0.36V$ and $0.47V$ (see Figures 4-8g–i).

These show that the maximum efficiency is reached when both inlet and outlet have maximum size at the same time, and the minimum efficiency is attained when the inlet size is maximum and the outlet size is minimum.

The effect of the wind at 45° with respect to the inlet is to promote a more efficient ventilation in all cases. Keeping the inlet size to $1/3L$ and increasing the outlet size from $1/3L$ to $2/3L$ and to full width L, the interior velocity varies, respectively, from $0.42V$ to $0.40V$ and to $0.44V$ (see Figure 4-9a–c).

With a 45° wind orientation with respect to the inlet, and keeping the latter to a size of $2/3L$ but increasing the outlet size from $1/3L$ to $2/3L$ and to full width L, the interior air velocity varies, respectively, from $0.43V$ to $0.51V$ and to $0.59V$ (see Figures 4-9d–f).

Still with a 45° wind orientation with respect to the inlet, and keeping the inlet size equal to L, but increasing the outlet size from $1/3L$ to $2/3L$ and full width L, the interior air velocity varies, respectively, from $0.41V$ to $0.62V$ and to $0.65V$ (see Figures 4-9g–i).

Note that even in the case of an oblique wind direction, the most efficient ventilation is attained when inlet and outlet are maximum. Vice versa, minimum efficiency takes place when the inlet size is equal to $1/3L$ and the outlet size is $2/3L$.

If the outlets and inlets are open in adjacent walls, which puts them at 90° from each other, the effect of the relative size can be described as follows. Assuming that the wind is perpendicular to the inlet, it is observed that in varying the proportion between the inlet and the outlet, the maximum interior air velocity equal to $0.51V$ is reached in two cases:

1. when the inlet is $1/3L$ and the outlet is equal to L
2. when the inlet is $2/3L$ and the outlet is $1/3L$.

On the other hand, the minimal efficiency takes place when the inlet size is $1/3L$ and the outlet size is $2/3L$, and the interior air velocity is $0.39V$ (see Figure 4-10a–e).

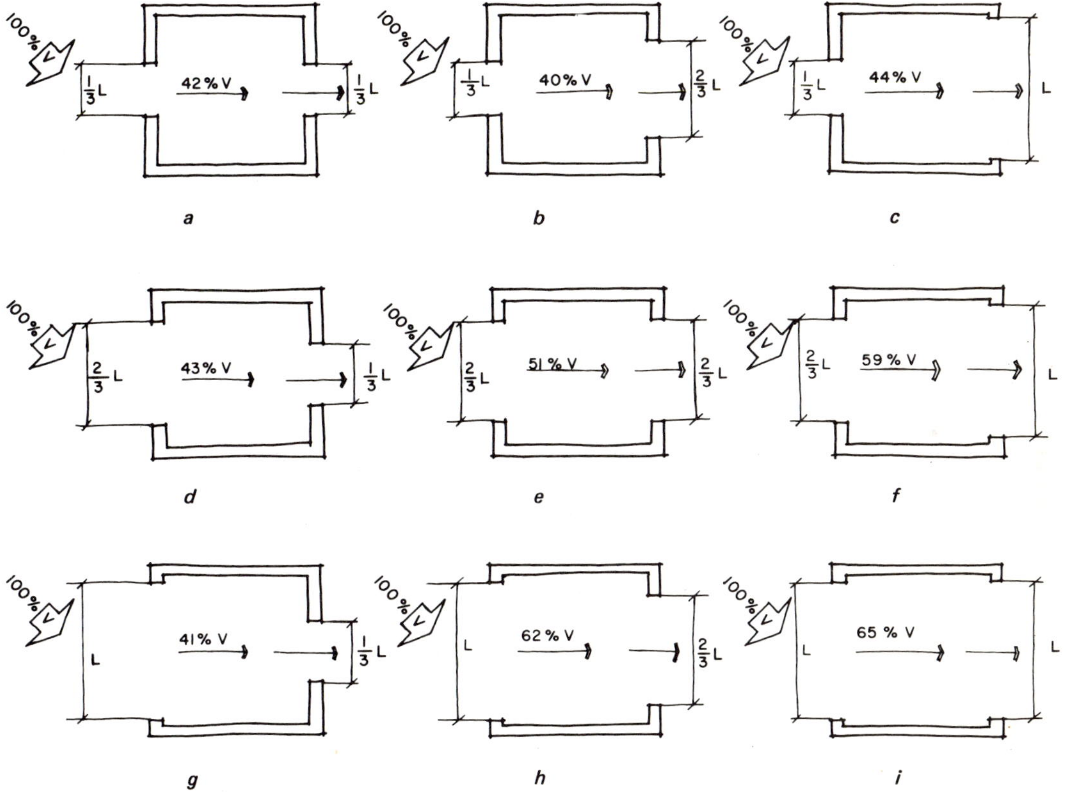

Figure 4-9. Openings on opposite walls; wind oblique to inlet.

In a condition similar to that of Figure 4-10, but with a wind orientation of 45° with respect to the inlet, the effect of varying the relationship between inlet and outlet size is as follows. The highest efficiency that produces an interior air velocity of $0.45V$ is attained when the inlet size is $1/3L$ and the outlet size is equal to L. On the other hand, the minimum efficiency takes place when the inlet size is $2/3L$ and the outlet size is $1/3L$. In this case, the interior air velocity is $0.36V$ (see Figure 4-11a–e).

The rate of airflow in the case of cross-ventilation in a given building can be computed from the following expression (Givoni 1965a):

$$Q = KAV$$

Q = airflow rate (cf/hr)
K = coefficient dependent on the ratio of inlet area to outlet area
V = velocity of undisturbed wind

In cross-ventilated spaces having a square floor plan and the inlet and outlet openings in opposite walls, the average indoor velocity of the moving air can be related to the ratio of opening area to wall area by means of the following formula:

$$\overline{V}_i = 0.45\,(1 - Ke^{-3.84})\,V_o$$

where

$\overline{V}_i$ = average indoor air velocity
K = ratio of window area to wall area
V_o = outdoor wind speed

The preceding formula developed at the Central Building Research Institute (1967) in Roorkee, India, derives from the experimental work originally conducted by Givoni (1965b).

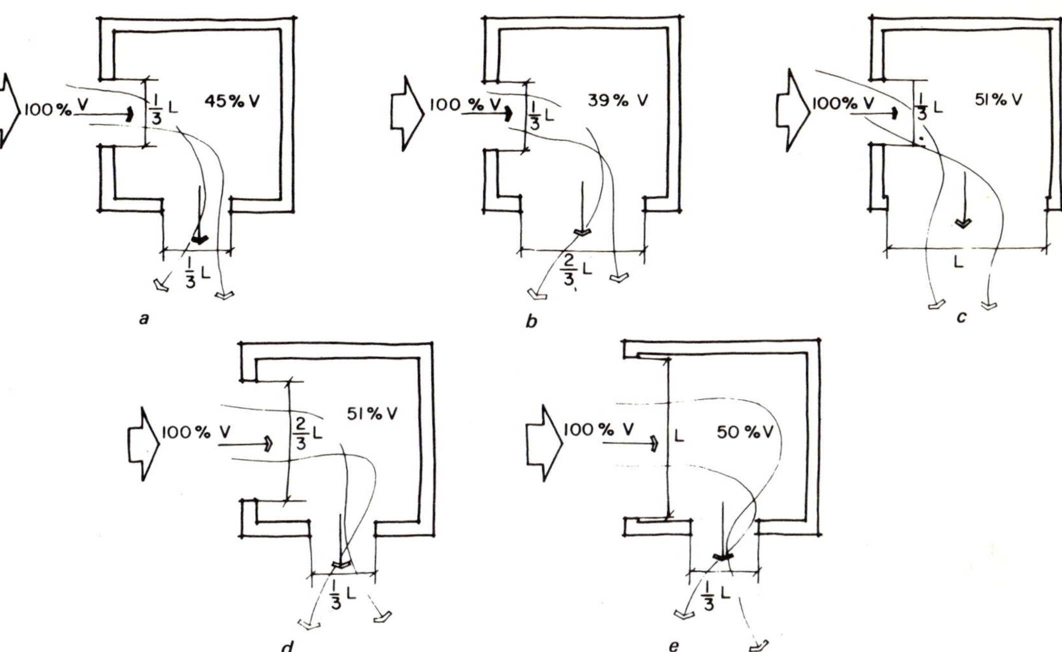

Figure 4-10. Effect of inlet and outlet sizes in cross-ventilated spaces: openings on adjacent walls; wind perpendicular to inlet.

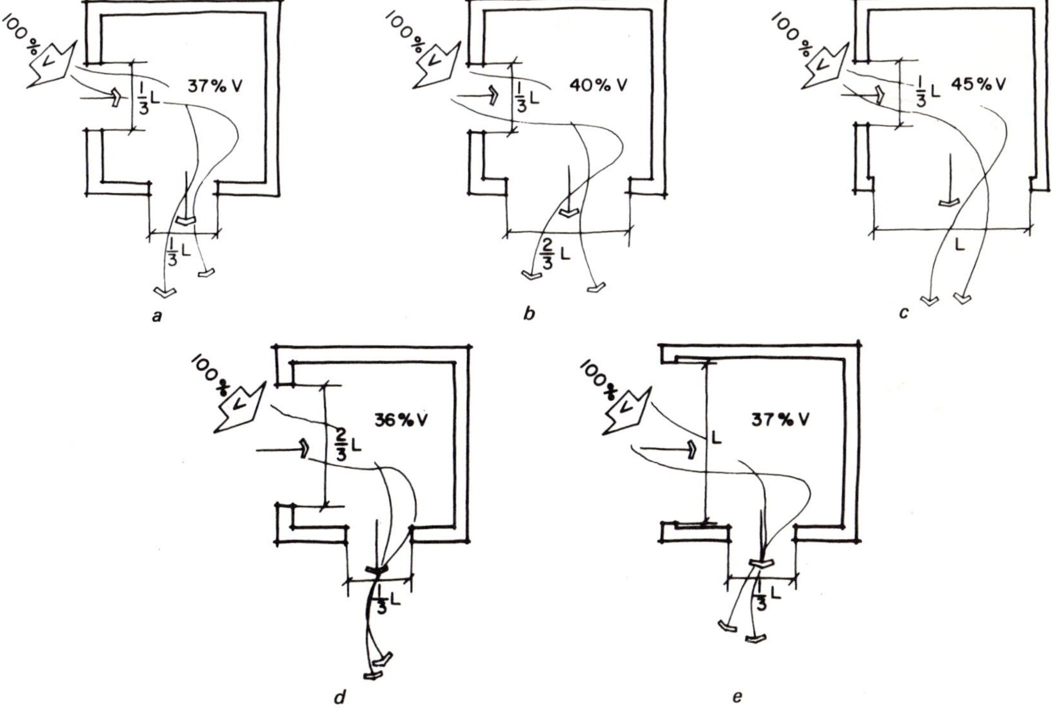

Figure 4-11. Openings on adjacent walls; wind oblique to inlet.

ANCIENT TECHNOLOGY

The development of passive systems for the efficient ventilation of buildings has a history that is centuries old, especially in the East and Middle East where unbearable climactic conditions forced people to produce inventive defensive strategies against the summer heat. For over five centuries, builders in Pakistan have been using masonry windscoops that extend above the roof line of buildings. The windscoops resembling large periscopes, are oriented in the direction of the prevailing winds. When the wind is opposite the face of the windscoop, the air enters it and is forced down into the building below. Eventually, it escapes through doors and windows, creating the desired ventilation. When the wind blows in the opposite direction, it produces a suction on the face of the scoop and thus air is pulled up from the building below. As the direction of the wind varies between these two extremes, the air either moves upward or downward. These opposite motions are due to the direction of the component of the wind velocity perpendicular to the scoop. Naturally, the efficiency of the system depends on the direction of the wind, and it is maximum when the wind blows in the prevailing direction. (See Figure 4-12.)

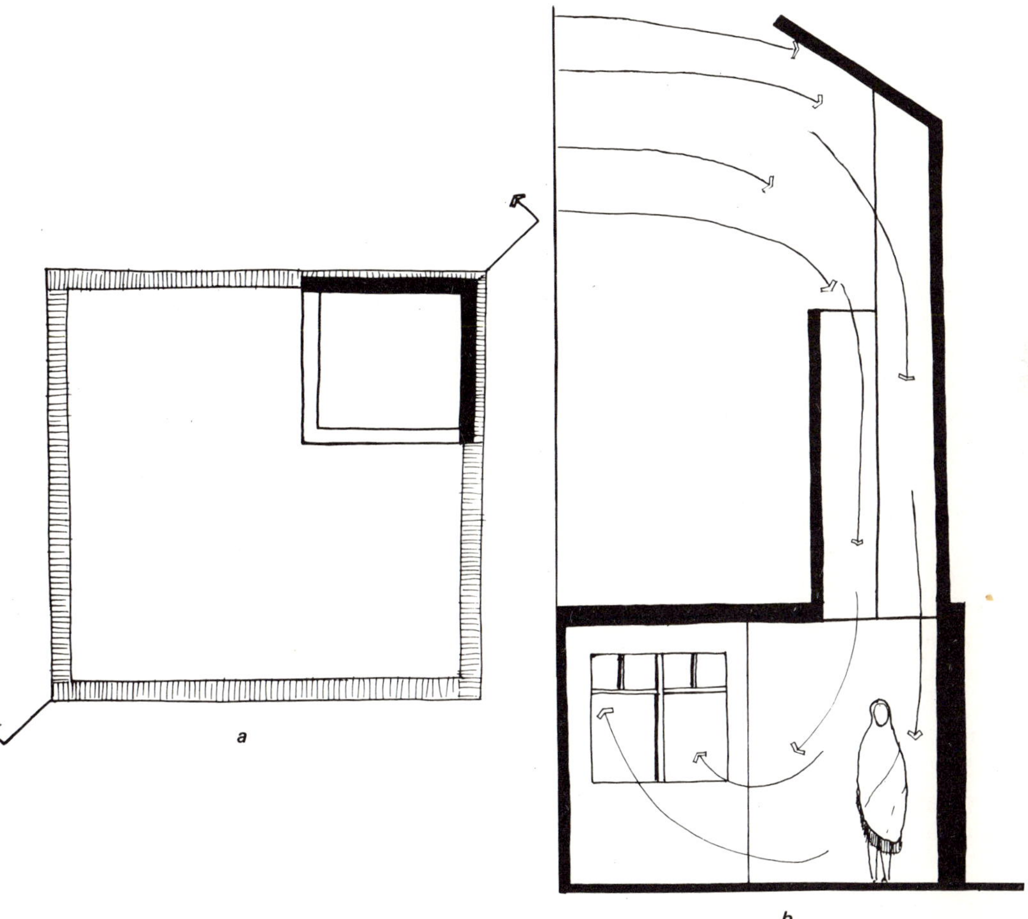

Figure 4-12. *a,* Passive systems for induced ventilation in buildings, used in Pakistan for the past five centuries. Windscoops oriented in the direction of the prevalent wind are placed above the roof. The windscoops catch the wind and force it to move downward into the rooms to be ventilated and finally to exit through the windows. However, when the wind blows in the opposite direction, the negative pressure in front creates suction that pulls up air from the building below. The efficiency of the system is well proved in the city of Hyderabad, Sind, Pakistan, where the 120°F outdoor temperature that prevails from April to June can be reduced to 95°F in the interior of buildings ventilated by this system. The roofscape of the city is dominated, of course, by the projecting windscoops, just as roofs of northern cities are characterized by chimneys. *b,* Schematic representation of the induced ventilation system in Pakistan. The wind, captured by the windscoop, is deflected downward into the space to be ventilated. Extending over the rooftop, the tower has a square or rectangular cross section up to a certain height, but in the upper part only two adjacent sides of the tower continue vertically. These two walls are covered at the top by a plate.

Like the chimneys of Western cities, the windscoops dominate the rooftops in Eastern city-scapes. Their efficiency, well proved by centuries of use, is such that temperature reductions as high as 25°F have been observed. Windscoops in Iran have a different design than those in Pakistan (Figure 4-13). Made of masonry, the windscoops have louvered openings on all faces, which allow the wind to flow horizontally through the top of the scoop, over the mouth of the scoop itself, for any wind direction. The wind, therefore, is not channeled through the mouth down into the building but flows always horizontally and perpendicularly to the flue, creating suction over the mouth. In all cases, then, air is always pulled up from the building. (See Figure 4-14.)

Ancient designers at times achieved efficient ventiliation in buildings, either accidently or intuitively. On walls or roofs, windows, doors, and apertures in general provided inlets and outlets for the circulation of air when they were properly placed in zones of positive and negative pressure, respectively. The Pantheon in Rome is the perfect example of an efficiently ventilated building. The round eye on the crown of the dome sucks the air from below as the wind blows over the dome. Air flowing under the portico and entering through the huge door rises through the space and exits from the top. (See Figure 4-15.) Note that regardless of the wind direction, the negative pressure (suction) on the top will always prevail and suck air up, in all wind conditions.

TALL BUILDINGS

In high-rise buildings, the combined effect of temperature-gradient and wind-pressure differentials activates a very strong wind-ventilation mechanism inside the buildings, which can create serious problems if not kept under control. On the other hand, a thoughtful design could capitalize on this phenomenon to use it advantageously as a passive ventilation system. The "stack effect" in a tall building takes place when a natural temperature gradient is formed: hot air rises and stratifies in layers of increasing temperatures on top of stairwells and elevator shafts. Consequently, cold air, pulled in from outside of the building, rushes in at the bottom of these shafts and rises as it warms up, setting in motion a strong, convective upward airflow.

In addition, apertures on top of the shafts and stairwells produce areas of negative pressure when the wind flows over them. A powerful airflow, as described previous-

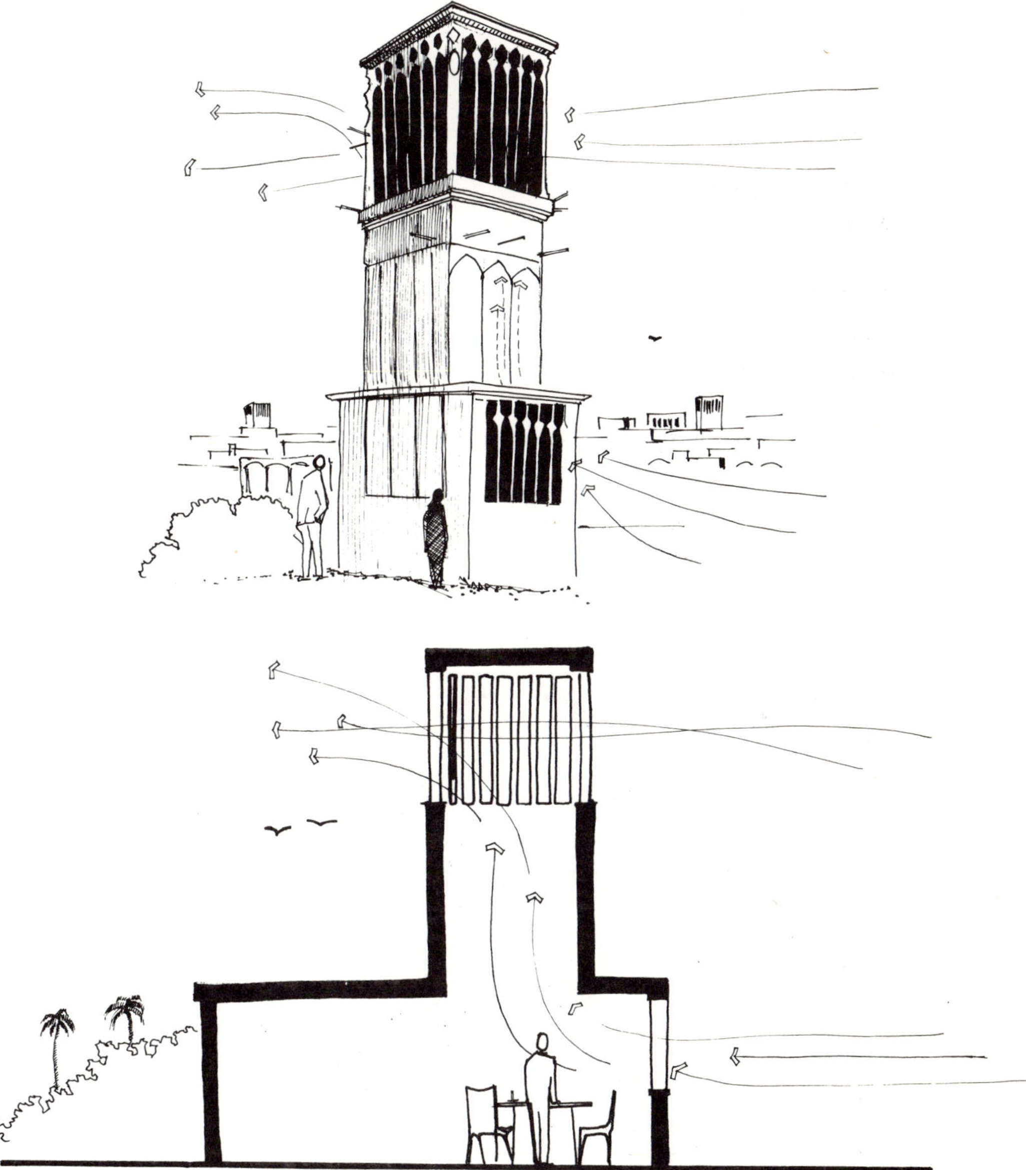

Figure 4-13. *a,* In Iran, the passive system typically used to induce ventilation in buildings is somewhat different than the one used in Pakistan. A venting tower above the roof sucks air from the space below. The tower in fact has vertical slots all around its sides so that the wind can flow horizontally from any direction through the tower and produce suction. *b,* Schematic illustration of air circulation in the passive venting system used in Iran.

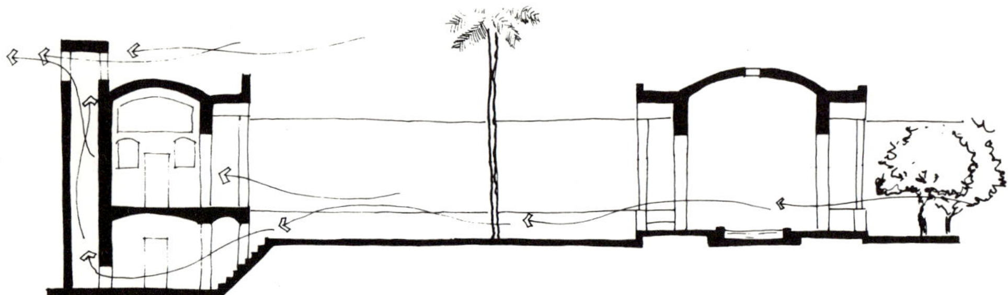

Figure 4-14. Most of the ventilation over a residential lot can be converged by proper design at the base of a wind tower where it will continue upward. As the air flows through cool shaded areas under trees, over open building spaces, or over water pools, considerable breezes can arise from very light air currents.

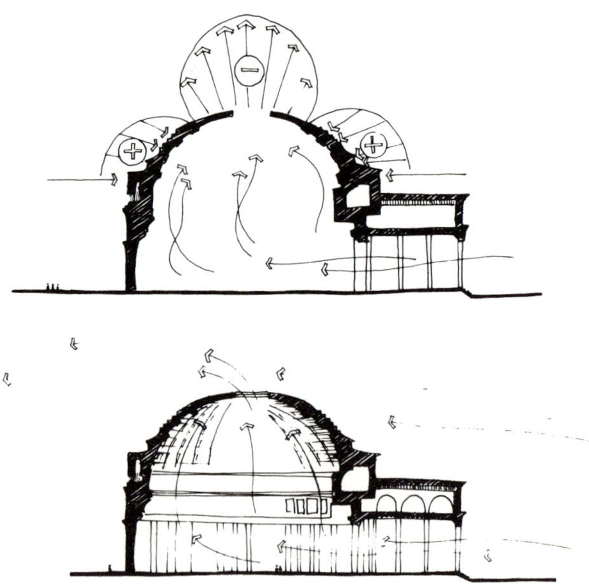

Figure 4-15. Maximum efficiency in the ventilation of interior spaces results when the air inlet is located where the wind pressure has a maximum positive value and the air outlet is located where the wind pressure has a maximum negative value (suction). An example of this, whether intentional or accidental, is the Pantheon in Rome. The circular opening at the crown of the dome coincides with the zone of suction with the result that, regardless of the positive pressure at the front door, the pressure differential and consequently the ventilation will always be effective, independent of the wind direction.

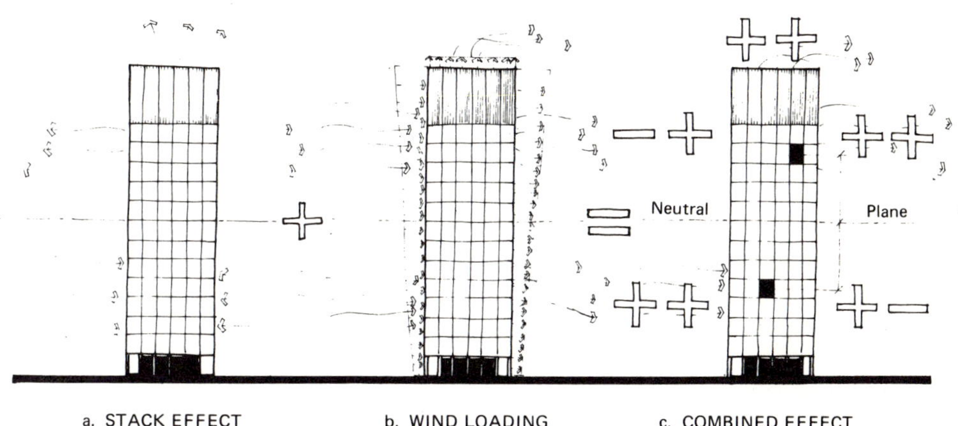

Figure 4-16. a, Because of the so-called stack effect, as air gets heated inside a building, it rises through elevator shafts, stairwells, etc., and escapes through openings and cracks in the upper floors and roof. At the same time, cold air is drawn in at the lower floors. *b,* The wind pressure around a building tends to push air into the building where the pressure is positive (windward areas) and to suck air out where pressure is negative (leeward and roof areas). *c,* Stack effect and wind pressure effect combine together vectorially, reinforcing each other on the windward zone in the lower floors and on the leeward zone in the upper floors and on the roof.

ly, is activated so that air is pulled through the doors and windows of a building and pushed out through the apertures on top, adding to the stack effect. (See Figure 4-16.) The problem is very common in tall buildings, and closing and opening doors is at times impossible, unless special devices are adopted to counteract the pressure, such as airlocks.

VENTILATION OF EXTERIOR SPACES

In the first century A.D., Vitruvius wrote of the importance of wind effects on the urban environment and gave suggestions for city planning. Twenty centuries later, our textbooks on environmental control do not add much more.

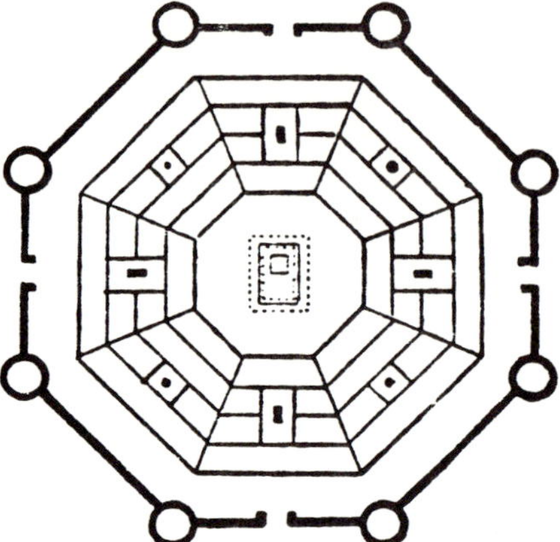

Figure 4-17. Vitruvius' plan for a typical town shows streets and alleys laid out with regard to the direction of the winds. (From *De Architectura* by Vitruvius)

Following the layout of the city walls, Vitruvius suggested that the lot subdivisions and the street pattern should be designed to block the wind in the alleys since cold, hot, and humid winds are, respectively, unpleasant, enervating, and unhealthy. Having identified eight principal directions from which winds blow, Vitruvius proposed a series of concentric octagons connected by eight radial lines as the main layout of the city streets. Each side of the octagon is perpendicular to the main wind directions, and the radial streets are at an angle from such directions. The concept, of course, is to keep the streets away from the wind paths as much as possible for all probable wind directions (Figure 4-17).

Current literature covering wind influence on the urban environment suggests basically: minimizing wind currents in the streets for cold climates; maximizing ventilation of streets in too hot and humid climates; and minimizing wind speed in hot and dry climates where soil erosion can be a problem. Building height and width, and the orientation of the streets are pointed out as major parameters for wind-velocity control in built-up areas. Channeled wind, obviously, can enter narrow streets, and, as in a natural canyon, its speed can increase considerably because of the Venturi effect. By the same token, the prevailing wind may be diverted before entering the streets by a proper arrangement of the layout of the streets, which could be determined by wind-tunnel experiments. The octagonal layout of Vitruvius' ideal city has a more rational counterpart in the prototypes designed following wind-tunnel testing.

In reality, it is improbable that the wind alone will be the generator of urban forms in planning. Other factors, such as the sun, must be considered concurrently, and the final forms will derive from a process of optimization. Furthermore, one could predict that between the two, the sun will probably have a dominant role in design as a form generator. However, these physical design parameters give way to socioeconomic factors that prevail in the overall layout of city streets.

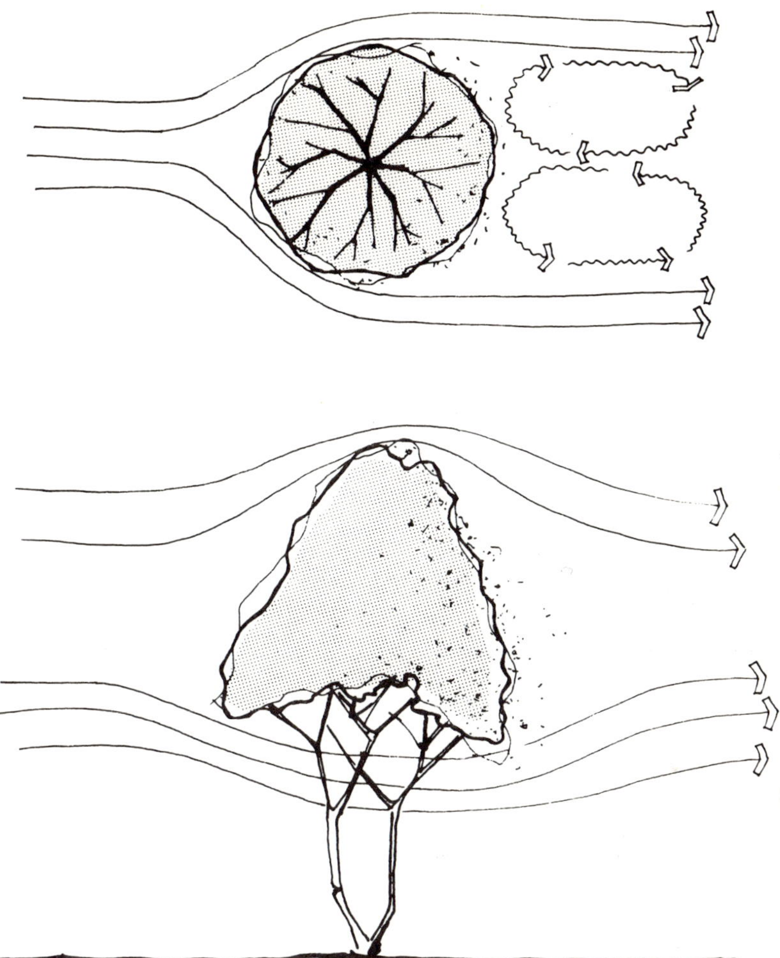

Figure 4-18. The wind deflection by an isolated tree creates a zone of higher speed under the tree and around its crown, while the sheltered area is just behind the foliage.

In low-density suburban areas, individual buildings stand alone in the free wind-fields, creating a different problem than exists in dense urban areas. Trees, shrubs, land formations, freedom in building orientation, and so forth are common factors for proper wind design.

If we look at the interaction of a single isolated tree in a natural wind-field, we can see how the deflected windflow produces a zone of reduced pressure on the leeward side, with some eddies being generated in this area. On the other hand, as the wind blows between the ground and the bottom of the tree foliage, it accelerates. This zone of higher speed can be taken into account if ventilation is desired, or, in the opposite case, when it has to be blocked, the phenomenon is equally relevant (Figure 4-18).

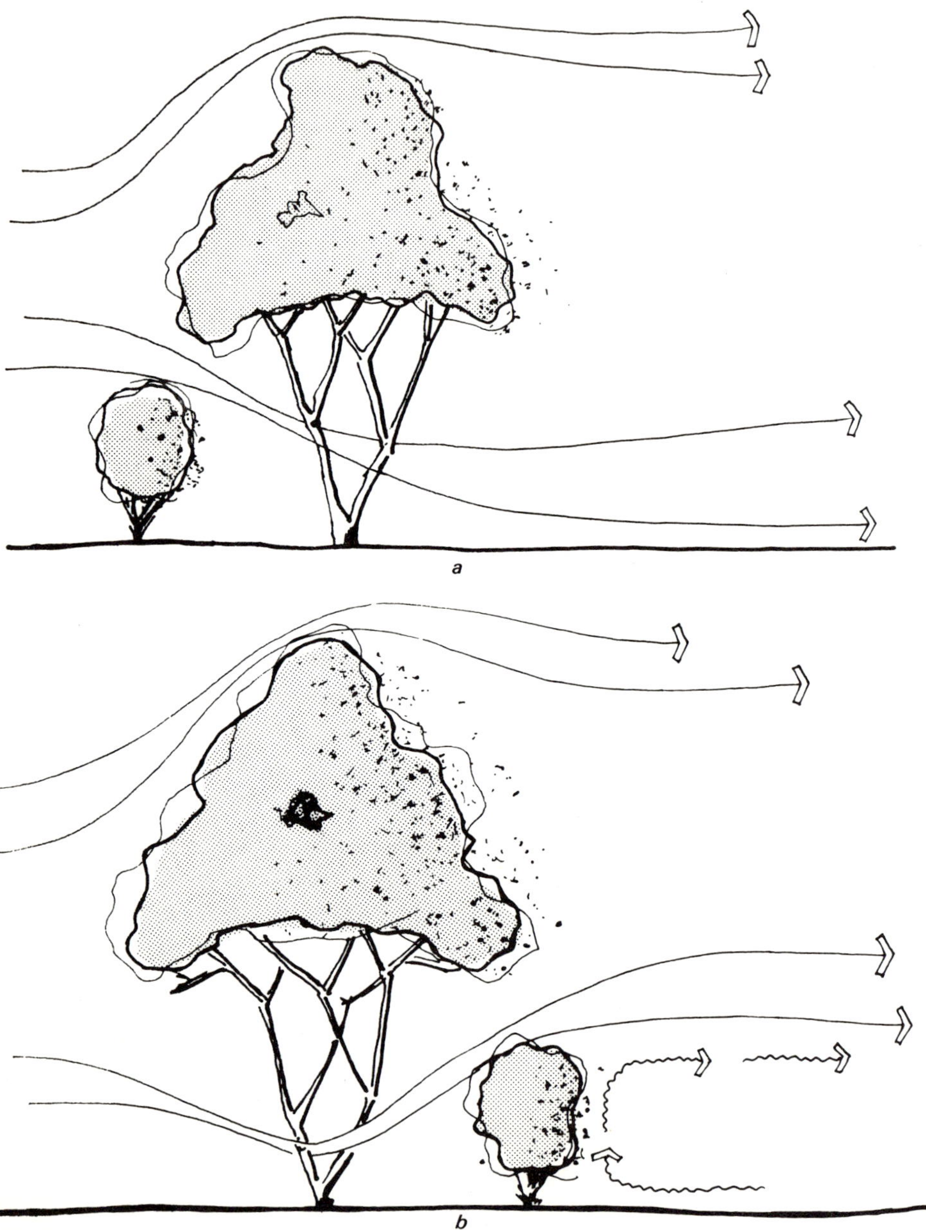

Figure 4-19. *a,* The addition of low shrubs in front of trees accelerates the air near the ground. *b,* By placing low shrubs behind trees, the accelerated windflow under the trees is deflected upward, and consequently a sheltered zone is created behind the shrubs.

The addition of simple shrubs to the isolated tree produces even a faster current under the tree if the shrubs are placed on the windward side of the tree, narrowing the space between the foliage and the ground. (Figure 4-19a). On the other hand, if the shrub is placed on the leeward side of the tree, the air current is deflected upward, creating a zone of reduced speed on the leeward of the shrub (Figure 4-19b).

The problem, of course, increases in complexity as the individual elements (tree and shrubs) increase in number and are used to produce a variety of effects that increase or decrease the wind speed. The environmental conditions of outdoor spaces can therefore be planned and designed on this basis for a large spectrum of outdoor activities. From the small scale of the residential backyard, the number of practical applications of wind control can include large areas for public uses, such as sports events and activities in general, or even larger-scale projects. In the latter case, we include the gigantic project accomplished in many parts of the world in which miles and miles of shelterbelts were built to reduce wind speed over crops. In particular, in the Soviet Union and the United States, shelterbelts and windbreakers have been used to protect livestock for many years.

SHIELDING EFFECT OF BUILDINGS

On the leeward side of a building, there is a shadow area of reduced wind speed where eddies are formed. The length of such an area varies with the geometry of the building: basically, the slope of the roof, the depth of the building in the direction of the wind, and the height.

For flat-roof buildings with a depth equal to the height, the shadow area is 3.75 times the building height. When the depth is 2 or 3 times the height, the length of the shadow area is, respectively, 3 or 3.25 times the height (see Figure 4-20).

For buildings with variable roof pitch that have a depth in the direction of the wind equal to the building's height, the lengths of the leeward shadow areas are $3.75A$, $3.75A$, $4.25A$, $4.75A$ for the respective roof pitch: 0, 0.33, 0.5, and 0.67 (where A is the building's height). See Figure 4-21.

For buildings with variable roof pitch that have a depth in the direction of the wind equal to 3 times the building height, the lengths of the leeward shadow areas are $3.25A$, $3.75A$, $4.25A$, and $4.25A$ for the respective roof pitch: 0, 0.33, 0.5, and 0.67. See Figure 4-22.

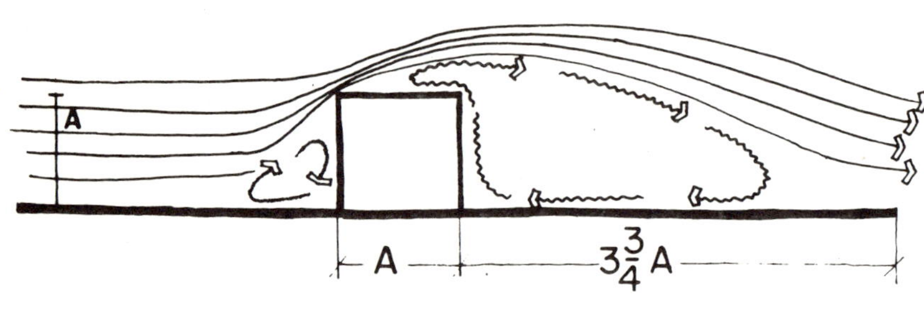

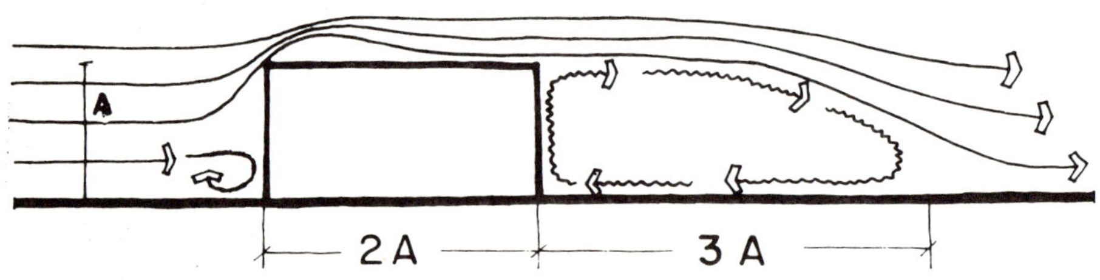

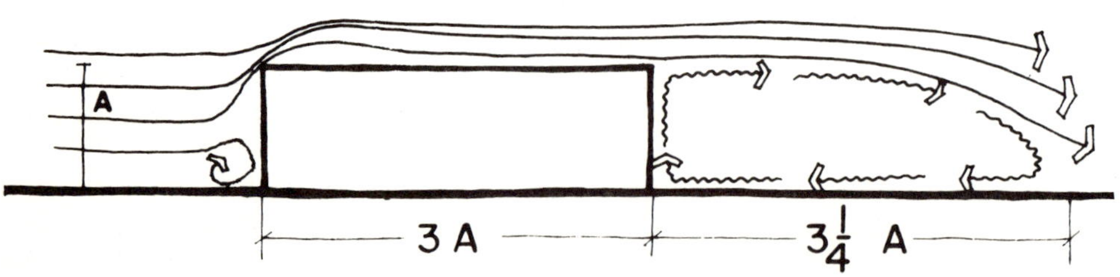

Figure 4-20. Length of leeward eddy area for flat-roof building.

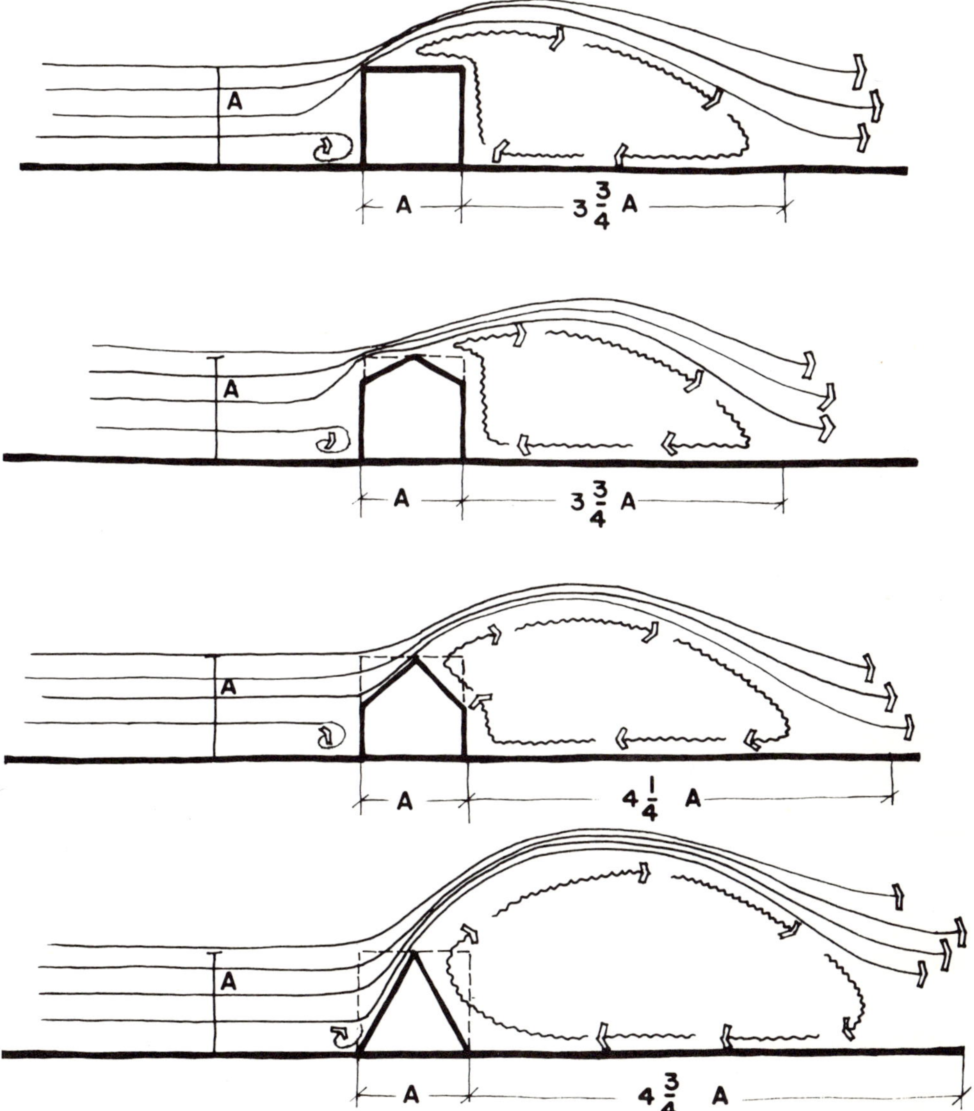

Figure 4-21. Length of the leeward eddy area for buildings with different roof pitch, having a depth equal to the height.

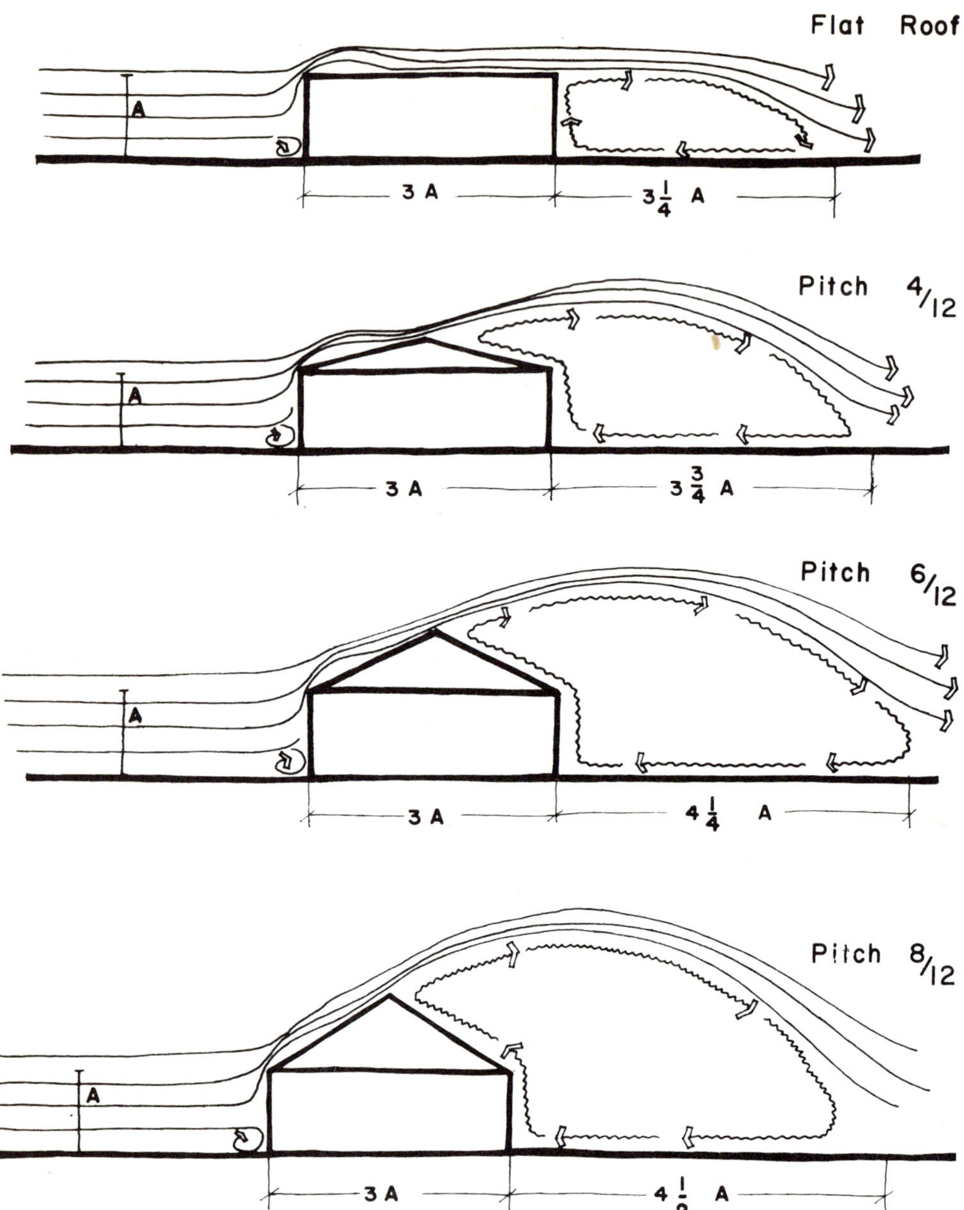

Figure 4-22. Length of the leeward eddy area for buildings with different roof pitch, having a depth equal to three times the building height.

Open structures exposed to the wind separate the streamlines so that the air flows not only over and around them, but also under them. Convex, flat or concave roof configurations generate zones of eddy formation over and under the structure and deflect wind in various ways, as illustrated in Figures 4-23a–g. As the air is deflected, the velocity also varies, increasing as the streamlines converge and decreasing as they diverge and their spacing widens.

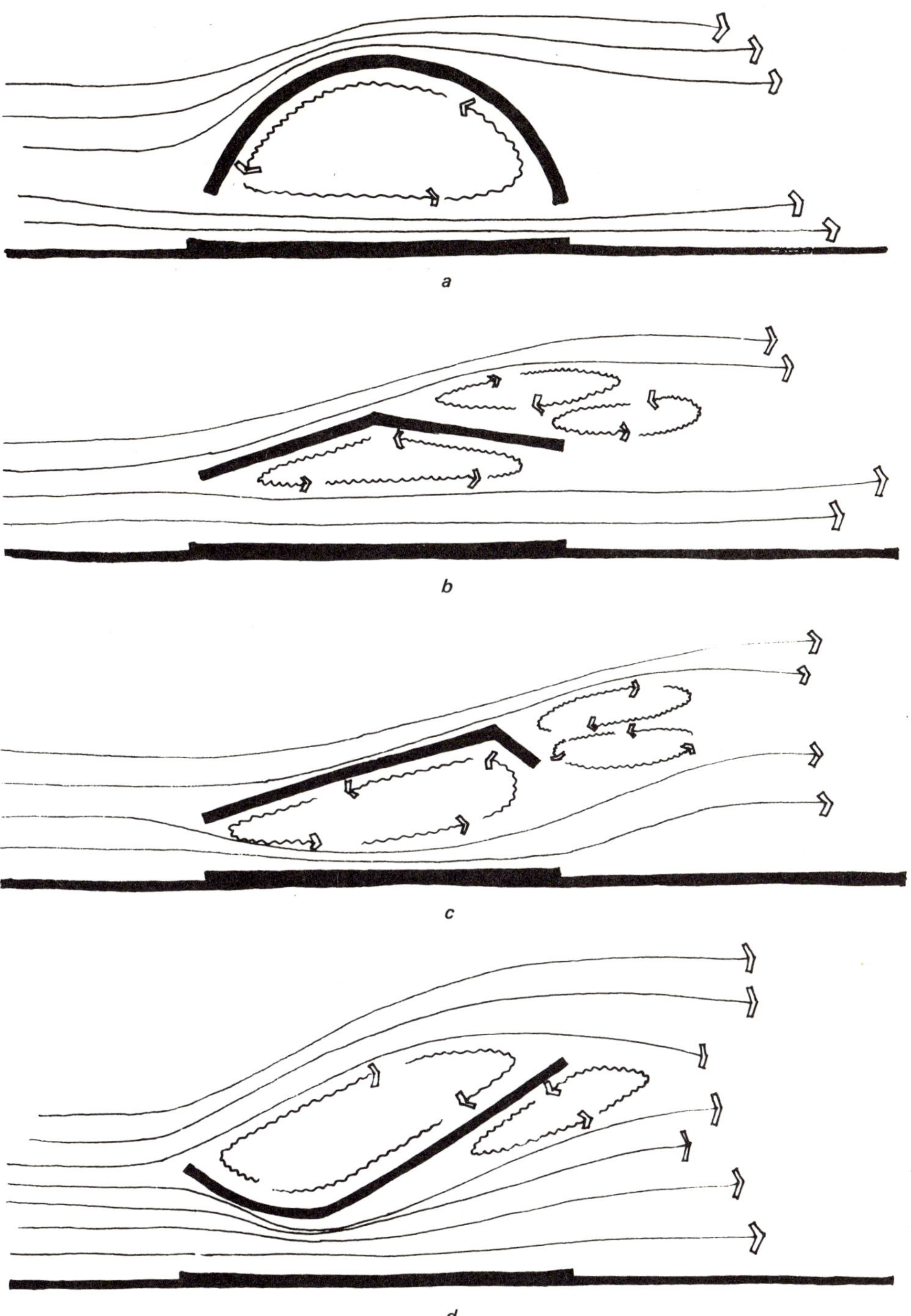

Figure 4-23. The deflection of the streamlines gives a visual representation of the airflow over and below open structures. (As the spacing between the streamlines increases, the velocity decreases; on the other hand, as the spacing gets smaller, the velocity gets higher.) Velocity changes and eddy zones produced by the interaction of wind and structures are shown for various configurations.

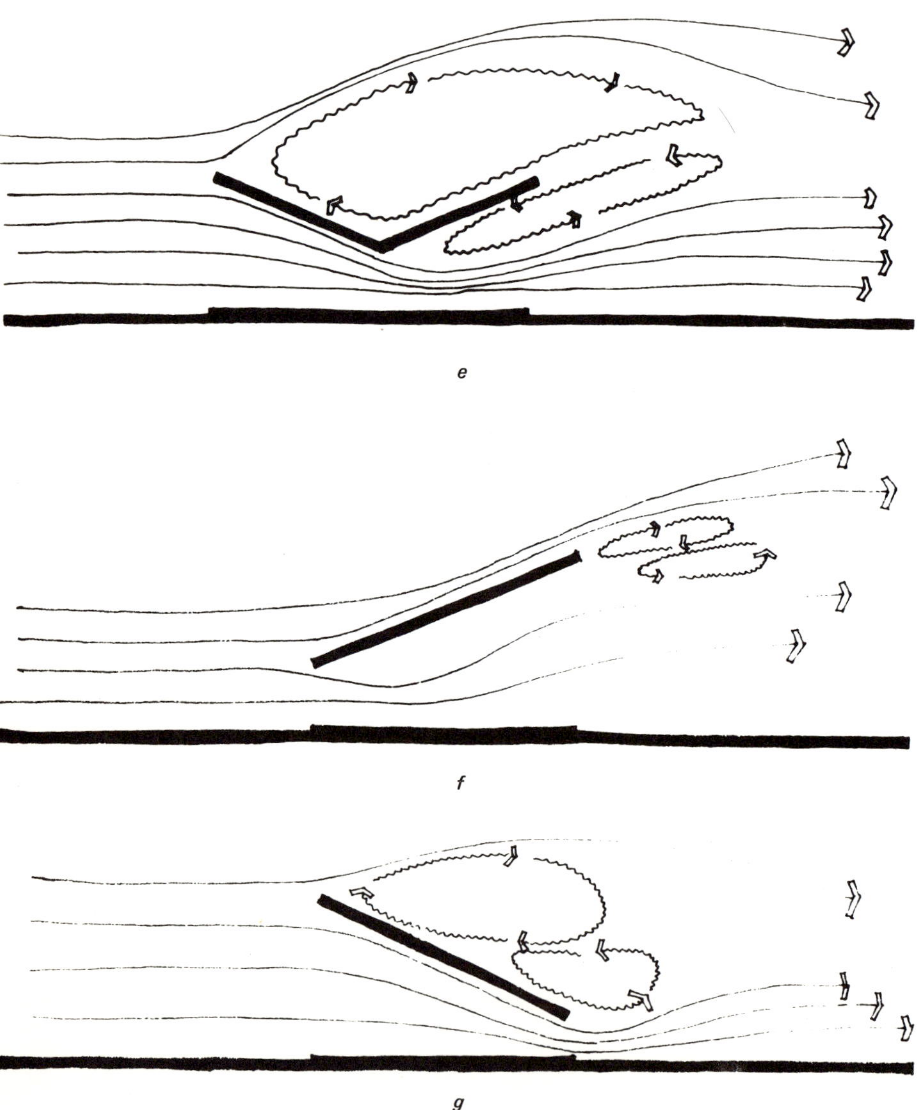

Figure 4-23. Continued.

Figure 4-24. Form response to windflow. Aerodynamic design and architectural composition have produced this exciting form for the smokestack of the San Sebastian. Responsible for this kind of achievement is the current trend to expand architectural design to all aspects of the physical environment affecting humanity, including naval architecture, which was once the exclusive area of engineers. Designed by the Italian architect Guarello, this structure solves the problem of accelerating the airflow and eliminating the falling of the ashes on the decks.

FORM RESPONSE TO WINDFLOW

Smokestack ashes falling on the decks of passenger ships is a problem of major importance. If the airflow due to the natural wind and to the relative motion of the ship with respect to the air could be accelerated, the ashes in the air would also accelerate and, with the increased speed, would fall in the water behind the ship. This is the problem facing the Italian architect Guarello, whose practice in Genova includes naval architecture, once the exclusive domain of engineers. Taking advantage of the Venturi effect, Guarello splits the conventional cylindrical smokestack into two streamlined vertical bodies, adds a horizontal airfoil on top, shapes the base into an inclined plane, and obtains an aerodynamic tri-dimensional structure that accelerates the wind passing through and around it. The result is a very exciting form that combines aerodynamic design and architectural composition.

WIND EFFECT ON BUILDING DAMPNESS

Wind-driven rain that penetrates buildings is a major cause of wetness and dampness that generate undesirable conditions for habitation. These include physical discomfort, illnesses (arthritis, rheumatism, etc.), distortion, alteration and decay of building materials (wood, metals, plaster, etc.), growth of fungi, and generation of bad odors.

As it rains, the wind pressure on the wet wall of a building forces the water to penetrate through the cracks that may be present on the surface. Cracks do not need to be very wide for the water to penetrate. From a minimum gap approximately 4×10^{-4} in. thick, the wind-driving force becomes effective in pushing water through such microscopic cracks. Naturally, as the gap increases, so does the penetrating effect of the water due to the wind. The wind action is not alone but is usually combined with the capillary action that reinforces the process. Other causes, such as diffusion and absorption, can also be present, but the wind pressure is still the dominant one.

THE ARTIFICIAL WIND

Air currents can also be artificially produced using mechanical power. Many processes that the wind could theoretically perform—such as ventilating, cooling of buildings, industrial drying of materials, and powering wind tunnels—are in practice accomplished by mechanical ventilation, which is easy to regulate. Natural and artificial air motion are very similar. The latter will be briefly discussed here so that the reader may visualize the problem of air motion more completely.

Artificial airflows at sustained speed became feasible with the advent of the electric motor, which presented an economic and regulable propulsion for mechanical fans. These fans are usually classified as follows:

1. The *centrifugal fan* consists of a wheel (rotor) with radial blades that can be straight, forward-curved, or backward-curved. The wheel rotates inside a volute, or scroll-shaped, housing whose cross section approaching the outlet gradually increases. The air enters parallel to the wheel axis, then turns 90°. Passing through the rotating blades, it enters the volute and exits at the outlet. The

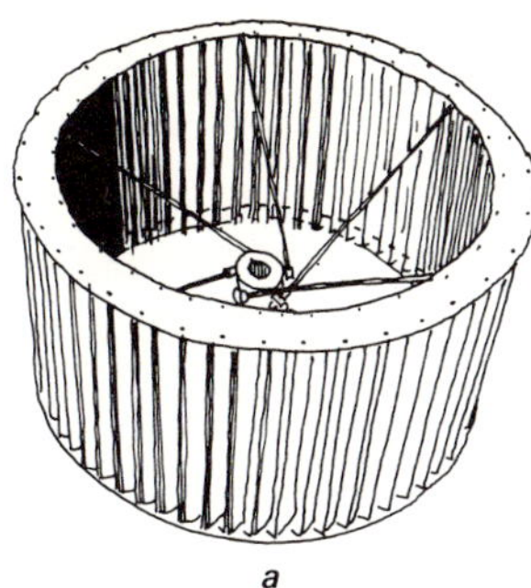
a

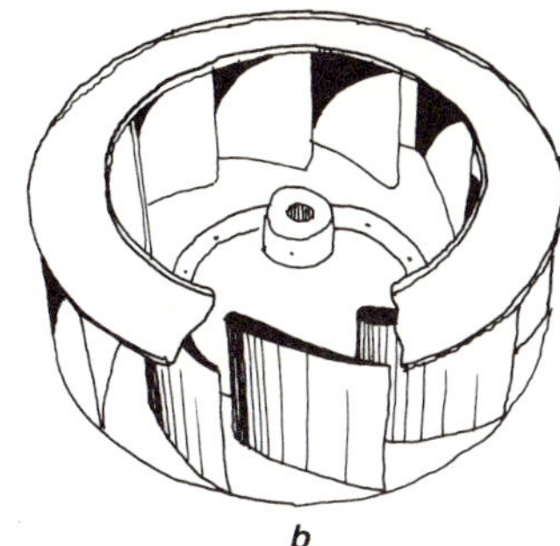
b

Figure 4-25. Vectorial sum *(V_1)* of the tip velocity *(V_t)* and the centrifugal velocity *(V_c)* for the three different types of centrifugal fans. *a,* radial straight blades; *b,* forward-curved blades; *c,* backward-curved blades.

centrifugal fan is the most popular and the most versatile for variable pressures and air volumes. (See Figure 4-25.)

Following is a description of the blades used in centrifugal fans:

a. *Radial straight blades* are more efficient than forward-curved blades but less efficient than backward-curved blades. The narrow wheel allows high velocities and pressures. Fans of this type perform well in parallel.
b. *Forward-curved blades* are the least efficient overall, yet they are the most efficient in weight, size, and cost. With these blades, air volumes can be dangerously reduced if resistance increases. In addition, horsepower increases at a much higher rate than air volume demand. They are difficult to work in parallel.
c. *Backward-curved blades* are the most efficient and dependable.

2. *Propeller fan* (axial flow) consists of a propeller or wheel mounted in a ring. Since the flow is parallel to the shaft, the air passes through the wheel perpendicularly to it. Although good for large volumes of air, it can only produce minimal pressure increases.
3. *Tube axial fan* (axial flow) is similar to the propeller fan but is mounted in a cylinder instead of a ring.
4. *Vane axial fan* (axial flow) is similar to propeller and tube axial fans. However, it includes a set of vanes placed either in front of or behind the wheel for the purpose of guiding the air.

Notice that the efficiency in terms of cost, weight, diameter, horsepower, and pressure is higher from type 2 through type 4.

FAN CHARACTERISTICS

The selection of a fan for a given system is made based on the manufacturer's data, which include the following:

- air volume (cfm)
- outlet velocity (fpm)
- angular velocity (rpm)

- tip speed (fpm)
- static pressure (in. water)
- brake horsepower
- wheel diameter (in.)
- outlet area (ft^2)

In addition, the following characteristics should be considered:

- efficiency
- stability to operation
- noise level
- compactness

VELOCITY IN CENTRIFUGAL FANS

As the air enters the inlet of a centrifugal fan and goes through the blades, it acquires a velocity V_1, which is the vectorial sum of the tip velocity V_t and the centrifugal velocity V_c. As indicated in Figure 4-26, V_i is greater than V_t in straight radial blades and even greater in the forward-curved blades. For backward-curved blades, V_i is smaller than V_t. Where the radius of the propeller and the revolutions per minute are known, the tip velocity V_t can be calculated:

$$V_t = 2\pi r\,(\text{RPM})$$

where

V_t = tip velocity (fpm)
r = radial length of the blade from its tip (ft)
RPM = revolutions per minute

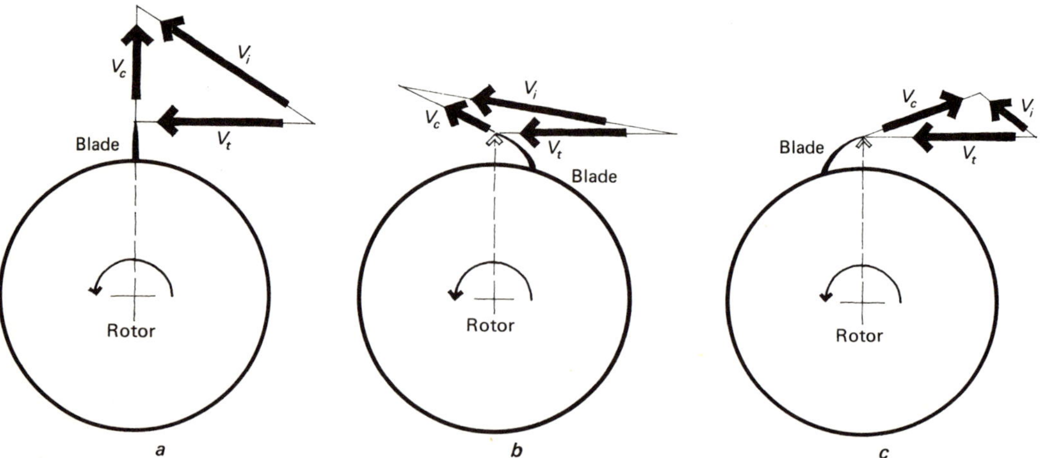

Figure 4-26. Static pressure, velocity pressure, and total pressure, with schematic illustration of various measuring techniques.

Example 4-1: Given a fan with blades 1.5 ft long that rotate at 500 rpm, calculate the tip speed.

Solution:

$$V_t = 2 \times \pi \times 1.5 \times 500 = 4710 \text{ fpm}$$

PRESSURE

The mechanical fan transfers energy received from its motor to the air, which is accelerated as it passes through the rotating blades. The energized air manifests this energy through an increase in pressure (total pressure *TP*). This total pressure is the sum of the static pressure (*SP*) and the velocity pressure (*VP*). As indicated in Figure 4-27, these three different quantities can be measured directly. The velocity pressure is the difference between (*TP*) and (*SP*).

The static pressure is considerably more significant than the velocity pressure in sustaining an airflow. Since the sum of the two is constant (neglecting the losses), it is convenient to convert (*VP*) into (*SP*). Centrifugal fans have divergent scrolls whose cross sections approaching the outlet increase. This reduces the velocity of the air and the velocity pressure, which in turn increases the static pressure of an amount ΔP. Such pressure increase can be expressed as follows:

$$\Delta P = K \frac{\rho}{2g} (V_1^2 - V_0^2)$$

where

ΔP = pressure rise (psf)
K = performance factor varying between 0.7 and 0.8
ρ = air density (lb/cf)
g = acceleration of gravity (ft/sec²)
V_i = air velocity as it leaves the blade (fps)
V_o = air velocity at the outlet (fps, $V_0 < V_1$)

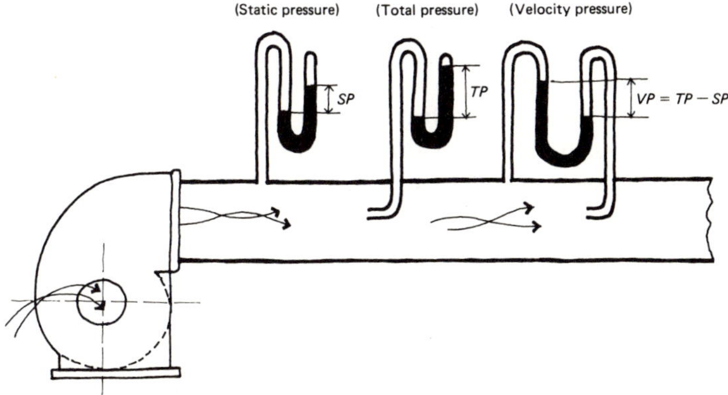

Figure 4-27. Rotors for centrifugal fans. *a*, rotor with forward curved blades; *b*, rotor with backward curved blades.

Example 4-2: As it leaves the blade, the air velocity (V_i) in a centrifugal fan is 100 fps. Because of a gradual enlargment of the volute cross section, the air velocity at the outlet (V_o) is 60 fps. Calculate the pressure increase (ΔP) assuming the air density at standard conditions, a performance factor (K) of 0.7, and an acceleration of gravity (g) of 32.2 ft/sec^2.

Solution:

$$\Delta P = 0.7 \times \frac{0.075}{2 \times 32.2}(100^2 - 60^2) = 5.2 \text{ psf}$$

HORSEPOWER

The following two expressions for fan horsepower are applicable whether or not air is assumed compressible. The first formula for horsepower is:

$$HP = \frac{(CFM)_i(TP_o - TP_i)}{6356\eta_t}$$

where

HP = horsepower (hp)
$(CFM)_i$ = air capacity at inlet (cfm)
TP_o = total pressure at outlet (in. water)
TP_i = total pressure at inlet (in. water)
η_t = total efficiency

Another useful formula for horsepower is:

$$HP = \frac{(CFM)_i(SP_o - TP_i)}{6356\eta_s}$$

where

$$HP = \text{horsepower (hp)}$$
$$(CFM)_i = \text{air capacity at inlet (cfm)}$$
$$SP_o = \text{static pressure at outlet (in. water)}$$
$$TP_i = \text{total pressure at inlet (in. water)}$$
$$\eta_s = \text{static efficiency}$$

Example 4-3: Calculate the necessary horsepower for a fan in order to provide 100,000 ft^3 of air at standard condition with a total pressure increase of 4 in. of water, assuming an efficiency of 90%.

Solution:

$$(CFM)_i = 100{,}000 \text{ cfm}$$
$$(SP_o - TP_i) = 4 \text{ in. water}$$
$$\eta_s = 0.90$$

The horsepower will be:

$$HP = \frac{4 \times 100{,}000}{6356 \times 0.90} = 69.9 \text{ hp}$$

FAN LAWS

The fan laws express the relationship between the performance parameters of two identical fans. Such parameters are:

- CFM: capacity of air delivered in ft^3/min
- SIZE: size of the fan expressed by a linear dimension, such as its radius or diameter

- RPM: rotative speed of the fan in rpm
- PRESS: air pressure, including the static pressure, the velocity pressure (kinetic pressure), or the total pressure
- HP: horsepower
- PWL: sound power level
- η: efficiency
- ρ: air density

The subscripts $_A$ and $_B$ for any of the performance parameters indicate, respectively, the parameters of fan *A* and fan *B*. In fact, when the parameters of one are known, the unknown parameters of the other can be derived by using the fan laws. Note that for the following laws, the efficiency of the fans is equal for both, and the fans operate at the same point of rating:

Law 1a (when $\rho_A = \rho_B =$ const)

$$\frac{(\text{CFM})_A}{(\text{CFM})_B} = \frac{(\text{SIZE})_A}{(\text{SIZE})_B} \times \frac{(\text{RPM})_A}{(\text{RPM})_B}$$

Law 1b

$$\frac{(\text{PRESS})_A}{(\text{PRESS})_B} = \frac{(\text{SIZE})_A{}^2}{(\text{SIZE})_B{}^2} \times \frac{(\text{RPM})_A{}^2}{(\text{RPM})_B{}^2} \times \frac{\rho_A}{\rho_B}$$

Law 1c

$$\frac{(\text{HP})_A}{(\text{HP})_B} = \frac{(\text{SIZE})_A{}^5}{(\text{SIZE})_B{}^5} \times \frac{(\text{RPM})_A{}^3}{(\text{RPM})_B{}^3} \times \frac{\rho_A}{\rho_B}$$

Law 1d

$$\frac{(\text{PWL})_A}{(\text{PWL})_B} = 70 \log_{10}\left(\frac{\text{SIZE}_A}{\text{SIZE}_B}\right) + 50 \log_{10}\left(\frac{\text{RPM}_A}{\text{RPM}_B}\right) + 20 \log_{10}\left(\frac{\rho_A}{\rho_B}\right)$$

Law 2a

$$\frac{(\mathrm{CFM})_A}{(\mathrm{CFM})_B} = \frac{(\mathrm{SIZE})_A{}^2}{(\mathrm{SIZE})_B{}^2} \times \sqrt{\frac{(\mathrm{PRESS})_A}{(\mathrm{PRESS})_B}} \times \sqrt{\left(\frac{\rho_B}{\rho_A}\right)}$$

Law 2b

$$\frac{(\mathrm{RPM})_A}{(\mathrm{RPM})_B} = \frac{(\mathrm{SIZE})_B}{(\mathrm{SIZE})_A} - \sqrt{\frac{(\mathrm{PRESS})_A}{(\mathrm{PRESS})_B}} \times \sqrt{\frac{\rho_B}{\rho_A}}$$

Law 2c

$$\frac{(\mathrm{HP})_A}{(\mathrm{HP})_B} = \frac{(\mathrm{SIZE})_A{}^2}{(\mathrm{SIZE})_B{}^2} \times \frac{(\mathrm{PRESS})_A{}^{2/3}}{(\mathrm{PRESS})_B{}^{2/3}} \times \sqrt{\frac{\rho_B}{\rho_A}}$$

Law 2d

$$\frac{(\mathrm{PWL})_A}{(\mathrm{PWL})_B} = 20 \log_{10} \frac{(\mathrm{SIZE})_A}{(\mathrm{SIZE})_B} + 25 \log_{10} \frac{(\mathrm{PRESS})_A}{(\mathrm{PRESS})_B} - 5 \log_{10} \left(\frac{\rho_A}{\rho_B}\right)$$

Law 3a (when $\rho_A = \rho_B = \text{const}$)

$$\frac{(\mathrm{RPM})_A}{(\mathrm{RPM})_B} = \frac{(\mathrm{SIZE})_B{}^3}{(\mathrm{SIZE})_A{}^3} \times \frac{(\mathrm{CFM})_A}{(\mathrm{CFM})_B}$$

Law 3b

$$\frac{(\mathrm{PRESS})_A}{(\mathrm{PRESS})_B} = \frac{(\mathrm{SIZE})_B{}^4}{(\mathrm{SIZE})_A{}^4} \times \frac{(\mathrm{CFM})_A{}^2}{(\mathrm{CFM})_B{}^2} \times \frac{\rho_A}{\rho_B}$$

Law 3c

$$\frac{(\mathrm{HP})_A}{(\mathrm{HP})_B} = \frac{(\mathrm{SIZE})_B{}^4}{(\mathrm{SIZE})_A{}^4} \times \frac{(\mathrm{CFM})_A{}^3}{(\mathrm{CFM})_B{}^3} \times \frac{\rho_A}{\rho_B}$$

Law 3d

$$\frac{(\mathrm{PWL})_A}{(\mathrm{PWL})_B} = -80 \log_{10}\left(\frac{\mathrm{SIZE}_A}{\mathrm{SIZE}_B}\right) + 50 \log_{10}\left(\frac{\mathrm{CFM}_A}{\mathrm{CFM}_B}\right) + 20 \log_{10}\left(\frac{\rho_A}{\rho_B}\right)$$

Law 4a

$$\frac{(\mathrm{CFM})_A}{(\mathrm{CFM})_B} = \frac{(\mathrm{SIZE})_A{}^{4/3}}{(\mathrm{SIZE})_B{}^{4/3}} \times \frac{(\mathrm{HP})_A{}^{1/3}}{(\mathrm{HP})_B{}^{1/3}} \times \frac{\rho_B{}^{1/3}}{\rho_A{}^{1/3}}$$

Law 4b

$$\frac{(\mathrm{PRESS})_A}{(\mathrm{PRESS})_B} = \frac{(\mathrm{SIZE})_B{}^{4/3}}{(\mathrm{SIZE})_A{}^{4/3}} \times \frac{(\mathrm{HP})_A{}^{2/3}}{(\mathrm{HP})_B{}^{2/3}} \times \frac{\rho_A{}^{1/3}}{\rho_B{}^{1/3}}$$

Law 4c

$$\frac{(\mathrm{RPM})_A}{(\mathrm{RPM})_B} = \frac{(\mathrm{SIZE})_B{}^{5/3}}{(\mathrm{SIZE})_A{}^{5/3}} \times \frac{(\mathrm{HP})_A{}^{1/3}}{(\mathrm{HP})_B{}^{1/3}} \times \frac{\rho_B{}^{1/3}}{\rho_A{}^{1/3}}$$

Law 4d

$$\frac{(\mathrm{PWL})_A}{(\mathrm{PWL})_B} = -13.3 \log_{10}\left(\frac{\mathrm{SIZE}_A}{\mathrm{SIZE}_B}\right) + 16.6 \log_{10}\left(\frac{\mathrm{HP}_A}{\mathrm{HP}_B}\right) + 3.3 \log_{10}\left(\frac{\rho_A}{\rho_B}\right)$$

Law 5a

$$\frac{(\text{SIZE})_A}{(\text{SIZE})_B} = \sqrt{\frac{(\text{CFM})_A}{(\text{CFM})_B}} \times \frac{(\text{PRESS})_B^{1/4}}{(\text{PRESS})_A^{1/4}} \times \frac{\rho_A^{1/4}}{\rho_B^{1/4}}$$

Law 5b

$$\frac{(\text{RPM})_A}{(\text{RPM})_B} = \sqrt{\frac{(\text{CFM})_B}{(\text{CFM})_A}} \times \frac{(\text{PRESS})_A^{3/4}}{(\text{PRESS})_B^{3/4}} \times \frac{\rho_B^{3/4}}{\rho_A^{3/4}}$$

Law 5c (when $\rho_A = \rho_B = \text{const}$)

$$\frac{(\text{HP})_A}{(\text{HP})_B} = \frac{(\text{CFM})_A}{(\text{CFM})_B} \times \frac{(\text{PRESS})_A}{(\text{PRESS})_B}$$

Law 5d

$$\frac{(\text{PWL})_A}{(\text{PWL})_B} = 10 \log_{10}\left(\frac{\text{CFM}_A}{\text{CFM}_B}\right) + 20 \log_{10}\left(\frac{\text{PRESS}_A}{\text{PRESS}_B}\right)$$

Law 6a (when $\rho_A = \rho_B = \text{const}$)

$$\frac{(\text{SIZE})_A}{(\text{SIZE})_B} = \frac{(\text{CFM})_A^{1/3}}{(\text{CFM})_B^{1/3}} \times \frac{(\text{RPM})_B^{1/3}}{(\text{RPM})_A^{1/3}}$$

Law 6b

$$\frac{(\text{PRESS})_A}{(\text{PRESS})_B} = \frac{(\text{CFM})_A^{2/3}}{(\text{CFM})_B^{2/3}} \times \frac{(\text{RPM})_A^{4/3}}{(\text{RPM})_B^{4/3}} \times \frac{\rho_A}{\rho_B}$$

Law 6c

$$\frac{(\mathrm{HP})_A}{(\mathrm{HP})_B} = \frac{(\mathrm{CFM})_A{}^{5/3}}{(\mathrm{CFM})_B{}^{5/3}} \times \frac{(\mathrm{RPM})_A{}^{4/3}}{(\mathrm{RPM})_B{}^{4/3}} \times \frac{\rho_A}{\rho_B}$$

Law 6d

$$\frac{(\mathrm{PWL})_A}{(\mathrm{PWL})_B} = 23.3 \log_{10}\left[\frac{(\mathrm{CFM})_A}{(\mathrm{CFM})_B}\right] + 26.6 \log_{10}\left[\frac{(\mathrm{RPM})_A}{(\mathrm{RPM})_B}\right] + 20 \log_{10}\left(\frac{\rho_A}{\rho_B}\right)$$

Law 7a

$$\frac{(\mathrm{SIZE})_A}{(\mathrm{SIZE})_B} = \sqrt{\frac{(\mathrm{PRESS})_A}{(\mathrm{PRESS})_B}} \times \frac{(\mathrm{RPM})_B}{(\mathrm{RPM})_A} \times \sqrt{\frac{\rho_B}{\rho_A}}$$

Law 7b

$$\frac{(\mathrm{CFM})_A}{(\mathrm{CFM})_B} = \frac{(\mathrm{PRESS})_A{}^{3/2}}{(\mathrm{PRESS})_B{}^{3/2}} \times \frac{(\mathrm{RPM})_B{}^2}{(\mathrm{RPM})_A{}^2} \times \frac{\rho_B{}^{3/2}}{\rho_A{}^{3/2}}$$

Law 7c

$$\frac{(\mathrm{HP})_A}{(\mathrm{HP})_B} = \frac{(\mathrm{PRESS})_A{}^{5/2}}{(\mathrm{PRESS})_B{}^{5/2}} \times \frac{(\mathrm{RPM})_B{}^2}{(\mathrm{RPM})_A{}^2} \times \frac{\rho_B{}^{3/2}}{\rho_A{}^{3/2}}$$

Law 7d

$$\frac{(\mathrm{PWL})_A}{(\mathrm{PWL})_B} = 35 \log_{10}\left(\frac{\mathrm{PRESS}_A}{\mathrm{PRESS}_B}\right) - 20 \log_{10}\left(\frac{\mathrm{RPM}_A}{\mathrm{RPM}_B}\right) - 15 \log_{10}\left(\frac{\rho_A}{\rho_B}\right)$$

Law 8a

$$\frac{(\text{SIZE})_A}{(\text{SIZE})_B} = \frac{(\text{HP})_B{}^{1/4}}{(\text{HP})_A{}^{1/4}} \times \frac{(\text{CFM})_A{}^{3/4}}{(\text{CFM})_B{}^{3/4}} \times \frac{\rho_A{}^{1/4}}{\rho_B{}^{1/4}}$$

Law 8b

$$\frac{(\text{RPM})_A}{(\text{RPM})_B} = \frac{(\text{HP})_A{}^{3/4}}{(\text{HP})_B{}^{3/4}} \times \frac{(\text{CFM})_B{}^{5/4}}{(\text{CFM})_A{}^{5/4}} \times \frac{\rho_B{}^{3/4}}{\rho_A{}^{3/4}}$$

Law 8c (when $\rho_A = \rho_b = \text{const}$)

$$\frac{(\text{PRESS})_A}{(\text{PRESS})_B} = \frac{(\text{HP})_A}{(\text{HP})_B} \times \frac{(\text{CFM})_B}{(\text{CFM})_A}$$

Law 8d

$$\frac{(\text{PWL})_A}{(\text{PWL})_B} = 20 \log_{10}\left(\frac{\text{HP}_A}{\text{HP}_B}\right) - 10 \log_{10}\left(\frac{\text{CFM}_A}{\text{CFM}_B}\right)$$

Law 9a

$$\frac{(\text{SIZE})_A}{(\text{SIZE})_B} = \sqrt{\frac{(\text{HP})_A}{(\text{HP})_B}} \times \frac{(\text{PRESS})_B{}^{3/4}}{(\text{PRESS})_A{}^{3/4}} \times \frac{\rho_A{}^{1/4}}{\rho_B{}^{1/4}}$$

Law 9b

$$\frac{(\text{RPM})_A}{(\text{RPM})_B} = \sqrt{\frac{(\text{HP})_B}{(\text{HP})_A}} \times \frac{(\text{PRESS})_A{}^{5/4}}{(\text{PRESS})_B{}^{5/4}} \times \frac{\rho_B{}^{3/4}}{\rho_A{}^{3/4}}$$

Law 9c (when $\rho_A = \rho_B =$ const)

$$\frac{(\text{CFM})_A}{(\text{CFM})_B} = \frac{(\text{HP})_A}{(\text{HP})_B} \times \frac{(\text{PRESS})_B}{(\text{PRESS})_A}$$

Law 9d

$$\frac{(\text{PWL})_A}{(\text{PWL})_B} = 10 \log_{10}\left(\frac{\text{HP}_A}{\text{HP}_B}\right) + 10 \log_{10}\left(\frac{\text{PRESS}_A}{\text{PRESS}_B}\right)$$

Law 10a

$$\frac{(\text{SIZE})_A}{(\text{SIZE})_B} = \frac{(\text{HP})_A{}^{1/5}}{(\text{HP})_B{}^{1/5}} \times \frac{(\text{RPM})_B{}^{3/5}}{(\text{RPM})_A{}^{3/5}} \times \frac{\rho_B{}^{1/5}}{\rho_A{}^{1/5}}$$

Law 10b

$$\frac{(\text{CFM})_A}{(\text{CFM})_B} = \frac{(\text{HP})_A{}^{3/5}}{(\text{HP})_B{}^{3/5}} \times \frac{(\text{RPM})_B{}^{4/5}}{(\text{RPM})_A{}^{4/5}} \times \frac{\rho_B{}^{3/5}}{\rho_A{}^{3/5}}$$

Law 10c

$$\frac{(\text{PRESS})_A}{(\text{PRESS})_B} = \frac{(\text{HP})_A{}^{2/5}}{(\text{HP})_B{}^{2/5}} \times \frac{(\text{RPM})_A{}^{4/5}}{(\text{RPM})_B{}^{4/5}} \times \frac{\rho_A{}^{3/5}}{\rho_B{}^{3/5}}$$

Law 10d

$$\frac{(\text{PWL})_A}{(\text{PWL})_B} = 14 \log_{10}\left[\frac{(\text{HP})_A}{(\text{HP})_B}\right] + 8 \log_{10}\left[\frac{(\text{RPM})_A}{(\text{RPM})_B}\right] + 6 \log_{10}\left(\frac{\rho_A}{\rho_B}\right)$$

Example 4-4: A fan delivers 22,390 cfm of air at a gauge pressure of ⅞ in. of water, rotating at a rate of 590 rpm using 5.4 hp. Calculate the air volume in cfm, the gauge pressure, and the horsepower required if the rate of revolution is increased to 678 rpm.

Solution:

From law 1a

$$\frac{(\text{CFM})_A}{(\text{CFM})_B} = \frac{(\text{SIZE})_A}{(\text{SIZE})_B} \times \frac{(\text{RPM})_A}{(\text{RPM})_B}$$

$$\frac{22{,}390}{(\text{CFM})_B} = 1 \times \frac{590}{678}$$

$$(\text{CFM})_B = 25{,}730 \text{ cfm}$$

From law 1b

$$\frac{(\text{PRESS})_A}{(\text{PRESS})_B} = \frac{(\text{SIZE})_A{}^2}{(\text{SIZE})_B{}^2} \times \frac{(\text{RPM})_A{}^2}{(\text{RPM})_B{}^2} \times \frac{\rho_A}{\rho_B}$$

$$\frac{7/8}{(\text{PRESS})_B} = 1 \times \frac{590^2}{678^2} \times 1$$

$$(\text{PRESS})_B = 1.16 \text{ in. of water}$$

From law 1c

$$\frac{(\text{HP})_A}{(\text{HP})_B} = \frac{(\text{SIZE})_A{}^5}{(\text{SIZE})_B{}^5} \times \frac{(\text{RPM})_A{}^3}{(\text{RPM})_B{}^3} \times \frac{\rho_A}{\rho_B}$$

$$\frac{5.4}{(\text{HP})_B} = 1 \times \frac{590^3}{678^3} \times 1$$

$$(\text{HP})_B = 8.19 \text{ hp}$$

REFERENCES

Central Research Building Institute. 1967. Building Digest No. 49. Central Research Building Institute, Roorkee, India.

Givoni, B. 1965a. *Man, Climate and Architecture.* New York: Elsevier Publishing Co. Ltd., p. 264.

———. 1965b. Laboratory study of the effect of window size and location on indoor air motion. *Architectural Science Review* 8(2): 42–46.

5
Wind in the Natural Environment

WIND OVER LAND

Wind Erosion

The erosive effect of the wind on soil that is not protected by vegetative cover is a major ecological problem. The wind picks up loose humus and other soil particles and carries them away, thus destroying farmland.

To what extent does wind erosion affect the agricultural economy and the quality of the environment in the United States? About 70 million acres of land are affected by wind erosion (Woodruff et al. 1972). Of these, 55 million are classified as cropland and 9 million as rangeland; the remaining 6 million are unclassified. Of the total 70 million, 46 million are in need of protection against wind erosion, which is achieved by stabilizing the loose soil. Wind erosion increases the amount of wasted land at a rate of 4.8 million acres per year in the United States, a significant figure that cannot be neglected. Furthermore, agricultural damage caused by the wind is not limited to the loss of cultivated land alone, but extends to the crops as well. More precisely, the abrasive action of windborne soil particles has a destructive effect on plants when driven by strong winds (Figure 5-1). For instance, corn, wheat, cotton, tomatoes, tobacco, and many other crops that differ in size and structure, and that have either been damaged or destroyed in various geographical areas, verify the phenomenon.

Wind erosion also causes air pollution. It is estimated that over the United States alone, over 30 million tons of dust particles per year are injected into the atmosphere by the wind. Of these, the largest percentage consists of soil particles.

Wind erosion is a global phenomenon that becomes more acute at times in areas where particular weather conditions occur. For instance, severe droughts in Africa in the 1970s created wind-erosion problems on the continent. Similarly, in the United States during the early 1930s, long periods of drought produced disastrous wind erosion of the soil that generated intense dust storms. Low visibility and particularly extensive road maintenance were characteristic of that period. In fact, in the 1930s, large-scale wind-protection projects using shelterbelts were initiated in the United States.

Figure 5-1. The steady cutting action of the wind has sheared the edge of the bush illustrated in the sketch.

What causes the phenomenon? When soil is bare and unprotected by vegetation and the soil particles are dry and loose, wind with its turbulent structure picks up the fine particles, whose gravitational properties are offset by the wind speed. Obviously, the higher the wind speed, the larger the weight of the particles that the wind can pick up. It is estimated that for dry, fine soil such as sand, a wind velocity of about 13 mph at an elevation of 1 ft above the ground is enough to initiate the injection of the particles into the air. The friction between particles and the cohesion, largely affected by moisture content, reduces the vulnerability of the soil to erosion.

The cloddiness of a soil is a major factor in determining its resistance to wind erosion. Soil clods offer some resistance to the wind forces that try to displace them. They also protect smaller particles under them from being picked up by the wind. The stability of the clods is affected considerably by their moisture, compaction, clay and lime content, organic matter content, and microbial bacteria. Tilling of the soil, the action of the weather, mechanical and animal traffic, and abrasion result in the breaking up of clods.

Certain types of soils are more susceptible to erosion caused by their breakdown because they have less organic matter content and lower amounts of clay and silt. Sands, sandy loams having a coarse structure, and loamy sands are in this category. To produce clodding, moisture and cohesion are needed during cultivation, but a variation in temperature, such as when the soil is freezing or thawing, or the excessive moisture present after rains, will break them up quite easily. Clods formed by soils having small particles, therefore, do not hold together very well under changing weather conditions. The best clodding soils and the most resistant to wind erosion are those with a clay content from 20% to 30% and a silt thickness from 0.005 to 0.01 mm (0.0002 to 0.0004 in.). Loams of different types, such as silty clay loams, clay loams, or silt loams, are generally the best clodding soils.

Other factors also determine the wind erosion of soils: whether the clods are large or small, their bulk density, and even the ratio of clods to the total amount of soil being considered. The erosion of soils by wind action can be determined by comparing among different soils the ratios of clods that exceed 0.84 mm in diameter. This ratio can be determined by dry sieving the soil and referring to tables made available for this purpose by the U.S. Dept. of Agriculture.

Specifying the cloddiness needed to prevent wind erosion of field soils can be done by commonly used methods; for instance, 50% of a given field surface should be covered with clods of diameters larger than 0.4 in., or two-thirds of the topsoil by weight should be of a size that will resist erosion (more than 0.03 in. diameter).

Absorption and deflection of wind forces can also prevent soil erosion. The uneven surface of a tilled field can offer protection to soil particles by slowing the wind speed and by trapping moving particles. In effect, the abrasive action of the wind is lessened by the rough surface of the field. However, the increasing wind turbulence due to the surface roughness can offset the advantages and increase the erosion of other areas exposed to the stronger wind action. Therefore, the maximum roughness is considered to be most effective between 2 and 5 in.

The moisture content of the soil and the velocity of the wind combine to increase or decrease soil erosion. It has been found that a difference of 10 mph in wind speed (30 mph versus 20 mph) can increase the rate of erosion by a factor of 3. This rate is less for moist soils and greater for dry soils. Soil dried by the air will erode 133% more than soil with moisture that is near the wilting point for most plants. Soil moisture and wind velocity are generally combined into a "local wind-erosion climatic factor"; the U.S. Dept. of Agriculture provides maps that give the monthly value of this climatic factor.

How much soil is lost from a particular field is based not only on the length of the field or on its width, since wind directions change frequently. Instead, the field length considered is based on the prevailing wind direction, and it is measured along

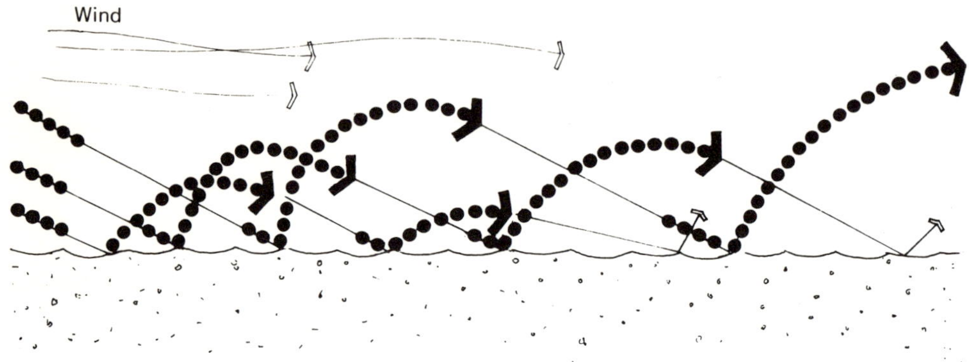

Figure 5-2. Saltation. Motion of solid particles picked up by the wind; they jump and spin as they rise, and break other particles loose as they fall and strike the ground.

this path. Wind barriers that shelter the field on the windward side must be included in this determination, and their amount of sheltering deducted from the total field length.

Wind erosion is best prevented by vegetative cover on the land surface. Organic matter present over the soil will shield those particles most likely to be displaced by direct wind action and act as a buffer against the wind speed. Orientation of the vegetative cover, with respect to the prevailing wind zone, as well as the amount, and how high it grows above the soil surface, all contribute to the prevention of erosion. Size of the residue is important; the smaller, finer residue is considered the most effective because it tends to reduce the wind speed. The type of residue available determines the size, and generally wheat stubble is preferred over corn stubble or sorghum. A single factor—combining type, orientation, size and amount of residue, and its effective cover against soil erosion—is normally used.

How does the wind carry soil particles aloft? Basically, there are two types of motions:

• *Suspension* occurs for small particles, such as dust, varying from 0.1 to 0.5 mm in diameter. Such particles are carried in a state of suspension and literally float in the air. They can therefore be carried for long distances.

• *Saltation* is characterized by the jumping action of soil particles, as the term implies (from Latin *saltus,* "jump"). As the soil particles are picked up, they rise vertically and rotate at the same time at a speed of several hundred revolutions per second. As the particles redescend, they also move forward for a distance varying between 10 and 15 times their height above the ground. Striking the ground as they fall, they break other soil particles loose, contributing to the erosion process. (See Figure 5-2.)

Figure 5-3. Natural rock formation in Utah, shaped like an arch by the eroding action of the wind.

Wind Erosion on Rock

The long-term effects of wind erosion on rock formations are evident (Figure 5-3). Although less intense than water and glacial action, wind has a significant effect. The abrasive action is mostly produced by the fine soil particles carried by the wind and suspended in the air. Examples of wind erosion include rocks with very interesting configurations that seem to resemble purposely shaped human artifacts, rather than being the result of the erratic actions of an uncontrolled force. (See Figures 5-4 and 5-5.)

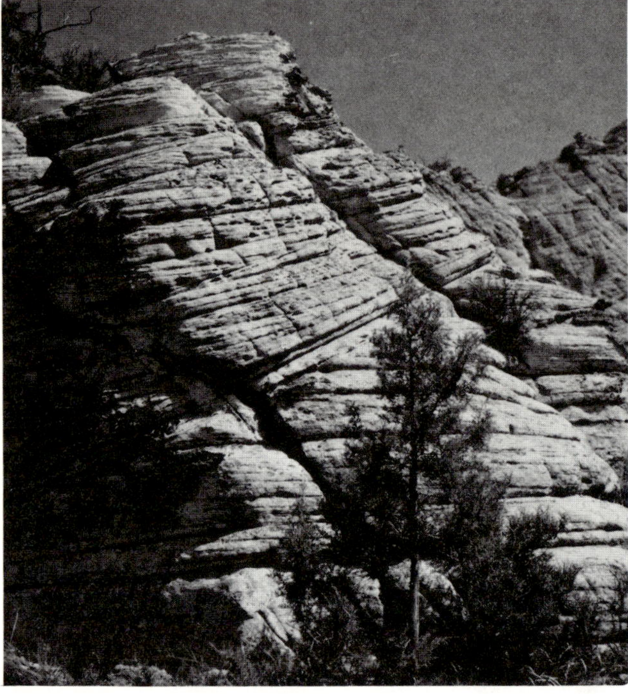

Figure 5-4. Wind erosion. Navajo sandstone at Finch Draw, about a mile west of Flaming Gorge, Daggett County, Utah. Characteristic wind-sculptured outcrop of back slope areas, showing bedding details and weathering habit. Sweeping tangential crossbedding indicates eolian origin. Circa 1953. (Courtesy of W. R. Hansen, U.S. Geological Survey)

Figure 5-5. Wind erosion. Jurassic sandstone (wind sculptured) on red beds, 7 miles south of Hot Springs, in the Black Hills of Fall River County, South Dakota. 1898. (Courtesy of N. H. Darton, U.S. Geological Survey)

From the quantitative point of view, wind erosion is usually expressed by a general equation that correlates six variables together. The equation is:

$$E = f(I, K, C, L, V)$$

where

E = average annual soil loss in tons per acre
I = soil erodibility index
K = soil surface roughness
C = climatic factor, including wind velocity and soil moisture
L = unsheltered field width measured along the direction of the prevailing wind
V = vegetative cover factor

Windbreaks

In addition to its general effects on climate and consequently on humans, wind also directly affects human habitats and such activities as agriculture, forestry, animal breeding, recreation, and navigation. People seek protection against wind damage through natural and artificial means. For instance, in siting settlements, choosing farmland, and selecting shelters for navigation, humans historically have not depended solely on the effects of natural land formations on wind, but have also constructed windbreaks, including the planting of shelterbelts.

Windbreaks are built using various materials and techniques. They can consist of trees properly selected and strategically planted (shelterbelts), or they can simply be fences (made of any suitable material), palisades, walls, earth berms, artificial sand dunes, etc. (Figures 5-6 and 5-7). Their basic function, of course, is to create an area relatively protected from the full force of the wind. The area of maximum protection is on the leeward side, but even on the windward side there is a zone of reduced wind speed. A schematic distribution of the wind velocity on either side of a windbreak is shown in Figure 5-8. In general, one could say that the area affected by the windbreak has a width of about 30 times its height on the leeward side and 10 times on the windward side (Figure 5-9). The speed of the wind varies gradually in this area and at a certain distance is back to its original value. It is interesting to note that at each end of the windbreak, there is a small area where the speed is higher than the speed of the unobstructed wind itself (Figure 5-10).

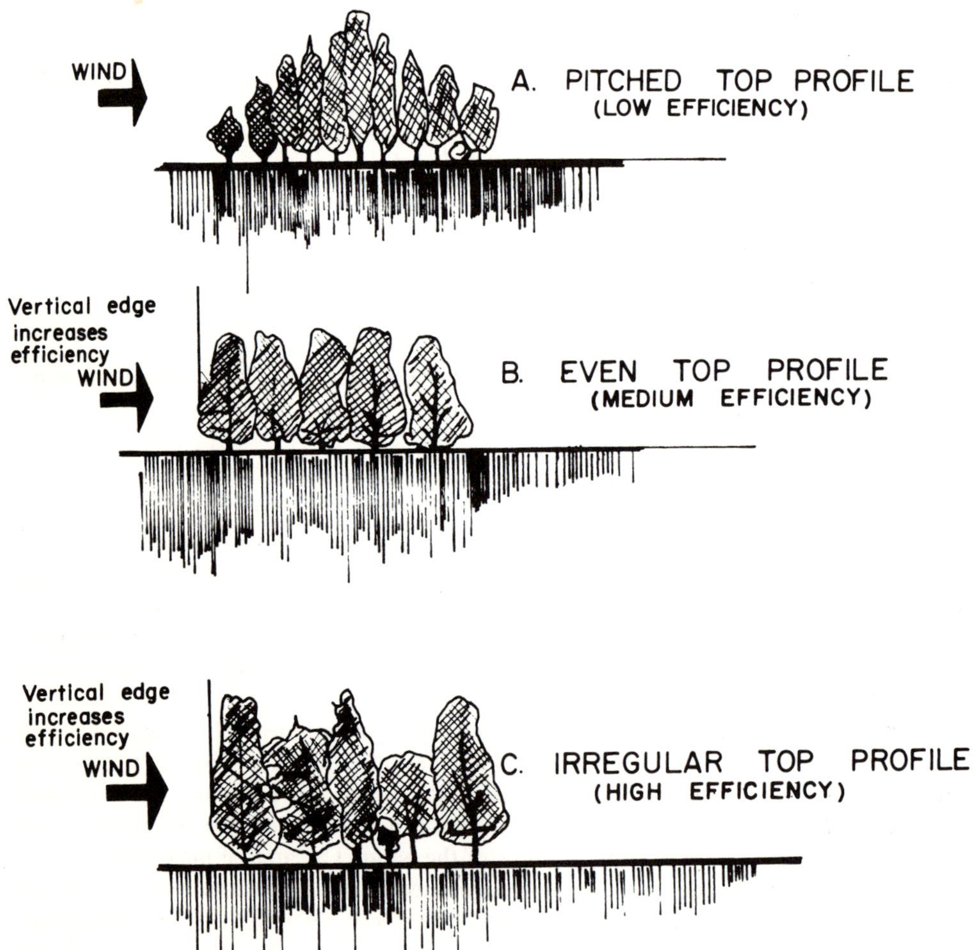

Figure 5-6. Efficiency of top profile for shelterbelts.

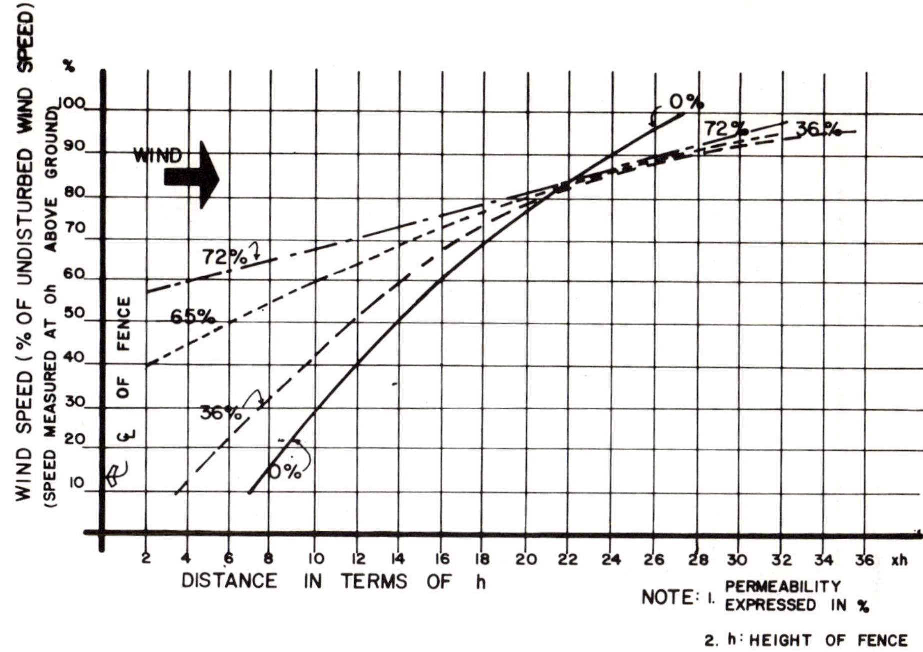

Figure 5-7. Reduction of wind speed for fences of various permeability.

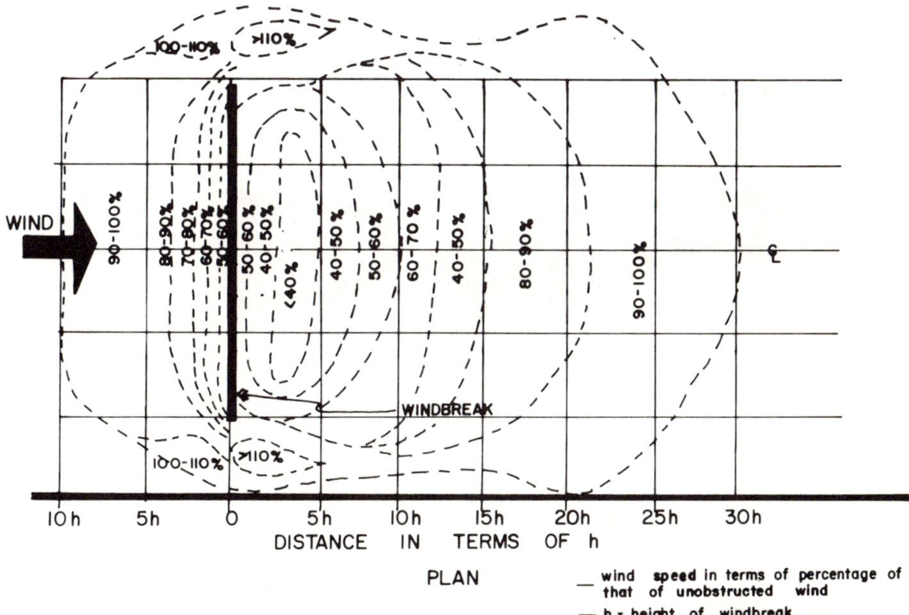

Figure 5-8. Distribution of the wind velocity around a windbreak whose permeability is defined as "moderately dense." The velocity is expressed as a percentage of the speed of the unobstructed wind. Notice that while the speed is reduced on the leeward and also on the windward side of the windbreak, it increases around the end edges up to 110%.

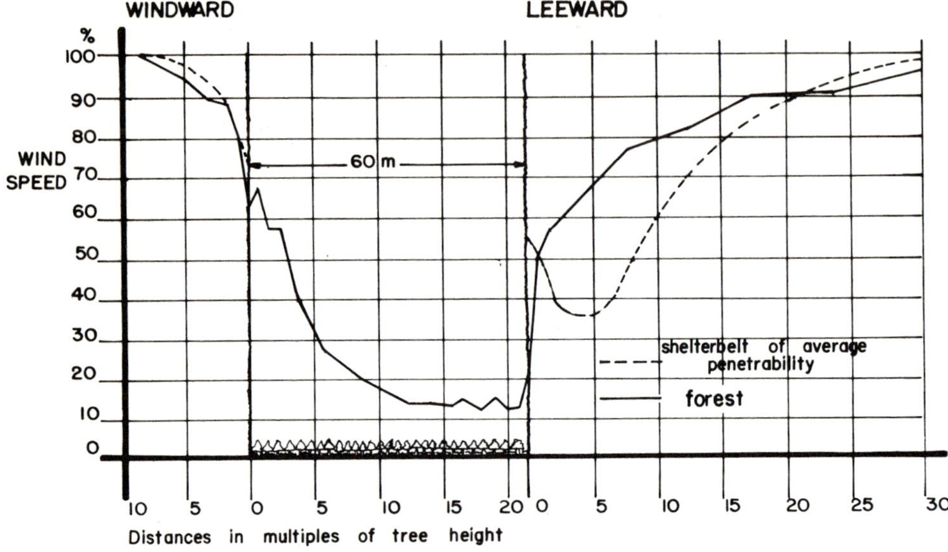

Figure 5-9. Profiles of the wind velocity produced by a forest and by a shelterbelt of average penetrability. As indicated, the shelterbelt is more efficient than the forest in reducing the wind speed, up to a distance of about 20 times the tree height on the leeward side.

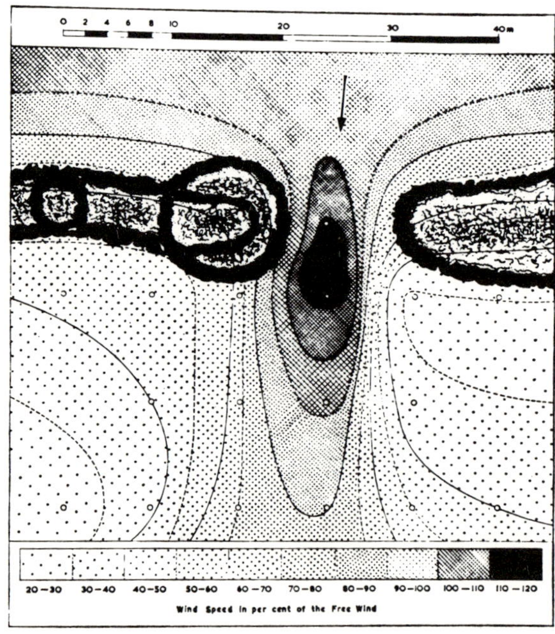

Figure 5-10. Effects of discontinuities in shelterbelts. An opening in a shelterbelt produces changes in the wind speed in the areas near the opening on the leeward and windward sides. Wind speeds through the opening can reach 120% the speed of the free wind. (After Nägeli, courtesy of J. M. Caborn)

Shelterbelts

A shelterbelt is a natural or artificial windbreak made out of trees and/or shrubs. The most effective orientation for a shelterbelt is perpendicular to the direction of the wind.

The width of the sheltered area near the sides of a shelterbelt is reduced, because conditions there are different. It is advisable not to make shelterbelts shorter than 12 times the height that they reach. For shelterbelt lengths up to 20 heights, the protected area increases at a sharp rate, but above this value, the rate of protection increases in direct proportion to the length of the shelterbelt.

The density of a shelterbelt is inversely proportional to the amount of air filtering through it. The denser a shelterbelt is, the lower the wind speed is (Figure 5-11). The density can be classified in terms of the lowest wind speed on the leeward side, as follows:

Density	*Lowest Wind Speed*
Very dense	15% of unobstructed wind speed
Dense	25% of unobstructed wind speed
Moderately dense	35% of unobstructed wind speed
Open	38% of unobstructed wind speed
Very open	68% of unobstructed wind speed

Notice that shelterbelts with different densities (when all other parameters are equal) have different wind-abatement profiles. The lowest wind speed occurs closer to the shelterbelt, as this is denser.

Shelterbelts with equal characteristics but different densities have different wind-abatement profiles. For instance, Figure 5-12 shows that the point of maximum wind abatement is closer to the shelterbelt when the shelterbelt is denser. A certain shelterbelt density can be achieved with a single row of tree planting or with multiple rows (Figure 5-13).

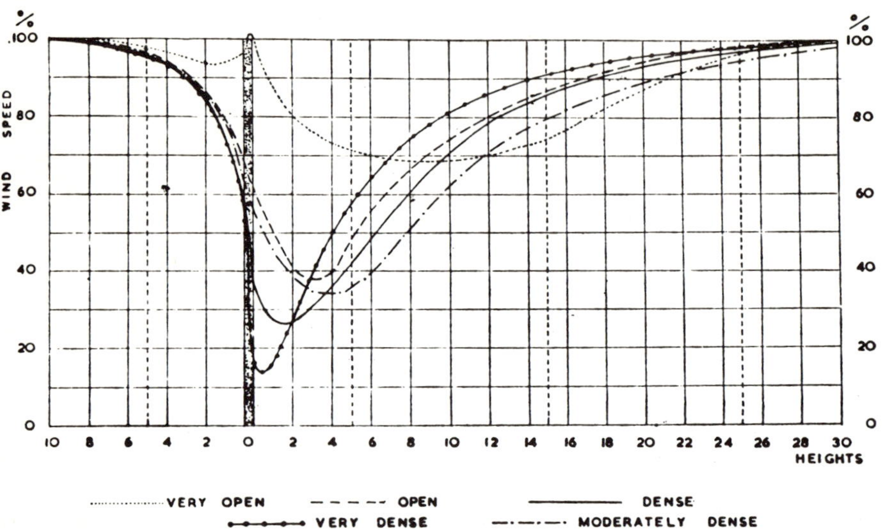

Figure 5-11. Profiles of wind velocity for various shelterbelts having different density. (After Nägeli, courtesy of J. M. Caborn)

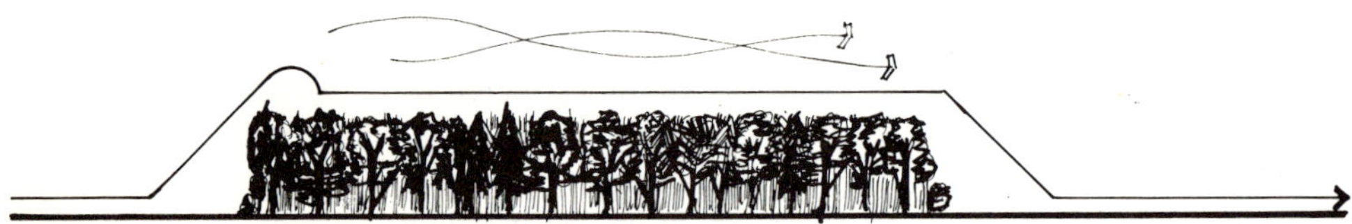

Figure 5-12. The depth of the sheltered area on the leeward side of a wide shelterbelt is very small, especially when compared with the depth of the area behind a narrow shelterbelt.

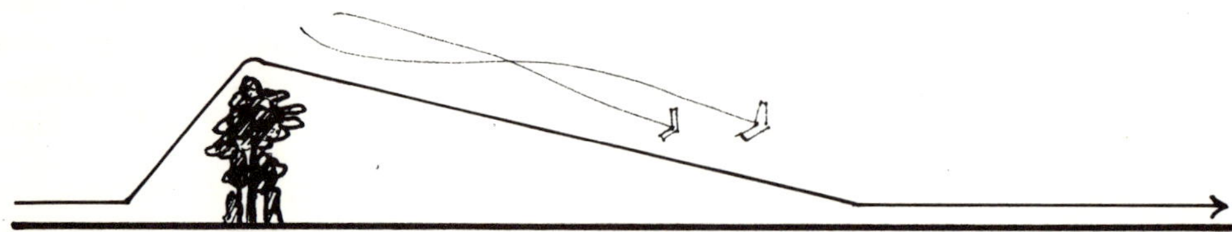

Figure 5-13. Profile of the sheltered area on the leeward side of a narrow shelterbelt shows the higher efficiency of narrow shelterbelts compared to thick ones.

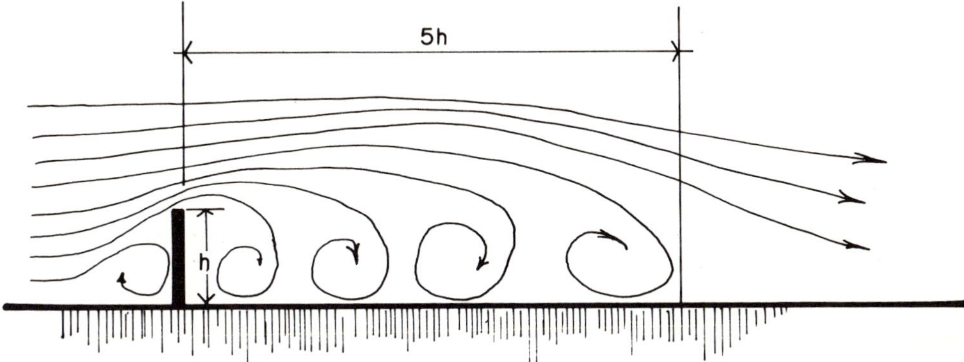

Figure 5-14. Eddy formation behind windscreen with 0% penetration.

Windscreens

In general, the efficiencies of windscreens (fences, palisades, walls, etc.) in reducing wind speed vary according to their permeability. Solid windscreens with zero permeability are not the most efficient. In fact, violent eddies are formed just behind the screen and extend for a distance of approximately five heights (see Figure 5-14). When a certain amount of air is allowed to go through the windscreen, such eddies are reduced. With a permeability of 30%, the eddies are practically eliminated, and the air flowing through and over the fence is steady and laminar. Up to a permeability of 35%, the efficiency increases, but for larger permeability, it starts to decline. Figure 5-7 shows the reduction of wind speed for fences with different permeabilities. (Note that the eddy effect is not indicated.)

EVAPORATION

Function of Wind Speed

Because the wind contributes to the evaporation of water, its action is one of the most important factors affecting the climate. At the same time, wind-caused evaporation influences human life by interfering with everyday tasks, from drying laundry on clotheslines to cooling the human body. In fact, evaporation from the oceans and other water bodies produces cloud formation and the consequential condensation and precipitation that occur within a system of energy transformation involving changes of state and latent heat. Moreover, on the microscale, evaporation with consequential heat absorption can be a factor in reducing high seasonal air temperatures in and around buildings. It can also be used in simple chores, such as drying clothes and foods, in parts of the world with low energy resources.

Let us see the correlation that exists between wind velocity and evaporation rate under certain given conditions.

Wind and Evaporation

As the wind blows over a liquid or a wet surface, it increases the rate at which the liquid evaporates. The wind is not the only factor; the temperature and nature of the liquid, the vapor pressure over the liquid, and the shape and area of the surface are also important in evaporation.

If we consider the evaporation from the oceans, lakes, and other large liquid surfaces and the consequent cloud formation and precipitation, we can see that the wind is a major factor, which, through the weather, affects the environment on a large scale. If we consider the evaporation from small surfaces, such as pools, wet pavement, and wet buildings, and the cooling that derives from the evaporation, we see that wind also influences the environment on the microscale.

How does wind affect evaporation? As heat energizes the molecules in a liquid, they escape from the liquid surface into the air above and collide with each other, producing a vapor pressure. Then, as the wind blows, the evaporation is accelerated because the wind carries away the vaporized molecules and thus reduces the vapor pressure, allowing more molecules to escape from the liquid.

Dalton Formula

The first mathematical equation on the phenomenon of evaporation was formulated by John Dalton, English chemist and physicist, in 1802:

$$E = C(e_w - e_a)$$

where

C = coefficient
e_w = maximum vapor pressure
e_a = actual vapor pressure in air

The Dalton formula, however, did not include the wind effect. In 1886, however, Fitzgerald modified Dalton's formula by including the wind factor, and later several other formulas that include the wind effect were developed.

Fitzgerald Formula

The Fitzgerald formula developed in 1886 is:

$$E = \psi\,(e_w - e_a)$$

where

$$\psi = 0.4 + 0.199w$$

and

E = rate of evaporation (in./24 hr)
ψ = wind factor
e_w = mean vapor pressure at water-surface temperature (in. of Hg)
e_a = mean vapor pressure of saturated air at temperature of dew point (in. of Hg)
w = mean wind velocity near surface of ground (mph)

Meyer Formula

The Meyer formula, developed in 1915, is as follows:

$$E = C\,(e_w - e_a)\,\psi$$

where

$$\psi = 1 + 0.1w$$

and

E = rate of evaporation (in./30-day month)
C = coefficient whose value is 15 for small, shallow-water areas and 11 for large, deep-water areas
e_w = maximum vapor pressure (in. of Hg) corresponding to monthly mean air temperature observed at nearby stations for small bodies of shallow water, or corresponding to water temperature instead of air temperature for large bodies of deep water

e_a = actual vapor pressure in air based on monthly mean air temperature and relative humidity at nearby stations for small bodies of shallow water, or based on information about 30 ft above water surface for large bodies of deep water
w = monthly mean wind velocity, in mph, at about 30 ft above ground
ψ = wind factor

Horton Formula

The Horton formula, developed in 1917, is

$$E = 0.4\,(\psi\, e_w - e_a)$$

where

$$\psi = 2 - e^{-0.2\,w}$$

and

E = rate of evaporation (in./24 hr)
ψ = wind factor
e_w = mean vapor pressure at water-surface temperature (in. of Hg)
e_a = mean vapor pressure of saturated air at temperature of dew point (in. of Hg)

w = mean wind velocity near surface of ground (mph)

For large areas, E is multiplied by

$$(1 - P) + P\frac{\psi - 1}{\psi - h}$$

where

P = fraction of time during which wind is turbulent
h = relative humidity

Rohwer Formula

The Rohwer formula, developed in 1931, is as follows:

$$E = 0.771\,(1.465 - 0.0186B)\,\psi\,(e_w - e_a)$$

where

$$\psi = 0.44 + 0.118_w$$

and

E = rate of evaporation (in./24 hr)
ψ = wind factor
e_w = mean vapor pressure at water-surface temperature (in. of Hg)
e_a = mean vapor pressure of saturated air at temperature of dew point (in. of Hg)
B = mean barometric reading (in. of Hg, at 32°F)
w = mean wind velocity near surface of ground (mph)

Lake Hefner Formula

The Lake Hefner formula (Harbeck et al. 1954) is as follows:

$$E = 0.00177w\,(e_w - e_a)$$

where

E = rate of evaporation (in./24 hr)
e_w = mean vapor pressure at water-surface temperature (mb)
e_a = mean vapor pressure of saturated air at temperature of dew point (mb)
w = mean wind velocity near surface of ground or water (mph)

Lake Mead Formula

The Lake Mead formula (Harbeck et al. 1958) is as follows:

$$E = 0.001813\, w\, (e_w - e_a)\, t\, [1 - 0.03\, (T_a - T_w)]$$

where

E = rate of evaporation (in./t days)
e_w = mean vapor pressure at water-surface temperature (mb)
e_a = mean vapor pressure of saturated air at temperature of dew point (mb)
w = mean wind velocity near surface of ground or water (mph)
t = number of days in period for evaporation
T_a = average air temperature (°C + 1.9°C)
T_w = average water-surface temperature (°C)

Evapotranspiration

Evapotranspiration is the combined effect of evaporation (from surfaces of water, snow, ice, soil, and vegetation, and from wet surfaces of buildings and pavements) and the transpiration process, i.e., the transfer of water from the moist soil to the atmosphere through the action of plants. More specifically, we note that a film of moisture is generally present on the underside of a leaf. Wind, even if gentle, can remove the moist air and leave dry air in its place. Surrounded by dry air, the plant releases more moisture because the vapor pressure inside the leaf is greater than that of the external air, which activates transpiration from the plant. As the speed of the wind increases, the transpiration rate also increases. However, this is only true for small values—such as 8 mph. Above this limit, transpiration decreases, and for strong winds, it may stop completely.

The relationship between wind and evapotranspiration expressed in analytical terms was first proposed by H. L. Penman in 1948:

$$U = \frac{AH + 0.27E}{A + 0.27}$$

where

$$E = 0.35\ (e_a - e_d)\ (1 + 0.0098\,W_2)$$
$$H = R(1 - r)\ (0.18 + 0.55\ S) - B[0.56 - 0.092(e_d)^{0.5}\ (0.10 + 0.90\ S)]$$

$A = \frac{de_a}{dt}$ in mm Hg/°F is derived as a function of the temperature (see Figure 5-13b)
B = coefficient depending on temperature mm of water per day (see Table 5-1)
E = daily evaporation in mm of water
H = daily heat budget at surface in mm of water
R = mean monthly solar radiation expressed in mm of water evaporated per day (see Table 5-2)
S = estimated ratio of actual bright sunshine time to the maximum theoretical time possible
e_a = saturated vapor pressure at mean air temperature in mm of Hg (see Figure 5-15)
e_d = actual vapor pressure in the air in mm of Hg (obtained as the product of e_a and the relative humidity in %)
r = estimated percentage of reflective surface
U = daily evaporation in mm
W_2 = mean wind speed at 2 m above ground in mi/day; or equal to $W_1 \left(\frac{\log 0.6}{\log h}\right)$, where W_1 is the measured wind speed in mi/day at height h in ft

at height h in ft

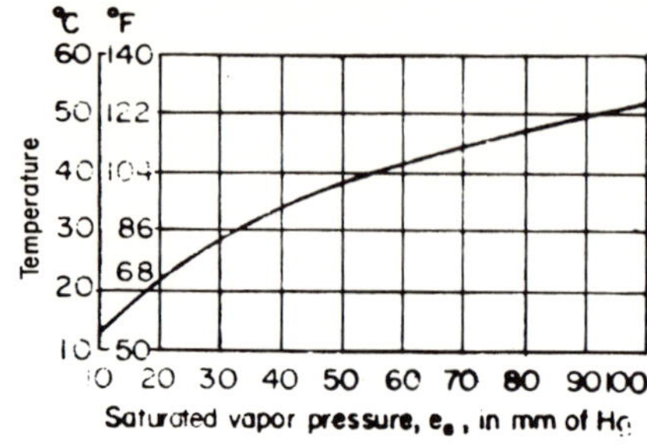

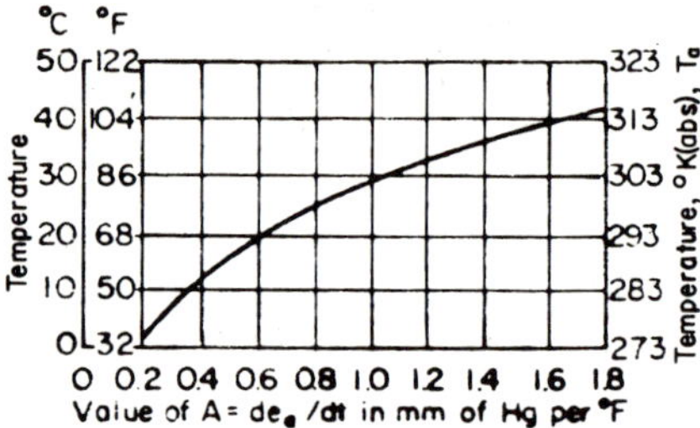

Figure 5-15.

Table 5-1. Values of *B* in the Penman equation.[a]

T_a, °ABS.	*B*, MM WATER/DAY	T_a, °F	*B*, MM WATER/DAY
270	10.73	35	11.48
275	11.51	40	11.96
280	12.40	45	12.45
285	13.20	50	12.94
290	14.26	55	13.45
295	15.30	60	13.96
300	16.34	65	14.52
305	17.46	70	15.10
310	18.60	75	15.65
315	19.85	80	16.25
320	21.15	85	16.85
325	22.50	90	17.46
		95	18.10
		100	18.80

[a] From Criddle (1958).

NOTE: $B = \sigma T_a^4$, where σ is Boltzmann constant, or 2.01×10^{-9} mm/day, and T_a is the air temperature. Heat of evaporation was assumed to be constant at 500 cal/g of water.

Table 5-2. Midmonthly intensity of solar radiation on a horizontal surface, in millimeters of water evaporated per day, or the value R in the Penman equation.[a]

LATITUDE (°)	J	F	M	A	M	J	J	A	S	O	N	D
North												
90	–	–	–	7.9	14.9	18.1	16.8	11.2	2.6	–	–	–
80	–	–	1.8	7.8	14.6	17.8	16.5	10.6	4.0	0.2	–	–
70	–	1.1	4.3	9.1	13.6	17.0	15.8	11.4	6.8	2.4	0.1	–
60	1.3	3.5	6.8	11.1	14.6	16.5	15.7	12.7	8.5	4.7	1.9	0.9
50	3.6	5.9	9.1	12.7	15.4	16.7	16.1	13.9	10.5	7.1	4.3	3.0
40	6.0	8.3	11.0	13.9	15.9	16.7	16.3	14.8	12.2	9.3	6.7	5.5
30	8.5	10.5	12.7	14.8	16.0	16.5	16.2	15.3	13.5	11.3	9.1	7.9
20	10.8	12.3	13.9	15.2	15.7	15.8	15.7	15.3	14.4	12.9	11.2	10.3
10	12.8	13.9	14.8	15.2	15.0	14.8	14.8	15.0	14.9	14.1	13.1	12.4
0	14.5	15.0	15.2	14.7	13.9	13.4	13.5	14.2	14.9	15.0	14.6	14.3
South												
10	15.8	15.7	15.1	13.8	12.4	11.6	11.9	13.0	14.4	15.3	15.7	15.8
20	16.8	16.0	14.6	12.5	10.7	9.6	10.0	11.5	13.5	15.3	16.4	16.9
30	17.3	15.8	13.6	10.8	8.7	7.4	7.8	9.6	12.1	14.8	16.7	17.6
40	17.3	15.2	12.2	8.8	6.4	5.1	5.6	7.5	10.5	13.8	16.5	17.8
50	17.1	14.1	10.5	6.6	4.1	2.8	3.3	5.2	8.5	12.5	16.0	17.8
60	16.6	12.7	8.4	4.3	1.9	0.8	1.2	2.9	6.2	10.7	15.2	17.5
70	16.5	11.2	6.1	1.9	0.1	–	–	0.8	3.8	8.8	14.5	18.1
80	17.3	10.5	3.6	–	–	–	–	–	1.3	7.1	15.0	18.9
90	17.6	10.7	1.9	–	–	–	–	–	–	7.0	15.3	19.3

[a] From Criddle (1958) and Shaw (1936).
NOTE: Values from Shaw (1958) multiplied by 0.86 and divided by 59 give the radiation in mm water/day.

WIND-FIRE INTERACTION

The influence of wind on fires is threefold:

1. It affects the burning rate through the rate of oxygen that the wind supplies to the burning fire.
2. It increases or decreases the rate of fire spreading according to whether the wind direction coincides with or is opposite to the direction of the fire spreading. The wind, in fact, tilts the flames forward or backward.
3. It increases the danger of potential forest fires by drying the vegetation and consequently increasing its flammability.

The wind's effects vary according to whether fires are stationary or moving. The wind can also increase the intensity of fires by affecting the condition of the convective air column that rises above them. This implies that the vertical profile of the wind's speed at low and high altitudes is an important factor. Furthermore, the wind's effect on the severity of potential fires makes the wind the fundamental parameter in the determination of fire danger.

Another detrimental aspect of wind action in combination with fires is its contribution to an increase in the general pollution of the air. According to statistics, throughout the world, large amounts of chemicals and solid particles are injected into the atmosphere by the smoke and the residual ashes of the numerous forest fires that occur annually.

High-Intensity Moving Fires

Moving fires, unlike the stationary type, consume fuel at a high rate and thus need to move to sustain themselves. Moving fires can reach high intensity if the wind speed above the fires either diminishes with height, or if it increases up to a maximum of about 1500 ft above ground and then decreases. Wind speed at the maximum point should be approximately 18 mph, in order for fires to reach the highest intensity.

Fire Spreading

The effect of the wind on fire is to increase the rate of spreading, not only when wind and fire move in the same direction, but even when they do not, or are exactly opposite. As the wind speed increases, the rate of fire propagation also increases. Naturally, when the wind and the fire move in the same direction, the fire advances at a higher speed than it would if wind and fire were opposite to each other. In Figures 5-16 and 5-17 the rate of fire spreading is plotted as a function of wind velocity. In both cases, when the wind velocity is positive (wind and fire moving in the same direction) or negative (wind and fire moving in opposite directions), the rate of spreading increases proportionally to wind velocity. One curve, however, is steeper than the other.

The fact that winds increase the speed at which fires advance, even when blowing in the direction opposite that of the fires, is a paradox that is easily explainable. As the wind's speed increases, the rate of oxygen fed to the fire increases, which in turn increases the rate of combustion forcing the fire to move forward where the fuel is. Figure 5-18 is based on small forest fires tested for experimentation, where wind velocities were measured at 3 ft over the ground level.

Another mechanism of fire spreading caused by the wind is "spotting." Smoldering ashes and burning embers can be picked up by the wind and carried forward, starting new fires ahead. Burning particles can also rise vertically thousands of feet where they are intercepted by high-level winds and carried several miles away, where they can start new fires at a considerable distance from the main fire.

Energy

We can visualize the wind mechanism associated with fires: air converging toward the low-pressure zone over the fire and the rising of a vertical convection column for several thousand feet. The energy sources involved for the existence of these winds can be either the heat generated by the combustion and then converted into kinetic energy, or the energy of the wind per se, independent of the fire.

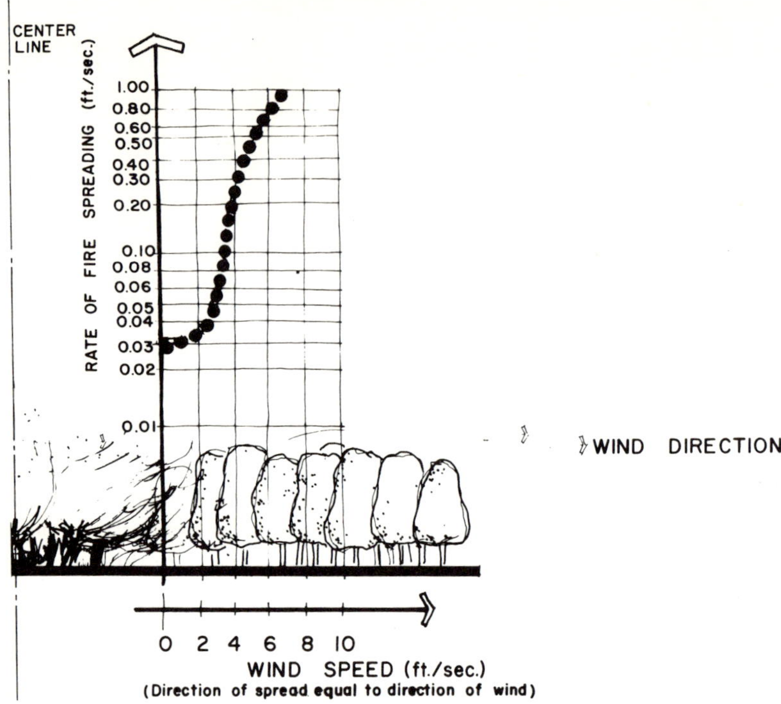

Figure 5-16. Rate of fire spreading when the directions of the wind and fire are the same.

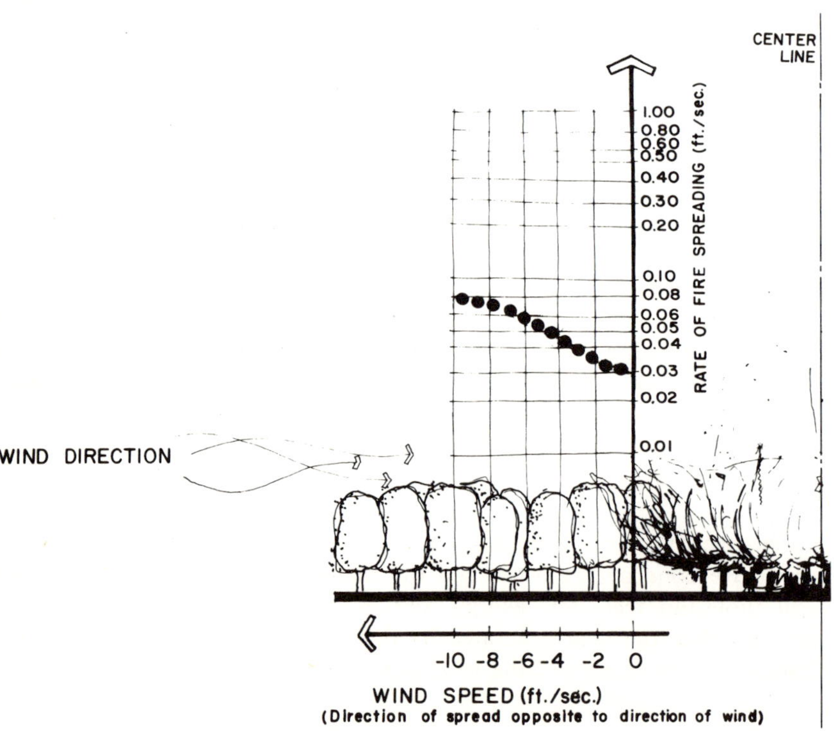

Figure 5-17. Rate of fire spreading when the directions of the wind and fire are opposite.

Figure 5-18. When large forest fires occur, large masses of air over the burning trees are heated and consequently rise because of their reduced density. This upward flow attracts cold air, which converges toward this low-pressure area. Air comes rushing (at times) from great distances, thus producing centripetal winds that can attain substantial speed.

The rate of thermal energy from the fire converted into kinetic energy is given by:

$$E_F = \frac{I}{c_p\,(T_o + 459)}$$

where

E_F = rate of thermal energy converted (ft lb/sec × ft²)
I = intensity of the fire (Btu/ft × sec)
c_p = specific heat of the air
T_o = air temperature (°F)

The rate of the kinetic energy of the wind is expressed by the following:

$$E_w = \frac{\rho\,(w - r)^3}{2g}$$

where

E_w = rate of kinetic energy in the wind (ft lb/sec × ft²)
g = acceleration of gravity (ft/sec²)
r = rate of fire spreading (ft/sec)
w = wind speed (ft/sec)
ρ = air density (lb/ft³)

Fire Danger

The possibility of fire ignition is affected by several factors, including wind. The rate of evaporation of the moisture in vegetation speeds up the drying process and consequently the flammability of the vegetation itself. Since the wind is a principal cause of evaporation, it is directly related to the danger of fire ignition. Wind, of course, is a factor of extreme variability; consequently, the possibility of a fire igniting also varies. Hour-by-hour evaluations of fire danger are necessary where special climatic conditions affect dense forests. These evaluations are done by forest rangers and related organizations.

There are several methods to measure fire danger, and all of them include the wind as an essential factor. For instance, the Southeastern Forest Experiment Station of the U.S. Forest Service classifies fire danger according to a burning index that uses a scale of 1 to 100 points, which are grouped into five classes. The burning index is based on four parameters, one of which is the wind.

Fire Fighting

The wind also has an important effect on fire-fighting operations. When blowing in the opposite direction of the water streams of firehoses, the wind can reduce the effective length of the water jets considerably. When wind speed is less than 5 mph, the wind's effect is negligible and can be ignored. At a speed of approximately 10 mph, the wind can reduce the range of the water streams up to 40%. This is indeed a significant value and should be taken into account for the computation of the effective range of a water stream.

On the other hand, when the wind's direction coincides with that of the water streams, their length increases. The direction of the water can be adjusted by the hose operators so that they can take full advantage of the wind-boosting action.

WIND EFFECT ON CLIMATE

The presence of wind in combination with other factors plays a major role in assessing the quality of climate for any given area around the world. The following relationship proposed by Maunder (1962) ties together 13 independent variables (R_1, R_2, R_3, S_1 . . .) concerning rainfall, number of sunshine days, temperature, humidity, the number of windy days, etc., in order to compute the dependent variable X, referred to as the *human climate index*.

Table 5-3. Rating values for evaluation of human climate index.

factor rating	1	2	3	4	5
R_1 = Mean annual rainfall (in.)	10.0–17.7	17.8-31.6	31.7-56.2	56.3-99.9	100.0-177.8
R_2 = Mean annual duration of rainfall (h)	355–446	447-562	563-707	708–891	892-1122
R_3 = Percentage of rainfall occurring at night (9 P.M.-9 A.M.)	53.0–54.9	51.0-52.9	49.0-50.9	47.0–48.9	45.0–46.9
S_1 = Mean annual duration of bright sunshine (h)	2400-2599	2200-2399	2000-2199	1800-1999	1600–1799
S_2 = Mean winter duration of bright sunshine (h)	450–499	400–449	350-399	300-349	250–299
T_1 = Mean annual degree-days (base 60°F)	1000–1349	1350-1905	1906-2630	2631-3630	3631-5011
T_2 = Mean number of days with screen frost per year [a]	0–7.9	8.0-23.9	24.0–47.9	48.0-79.9	80.0-119.9
T_3 = Mean daily maximum temperature of coldest month (°F)	56.0–59.9	52.0-55.9	48.0-51.9	44.0–47.9	40.0–43.9
T_4 = Mean annual maximum temperature (°F)	74.0–77.9	78.0-81.9	82.0-85.9	86.0–89.9	90.0–93.9
T_5 = Mean number of days with ground frost per year [b]	0–14.9	15.0–44.9	45.0–89.9	90.0-149.9	150.0-224.9
H_1 = Humidity index expressed as dew point (°F)	50.0–51.9	52.0-53.9	54.0-55.9	56.0–57.9	58.0–59.9
W_1 = Mean number of days with wind gusts ≥40 mph per year	0–14.9	15.0–44.9	45.0–89.9	90.0-149.9	150.0–224.9
W_2 = Mean number of days with wind gusts ≥60 mph per year	0–2.4	2.5-4.9	5.0-12.4	12.5-22.4	22.5-35.0

Source: W. J. Maunder, Weather **17**, No. 1, Jan. (1962), p. 9.

[a] Days when the meteorological instrument shelter falls below 0°C (32°F).
[b] Days when the minimum temperature of the grass falls below −0.8°C (30.5°F).

$$X = (3R_1 + 3R_2 + 2R_3) + (4S_1 + 3S_2) + (2T_1 + T_2 + T_3 + T_4 + T_5) + (5H_1) + (2W_1 + 2W_2)$$

The definition of the variables and their values are given in Table 5-3. Note that each of the independent variables is rated from 1 to 5, with 1 the best rating and 5 the worst. Therefore, the best value for X is 30. This value is obtained when all variables are rated 1. The worst value for X is 150, which is obtained when all variables are rated 5.

A significant aspect of Maunder's human climate index is that it takes into consideration not only the mean annual condition of the location, but also the winter and summer conditions independently, and correlates all three into one index. Among sites where the 13 independent variables can be measured, a comparative evaluation is now possible due to Maunder's criteria.

Notice that the wind factors (W_1 and W_2) in the formula only take into consideration the negative effects of the wind on the climate; the positive effects of the wind are not considered. These limitations, however, could eventually be removed by reevaluating and correcting the basic formula, which in itself constitutes a significant beginning for the inclusion of wind in climate classifications. The practicality of the formula can at times be reduced by the difficulty in collecting the necessary climatological data. Examples of available measurements for a given geographical locality in the United States are shown in Appendix C.

Wind Influence on Fog

Fog is like a cloud, which, rather than being aloft, is formed in a layer within 50 ft of the ground and remains there. It is composed of either small droplets of water or small crystals of ice.

The formation of fog is greatly influenced by the wind speed. If the wind speed is low, the layer of turbulence over the ground where the air continuously mixes is very thin. The air in this layer is cooled by the cold surface of the ground and makes the water vapor condense and become dew. On the other hand, if the wind speed is too high, this layer of turbulence is very thick so that the fog being formed at the ground is lifted by the turbulent air and changed from fog to what we can call a "low cloud."

Windchill

The blowing of the wind against an object increases the rate of heat flow from the object to the surrounding air. The rate increases as the wind speed rises. Siple (1951) developed an index that combines temperature and wind speed into one value called the *windchill*. This value is equivalent to the temperature one would feel without the wind. For instance, if the actual measured temperature is 90°F, and the wind speed is 45 mph, the windchill is 83°F. A compilation of windchill temperatures is given in Tables 5-4 and 5-5.

The amount of heat loss produced by the windchill effect is given in metric form by the following equation (Siple 1951):

$$\text{Windchill} = (10.45 + 10\sqrt{v} - v)(33 - t)$$

For example, at a temperature of 32.22°C (90°F) and a wind speed of 20.11 m/sec (45 mph), a person will experience a heat loss as given by the following equation:

$$\text{Windchill} = (10.45 + 10\sqrt{20.11} - 20.11)(33 - 32.22) = 27.44 \text{ kcal/m}^2\text{/hr}$$

See Figure 5-19.

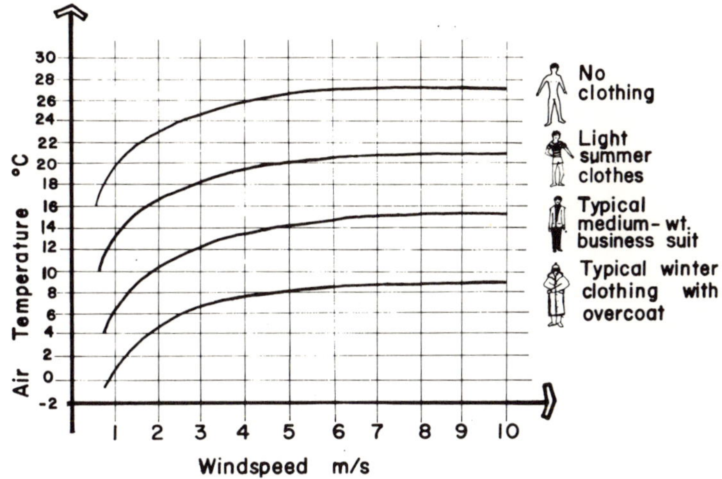

Figure 5-19. Windchilling effect.

Human Comfort

The effect of the wind on the human body is basically twofold:

1. It affects the heat exchange between body and air proportionally to the air speed through convection.
2. It increases the evaporation of sweat and therefore the cooling of the skin.

Table 5-4. Equivalent temperatures on exposed flesh at varying wind velocities.

WIND VELOCITY (MPH)										
45	35	25	20	15	10	5	3	2	1	EQUIVALENT TEMPERATURE (°F)
TEMPERATURE READING (°F)										
90	89.5	89	88.5	88	88.8	87.5	87	86	84.5	83
82	81	80.5	80	79.5	78	76	74	72.5	70	60
72	71	69.5	68	67	65	60	57	53.5	47.5	23
63	61	59	57	55	52	44.5	39	34.5	20	−11
51	49	47	45	42.5	38	28	18.5	11	0	−27
41	39	36	34	30.5	24	11	0	− 9	−23.5	−38
30	28	25	23	18	11	− 5	−16.5	−40	–	–
20	18	14	11	6	− 2	−19	−40	–	–	–
10	7.5	3	0	− 6	−15	−35	–	–	–	–
0	− 2.5	− 8	−12	−18	−29	–	–	–	–	–
−11	−14	−18	−23	−30	–	–	–	–	–	–
−21	−24	−30	−35	–	–	–	–	–	–	–
−32	−35	−40	−40	–	–	–	–	–	–	–

Table 5-5. Windchill factors. (Courtesy U.S. Dept. of Commerce, Environmental Science Services Administration)

	dry-bulb temperature (°F)																
	35	30	25	20	15	10	5	0	-5	-10	-15	-20	-25	-30	-35	-40	-45
mph	*equivalent temperature*[a] *of wind chill index* (°F)																
calm	35	30	25	20	15	10	5	0	-5	-10	-15	-20	-25	-30	-35	-40	-45
5	33	27	21	16	12	7	1	-6	-11	-15	-20	-26	-31	-35	-41	-47	-54
10	21	16	9	2	-2	-9	-15	-22	-27	-31	38	-45	52	-58	-64	-70	-77
			very cold														
15	16	11	1	-6	-11	-18	-25	-33	-40	-45	-51	-60	-65	-70	-78	85	-90
				bitterly cold													
20	12	3	-4	-9	-17	-24	-32	-40	-46	52	-60	-68	-76	-81	-88	-96	-103
25	7	0	-7	-15	-22	-29	-37	-45	-52	58	67	-75	-83	-89	-96	-104	-112
						extreme cold											
30	5	-2	-11	-18	-26	-33	-41	-49	-56	-63	-70	-78	-87	94	-101	-109	-117
35	3	-4	-13	-20	27	-35	-43	-52	-60	-67	-72	-83	-90	-98	-105	-113	-123
40	1	4	15	22	29	-36	45	54	62	69	76	87	-94	101	107	116	128
45	1	-6	-17	-24	-31	-38	-46	-54	-63	-70	-78	-87	-94	-101	-108	-118	-128
50	0	-7	-17	-24	-31	-38	47	56	-63	70	-79	-88	-96	-103	-110	-120	-128

[a] Equivalent in cooling power or exposed flesh under calm condition.

WIND AND NATURE

Environment-Organism Energy Exchanges

The wind is one of the factors that regulate the heat exchange between animal organisms and the physical environment that surrounds them. The exchange mechanism involving only the wind is the heat flow due to convection. The total energy exchange, instead, also includes other mechanisms as will be shown hereafter. Notice that the energy transfers either from the organism to the environment or from the environment to the organism, until a condition of equilibrium between the two is reached.

The energy cycle can be seen as a transformation process that converts the chemical energy of food into the mechanical energy necessary for work, for breathing and blood circulation, for metabolic processes, and for heat. One can also note the energy present in the environment that comes from the sun, and its transformation into the kinetic energy of the wind, chemical energy, and other forms supporting organic processes, photosynthesis, etc. The two-way energy exchanges are also easily visualized. For instance, heat radiation from the surface of an animal body to the air around it, or vice versa, and heat radiation from objects of a given emissivity to an animal body, are forms of such exchange. Heat conduction or convection are other possible ways of energy exchange.

The expression for energy equilibrium between the body surface of an animal organism and the environment around it is:

$$M - E + R = \epsilon\sigma(T_s)^4 + \frac{K\sqrt[3]{V}}{\sqrt[3]{D^2}}(T_s - T_a) - C$$

where

M = metabolic energy (heat); chemical energy from food transformed into other forms and finally rejected as heat.
E = energy loss due to the evaporation of liquids from the organism through respiration and perspiration. This can be expressed in cal/gram of evaporated liquid.
R = radiant energy absorbed by the surface of the animal organism
ϵ = emissivity of the surface
σ = Stefan-Boltzmann constant
T_s = absolute temperature of the surface
K = constant of proportionality
V = wind velocity (cm/sec)
D = diameter of the body (cm)
T_a = absolute temperature of the air (°K)
C = energy exchanged by conduction

Note that the term $\epsilon\sigma(T_s)^4$ is the radiant energy emitted per unit area, and $\frac{K\sqrt[3]{V}}{\sqrt[3]{D^2}}$ $(T_s - T_a)$ is the heat transferred by the wind due to convection.

Animal Structures

A bird's nest set on a tree branch would fall with a gust of wind if special fastening had not been provided. A bird somehow evaluates the wind force on the light, intricate nest as it struggles to construct it. The job is not easy, considering that a bird can only use its beak and claws to hold and fasten and weave the little components it finds. The inverted dome, which gradually acquires tensile and compressive strength, is built in a sequence of operations that makes the portion already built stable and resistant against external forces, of which the wind is the most severe. Behavioral studies of certain birds show some of the extraordinary techniques employed during the construction of the nest.

Plant Structural Resistance to Wind

The omnipresence of wind on earth is such that any form of life is to some extent compatible with its effects. Plants must be able to survive the force of conventional winds by means of some structural mechanism. For example, the root system of a tree spreads widely underground to obtain nourishment and thus achieves the structural requirement for wind resistance. Of course, superficial root development due to a high water table, although sufficient for physiological requirements, is often inadequate for resisting wind forces. Wind forces on trees are proportional to the surface of their crowns and are applied as in any other structure as positive pressure and suction, with a center of gravity at a certain height from the ground where the resultant is located. Proportional to this resultant, of course, there is a reaction applied at a point underground, at the center of gravity of the tensile forces developed in the root system. The root filaments can only react with tensile forces to the horizontal wind action over the crown and trunk of the tree. For the stability of the tree, the sum of the forces and reaction must be equal to zero, and the sum of the moments of said forces and reaction with respect to any point must also be equal to zero. Assuming a fulcrum somewhere underground near the surface, the mechanism for resisting the overturning moment can be visualized (see Figure 5-20).

Airborne Pollen

In the complex mechanics of nature, the wind performs a vital role in the reproduction of plants. Plants with bright colored flowers that have fragrant perfumes, nectar, and moist pollen are fertilized by insects that carry the pollen to the female cones or to the eggs in the pistils of the flowers they visit. Plants that do not attract insects

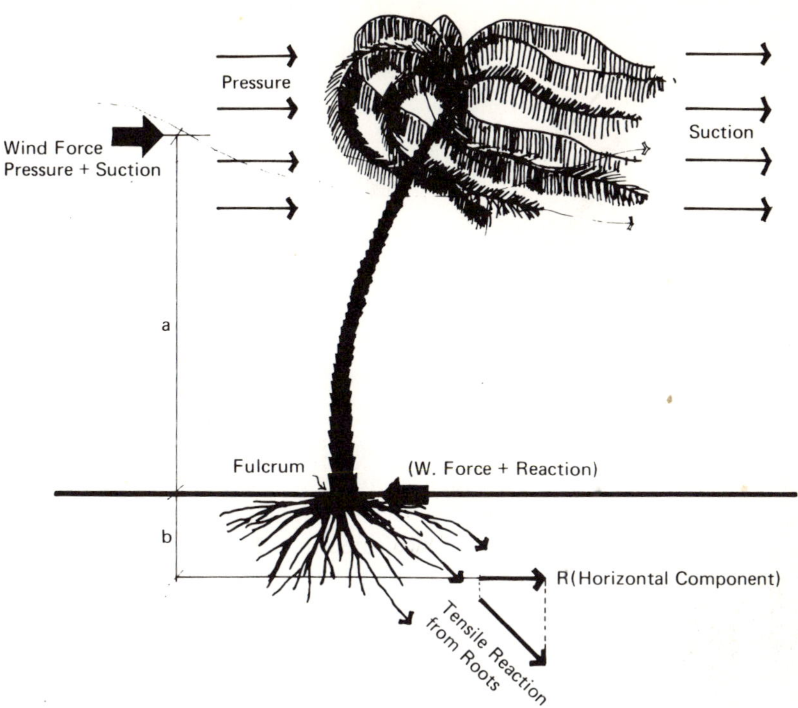

Figure 5-20. Structural strength of trees to wind loads is developed by tensile forces applied to the root system. Adequate resisting moments usually counteract overturning moments caused by the wind, and structural failures generally occur in the trunk and at the fulcrum due to bending. Uprooting (inadequate tensile resistance) is typically more rare.

Figure 5-21. Carried by the wind, tumbleweeds roll and roll over the plains. They are typical of the deserts and prairies of the United States and Russia. A tumbleweed is like a dry, light ball up to 3 ft in diameter. Broken off at the ground line by the wind when their seeds have matured, tumbleweeds spread their seeds on the land as they roll.

are pollinated by the wind. Light, dry pollen produced by these plants is carried in great quantities by the wind to the female organs of other plants. Among these plants are pine, oak, birch, and grasses. Such an important and inevitable natural phenomenon as wind-aided pollination should not be neglected in the design of environmental conditions for human comfort.

In addition to its positive role in nature, wind also has a negative aspect; that is, it carries aeroallergens that can cause allergic reactions in some individuals when they are exposed to them. The airborne allergens are deposited in a person's upper air passages when they are inhaled, or they can irritate the mucous membranes of the eyes if the wind speed is high. However, if a person breathes through the nose, most of the pollen is trapped in the nasal passages and does not reach the lungs. Aeroallergens are small and have an average diameter of 1 to 80 μ. Their density also varies; for most types of pollen, it is between 0.4 to 1.2 grams/cm^2 (0.014 to 0.043 lb/in.^{3}).

The release of pollens and fungi spores into the air is mostly due to the turbulence of the wind that detaches them from the mother plants. Sometimes, in rarer cases, such as with the mulberry, the anther sacs burst open and eject the pollen into the air. The release of pollens and fungi spores by the wind turbulence increases quantitatively in proportion to the wind speed.

See Figure 5-21.

WIND OVER WATER AND SHORES

Wind Over Water

The effect of the wind on water is no longer a specific engineering topic but now involves architecture as well, since urban expansion is pushing us toward the sea. Lakes, lagoons, sound waters, and even the open ocean waters have become objects of study for potential urban sites. Such interest is not new, as evidenced from the lacustrine settlements of the neolithic age and the example of Venice in its fifteenth century of glorious existence. (See Figures 5-22 and 5-23.)

Figure 5-22. The action of moist winds combined with the sea has been tested in Venice for over 15 centuries. For the most part, the buildings have resisted the battering of the wind-generated waves against the foundations. (Photographs by Ed. A Storti, Venice)

Recent studies on this topic include:

- "Triton City," a project carried out in 1966 by Buckminster Fuller for the United States Department of Housing and Urban Development exploring the possibility of floating cities
- "Hawaii's Floating City," a reinforced concrete floating island designed in 1970 at the University of Hawaii for a site in the open ocean approximately 6 mi from the coast (Figure 5-24)
- "Sea City," a study by Pilkington Glass Co. of London for a city supported on piles in the shallow waters of the North Sea 15 mi off the shores of Yarmouth (Figure 5-25)
- "Aquapolis," the floating city of the future, represented by the small prototype built for Expo '75 at Okinawa

Figure 5-23. In Amsterdam, even in cold weather, water and wind do not detract from the beauty or comfort of living in a water-based city.

Figure 5-24. Model of Hawaii's Floating City Project, by the University of Hawaii Department of Architecture. A modular type system using concrete designed for the open ocean, 5 miles south of Honolulu. (Research sponsored by NOAA)

Figure 5-25. "Sea City," a study by the Pilkington Glass Co., in London, for a city supported on piles in the shallow waters of the North Sea, 15 miles off the shores of Yarmouth. Shaped like an amphitheater, the model indicated in the figure shows a surrounding wall enclosing 16-story structures. It is 4700 ft long and 3300 ft wide. (Architect, Hal Moggridge)

Figure 5-26. Housing units in the Tokyo Bay. (Architect, Kenzo Tange)

Of course, a major obstacle to the practical realization of building on water is the difficulty of providing the structural stability needed against the action of wind-induced waves and the wind per se. A suggested method for reducing this action is to provide floating breakwaters consisting of cylindrical plastic-coated bags filled 90% with fresh water and 10% with air. These partially submerged bags counteract the wave motion by the action of secondary waves within the bags. Such a method is also applicable in deep waters, where no other systems are available. Another practical system that seems to be effective in reducing wind action on water is the release of compressed air bubbles from an undersea pipeline. Such a system, however, has only limited application and is suggested only for critical spots of limited size, such as the entrances to ports.

Wind-Induced Waves

Whether floating, or bearing on piles at an offshore site, or leaning out over water from a rocky edge, buildings must be designed for wave action. Avoiding, minimizing, or resisting the powerful force of waves employs a certain knowledge of how these wind-induced wave phenomena are generated.

Water waves are basically characterized by height, length, period, and speed. Most important for the design purposes considered herein is the height, which is proportional to:

- length of time the wind has blown
- the fetch (unobstructed length over which the wind has an approximate constant speed and direction)
- wind speed
- water depth

A group of empirical formulas particularly simple and very approximate are presented here first. More specific and exact types will be introduced later.

Stevenson Formula

The formula proposed by Thomas S. Stevenson in 1864 applies to small ponds with a minimum fetch of 600 ft and to the ocean with fetches of hundreds of miles:

$$H = 1.33h \text{ on vertical surfaces (see Figure 5-27)}$$

and

$$H = 1.5h \text{ on inclined surfaces}$$

where

H = maximum wave height (ft)
h = wave height (ft)
F = fetch (mi)

Modified Stevenson Formula

The modified Stevenson formula takes into account not just the fetch but also the wind speed. Furthermore, the formula is expressed differently according to the fetch length:

1. for fetches less than 20 mi:

$$H = 0.17\sqrt{UF} + 2.5 - F^{1/4}$$

2. for fetches longer than 20 mi:

$$H = 0.17\sqrt{UF}$$

where

H = wave height (ft)
F = fetch (mi)
U = wind speed (mph)

Scripps Formula

The Scripps formula, proposed by the Scripps Institution of Oceanography in La Jolla, California, is applicable to ocean waves. The relationship is limited only to wave height and wind velocity.

$$H = 0.026\ U^2$$

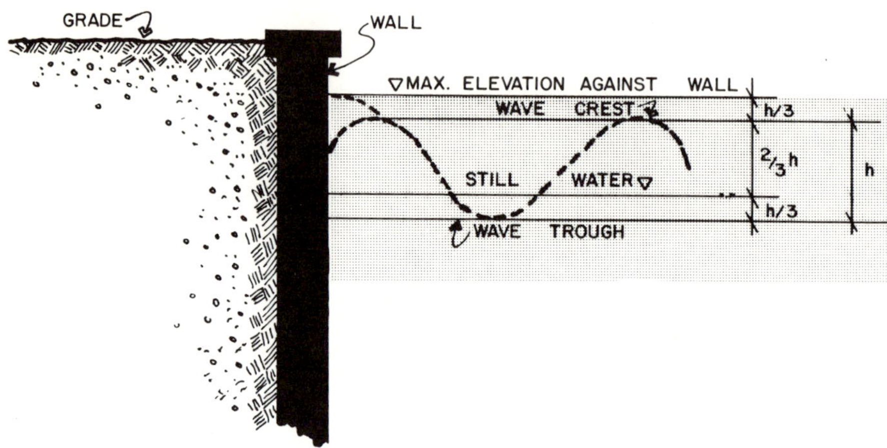

Figure 5-27. Height of freeboard wave action using the Stevenson formula based on wind velocity and other parameters.

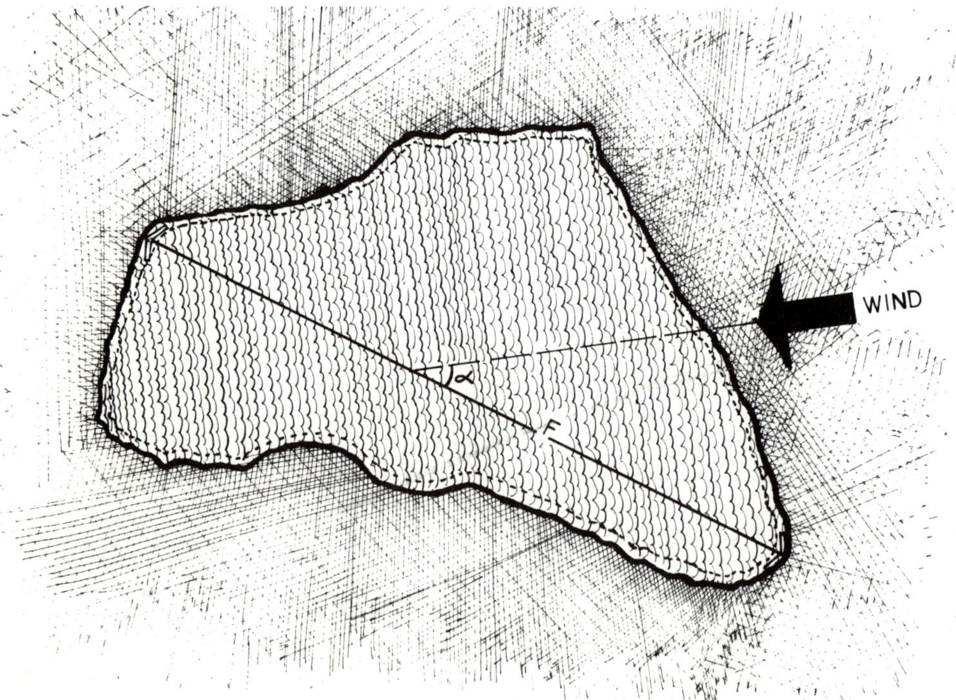

Figure 5-28. For predicting the rise of the water level in lakes due to wind, the Zuider Zee formula considers the maximum possible fetch F in the lake and the angle α between the fetch and the wind direction.

where

H = wave height (ft)
U = wind speed (knots) (Note: to convert mph to knots, multiply mph $\times$ 0.8689)

Zuider Zee Formula

The Zuider Zee formula combines not only the wind speed and fetch to the rise of water level, but also the water depth and the angle between the wind direction and the fetch that are considered here to be different (Seelye 1962). The fetch is assumed to be the largest dimension of the lake. (See Figure 5-28.)

$$S = \frac{U^2 F}{1400\, d} \cos a$$

where

S = rise of water level above normal (ft)
F = fetch (unobstructed largest dimension of the lake) (ft)
U = wind speed (mph)
d = average water depth (ft)
α = angle between wind direction and fetch

More accurate formulas for the prediction of wave height, together with time-saving charts, follow.

Sverdrup–Munk–Bretschneider Formula

The formula originally produced by Sverdrup and Munk in 1947 and later modified by Bretschneider in 1952 applies to deep water waves (U.S. ACERC 1973):

$$\frac{gH}{U^2} = 0.283 \tan h \left[0.0125 \left(\frac{gF}{U^2} \right)^{0.42} \right]$$

$$\frac{gT}{2\pi U} = 1.20 \tan h \left[0.077 \left(\frac{gF}{U^2} \right)^{0.25} \right]$$

and

$$\frac{gt}{U} = K \exp\left(\left\{A\left[\ln\left(\frac{gF}{U^2}\right)\right]^2 - B\ln\left(\frac{gF}{U^2}\right) + C\right\}^{1/2} + D\ln\left(\frac{gF}{U^2}\right)\right)$$

where

$\exp\{x\} = e^{\{x\}}$
$\ln = \log_e$
$K = 6.5882$
$A = 0.0161$
$B = 0.3692$
$C = 2.2024$
$D = 0.8798$
$F =$ fetch
$U =$ wind speed
$T =$ period
$g =$ acceleration of gravity

This formula has been used in Figures 5-29 and 5-30 to facilitate its use. Knowing the wind speed in knots or mph, the fetch length in statute or nautical miles, and the duration of the wind in hours, it is possible to read directly the corresponding wave height in feet and the period in seconds.

Example: (U.S. ACERC 1973)
Given: A wind speed $U = 35$ knots (40 mph), and a duration $t = 10$ hours.
Find: The significant wave height H_F, and the significant period T_F.

1. A fetch length, $F = 200$ NM.
2. A fetch length, $F = 80$ NM.

Solution:

1. Read Figure 5-29 from the left side at $U = 35$ knots, and move horizontally across the figure from the left toward the right, until you intersect the dashed line

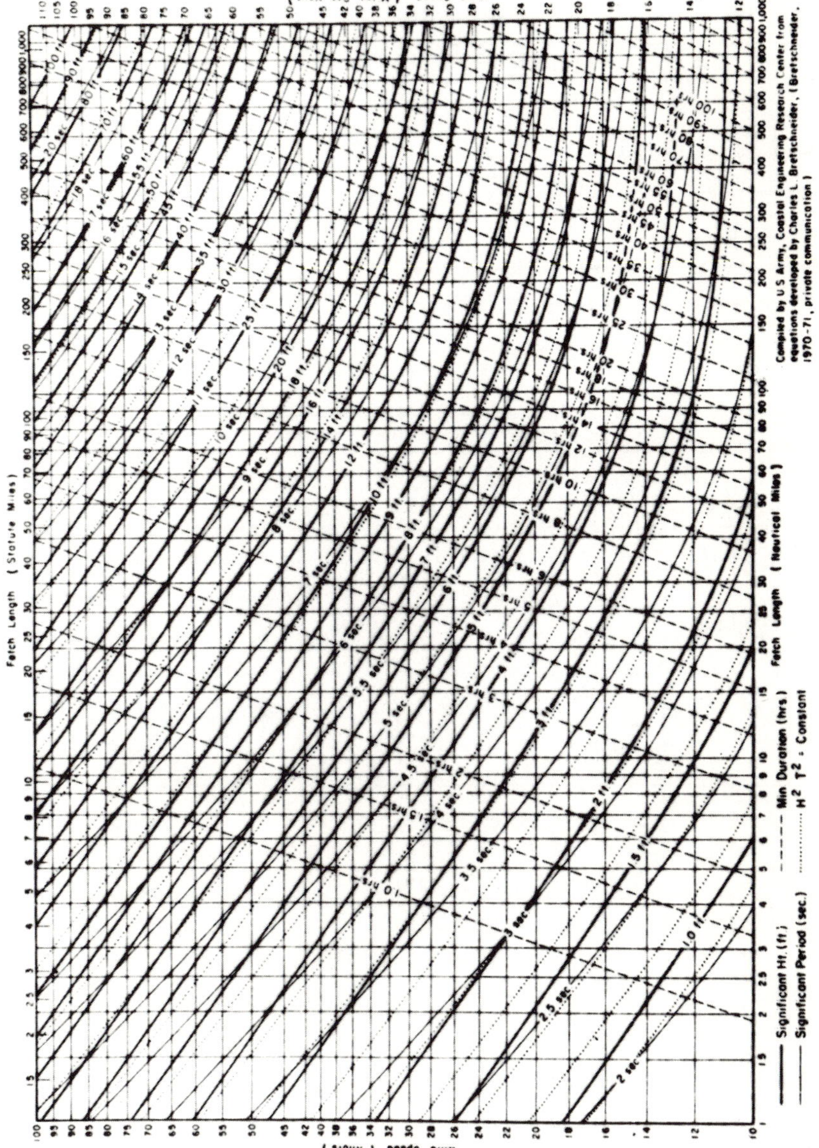

Figure 5-29. Deep-water wave forecasting curves as a function of wind speed, fetch length, and wind duration (for fetches 1 to 1000 nautical miles).

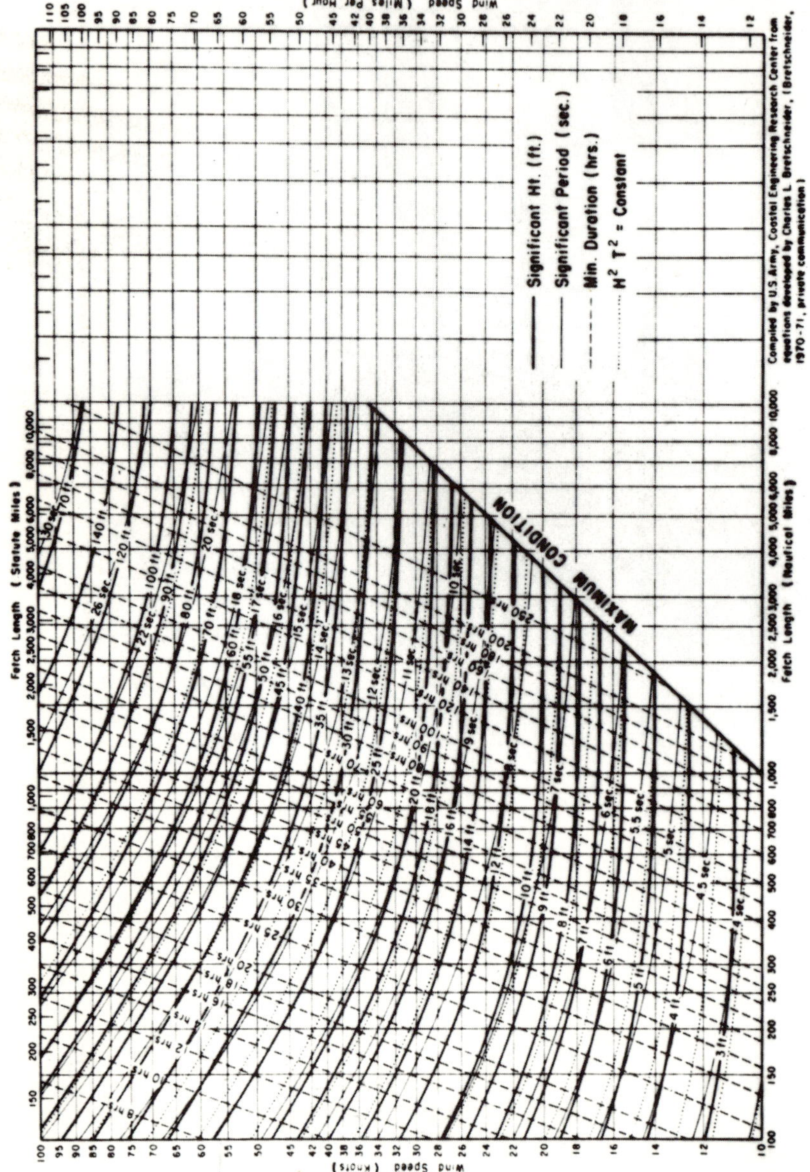

Figure 5-30. Deep-water wave forecasting curves as a function of wind speed, fetch length, and wind duration (for fetches 1 to 10,000 nautical miles).

representing a time duration of 10 hr that comes before the line indicating a fetch length of 200 NM. At the 10-hr duration line, $F = 92$ NM; this is the minimum fetch F_m for this case. With $U = 35$ knots, $t = 10$ hr, and $F = 200$ NM, then $H_F = 13.1$ ft, $T_F = 8.0$ sec, t_m equals 10 hr, and $F_m = 92$ NM.

2. Looking at Figure 5-29, when $F = 80$ NM and $t = 10$ hr, then the heights, periods, minimum duration, and fetch would be $H_F = 12.6$ ft, $T_F = 7.8$ sec, and $t_m = 9.0$ hr. The minimum duration t_m is 9 hr, corresponding to the miles which limit generation, and comes before a duration of 10 hr.

In this example, the wave pattern in (1) is limited by the duration; the wave pattern in (2) is limited by the fetch.

Shallow-Water Wave Formula

The shallow-water wave formula is a modification of the previous formula used for deep water. Considerations of energy losses due to bottom friction and percolation are added. The first form of the formula (U.S. ACERC 1973) is as follows:

$$\frac{gH}{U^2} = 0.283 \tan h\left[0.578\left(\frac{gd}{U^2}\right)^{0.75}\right] \tan h\left\{\frac{0.0125\left(\frac{gF}{U^2}\right)^{0.42}}{\tan h\left[0.578\left(\frac{gd}{U^2}\right)^{0.75}\right]}\right\}$$

and

$$\frac{gT}{2\pi U} = 1.20 \tan h\left[0.520\left(\frac{gd}{U^2}\right)^{0.375}\right] \tan h\left\{\frac{0.077\left(\frac{gF}{U^2}\right)^{0.25}}{\tan h\left[0.520\left(\frac{gd}{U^2}\right)^{0.375}\right]}\right\}$$

Figure 5-31a–j allow the reader, knowing *the* wind speed in mph and the fetch in feet, to find, for a given water depth, the wave height in feet and the period in seconds.

Example:
Given: Fetch $F = 80{,}000$ ft, wind speed $U = 50$ mph, water depth $d = 35$ ft (average constant depth), bottom friction factor $f_f = 0.01$ (assumed).

Find: Wave height H and wave period T.

Solution: From Figure 5-31 for constant depth, $d = 35$ ft, for $F = 80{,}000$ ft, and $U = 50$ mph. Then, $H = 6.2$ ft (say, 6 ft) and $T = 4.1$ sec (say, 4 sec).

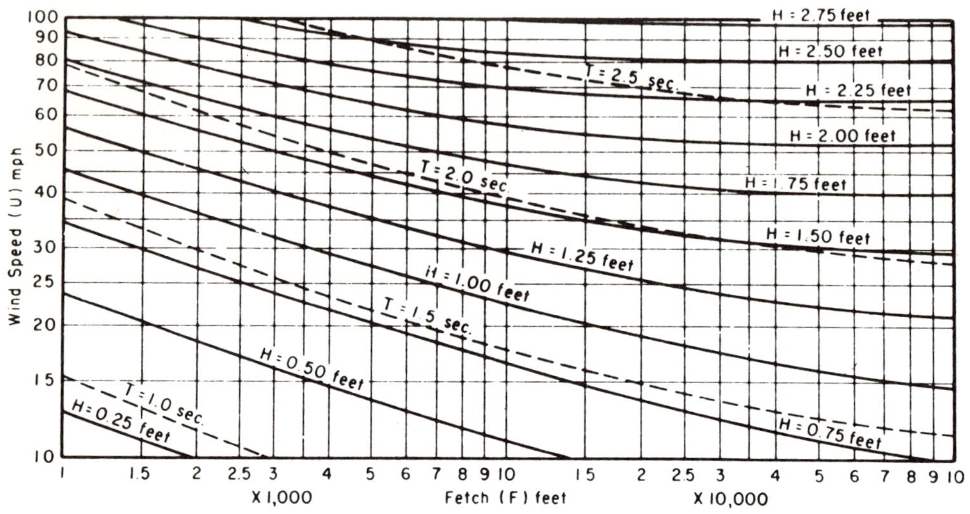

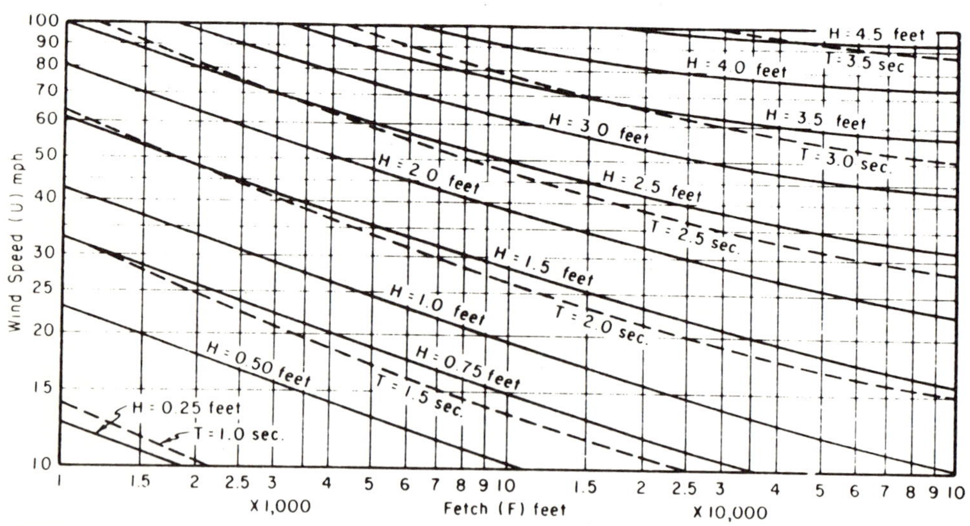

Figure 5-31. Forecasting curves for shallow-water waves. *a,* Constant depth = 5 ft; *b,* 10 ft; *c,* 15 ft; *d,* 20 ft; *e,* 25 ft; *f,* 30 ft; *g,* 35 ft; *h,* 40 ft; *i,* 45 ft; *j,* 50 ft. (Courtesy of U.S. Army Coastal Engineering Research Center, *Shore Protection Manual,* vol. II, Dept. of the Army Corps of Engineers, 1973)

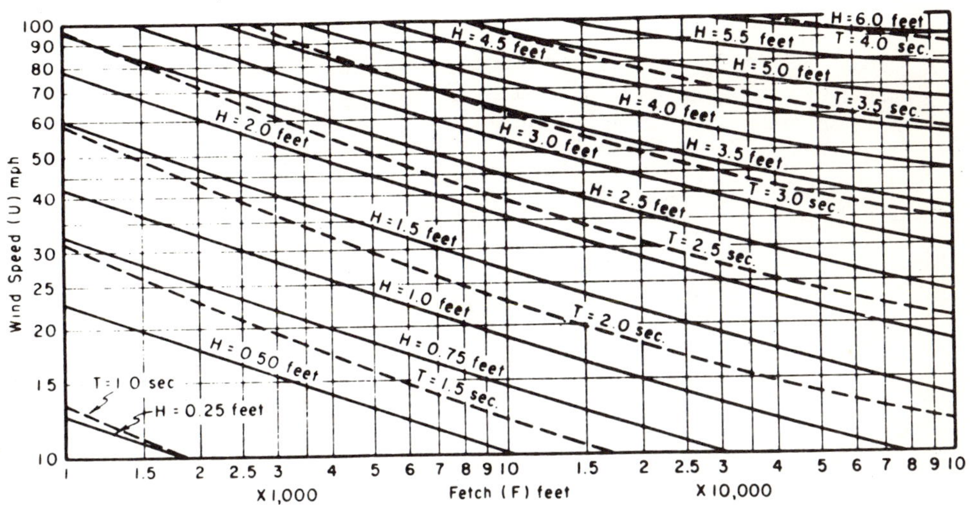

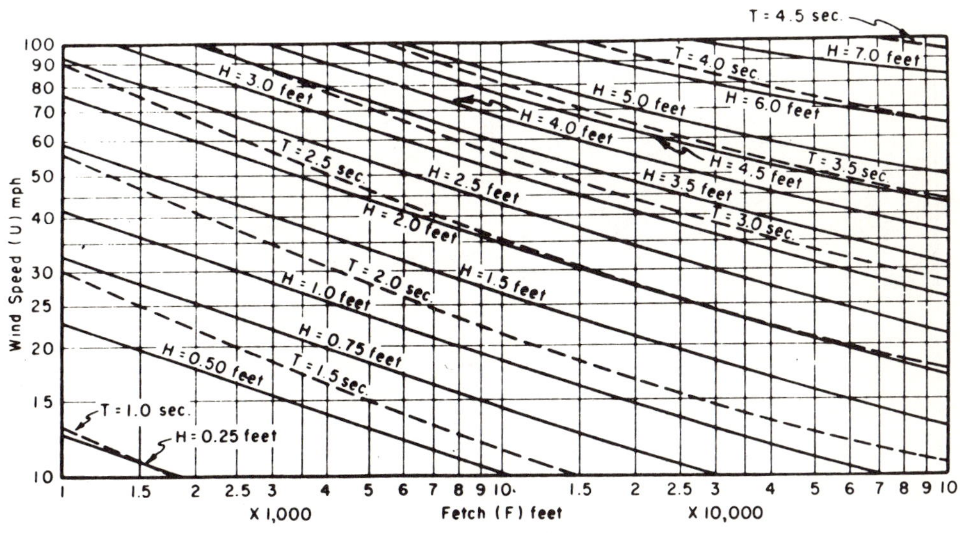

Figure 5-31. Continued.

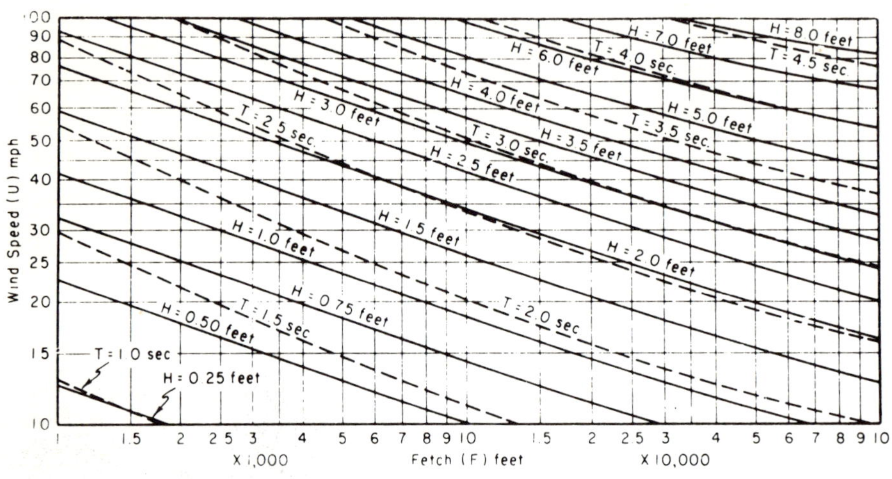

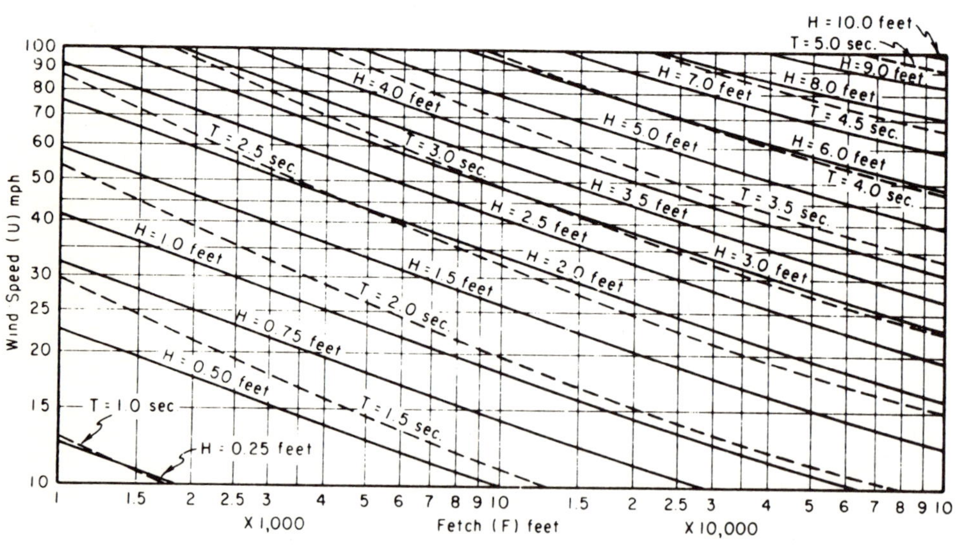

Figure 5-31. Continued.

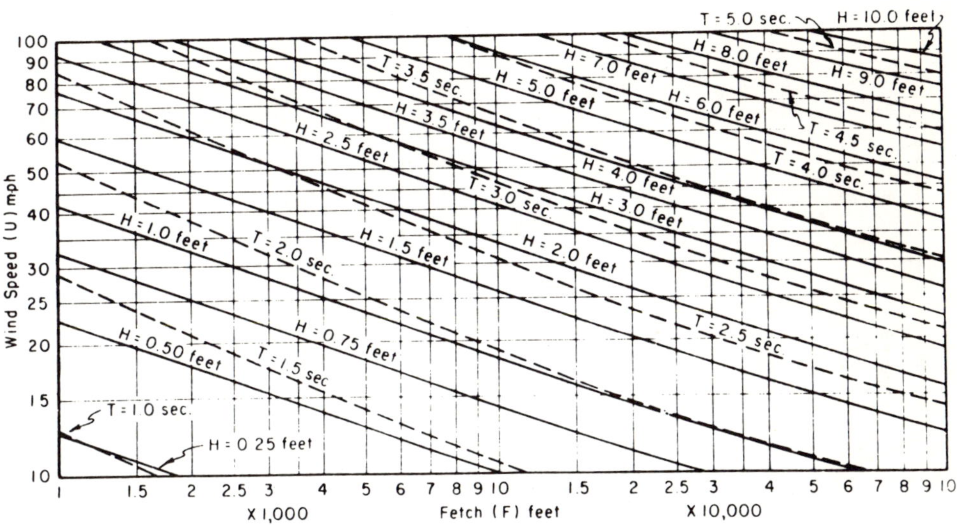

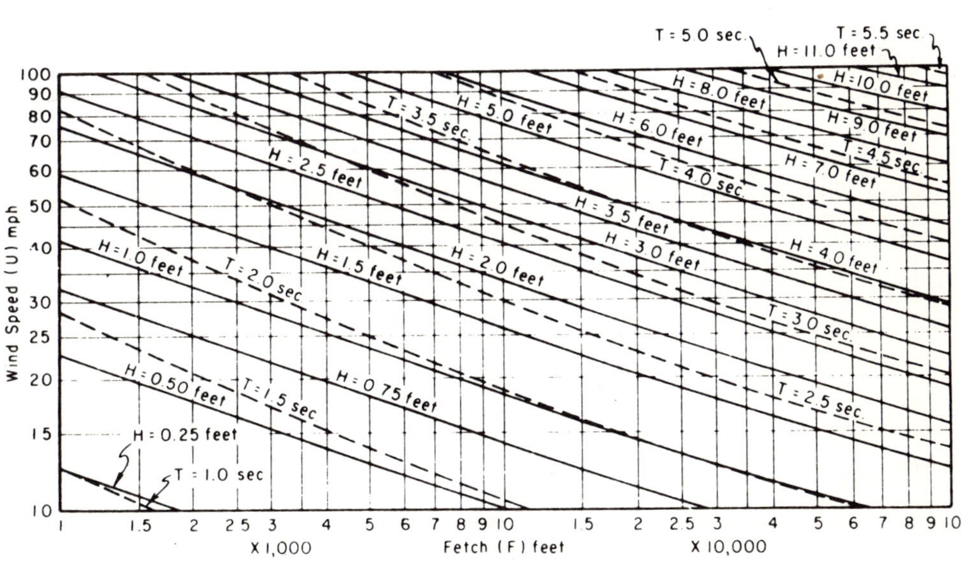

Figure 5-31. Continued.

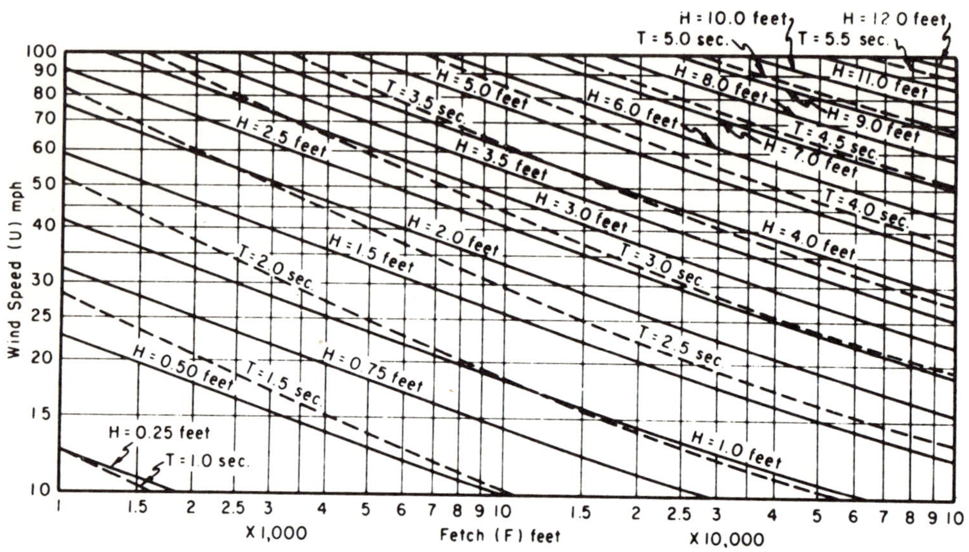

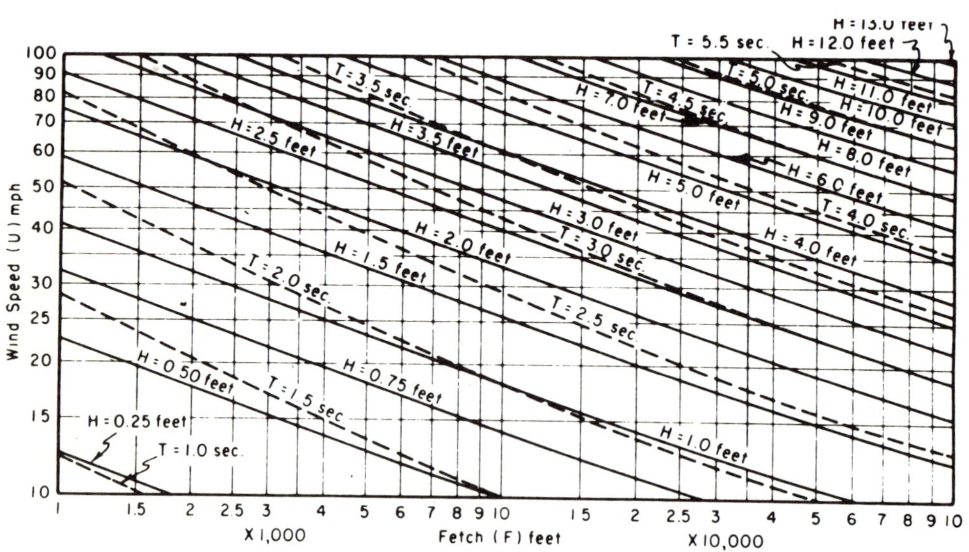

Figure 5-31. Continued.

Wind Influence on Shore Erosion and Sedimentation

Together with river discharge and the tidal fluctuation, the wind is a major factor influencing the circulation of waters and the suspended solid particles in estuarine zones. The action of the wind, which is expressed basically through the dynamics of the water waves that it creates, produces currents and affects the suspension of sediments, with consequential deposit or erosion. The effect of sediment transport depends on the wind speed and direction, as well as on the fetch and the duration.

In an experiment on the transport of sediments, wind speed and direction were recorded at Dover Air Force Base and Delaware Breakwater in Delaware and at the lighthouses at Miah Maull Shoal, New Jersey, and Harbor of Refuge, Delaware. A regression analysis of the data collected shows that the wind speed is more significant than tides and river discharge on sediment transportation during tidal ebb and flood.

Sand Dunes

Along shorelines, where land emerges from the oceans and seas, the interaction between land and water is a timeless dynamic state that affects people everywhere. The equilibrium of sandy shores is basically dependent on the sand dunes, which are generated and destroyed by the individual or combined action of wind and waves. Sand dunes constitute the essential defense system of the shore, whose fragile existence must be preserved.

Dunes can be formed by wave action that creates a submerged sandbar near the shore when the slope of the ocean floor is gentle (no more than 10°). Then, as the sandbar emerges, the wind builds it up by depositing sand particles against it. Dunes can also be caused directly by wind action when an obstacle on the shore initiates the deposit of sand, which gradually increases. In fact, the artificial formation of dunes is accomplished with artificial fences, similar to snow fences, placed on long lines parallel to the shore. Their stabilization is usually provided by adequate vegetation. The destruction of dunes is usually caused by waves or tides of several storms, or by wind erosion when overgrazing, drought, or other causes kill the stabilizing vegetation.

Dunes made with sand fences that trap the sand carried by the wind have been constructed using fascines, brush, picket fences, fences of plastic fabrics or jute-mesh, and long inflated plastic hoses. Sand fences are placed parallel to the shore, rather than perpendicular to the prevailing winds, as one might think, for, if they are at an angle with the wind, they are filled much faster. The filling time gets shorter as the wind speed increases. When the dune reaches the top of the fence, the fence can be raised to make the dune higher. Such an operation can be repeated several times a year. The base of the dune can be thickened by moving the fence on a parallel line. Fences should be moved about four times their height to achieve maximum efficiency. The expected slope of the dune can vary from 1:4 to 1:7, according to the size of the sand grains. Arranging the fence on a zig-zag line or any other geometrical configuration rather than straight does not increase the efficiency; therefore, straight lines are more economical. Optimum porosity of the fence (area of voids per total area) is about 50%. To trap the sand, the width of solid areas and areas of voids should be less than 2 in. Fences have been found to be the most economical means of building artificial dunes. (See Figure 5-32.) At times, even junked automobiles have been used to build dunes, but they cost more than fences, and the results were poor.

Stabilization is of great importance because dunes can be destroyed very easily. Planting is the best form of stabilization; it can be done either by seeding an appropriate grass or by transplanting. Seeding is difficult to do on a slope and is ineffective in the presence of high wind and without proper irrigation. Seeds are also difficult to purchase commercially, but must be procured by harvesting the grass, collecting the seeds, and allowing them to dry. Transplanting, on the other hand, is much simpler and more common. The types of grass usually planted are:

American beach grass
Australian veldt grass
Bamboo grass
European beach grass
Marram grass
Pampas grass
Perennial rye grass
Star grass
Weeping love grass

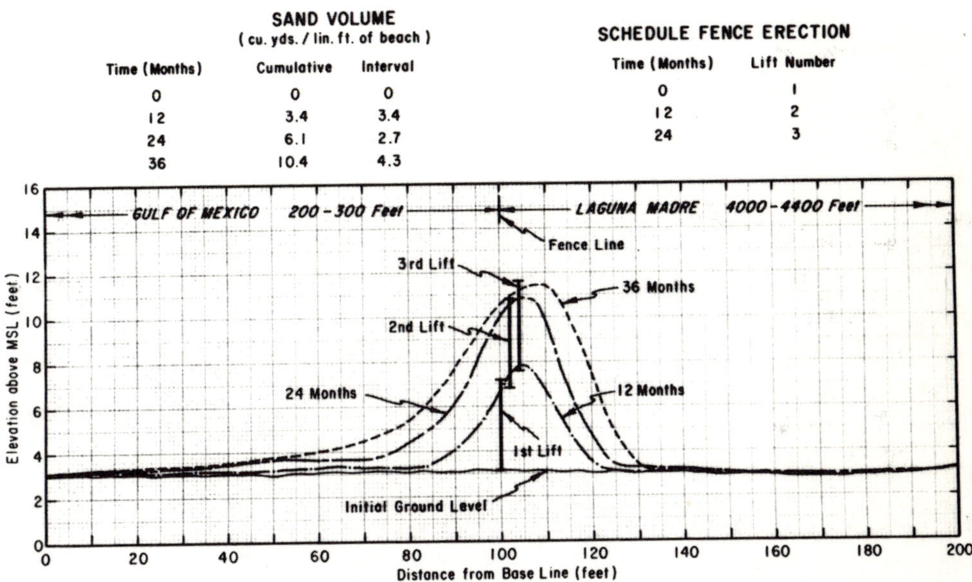

Figure 5-32. Sand fence dune, Padre Island, Texas. (From: U.S. Army Coastal Engineering Research Center, *Shore Protection Manual,* vol. II, Dept. of the Army Corps of Engineers, 1973)

REFERENCES

Criddle, W. D. 1958. Methods of computing consumptive use of water. *Proceedings of the American Society of Civil Engineers, Journal of Irrigation and Drainage Division* 84(IR1): 1–27.

Dalton, J. 1802. Experimental essays. *Manchester Literary and Philosophical Society Memoirs Proceedings* 5: 536–602.

Fitzgerald, D. 1886. Evaporation. *Transactions of the American Society of Civil Engineers* 15: 581–646.

Harbeck, G. E., Jr., M. A. Kohler, G. E. Koberg, et al. 1958. Water-loss investigations, Lake Mead studies. *U.S. Geological Survey Professional Paper 298.*

Harbeck, G. E., Jr., et al. 1954. Water-loss investigations, vol. 1, Lake Hefner studies. Technical report. *U.S. Geological Survey Paper 269.*

Horton, R. E. 1917. A new evaporation formula developed. *Engineering News Record* 78(4): 196–199.

Maunder, W. J. 1917. *Weather* 17(1): 9.

Meyer, A. F. 1915. Computing run-off from rainfall and other physical data. *Transactions of the American Society of Civil Engineers* 79: 1056–1155.

Penman, H. L. 1948. Natural evaporation from open water, bare soil, and grass. *Proceedings of the Royal Society of London* 193: 120–145.

Rohwer, C. 1931. *Evaporation from Free Water Surfaces.* U.S. Department of Agriculture, technical bulletin 271.

Seelye, E. E. 1962. Design. Data Book for Civil Engineers. John Wiley & Sons, p. 16.

Shaw, N. 1936. Comparative meteorology. *Manual of Meterology.* London: Cambridge University Press.

Siple, P. A. 1951. Regional climate analyses and design data. *Bulletin of the American Institute of Architects,* Washington, D.C., Sept. 6.

U.S. Army Coastal Engineering Research Center. 1973. *Shore Protection Manual.* Vol. 1.

Woodruff, N. P., L. Lyles, F. H. Siddoway, and D. W. Fryrear. 1972. *Soil and Water Conservation.* Research Division, Agricultural Research Service.

———. 1972. *How to Control Wind Erosion.* Agricultural Information Bulletin No. 354. Washington, D.C.

Woodruff, N. P., Leon Lyles, F. M. Siddoway, and D. W. Fryrear: Soil and Water Conservation, Research Division, Agricultural Research Service, How to Control Wind Erosion, Agriculture Information Bulletin No. 354, June 1972, Washington, D.C.

6
Wind in the Urban and Regional Environment

HEAT LOSSES BY WIND

A very thin layer of air adheres to all surfaces in contact with air and acts as a resistance to the flow of heat between the surfaces and the air itself. This is true, of course, if air is stationary; however, if air moves, the resistance is diminished and the reverse is true. In other words, whereas still air acts as a thermic insulator, an air current is an active factor in accelerating the heat flow. Just like in the radiator of an automobile, the higher the speed of the moving vehicle, the greater the amount of heat that goes from the water in the radiator to the air. In fact, at higher speeds, more air comes in contact with the radiator surface and because each particle of air can extract some heat from the surface, more heat is taken from the radiator. This phenomenon is very important for understanding why heat losses increase in buildings exposed to the wind. The higher the wind speed, the greater the amount of heat that is wasted through the exterior walls and roof. (See Figure 6-1.)

To consider the quantitative aspects of the phenomenon, note that the film or surface conductance for still air is:

$$f_o = 1.46 \text{ Btu/ft}^2 \times \text{hr} \times {}^\circ\text{F}$$

For a 7.5-mph wind, it is:

$$f_i = 4.0 \text{ Btu/ft}^2 \times \text{hr} \times {}^\circ\text{F}$$

For a 15-mph wind, it is:

$$f_i = 6.0 \text{ Btu/ft}^2 \times \text{hr} \times {}^\circ\text{F}$$

This applies to a vertical wall and a horizontal heat flow through it.

Corresponding to these conductances, we have the reverse—that is, the heat resistances. Thus, for $f_o = 1.46$, $R = 1/f_o = 0.68$ and for a 7.5-mph wind:

Figure 6-1. Even when buildings are heavily insulated, a certain amount of heat flows outward during the cold season. The rate of heat flow through walls and roof is increased by the wind striking these exterior surfaces. The higher the wind speed, the higher the rate of heat loss. Typical procedures for calculating heat losses in temperate climate assume an average wind speed of 15 mph.

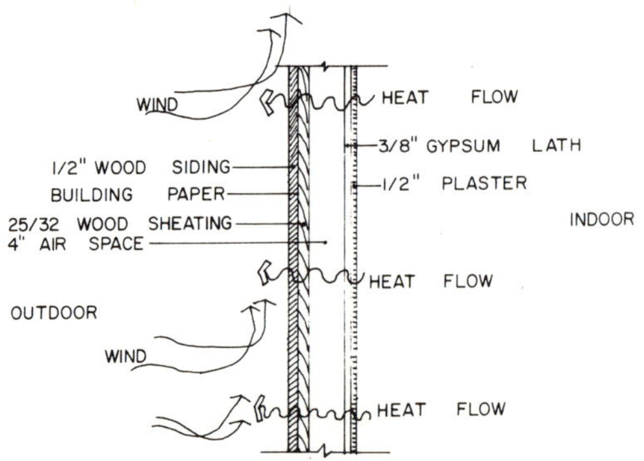

Figure 6-2. Section through an exterior wall.

$$f_i = 4.00, \quad R = 1/f_i = 0.25$$

and for a 15-mph wind:

$$f_i = 6.00, \quad R = 1/f_i = 0.17$$

Table 6-1 lists some values of film conductance for various wind speeds.

Table 6-1. Film or surface conductance for various wind speeds.

FILM OR SURFACE CONDUCTANCE (F_o IN BTU/FT^2 × HR × °F)	WIND SPEED (MPH)	SOURCE
4.00	7.5	Experimental
6.00	15.0	
7.00	20.0	
8.00	25.0	Extrapolated
9.00	30.0	
10.65	40.0	Extrapolated
12.13	50.0	
13.50	60.0	Extrapolated

Let us consider an exterior wall consisting of: 0.5 wood siding; building paper; 25/32 wood sheathing; 4-in. airspace; 3.75-in. gypsum lath; and 0.5-in. plaster. (See Figure 6-2.) The thermal resistance of the materials along the wall is:

$$R = 3.27 \text{ ft}^2 \times \text{hr} \times \text{°F/Btu}$$

to which the resistance due to still air on each side must be added. Thus, the total resistance for the wall with no wind is:

$$R = 3.27 + 0.68 + 0.68 = 4.63 \text{ ft}^2 \times \text{hr} \times \text{°F/Btu}$$

and the conductance per unit surface area is:

$$U = 1/R = 1/4.63 = 0.216 \text{ Btu/ft}^2 \times \text{hr} \times {}^\circ\text{F}$$

The resistance and the conductance for the same wall with a 7.5-mph wind are, respectively:

$$R = 3.27 + 0.68 + 0.25 = 4.20 \text{ ft}^2 \times \text{hr} \times {}^\circ\text{F/Btu}$$
$$U = 1/4.20 = 0.238 \text{ Btu/ft}^2 \times \text{hr} \times {}^\circ\text{F}$$

The resistance and conductance for the same wall with a 15-mph wind are, respectively:

$$R = 3.27 + 0.68 + 0.17 = 4.12 \text{ sf} \times \text{hr} \times {}^\circ\text{F/Btu}$$
$$U = 1/4.12 = 0.24 \text{ Btu/ft}^2 \times \text{hr} \times {}^\circ\text{F}$$

The increase in heat losses due to a 7.5-mph wind is

$$U = 0.238 - 0.216 = 0.022 \text{ Btu/ft}^2 \times \text{hr} \times {}^\circ\text{F}$$

which amounts to an increase of 10.2%. But the increase in heat losses for a 15-mph wind (twice, as in the previous case) is:

$$U = 0.24 - 0.216 = 0.024 \text{ Btu/ft}^2 \times \text{hr} \times {}^\circ\text{F}$$

which amounts to only 11.1%.

WIND DISPERSION ACTION AND AIR POLLUTANTS

Low-speed winds, or lack of wind when combined with a temperature inversion over an industrial urban center, can produce a severe air-pollution problem that could endanger human health, as previous experiences have proved. How does the phenomenon occur? Under ordinary conditions, air temperature decreases with height; thus, hot gases and smoke from different sources (automobiles, industries, etc.) rise high because of their lower density with respect to the colder air. The higher they rise, the more the dispersion of the pollutants in the air. On the other hand, during thermal inversions, hot polluted air meets warmer air as it rises, and therefore it cannot rise any longer and it accumulates in layers over the city. Moreover, if there is not sufficient wind to carry this air away, the accumulation of pollutants increases until the conditions change. (See Figure 6-3.)

The cleansing action of the wind, therefore, disperses the pollutants existing in the air from a limited area of high concentration to a wider zone determined by the fetch. From a point source, the dispersing action of the wind is basically a two-dimensional phenomenon—the diffusion spreads horizontally, fanning out with the wind, and spreads vertically as the different layers of the atmosphere mix. However, the scale in a vertical direction is much different from the horizontal one. More specifically, in the vertical direction the depth is approximately 10 km, while horizontally the area potentially involved is the entire globe. Therefore, there is a great difference in the two scales, so that the critical aspect of the distribution is in the vertical direction. The 10-km depth is actually the upper limit of vertical diffusion that will occur only when the vertical mixture of the atmospheres is active at its peak; under certain weather conditions, it could be reduced considerably. At times, such as during thermal inversion, the vertical depth of this active layer of air can be zero, with the consequential absence of pollutant dispersion.

The following definitions are important to the present discussion.

Lapse rate: Thermal profile; the change of air temperature with height.

Positive lapse rate: When the air temperature decreases with height.

Unstable air conditions: When the lapse rate is positive; air temperature decreases with height.

Negative lapse rate: When the air temperature increases with height.

Thermal inversion: When the air temperature increases with height (negative lapse rate).

Stable air condition: When the air temperature increases with height (thermal inversion, negative lapse rate).

Pollutant dispersion: It is good when conditions include positive lapse rate and unstable air conditions because air rises and is dispersed by winds. It is bad when conditions include negative lapse rate, thermal inversion, and stable air conditions.

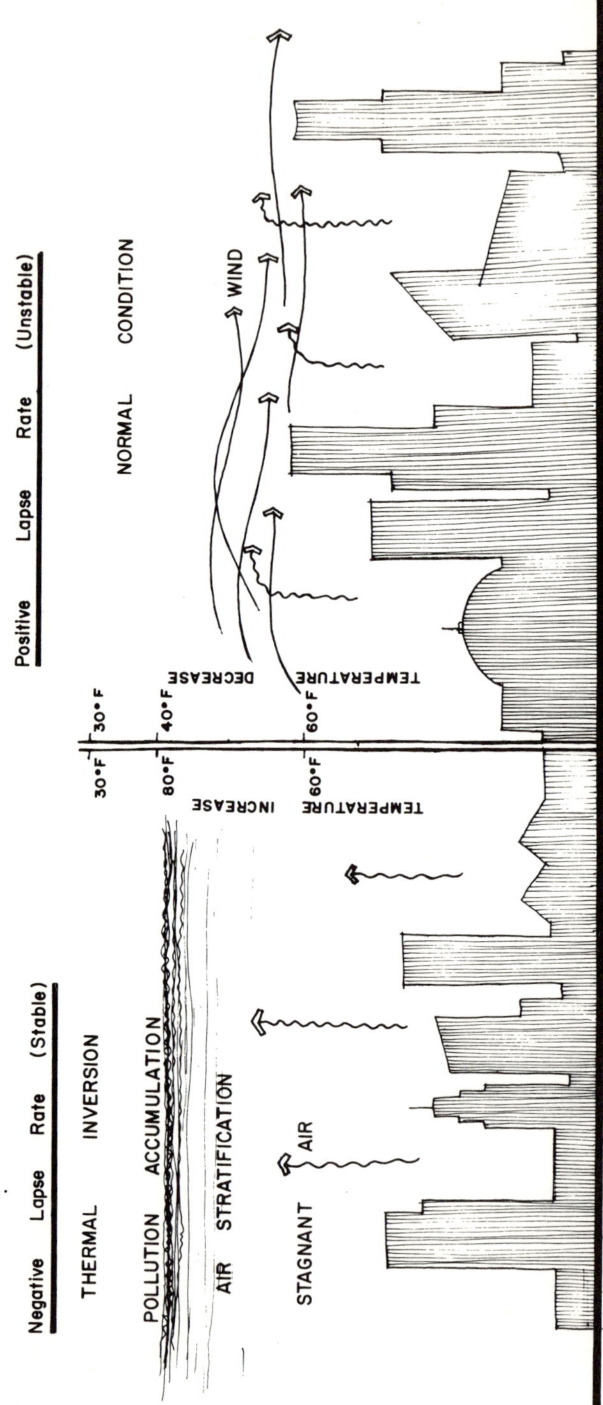

Figure 6-3. Combined effect of thermal inversion and low-speed wind or lack of wind over urban centers having air pollution factors.

AEROALLERGENS

Among the many types of airborne particles carried by the wind, there is a group that can generate allergic reactions to a spectrum of sensitive subjects in the human population. Such reactions include acute asthmatic attacks, ocular pruritis, nasal congestion, and hay fever. The particles referred to as *aeroallergens,* are solid particles with diameters that vary from 1 to 80 μ. The term is usually used to indicate not only those specific molecules that generate allergies, but also those inert particles that are attached to the molecules. For instance, dust, spores, and pollens are often mistaken for allergens, but in reality they only serve as vehicles that carry the true allergens; however, they are generally included with the aeroallergens.

In general, there is a correlation between the particles suspended in the air and the hazard they represent to human health. The major factor supporting the correlation is the size of the actual particles. In fact, when airborne particles are inhaled, they are deposited in different parts of the respiratory tract according to their size. Usually, the largest particles are trapped in the nasal cavities, when the subject breathes through the nose, or are caught by the mucus in the tracheobronchial system. The smaller particles (radius about 1 μ) reach the alveolar region of the lungs, where the process of elimination is much slower. Therefore, their irritating action is more prolonged and potentially more dangerous. (See Figure 6-4.)

Airborne particles vary considerably in size. The examples in Figure 6-5 show, at the low end of the spectrum, the gas molecules whose sizes range from 0.0001 to 0.001 μ. From the low end onward, the figure shows a variety of particles whose size is below the 0.001μ upper limit, above which particles are visible to the naked eye.

A: Visible to the naked eye
B: Diameter of human hair
C: Fly ash
D: Oil smoke
E: Diameter of gas molecule
F: Carbon dioxide
G: Ammonia, water
H: Nitrogen gas
I: Hydrogen gas
K: Sneezes
L: Tobacco smoke
M: Pollens
N: Bacteria
O: Insecticide dust, plant spores
P: Pulverized coal
Q: Viruses and vitamins
R: Cement
S: Mist of sulfuric acid
T: Fumes, smelter dust
V: Mist
W: Fog
X: Aerosols

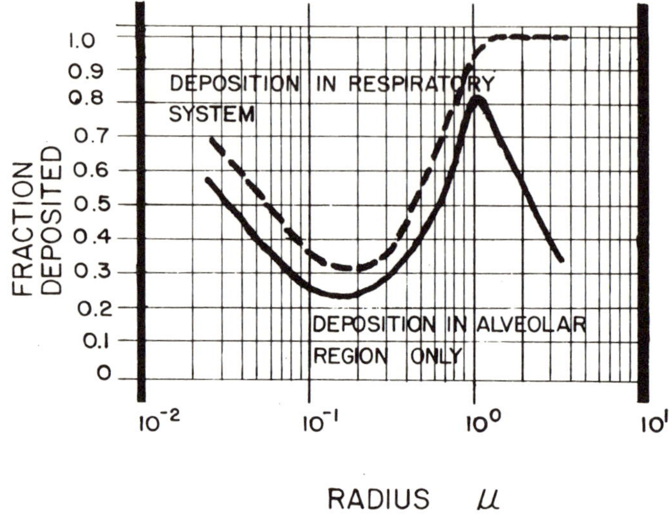

Figure 6-4. Distribution of particles in various parts of the respiratory system as a function of the particle size. (From National Air Pollution Control Administration, 1969, *Air Quality Criteria for Particulate Matter,* publication AP-49)

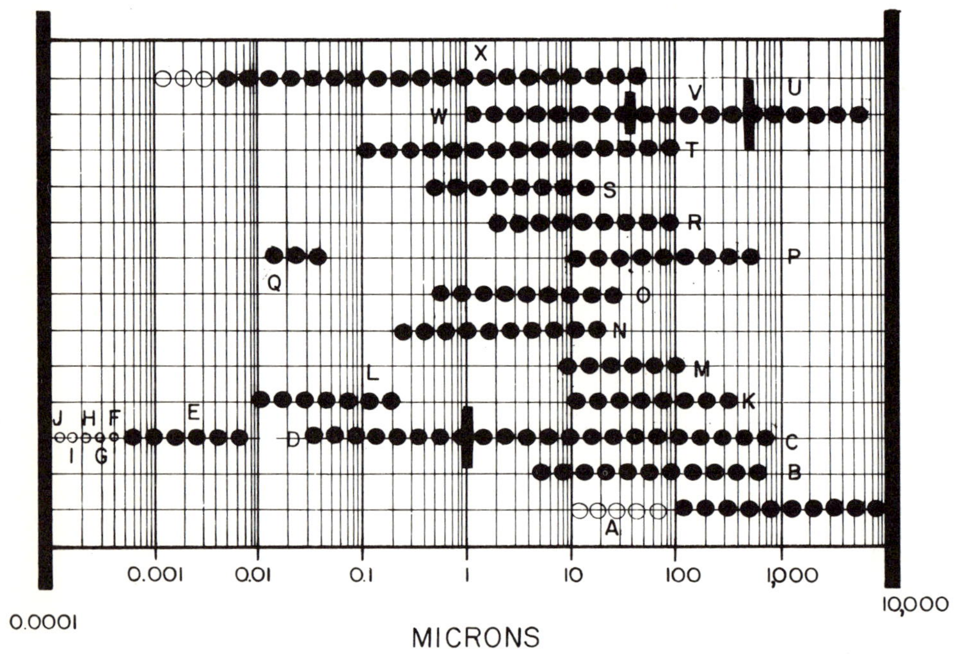

Figure 6-5. Spectrum of airborne particles potentially present in the air. (From T. F. Hatch and P. Gross, *Pulmonary Deposition and Retention of Inhaled Aerosols,* Academic Press, New York, 1964)

Because of their size, which is about 10μ, aeroallergen pollens and spores are usually trapped in the upper part of the respiratory system, as previously mentioned, and do not reach the lungs. Notice that many particles of organic nature are hygroscopic and therefore swell as they absorb moisture, thus increasing considerably in size. Pollens, spores, fungi, insect products, animal danders, and vegetable gums are included in this category.

The suspension of particles in the air depends, of course, upon several factors. In still air, particles in the 10-μ category fall and are deposited by gravitation. Smaller particles, on the other hand, remain airborne. The settling in still air is related to a certain typical velocity of the particle, referred to as the *settling velocity,* which for pollens varies from 1 to 10 cm/sec (0.0328 to 0.3280 ft/sec). According to Stokes'* law, however, the settling velocity is also proportional to the density and the square root of the equivalent radius. Such settling velocities apply in still air; however, if the air is in motion, these particles will still be airborne. In fact, average wind velocities carry aloft particles of large size, and, as the speed increases to the extremes of tornadoes, extremely heavy objects are also picked up by the wind.

How far can the wind carry airborne particles? Long-range transport of organic windborne particles is verified by observed pollen concentrations in the atmosphere over polar regions. Therefore, since considerable distance exists between the poles and the closest sources of certain types of pollens, the evaluation of long-range transport is obviously proven.

WIND-RELATED CATASTROPHES

Prior to the industrial era, the emission of pollutants in the air had been historically inconspicuous. In fact, the normal amount of waste that is produced by humans in their habitat without their machines is nominal and is easily dispersed by wind in the vast atmosphere. However, the high level of present industrial activity in technologically developed countries is causing the injection of enormous amounts of volatile refuse in the air. Aggravating the problem of air pollution in terms of its consequences on the health of the population, we notice that the major points of waste discharge

* George Gabriel Stokes (1819–1903), Irish physicist and mathematician.

into the air are not over open country or deserted areas but coincide with urban centers where the population is densest. In abnormal atmospheric conditions with stagnant air, the danger of serious effects on human health is potentially high as proved by specific documented cases.

In December 1930, an abnormal atmospheric condition in the Meuse River Valley in Belgium, between Liège and Huy, produced a stratification of stagnant air that lasted 5 days. The lack of wind with its cleansing action caused an accumulation of pollution that affected people as well as animals. Respiratory complications caused the deaths of 63 people, while a very large number were affected by such symptoms as chest pain, shortening of breath, coughing, and many disorders related to the heart and circulatory system.

In December 1952, in London, England, a severe catastrophe warned the world of the tragic potential of air pollution. A lack of wind caused a concentration of air pollutants, which, combined with heavy fog, caused much of the population to suffer respiratory illness, and circulatory and cardiac disorders. Illnesses due to the conditions alone were twice the number normally reported. About 4000 deaths were attributed to the phenomenon. It is important to note that illnesses and deaths were sharply higher in people under 5 years of age and those over 45. Statistical analysis of records of health conditions in the London population, conducted after this episode, revealed that sharp increases in sickness and death attributable to air pollution had occurred several times since 1873. For instance, in February 1880, a similar situation caused an estimated 1000 air-pollution–related deaths. In addition, high concentrations of pollutants in London caused hundreds of deaths in December 1957, and in December 1962.

In October 1948, in Donora, Pennsylvania, south of Pittsburgh, the lack of wind resulted in stagnant air that combined fog and a high concentration of pollutants in the air over the small industrial town. A total of 40% of the population were affected by the condition; 10% of these were affected by respiratory difficulties, including coughing and complications in breathing. Some of the symptoms, such as coughing, continued for as long as 4 months after the episode and affected a segment of the population that included 12% over 45 years of age. Other illnesses reported during the incident included gastrointestinal symptoms, vomiting, and diarrhea. In a population of 14,000, a total of 21 people died. The animal population was also affected. Many dogs, cats, canaries, and fowls died.

In November 1966, in New York City, lack of wind for 3 days created an accumulation of pollutants over the city. An acute increase in deaths was reported for a period of 7 days. Thus, the effects of the 4-day weather episode continued even after the physical conditions had returned to normal. The number of deaths attributed to the air pollutants was reported to be 168. This episode was preceded by an even more severe case in January and February of 1963, when the number of deaths was estimated to be from 200 to 400, due to the high concentration of pollutants in the air. In November 1963, a similar condition occurred in New York. The estimated number of deaths was 250.

It seems ironic that the lack of wind in cities is more lethal than the devastating fury of tornadoes and hurricanes. No hurricane has ever killed 4000 people, but zero wind velocity in London did.

Table 6-2. Major catastrophes due to high concentration of air pollutants caused by insufficient wind circulation.

YEAR	MONTH	LOCATION	DEATHS
1880	February	London	1000
1930	December	Meuse Valley, Belgium	63
1948	October	Donora, Pennsylvania	20
1952	December	London	4000
1953	November	New York City	250
1956	January	London	1000
1957	December	London	700–800
1962	December	London	700
1963	January–February	New York City	200–400
1966	November	New York City	168

WIND-CARRIED ODORS

Suspended in the air, odors are carried by the motion of the air itself. In fact, the particulates that produce odors can be transported for great distances as they remain aloft. The vehicle that carries odors and spreads them over vast territories is, of course, the wind. For such odor-transporting action, the wind is naturally the most important factor upon which the quality of the air depends.

Until recently, in observations of the impact of wind on the environment, odors have never been sufficiently emphasized because of the limited research work that had been done in this field. However, work undertaken in the last few years has broadened our understanding of the topic.

In the animal kingdom, the sense of smell is a major means of communication between individuals and between the individual and the exterior world. This sense is also very refined in humans. In fact, it enables people to perceive both the natural and the artificial environment in more complex and subtle ways.

In addition, unpleasant odors play a major role among pollutants, whereas fragrant odors can enhance the environment considerably. Negative and positive effects, however, do not have equal potential impact since the negative effects of bad odors can have deeper repercussions than those of pleasant nature.

As the wind blows away from the source where odorous particulates have been picked up, the concentration of the particulates diminishes with the traveled distance. The scent's effectiveness persists, however, even when the concentration is very low. Just 42 molecules of a substance, in fact, are enough to be detected by the human olfactory senses.

How does the human body perceive odors? Deep inside the cavities of the nasal passages, the smell receptors are activated by particulates that are carried by the air as it flows through the nose up to the scent sensors. An odor is perceived when the molecules reach the cells of the olfactory organs. It is interesting to note that seven types of olfactory cell hairs have been identified in the human smell receptors, and some researchers have identified seven basic classes in which to group odors. The implication, of course, is that each of the different cell types of the olfactory system may be sensitive to one of the seven classes. However, other researchers, disregarding this concept, have subdivided the field of odors into a different number of classes.

In practical terms, odor intensity or strength can be quantified by measuring the amount of odor-free dilutant (air, water, or other substance) that must be added to an odorous substance to reach the point when the odor can barely be perceived. This is called the *threshold value.* It is measured in terms of the number of dilutions that are necessary to reach the threshold concentration.

In 1968, the American Society of Testing Materials (ASTM) established the basic rules for measuring the intensity and quality of odors.

Once measured, odor intensity can be expressed in two equivalent ways:

1. by the *threshold odor number (TON),* which indicates the number of parts of odor-free dilutant to be added.
2. by the *odor intensity index (OII),* which indicates how many times the odorous substance must be diluted in half to reach the threshold point.

The two expressions are related by the relationship:

$$TON = 2^{(OII)}$$

For example, if the OII is 6, the TON will be 64 ($TON = 2^6$).

For measuring the concentration at the threshold point, the ASTM specifies a minimum of five individuals to assure sufficient objectivity in the tests. Furthermore, the participants must be examined for their ability to respond normally to odor stimuli.

To assist the human nose during the testing, the gas-liquid chromatograph can help considerably by subdividing a complex odor in its individual components and quantifying them proportionally. Other apparatuses normally used include the odormeter and the osmometer that help in the preparation of the dilutions of the substance with the odor-free medium (usually air or water). It is important to point out that, despite their name, these machines do not measure odors directly. They are just an aid to the operator who depends on his or her olfactory sensitivity in the final analysis.

Another apparatus that is very helpful for field measurements is the so-called scentometer. It consists of a small plastic box containing two beds of activated charcoal, one on top and the other at the bottom. Each bed is pierced by one ½-in.-diameter hole acting as an air inlet. On one side of the box there are four holes (odor inlets) whose diameters are 1⁄16, ⅛, ¼, and ½ in. On the other side there are two sniffing holes.

The operators in the field adjust their sense of smell to odor-free conditions by breathing through the sniffing holes while they keep the odor inlets closed with their fingers. The air that they are breathing comes from the two ½-in.-inlets and is odor-free because it is purified by the charcoal. When well adjusted to odor-free conditions, the operators open the smallest of the four odor inlets while the two ½ in.-inlets carrying odor-free air are always left open. The odorous air is mixed with odor-free air at its maximum dilution, and if no odor is perceived the operators will close this inlet and open the next larger one. They will continue until the threshold point is reached. Note that the operators can also choose to have more than one inlet open simultaneously in any combination so that the desired dilution is reached. The correlation between the open inlets and the dilution is indicated in Table 6-3.

Table 6-3. Correlation between the odor inlets left open to the passage of odorous air and the number of dilutions in the use of the scentometer.

DILUTIONS TO THRESHOLD	ODOROUS AIR INLETS[a] ½ IN.	¼ IN.	⅛ IN.	1/16 IN.
1.47	0	0	0	0
1.49	0	0	0	x
1.55	0	0	x	x
1.88	x	0	x	0
2.0	0	x	x	x
5.55	x	0	0	0
5.75	x	0	0	x
6.75	0	x	0	x
7.0	x	0	x	x
27.0	x	x	0	0
31.0	x	x	0	x
170.0	x	x	x	0

[a] x = port is covered.
0 = port is open.

Naturally, not all substances carried by the wind produce odors that humans can detect. Only those with a molecular weight of more than 15 times the weight of hydrogen can generate scent. So, just as we can see the wind, feel it, and hear it, we can also perceive it with our olfactory senses when it carries odors. Several classifications of odors carried by the wind have been proposed since the end of the nineteenth century. In 1966, Moncrieff (1966) compiled a survey of the classification systems that had been published by various authors.

Zwaardemaker (1895) proposed a subdivision that consisted of nine basic odors: ambrosial, balsamic or fragrant, ethereal, aromatic, empyreumatic, alliaceous, caprylic (goaty), repulsive, and nauseating or foetid. Henning (1916) developed a classification of odors divided into five types: flowery or resinous, fruity, spicy, burnt, and foul. Crocker and Henderson (1927) subdivided odors into four categories: fragrant, burnt, caprylic (goaty), and acid. Amoore (1952) classified odors into seven classes: musky, floral, ethereal, camphoraceous, minty, putrid, and pungent. Finally, Davies (1965) formulated a system of identification of odors that included: musky, floral or cedary, ethereal or fruity, alcoholic, camphoraceous or aromatic, pepperminty, and almond.

Many animals depend on smells—and upon the wind to carry these smells—for food procurement, defense, reproduction, and orientation. For humans, on the other hand, smell is not as important on the physical level, as it is on the psychological level. Freud, for instance, found that in many of his patients there existed correlations between odors and psychological pathology. In addition, odors transported by wind enhance human experiences and aid the memory of them: "Smells have remarkable power in calling up remembrances of past scenes" (Jackson 1899).

The large variety of scents carried by the wind affect humans with different intensity and at various levels of consciousness. Since antiquity, humans have expressed their aesthetic values with perfumes of a complex nature that appealed to the olfactory sense. In practical terms, however, they experienced both the positive and negative aspects of smell in their environment with a relatively low concern. The problem of worldwide pollution has also raised the human concern with the negative aspect of odors. Sometimes associated with the toxicity of poisonous gases and at other times acting independently, odors of high density permeate the atmosphere over the congested urban centers of the world. The negative effect of many of these odors can be recalled vividly by mentioning a few typical offensive sources: paper mills with their nauseating smells of decomposing wood pulp, fish processing plants, poultry processing plants, sewage treatment plants, chemical plants, oil refineries, etc.

On the positive side, the presence of scents in our everyday experience enriches the quality of life. This consideration, of course, is extremely important in fields concerned with the design of the human environment such as planning, architecture, civil engineering, environmental sciences, etc.

As the wind blows over the land, it diffuses the scents that it has picked up and allows them to be perceived. For example, fruit trees, flowers, herbs, and even the oceans contribute fragrances to the wind. In certain geographical areas, one could experience many of these odors combined. For instance, odors from lemon and orange trees, magnolias, honeysuckle, rosemary, thyme, and the sea are such that even the mental recollection of them can produce pleasant feelings. This has been demonstrated by the many literary evocations used in prose and poetry.

Thus, we can see how wind characteristics can greatly affect the selection of the site for any center of activities. Existing and potential sources of odors upwind must be carefully scrutinized for assessing the expected quality of the site under consideration. For instance, a site located downwind from any polluting industry is drastically affected in its economic value. Therefore, in terms of prevailing directions, wind conditions affect the desirability and the economics of land. Notice that the negative effect of wind on the value of land is a relatively new phenomenon that started when industries concentrated specific activities in areas where odors originate at high concentration. Never in human history have wind-carried odors had such an impact on the environment.

SMOKESTACKS

As smoke leaves a chimney, it acquires a certain configuration well distinguishable in the air which is referred to as the *stack plume.* Air temperature and wind speed profiles (the variation of temperature and speed with respect to the height above the ground) are the most important parameters that influence the plume configuration and consequently the transport, diffusion, and dilution of the polluting smoke in the atmosphere.

The major types of plume configurations that can be observed are characterized by the following kind of behavior: looping, coning, fanning, lofting, trapping, fumigation (inversion), and fumigation (strong winds). See Figure 6-6.

To predict the stack plume configuration and its behavior, it can be said in general that:

1. If the outlet stack velocity is less than the wind velocity, the plume will be deflected toward the ground (Δh is negative). The phenomenon is referred to as *aerodynamic downwash,* which causes pollution at the ground level (fumigation).
2. If the outlet stack velocity is higher than the wind velocity, the plume rises as in the "lofting" behavior.

See Figure 6-7.

Moses-Carson Formula

A general formula for calculating the rise of the plume above the stack top is the Moses-Carson formula (1969):

$$\Delta h = A\left(-0.029\frac{V_s d}{W} + 5.35\frac{\sqrt{Q}}{W}\right)$$

where
Δh = plume rise (m)
A = 2.65 for unstable conditions; 1.08 for neutral conditions; 0.68 for stable conditions
V_s = stack velocity (msec)
d = stack diameter (m)
Q = heat emission rate (kcal/sec)
W = mean wind speed

See Figure 6-8.

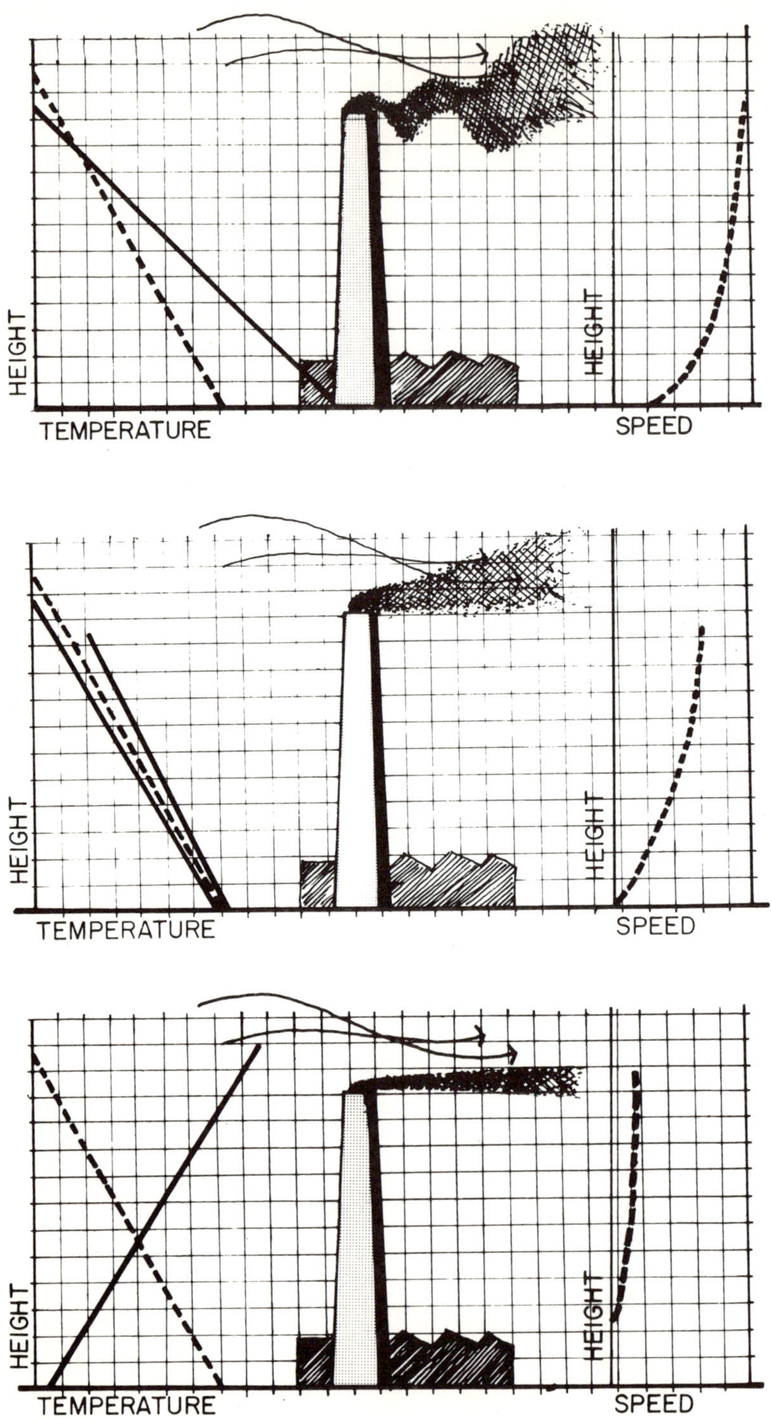

Figure 6-6. Types of plume configurations.

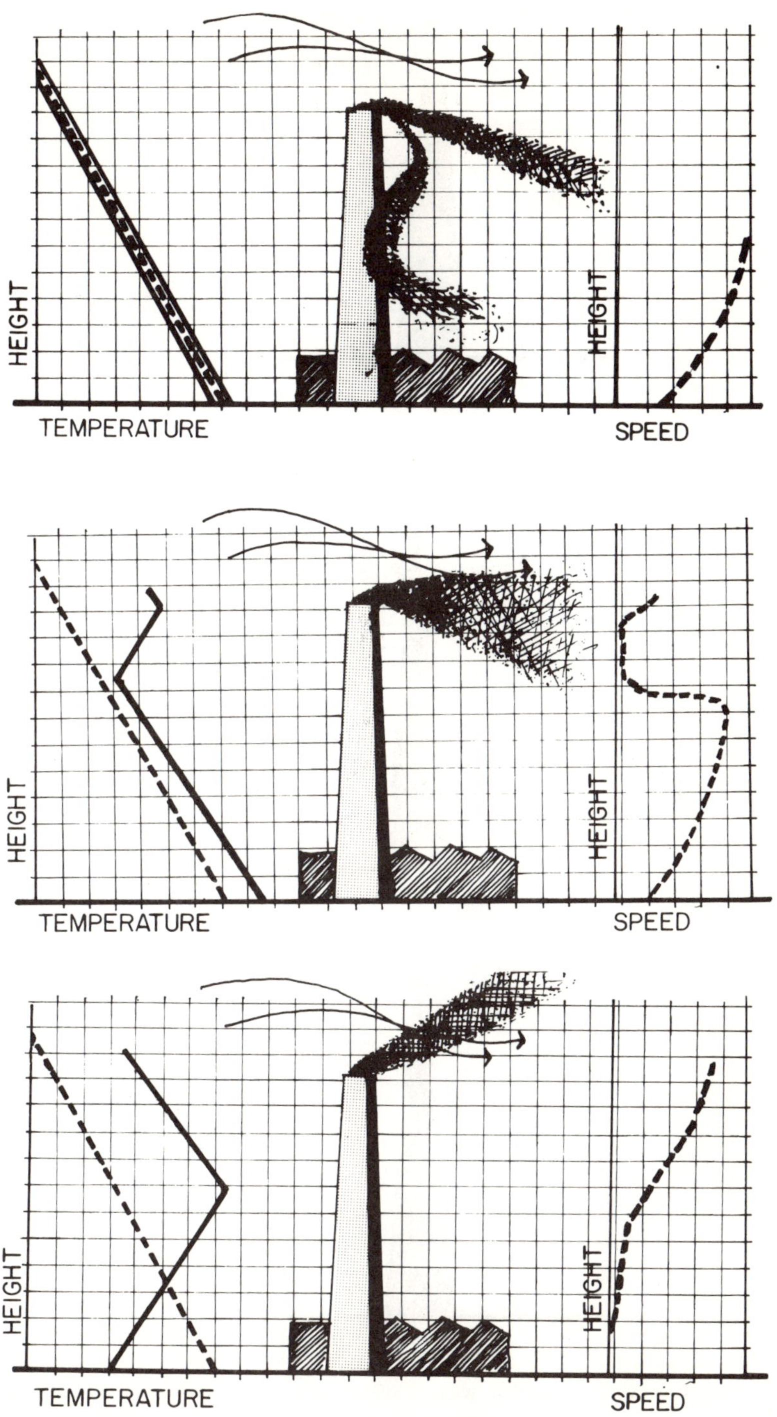

Figure 6-6. Continued.

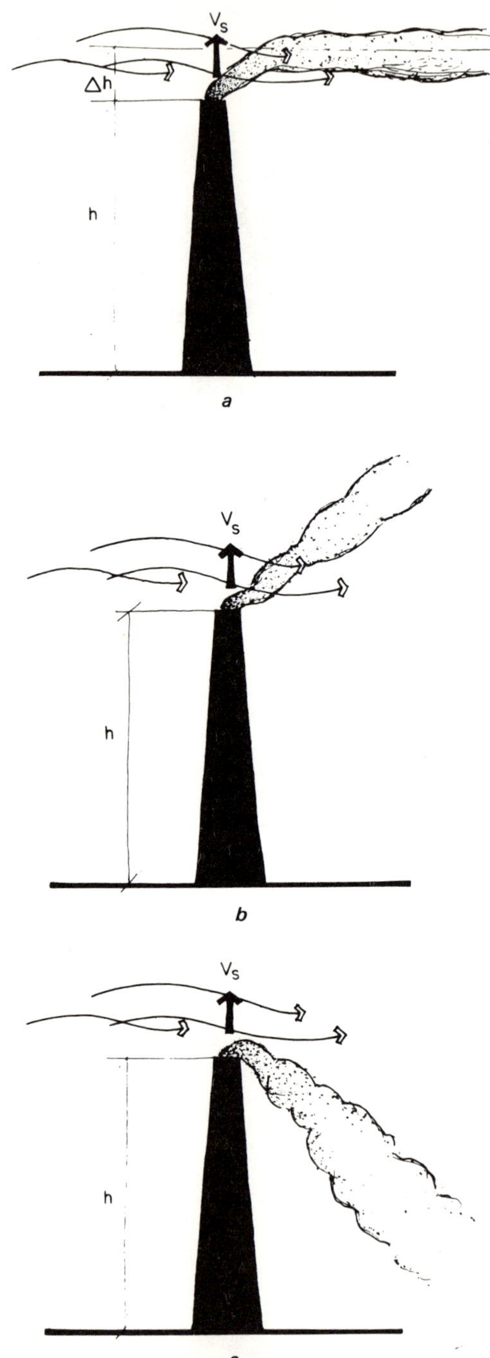

Figure 6-7. *a,* Stack velocity is less than wind velocity: aerodynamic downwash (fumigation). *b,* Stack velocity is higher than wind velocity (lofting). *c,* Rise of the plume above the stack top (Moses-Carson formula, 1969).

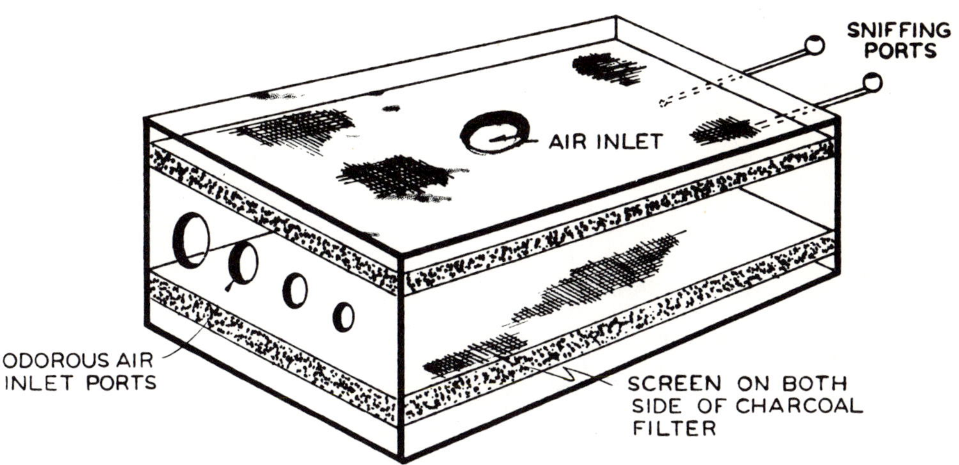

Figure 6-8. Scentometer, a useful aid for field measurements of odor intensity. Dilutions with odor-free air and determination of the threshold point are easily found.

Table 6-4. Stack plumes.

PLUME BEHAVIOR	THERMAL PROFILE	CONDITION OF AIR	POLLUTION CHARACTERISTICS	WIND
A	Lapse rate: positive	Unstable	–	Light winds, intense thermal turbulence
B	Lapse rate: positive	Neutral or stable	–	Moderate to strong winds, mechanical turbulence
C	Lapse rate: negative (inversion)	Very stable	–	Light winds, little turbulence
D	Lapse rate: positive above stack top; negative below stack top (inversion)	Unstable above stack top, stable below stack top	–	Moderate wind and strong turbulence aloft, very light wind and no turbulence at low levels
E	Lapse rate: positive	Unstable	Short duration at sunrise	Strong winds, high mechanical turbulence
F	Lapse rate: negative above stack top (inversion); positive below stack top	Stable aloft, unstable in lower levels	–	Light wind aloft, strong wind at low levels

Wind Dispersion of Radioactive Particles

Among the psychological and physical effects of the wind on humans and their environment, new aspects have recently emerged. These transcend for the most part the original fear of the wind god and his violent manifestations that the tribes of the ancient world had. The present manipulation of nuclear energy makes the wind a potential carrier of deadly radioactivity. Thousands upon thousands of human lives and other living organisms could be lost in a nuclear accident as the wind spreads radioactive particles. Comparisons of destruction caused by hurricanes and tornadoes with that which would be caused by contamination spread by the wind after a nuclear accident make any violent storm seem negligible.

The accident in 1979 at the Three Mile Island nuclear plant in Harrisburg, Pennsylvania, revealed the danger of nuclear power disasters. The spreading of radioactive materials, whether gas or solid particles erupting from a leaking vessel, can occur in different ways, but its dispersion by wind is by far the most dangerous. The speed at which it spreads and the range affected are of great proportions. Should an accident occur, wind speed and direction will be the first factors scrutinized in making emergency safety plans.

In 1979, there were 72 active power plants in the United States alone, and in recent years many others have proliferated all around the world. The risk of accidents and the consequential spread of radioactivity also rise. The potential danger makes one wonder about the desirability of sites located downstream from a nuclear power plant in the direction of the prevailing winds.

WIND EFFECT ON SOUND PROPAGATION

Sound Propagation through Air

Significant values of the effective sound pressure in the human environment vary from a minimum of approximately 0.0002 μbar ($1/10^6$ atm) when sound is perceived to a maximum of 100,000 μbars that could be produced by a blast.

To convert the effective sound pressure to the sound pressure level, we assume that:

$$\text{Sound pressure level} = 20 \log \left(\frac{p}{p_{\text{ref}}}\right)$$

where

p = effective sound pressure (μbar)
p_{ref} = a reference value for p equal to 0.0002 μbar

Substituting the value of p_{ref} we obtain:

$$\text{Sound pressure level} = 20 \log_{10} p + 94$$

where

sound pressure level is in db relative to 0.0002 μbar
p = effective sound pressure in N/m^2

Converting these two limits into sound pressure levels, we find, respectively, 0 and 174 db.

The propagation of sound from the source to the receiver reduces the sound pressure level proportionally to the distance due to the spherical propagation of sound (see Figure 6-9). This can be calculated as follows:

$$P_L = W_L + I - 20 \log_{10} d - 11$$

where

p_L = sound pressure level at the receiver located at a distance d(ft) from the sound source (db relative to 0.0002 μbar)
W_L = sound power level (db relative to 10^{-13} W)
I = directivity index from the source to the receiver (db)
d = distance from source to receiver (ft)

In addition to the sound reduction due to distance, there are additional attenuation factors, one of which is the wind. The sound pressure level inclusive of these factors is given by:

$$p_L = p_{L,o} - \log \frac{d}{d_o} - (R_1 + R_2 + R_3 + R_w)$$

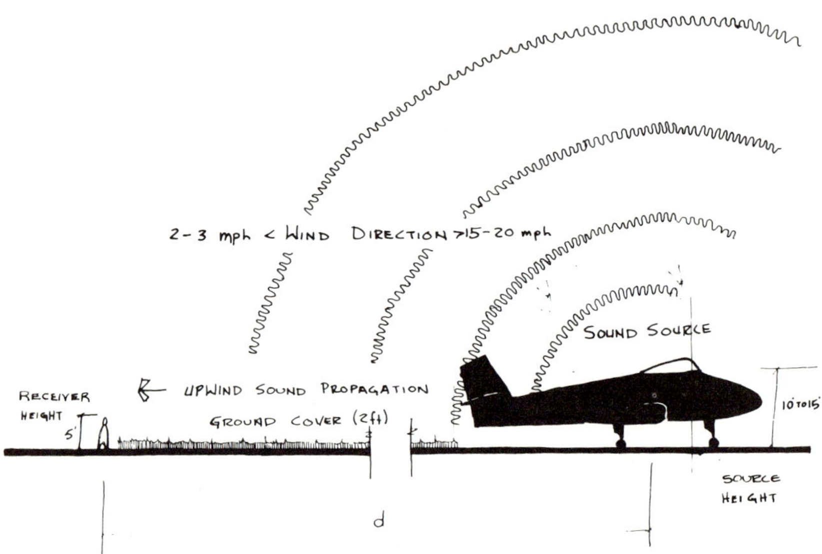

Figure 6-9. Propagation of sound from the source to the receiver.

where

p_L = sound pressure level at the receiver located at a distance d (ft) from the sound source (db relative to 0.0002 μbar)

$p_{L,o}$ = sound pressure level at a distance d_o (ft) from the sound source; d_o is an arbitrary distance where the sound pressure level is measured.

d = distance between receiver and source (ft)

d_o = distance where the reference measurement is made

R_1 = reduction factor due to the air

R_2 = reduction factor due to atmospheric conditions

R_3 = reduction factor due to obstacles

R_w = reduction factor due to wind and temperature

Wind Effect

Wind bends sound waves upward when the sound is traveling against the wind and downward when it travels with the wind. When the waves are deflected upward, there is a shadow area over the earth's surface where there is no sound. When the deflection is downward, of course, such shadows do not occur. The edge of these shadow areas, although a little fuzzy, are clearly identifiable.

For both upward and downward zones, the estimation of the sound attenuation factor R_w for certain conditions can be made as follows.

Upwind Zone

When the sound waves travel against the wind, the value of R_w can be determined if conditions are within the following limits:

- wind speed should be within 2 to 20 mph
- ground cover should be no more than 2 ft high
- source height above ground should not be more than 15 ft
- receiver height above ground should not exceed 5 ft
- distance between sound source and receiver should be less than 1 mi

Table 6-5. Distance between the sound source and the shadow area.

CONDITION	D (ft)[a]
Wind speed	
2–4 mph	2000
10–15 mph	400
10–18 mph	250
Temperature profile	
Lapse	250
Neutral	400
Inversion	2000
Sky	
Cloudy	400
Clear	2000
Nighttime	2000
Daytime	250–400

[a]The distance D between the source and the shadow area. (Select D according to the condition that generates the smallest value.)

To find R_w (db) establish the distance d (ft) between source and receiver. Determine the distance D (ft) between the source and the edge of the shadow area, either by direct field measurement or from Table 6-5. Calculate the ratio d/D, and read R_w from Figure 6-10.

For example, consider an aircraft operating in an airport. The ground cover is grass less than 2 ft high. The distance d between source and receiver is 500 ft. The environmental conditions are: daytime, clear sky, lapse temperature profile, wind at 15 mph opposite to the direction of the sound. From these conditions, we find from Table 6-5 that $D = 250$ ft. Source height above ground is 10 ft. Receiver height above ground is 5 ft. Compute $d/D = 500/250 = 2$. From Figure 6-10, find $R_w = 20$ db.

Downwind Zone

When the sound waves travel in the same direction of the wind, R_w can be found if the conditions are within the following limits.

- wind speed to be within 2 to 30 mph
- ground cover to be no more than 2 ft high
- sound-source height above ground to be no more than 15 ft
- receiver height above ground to be no more than 5 ft
- distance between sound source and receiver to be less 1 mile
- sound frequency to be within 300 to 5000 cps

To determine R_w (db), establish first the distance d (ft) between source and receiver. Establish then the mean frequency f_m of the sound and compute the product $(f_m)(d)$. Find the value of R_w on the diagram of Figure 6-11.

For example, consider the same condition as in the previous example except that the receiver is downwind with respect to the sound source. Assume also that the mean frequency of the aircraft noise is: $f_m = 2000$ cps. The product $(f_m)(d)$ is: $500 \times 2000 = 1{,}000{,}000$. From Figure 6-11, we find $R_w \cong 4$ db.

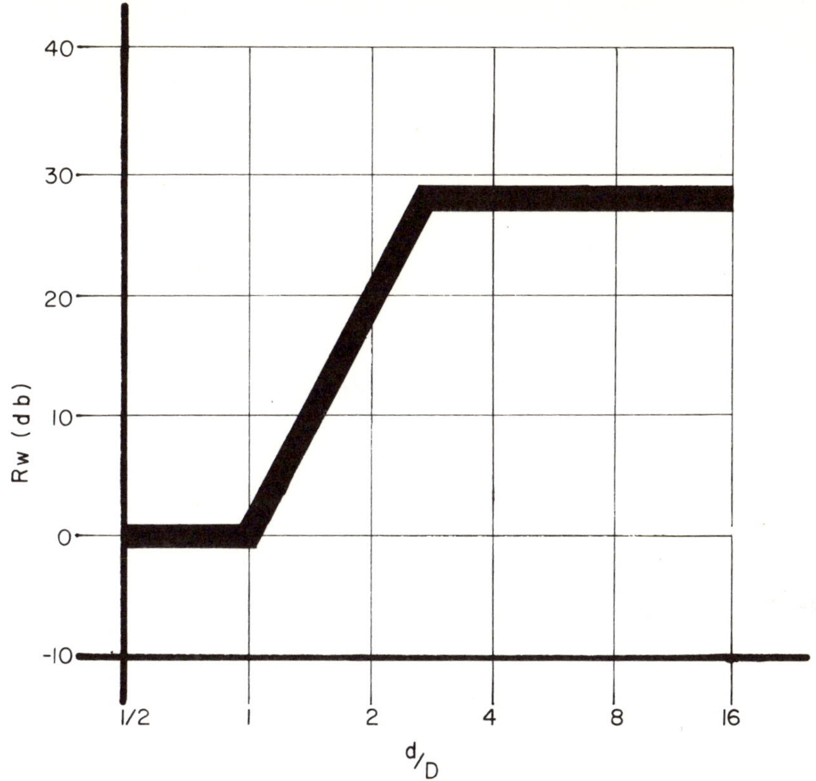

Figure 6-10.

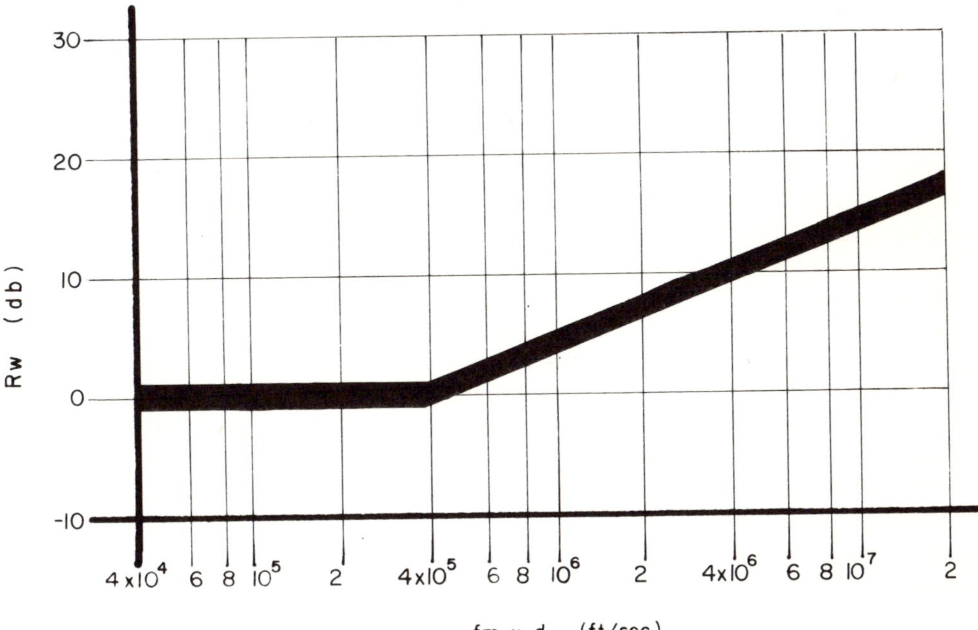

Figure 6-11.

Table 6-6. Equivalent runway length with respect to wind (Hornjeff 1962).[a]

TYPE OF AIRPLANE	TAKEOFF LOAD (LB)	RUNWAY LENGTH (FT)	
		STILL AIR	20-KNOT HEAD WINDS
Boeing 707–320	300,000	10,500	9,500
DC-8 (domestic)	250,000	8,400	7,650

[a] Standard daysea level, no slope.

EFFECT OF WIND ON AIRPORT DESIGN

The influence of the wind, alone or combined with atmospheric temperature, greatly affects the design of airports. In fact, the wind is a factor that determines the length and the orientation of the runways, as well as their number.

The general rule relating wind to runway length is that head winds decrease the necessary length of runways, whereas tail winds increase it. This is true, of course, for both takeoffs and landings. In fact, what counts in flying is the velocity of the plane with respect to the air, not the ground. However, for an aircraft, air speed and ground speed coincide when the air is still. With frontal winds, the air velocity is the sum of the ground speed and the wind speed. In this case, the air speed is higher, which consequently reduces the length required for the runway. On the other hand, tail winds reduce the air speed, which is equal to the difference between the ground speed and the speed of the wind. In this case, obviously, tail winds have the effect of reducing the length of runways.

Thus, the equivalent length of a given runway under a certain load condition varies according to the wind speed: that is, takeoff and landing lengths have to be established for each wind velocity. Since head winds are beneficial to the operation of an aircraft, it is customary to assume the value of the wind speed in the engineering calculations to be no more than half the real value. Such a safety precaution is also followed when the wind blows in the same direction of the aircraft motion (tail wind). The wind speed assumed in the engineering calculations is in this case 1.5, the actual velocity of the wind.

Table 6-6 gives some indications of the relationship between the length of runways and such parameters as the type of aircraft, the takeoff load, the absence of wind, and the effect of 20-knot head winds.

In general, the primary runways are oriented as close as possible to the direction of the prevailing wind. Any divergence between the orientation of the runway and the direction of the wind creates a wind component that is perpendicular to the runway, which affects considerably the safety of airplanes in both takeoff and landing. Such perpendicular wind components are usually referred to as *crosswinds.* The maximum crosswind speed allowed for an aircraft depends on many factors, including the size of the airplane and the configuration of the winds. Single-engine planes or other lightweight craft cannot operate safely in crosswinds greater than 15 mph. On the other hand, transport airplanes, which are much less sensitive to crosswinds, can operate in winds up to 35 mph. However, 15-mph crosswind speed is usually adopted for all cases; consequently, the orientation of runways is such that 95% of the time airplanes do not experience crosswinds in excess of 15 mph in either takeoff or landing. But, while this is true for mixed-traffic airports, those airports with predominantly cargo operations have runways oriented to allow crosswinds of up to 23 mph. By the same token, at small airports, used mostly for light planes, the maximum crosswinds allowed are approximately 10 mph.

In determining the final orientation of a runway, the wind data for the site must be gathered. For instance, see the summary of hourly observations for the Boston area in Appendix B. If we wanted to design a runway for that area, we would consider the percentage frequencies of wind direction and speed (annual values) given in Table B-3, and we would then draw a wind rose. Essentially this is a graphic representation of the data in Table C-3 that illustrates the percentage of the wind duration from 16 major directions and for the velocity ranges of 0 to 4, 4 to 15, 15 to 31, and 31 to 47. After the wind rose has been drawn and the maximum allowed crosswind speed has been determined (based on the type of aircraft using the facility), the orientation of the runway is determined as follows. On a strip of transparent paper, three parallel lines are drawn. The distance between the centerline and each of the two adjacent lines represents in any arbitrary scale the maximum allowed crosswind speed. The transparent strip is then placed over the wind rose so that the centerline goes through the center of the wind rose. A pin is put through the center of the wind rose, and the transparent strip with the three lines is rotated. As it rotates, one can observe that different sectors of the wind rose are contained within the area delimited by the three lines. Notice that each sector of the wind rose indicates a certain percentage of the duration of the wind. For each orientation of the strip, the percentages of the wind direction in the sectors being included are added. The orientation of the strip that produces the highest value of such sum is the desired direction for the runway.

VISUALIZATION OF THE WIND

One cannot escape the fact that in the immensity of a dark universe, our atmosphere with its luminous skies and lively winds is a unique spatial entity. From this point of view, the winds seem to have a special value in their role of creating motion around the globe. Using the inexhaustible energy of the sun, they produce such images as passing clouds, waving oceans, and rippling grasses when they are moderate, and generate terrifying scenes of violence over land and water when they are stormy.

Besides the global circulation of the winds from near the earth surface up to altitudes of several miles, the dynamics of the wind is also particularly interesting on a smaller scale, such as on the level of the urban environment. The wind is in reality a form of energy that permeates the city with subdued effects that give life to inanimate objects. These visible kinetic effects appear in such familiar scenes as waving trees, oscillating lamps and street signs, water gushing in fountains, deflected smoke plumes, and drifting snow. In a sense, the wind can be visualized as an invisible force flowing around buildings or other obstacles, through the intricate patterns of the narrow streets of a medieval European town, or through the large avenues of a modern metropolis. As this force finds its way through the streets, flowing straight or turning corners, one can see the speed increasing as the path narrows or decreasing as it expands. By the same token, it is simple to visualize zones of turbulence or smooth flows according to the shape and roughness of the path.

Kinetic expressions generated by the wind, whether in nature or in human-built spaces, are almost totally accidental. However, a significant exception is the modern art movement in sculpture known as kinetic art. In the wake of the numerous art movements of the early 1900s that broke long-standing traditions, the kinetic art appeared with its revolutionary concepts in the early 1930s. Its basic idea consisted of incorporating a dynamic component in sculpture by actually including physical motion in the artistic composition. Parts of the sculpture were made to move, creating something never attempted before in the various media used by sculptors. The dramatic effects have gradually increased, and their momentum was such that up to the present the interest in kinetic art is still alive. See Figures 6-12 through 6-14.

Motion was achieved through several means, but the predominant sources were the force of the wind (for the outdoor prototypes) and light air currents (for the small indoor pieces). Placed on museum grounds for outdoor display or outside buildings to complement the architectural composition, the large kinetic sculptures appear like silent dancers moving to the rhythm of an inaudible tune. In city parks, in pedestrian plazas, in the middle of public squares, the moving limbs of the sculptures react to even the most gentle breezes.

The mechanisms used to produce the desired kinetic effect vary tremendously, from simple suspended masses, to intricate mechanical assemblies capable of generating individual basic motions as well as any combination of them. The fundamental parameters used to create a kinetic theme are essentially geometry and velocity. For the most part, the materials used are noncorrosive metals including copper, brass, and stainless steel.

Naturally, the fundamental parameters for any work dealing with motion must include the same basic principle of kinematics such as path geometry, velocity, and acceleration. Kinetic sculpture, however, goes even farther, and like other art forms based on time functions, such as music or choreography, it depends also on cadence or rhythm, periodicity, and other parameters. Probably no other art form has such a potentially large spectrum of expression as kinetic art. Even the repertoire of motion per se is wider for kinetic sculpture than for choreography itself. The geometry of the paths of the moving parts can include such patterns as orbital, pendular, looping (repetitive or irregular), linear (back and forth), reversible, planar, spatial, and helicoidal. In other words, any linear configuration in any plane or in space is actually possible using a geometric directrix as one chooses. From the point of view of playing with variable speed, the range of possibilities is unlimited. For instance, some of the characteristics of motion that have been actually used in kinetic sculpture can be described as lulling, frenzied, vibratory, swaying, pitching, or rolling. In terms of motion combinations that potentially create choreography, one can visualize, for instance, a sequence of this type: a moving element that comes slashing as a dagger and then is softly decelerated to an extremely slow motion with sudden changes

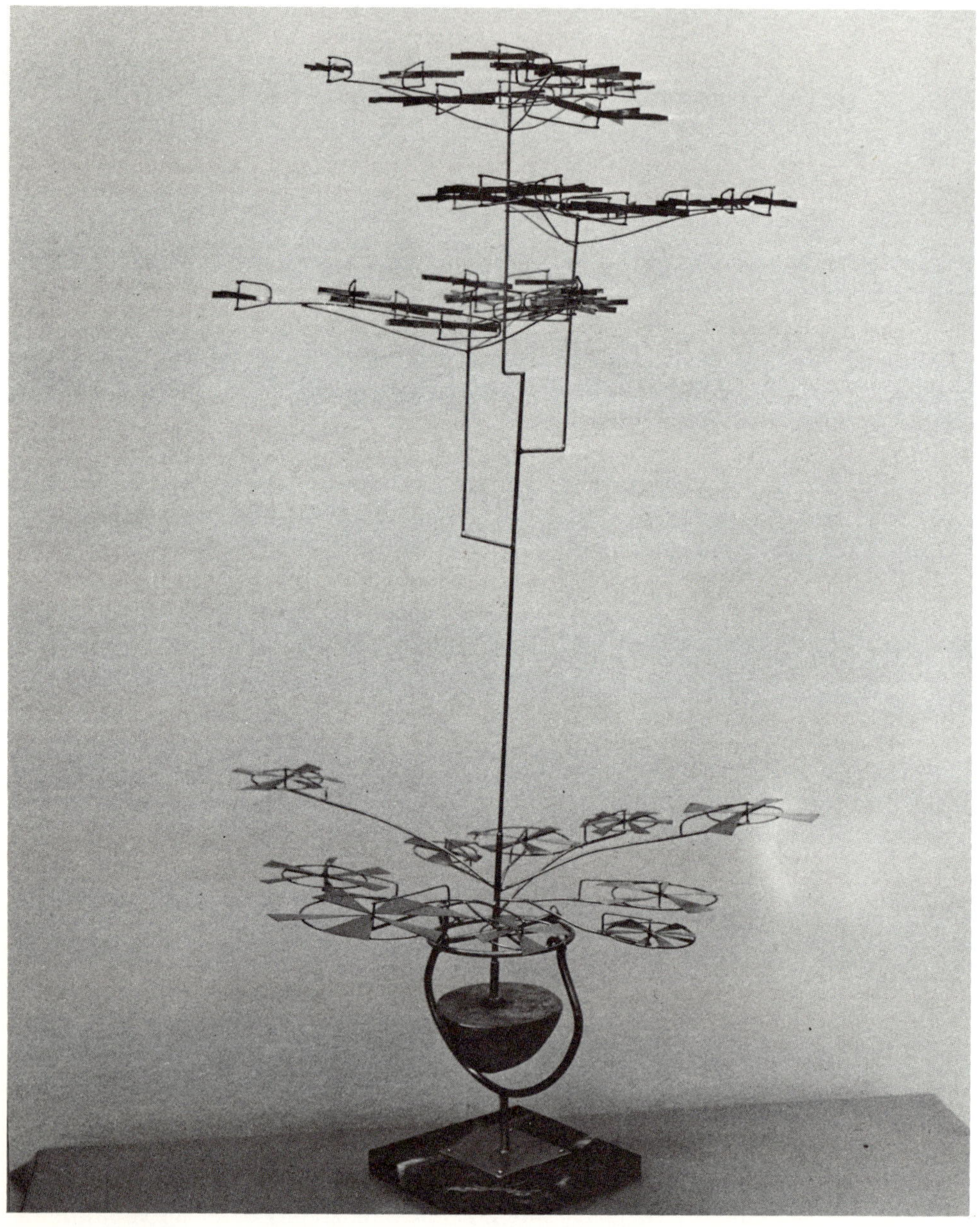

Figure 6-12. Windflower I (1959) by George Rickey. Stainless steel, brass, and copper—34 × 18 × 16 in. (Courtesy of Montclair Art Museum, New Jersey)

Figure 6-13. Crucifera IV (1965) by George Rickey. Stainless steel, 21 × 32 × 7 ft. (Courtesy of Great Western Savings and Loan Association, Oakland, California)

Figure 6-14. Four Squares in a Square (1969) by George Rickey. Stainless steel—19 × 11 × 1 ft. (New National Gallery, West Berlin; Courtesy of George Rickey)

that, even if abrupt, always maintain a sense of continuity and smoothness as the flow of a fluid. To achieve effects such as these, several dampers of various types are necessary. A damper, which can be imagined as an energy eater, is actually a mechanism that absorbs and dissipates energy in different ways. Therefore taking advantage of such variety, an artist can create complex orchestrations of motions.

Kinetic art originated in the United States from the pioneering work of the American sculptor Alexander Calder. His famous mobiles have gained great popularity since the 1930s, gradually obtaining worldwide recognition that inspired artists in many countries. Among Calder's early mobiles was the *Calderberry Bush,* built in 1932. Using steel, wood, and aluminum, the 7-ft composition moves elegantly as the air flows through it.

In England, Lynn Chadwick emerged as one of the outstanding artists in kinetic sculpture. The *Fisheater* (1952), part of the collection of the Art Council of Great Britain, is a significant example of his work. This outdoor piece made of copper and brass stands 8 ft high and is a light elegant structure animated by the wind.

Another American sculptor who devoted most of his work to the cause of kinetic art is George Rickey. His exciting and inspiring production includes a large number of indoor and outdoor pieces. Among the indoor sculptures are *Trastevere Flower* (1957), *Windflower I* (1959), and *Sunflower* (1960). Intriguing in these examples are the simplicity and precision of the design, construction details, and the type of material used. The preciousness of jewelry and the precision of toolmaking seem to be combined in Rickey's work. The large outdoor artworks are astonishing prototypes of constructivism and exciting designs of structural and dynamic engineering in the interpretation of an artist. *Two Red Lines* (1963), *Two Lines Temporal I* (1964), *Two Rectangles Vertical Gyratory* (1969), *Space Churn with Sphere* (1969), and *Four Square in a Square* (1969) are among his kinetic sculptures.

In Italy, Arnaldo Pomodoro is part of the kinetic art movement. His *Grande Disco* (1974) is a large and powerful bronze casting that spins around a vertical axis under the action of the wind. At present, however, the piece is immobile for the protection of passersby.

THE MONROE PHENOMENON

Over the windward facade of tall buildings, a portion of the wind, on the upper levels, rises vertically and then goes over the rooftop, while another part descends toward the street and bounces up as it hits the sidewalk, creating eddies and turbulence. An article in *New York* described this effect as the "Monroe phenomenon." The expression comes from the film, *The Seven Year Itch,* in which Marilyn Monroe's skirt was repeatedly blown up by the wind as she was walking on New York's sidewalks. Such winds are particularly strong around buildings that have smooth surfaces and less so around buildings that use setbacks.

Identified first by A. F. E. Wise, this phenomenon creates turbulent zones at the ground where the comfort parameter ψ can be as high as 1.5 for buildings approximately 180 ft high, and it can reach values up to 1.8 under certain conditions. For instance, if the wind is perpendicular to a building, and if, at a distance *n* upstream, there is another building *n* ft high, then such an arrangement of buildings is considered critical and ψ can be as high as 1.8, as previously mentioned.

THE COMFORT PARAMETERS (ψ)

In describing the effect of the wind speed and the wind turbulence around single buildings and groups of buildings, it is convenient to relate the wind effect to a parameter called ψ (the *comfort parameter* of the wind). It is defined as the ratio between the sum of the average speed and the turbulence at any given point, and the speed and the turbulence at a specific reference point. Thus, an increase of wind speed and turbulence corresponds with an increase of ψ.

THE ROW EFFECT

The row effect is a phenomenon caused by a row of buildings on the windflow, when, for instance, the width of the row is approximately 30 ft, the height is not over 90 ft, and the length is not less than 700 ft. If the wind is at a 45° angle with the row, and the height of the row averages from 45 to 55 ft, most of the wind goes over the row and creates eddies behind the buildings. There, the comfort parameter $\psi = 1.4$. If the row includes some openings, of which the width is between 1 and 2 times the height of the row, the comfort parameter is $\psi = 1.3$. Such an effect, however, can be eliminated if the integrity of the row is altered by some nearby building placed at 90° with the row itself.

THE VENTURI EFFECT

If two buildings or two rows of buildings are arranged so as to funnel the wind, the wind speed increases up to a critical zone that is the narrowest. Under this condition, several cases can be identified. If the shape of the space between buildings is such that it converges toward the narrowest section and then diverges, the comfort parameter ψ can be as high as 2 if the building height is about 150 ft. In general, for the Venturi effect to occur, the buildings must be about 45 ft high (or higher) and the length of the two sides should be not less than 300 ft. If the inlet and outlet have widths from 2 to 3 times the building height, the comfort parameter ψ is 1.3 for buildings 75 ft high, and 1.6 for 90-ft-high buildings.

EFFECT OF OPENINGS UNDER BUILDINGS

Many modern structures have open spaces in between the columns at ground level. The wind flowing under the building creates a zone of discomfort on the leeward side that is more pronounced as the building height increases. For building heights up to 45 ft, the phenomenon is negligible. For heights of approximately 60 ft, the comfort parameter ψ is 1.2, and for heights of about 150 ft, the comfort parameter can reach 1.5 independently from the length of the building.

EFFECT OF CORNERS

Turbulence at the corner of buildings is a local phenomenon that does not extend much farther away from the building itself. For prismatic buildings approximately 45 ft high, the comfort parameter ψ is about 1.2. On the other hand, for buildings higher than 100 ft, ψ is equal to 1.5. For building heights of 300 ft, ψ increases up to 2.2.

THE CELL EFFECT

The *cell effect,* a common phenomenon connected with the grouping of buildings, is typical of many urban centers. The term indicates open spaces delimited by buildings all around the perimeter. The phenomenon consists of a wind speed abatement over the open spaces due to the shielding of the building belt.

For a space to qualify as a cell, the number of sides forming its perimeter is not relevant, but the openings cannot exceed 25% of the perimeter itself. In this respect, we distinguish between "open cells," which have openings facing the wind either directly or within a 45° sector centered in the direction of the wind, and "closed cells" in which the openings, if any, are facing the downwind direction. Large courtyards, squares with and without converging streets, and plazas are examples of where the cell effect can occur.

The major parameters affecting the wind abatement in the cell are the average height of the building (h_m) and the quantity $\frac{S}{(h_m)^2}$, where S is the area of the cell. The effect of the phenomenon in reducing the wind effect is measured in terms of the comfort parameter ψ.

For instance, when h_m varies within 45 and 75 ft and $\frac{S}{(h_m)^2}$ is not larger than 10, the parameter ψ varies from 0.4 to 0.8. Note also that the apertures can be h oriented anywhere along the perimeter without affecting the extent of wind abatement. For example, assuming $h_m = 50$ ft, the maximum value for S would be $S = (50)^2 \times 10 = 25{,}000$ ft^2 (approximately 158×158 ft).

In the case of open cells with $\frac{S}{(h_m)^2} < 20$, the comfort parameter ψ varies from 0.7 to 1.1. On the other hand, if the orientation of the opening is equal to the direction of the wind, and h_m and $\frac{S}{(h_m)^2}$ have the values mentioned previously, the comfort parameter ψ is either 1 or less than 1.

In the case of closed cells with $\frac{S}{(h_m)^2} < 30$, the average value for the comfort parameter ψ is 0.5 or less.

THE SETBACK EFFECT

In high-density urban areas, it is very common to observe tall buildings in which the width is gradually reduced with height, which generates an overall effect of truncated pyramids. In New York City, the "setback" laws determined this characteristic shape for the first skyscrapers. As the wind engulfs these structures, the stepping of the stories creates a roughness that dissipates the wind energy and produces a condition that can be referred to as the *setback effect.*

Because of this effect, a high turbulence is observed at the street level, especially at the windward corners. The comfort parameter ψ is 0.6, relatively low, but the turbulence component is particularly high. For buildings of approximately 13 stories, it is observed that at the windward corners, at a certain height, the parameter ψ can reach 1.6. Balconies, when present, also contribute to turbulence. For instance, for a 17-story building, the turbulence caused by the balconies on the windward side, at the top of the building, generate a comfort parameter ψ that can be as high as 1.6. However, except for these top balconies, the value of ψ is only 0.5 for the majority of them (±80%).

SOUND

Whenever the wind blew through the reeds in the valley, the Greeks said it was the cry of Syrinx, the beautiful nymph whom the gods changed into reeds when she was pursued by Pan. We in the twentieth century talk instead about the aerodynamic noises that surround us in our cities.

What are these noises, and what causes them? We know by experience that as the wind blows over and around tall buildings, aerodynamic noises arise and propagate through the interior spaces, creating a real nuisance to the occupants. Exterior projections from the facade, such as cornices, sills, and mullions, can be sound sources, but their role is secondary. The most important factor is the air that penetrates the building and passes through elevator shafts, stairwells, and chutes to produce moaning and wailing sounds as it escapes from the vents on the top. The phenomenon can be compared to the effect that one can get by blowing over the top of an open bottle. Practically no buildings are absolutely airtight. Air enters them through gaps or as the exterior doors are opened. The combined effects of pressure and suction on a building surface facilitate airflow through the structure. As the air enters the shafts, it is sucked in at the base and accelerated as it moves upward, partially because of thermal effects that make warm air rise and partially because of the wind suction over the vents. The phenomenon is usually called the *chimney* or *stack effect* since it is similar to what happens in a chimney or smokestack. Warm air heated by the fire rises, and it is further accelerated by the shape of the stack due to the Venturi effect. In fact, the elevated air speed is evident by the rise of the smoke over the chimneytop. To stop the phenomenon a simple solution would be to close the air vents; however, building codes require open vents for sucking smoke out in case of fire. Another solution would be to reduce the penetration of air through exterior doors by using several doors, thus creating a system of airlocks. This was actually done in the World Trade Center in New York City. Most effective, however, are revolving doors, which also function as airlocks. A typical example of stack effects in elevator shafts occurred in the Pan Am Building in New York City. The pressure generated by the air currents in the elevator shafts was so great on windy days that the elevator doors were vibrating and would not close. As a result, all the motors operating the doors had to be replaced shortly after the completion of the building.

Another source of noise in tall buildings on windy days is caused by building motion due to powerful wind forces. The squeaking and creaking that occur are due to the actual distortion of the building frame. Efficient structural systems of recent steel buildings have considerably reduced building weight, making them light and flexible so that relatively large deflections are to be expected. However, even the Empire State Building, whose stiffness and damping qualities are conservative, produces creaking noises when the wind speed is particularly high. Other noise sources in wind-sensitive buildings are the rattling of elevator cabs as they hit the tracks in the shafts, as well as the vibrations of the elevator doors.

A *noise canyon* is a term that describes a narrow urban street lined on both sides by tall buildings. These buildings produce reverberating sound effects of such intensity that they create a real problem for anyone who lives or works there. In fact, when buildings of equal height face each other on opposite sides of a street, the street sounds are carried to the upper stories by the accelerated wind flow. Screeching tires, sirens, backfiring automobiles, and screams emerge over the uniform street sounds and are quite disturbing, especially when the wind gathers them from a large contributing area.

WHERE THE WIND DOES NOT BLOW

There is no place on the surface of the earth where the wind does not blow. The atmosphere, crossed by heat radiations from the sun and the earth, is in constant motion. Therefore, anything we build must respond to the wind. Cold or hot, moist or dry, violent or mild, the wind influences building forms. But how would the form respond to a windless environment? What would evolve in the absence of any air motion? Although such a condition does not exist on earth, it can be found in outer space.

Since the beginning of the Space Age, orbiting satellites, the landing of men on the moon, and the success of probes sent to other planets have all paved the way for the possibility of temporary and permanent space colonies. (See Figure 6-15.)

Starting in the 1980s, plans for the future include, in progressive order:

- shuttle operations
- space transportation system payloads
- low-earth-orbit space stations
- sortie missions to geosynchronized orbit
- use of large space structures

The next phase, characterized by permanently occupied space colonies orbiting the earth, includes:

- space manufacturing
- geosynchronous space station
- orbital launch facility
- large structural applications (SSPS)
- space-base operations

The third phase, which projects human occupancy of space, is based on self-sufficient colonization systems that include the following progressive stages:

- close ecology
- lunar base
- space industrialization
- space construction
- exploitation of extraterrestrial materials and energy

Figure 6-15. Artist's concept of outside view of lunar settlement under construction (200 to 300 people). (Courtesy of NASA)

Regardless of science fiction fantasies, the extension of human habitats to extraterrestrial bases may be necessary if humans are to survive the crisis produced by the population growth and the scarcity of energy and other vital resources. Thus, architects of the near future will be concerned with space architecture, dealing with quiet motionless environments.

The differentiation of building forms in wind-containing and windless environments, in terms of the wind influence alone, is not simple because of many contributing factors. Reduced and even total absence of gravity, pressure differences between the inside and outside of structures, and usage of earth-built components versus usage of materials from other planets are among the major factors that produce differences in form.

Windless environments can produce vertical structures more economically than can earth-based structures, which must pay a high premium for wind. The cost per square foot of building area for high-rise buildings is known to increase as the height increases because of the wind action and the difficulty in resisting it. In contrast, vertical expansion of structures in the absence of wind will have practically no structural limits. Some limits, however, will be imposed by other parameters. The super-skyscraper, in fact, could be a prototype design in the windless lunar landscape. It would allow a long distance view over the moon surface. The concept of vertical extension, which the absence of wind already permits, is significantly reinforced by the reduced gravitational field. Large bubbles constituted of thin membranes can enclose immense spaces when unaffected by wind forces. An enclosed artificial atmosphere is probably a very realistic design concept. However, a practical limitation to the size of inflatable bubbles arises from the possibility of induced air circulation within the enclosed space itself. Air currents due to thermal differences might produce forces against the enclosing membrane.

Pressurized tubular components could generate a rigid system of columns and beams that could form frames or space structures. Still in the realm of pneumatic structures, the inflated-tube concept uses only light membranes in tension. The lightness, of course, is the key factor in any structure in outer space that depends on material transported from earth.

WIND AS A RECREATIONAL RESOURCE

In the twentieth century, the concept of mental and physical recreation has been reassessed after two millennia of neglect and has been given a level of priority that surpasses classical splendor of Greek and Roman times. Therefore, we must not overlook those recreational activities based on the use of wind power, since they can be of particular interest for the planning of regional developments. Winter and summer sports based on wind power can generate additional sport resorts in new areas with favorable conditions that would tap a natural resource never used before: the wind. The wind is just as vital to economics and to human welfare as other natural resources. On this basis, we can classify wind power as a natural resource for recreation and assess geographical areas according to how favorable wind conditions are, just as it is done for recreational activities based on water and snow.

Iceboating

In its original form, an iceboat consisted of a sailboat equipped with runners that allowed it to slide on ice. Iceboats were used in the 1700s in Scandinavia and also in the United States. However, as a real sport, iceboating was not established until the end of the 1800s. Over the years, several refinements have improved the iceboat's design; the modern version can reach speeds of 160 mph, making the sport risky, although quite exciting. Modern iceboats include smaller, more economical crafts than those of the past, thus contributing to the increasing popularity of this sport. (See Figure 6-16.)

Skate Sailing

Skate sailing is a Scandinavian recreation that dates back to the Middle Ages. In skate sailing, an ice skater holds a frame with a sail on his or her shoulder and maneuvers it, always keeping it at an angle with the wind. With up to 70 ft^2 of nylon or Dacron stretched over a light frame, the skater can reach speeds of 60 mph. (See Figure 6-17.) The sport spread to the United States in the late 1800s and became particularly popular in the New York and Great Lakes areas.

Kite-Flying

As history shows, the origins of kite-flying included social and religious aspects in Oriental countries and extended much later to scientific and technological applications in the Western world. Although some legends credit the Greeks as the inventors of the kite (with one in particular referring to Archytas of Tarentum in the fourth century B.C.), kites were originally used in Asia since prehistoric times. Carried aloft by the wind, the fragile kite must have inspired a variety of symbolic expressions in primitive people. (See Figures 6-18 through 6-20.)

During religious ceremonies, the Maoris, a primitive tribe in New Zealand, used to send up a kite while singing traditional chants. China, Japan, Korea, Vietnam,

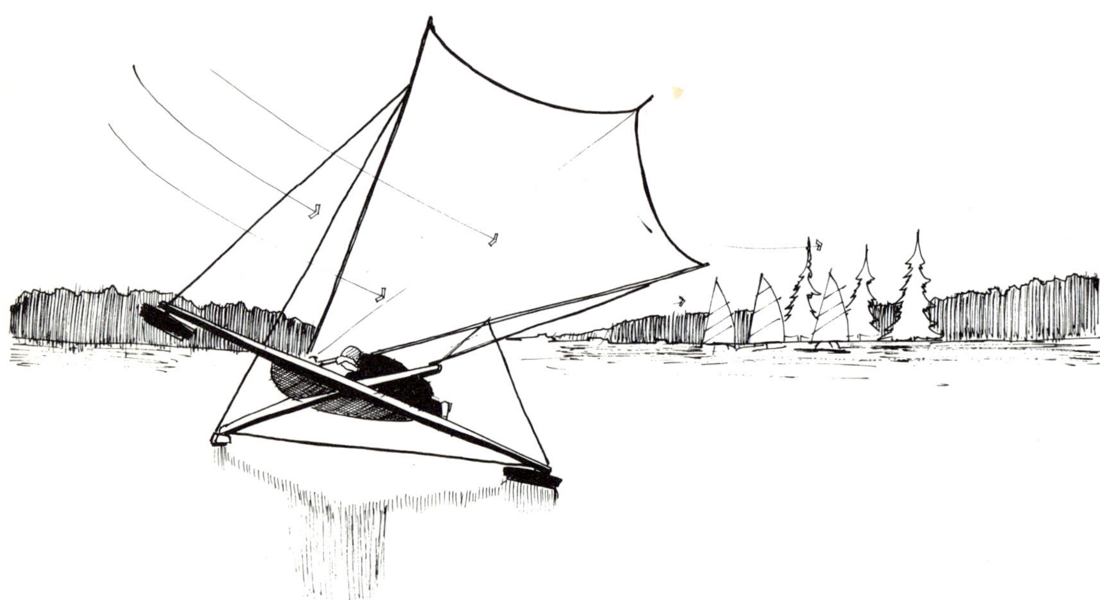

Figure 6-16. Iceboating dates back to the 1700s. Modern iceboats can reach 160 mph with sails of approximately 70 ft^2.

Figure 6-17. Skate-sailing on a frozen lake. This sport, which originated in Scandinavia, dates back to the Middle Ages. It was introduced in the United States during the 1800s. The 50-to-70-ft^2 sail is kept at an angle to the wind—never perpendicular. Holding the sail by its metal frame and using racing skates, speeds up to 60 mph can be reached.

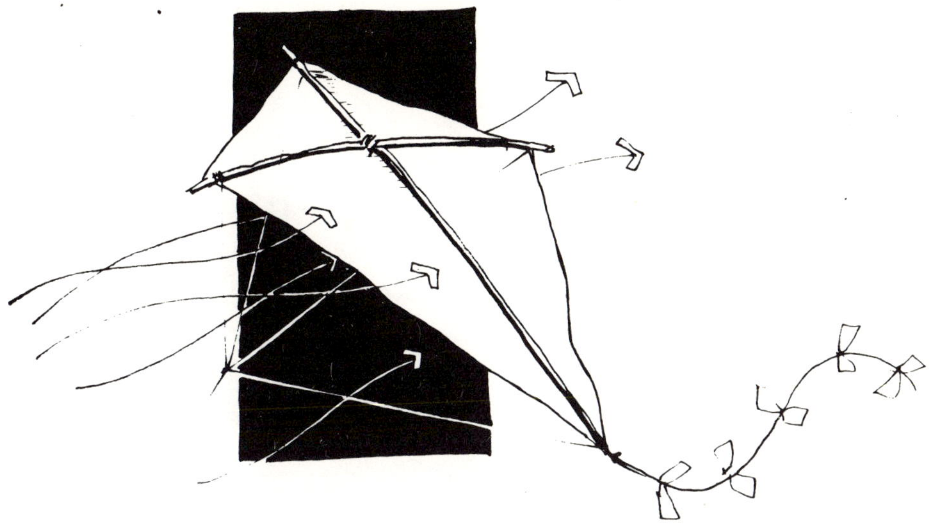

Figure 6-18. Conventional kite built with a planar cross frame. It requires a tail for stability.

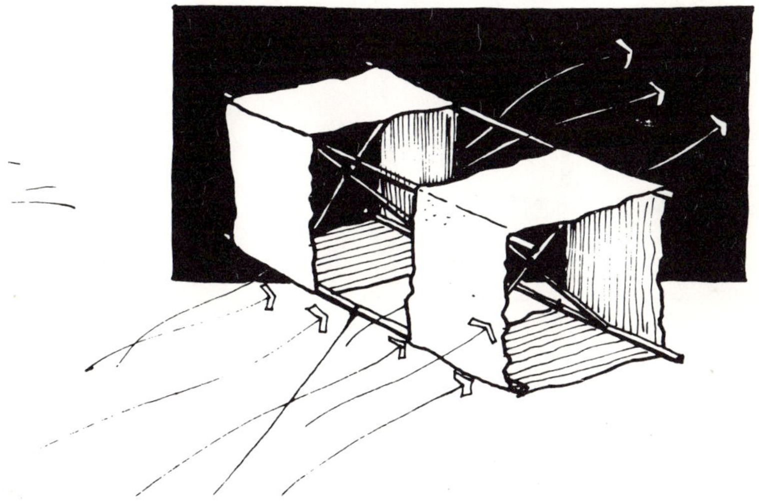

Figure 6-19. Box kite. Box kites were used for experimentation by Orville and Wilbur Wright.

Malaya, and India consider kite-flying a national pastime. In China, the ninth day of the ninth month of the year is "kite day." In India, kite-fighting is a popular competition in which the opponents try to cut each other's line by using a string covered with glue and glass fragments. No less competitive is kite-fighting in South America, where a rotating small disk sliding aloft along the line is used to cut the

Figure 6-20. Maneuverable kites are a combination of the Rogallo aerodynamic shape, sturdy construction, and controls with special bridle connections.

adversary's line. In Asia, peddlers are often seen flying kites to pass the time while they wait for customers to stop by.

The kite, utilizing the kinetic energy of the wind, can lift and keep aloft a mass whose weight is heavier than that of an equal volume of air. Such a machine, conceived prior to any knowledge of aerodynamics, was then capable of lifting a man when, in the nineteenth century its design was refined without adding any complexity to its original simplicity enabling humans to fly for the first time in history. Thus, since it was discovered centuries ago how to overcome gravity by using the wind, it is hard to explain why it was only in the nineteenth century that the idea was applied to human flight. Moreover, it does not seem logical that after such success in the 1890s, the idea was applied to hang-gliding only a few years ago.

For scientific purposes in general, kites have been employed since the middle of the eighteenth century. In 1749, Alexander Wilson and Thomas Melville measured the air temperature at various heights above ground by flying a kite that carried a thermometer. Benjamin Franklin demonstrated in 1752 the electrical nature of lightning by using a kite to lift a metal conductor in the air during stormy weather. Kites were commonly used for meteorological measurements until the 1920s when they were gradually replaced by balloons and airplanes. The kites used in meteorology were mostly of the box type, with a sail area of up to 80 ft^2. Typical altitudes reached with single kites were over 1 mi, but with trains of several kites attached in series, several miles have been reached. On November 29, 1905, an assembly of six kites in series was raised up to an altitude of 4 mi at the observatory in Lindenberg, Germany.

Prior to the use of aircraft in warfare, the possibility of hoisting strategic apparatuses over enemy lines or even lifting human observers at a certain altitude was indeed intriguing. Less vulnerable than balloons to enemy fire, kites seemed to be more feasible, which, in fact, was proved by several experiments at the end of the nineteenth century. In Australia, Lawrence Hargrave designed an innovative kite form usually referred to as the "box kite," with which it was possible to hoist a man 40 ft above the ground. It was later adopted for meteorological uses. Soon after, B. F. S. Baden-Powell, an officer of the British Army, also succeeded in lifting a man with a 36-ft-high kite a number of times. With an assembly of six smaller kites, he was also able to lift and keep aloft a man at an altitude of 100 ft. The technique was then adopted by most European armies. By 1897, similar experiments were also conducted by the United States Army using box kites.

Other names are associated with the evolution of kites. The most prominent include: Octave Chanute (1832–1910) in the United States; William A. Eddy (1858–1909) who invented a kite without a tail; Samuel P. Langley (1834–1906) in the United States; Charles F. Marvin (1858–1943) in the United States; Paul Garber in the United States; and Francis Rogallo in the United States.

The kite is no longer a static apparatus at the end of a line, but a dynamic flying device that can perform exact maneuvers and evolutions under the control of the flyer. The kite, in fact, can follow complex patterns moving on vertical, horizontal, or inclined planes, inverting its direction such as from upward to downward and vice versa and from left to right or right to left. It can also perform single or multiple loops, and like a sailboat, it can fly in almost any direction with respect to the direction of the wind. The controls, two lines held by the operator and extending to the kite, are connected to the kite by means of a bridle of simple design.

At its present stage of development, the kite has potential as a new sport. A few changes are necessary. The tensile force on the control lines, for instance, must be increased in comparison to that required in conventional kite-flying. Muscular effort equal to that required in most sports must be imposed on the operator by enlarging the sail area and reinforcing the frame as necessary. The suggested level of action should be equal, for example, to that required in deep-water fishing when reeling in a large catch.

Although the idea of controllable kites is relatively simple, it did not appear until very recently. During World War II, in fact, the United States Navy applied the concept of maneuverable kite-flying to artillery target practice through the experimental efforts of Paul Garber, a naval officer at the time. The trials were successful, and the use of kites became standard practice. Almost forgotten for years, only recently has maneuverable kite-flying reemerged as a recreational activity and a sport, for the most part in England and the United States.

Hang-Gliding

Midway between gliders or sailplanes and kites, the hang-glider has become a new sport that has gained incredible popularity in the past few years. Hang-gliders are very inexpensive and therefore have a large economic potential. These crafts can do almost everything that gliders and sailplanes do, although with less aerodynamic efficiency. They can glide for over 100 miles. They soar by gaining altitude by riding over thermal currents. However, they differ from gliders and sailplanes in the following ways: (1) absence of fuselage that leaves the pilot totally exposed to the wind and unprotected landing; (2) absence of mechanical controls because governing the craft is done by shifting the pilot's weight; and (3) potential inability to control the craft in case of turbulence or wave lifts with dangerous consequences.

Hang-gliders look like large kites (Figure 6-21). Although they can be made of rigid material, they are typically made of a plastic membrane, usually dacron, and a frame of tubular aluminum. Their construction is similar to that of a kite, but their performance is like that of a sailplane.

The most popular type of hang-glider is based on Rogallo's patents of 1949–1951, which featured a special airfoil (an aerodynamic shape capable of obtaining an uplifting

Figure 6-21. Typical hang glider.

component from an air current) produced out of a flexible material–in other words, a tensioned membrane. The surface area usually ranges from 120 to 280 ft². The frame of aluminum alloy includes four tubes, 1 to 2 in. in diameter; two form the leading edge, one is the cross tube, and one is the keel. A model with 18 ft of leading edge weighs approximately 40 lb.

The initial motion in gliding is by gravity, which the hang-glider obtains by jumping off a hilltop or a cliff. In gliding, in fact, the craft uses gravity and the air resistance in a controlled descent. The other component of hang-gliding is soaring. In soaring, the operator uses vertical air currents originated by thermal sources or by the characteristics of the land below (wave and ridge flying).

The wind speed for taking-off should be from 10 to 25 mph in smooth terrain with no turbulence, and from 10 to 15 mph in rough terrain with turbulence. In higher wind velocities, the drag force of the wind prevails, pushing the hang-glider backward with respect to the wind direction rather than providing lift. Thus, hang-gliding is extremely dependent on the characteristics of the terrain. Small waves from ridges and thermal sources have a critical influence, not just for taking-off and landing, but also for in-flight conditions. A classification of geographical characteristics varying from optimum to poor should be established. For example, in the United States, the best hang-gliding conditions are usually found on the West Coast, the western deserts, and the outer banks in the Carolinas. (See Figures 6-22 through 6-25.)

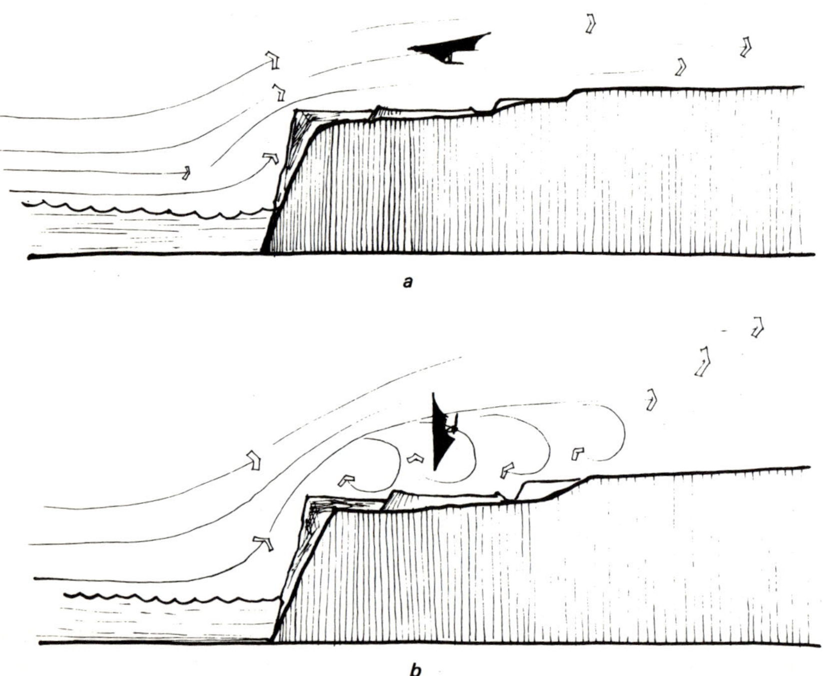

Figure 6-22. Cliffs deflect the windflow upward creating a lifting current that could be advantageously used by hang-gliders. *a,* When winds are light, the flow is smooth (laminar) and there are no dangers. *b,* When winds are moderate or stronger, turbulence forms on top of the cliff. The situation is now dangerous, and no attempt should be made at landing or takeoff.

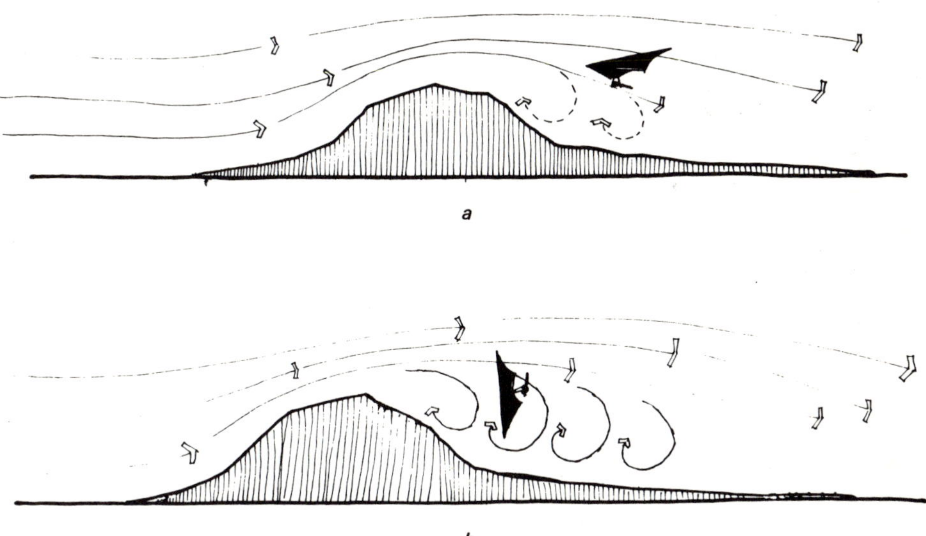

Figure 6-23. *a,* If winds are light, the turbulence is weak. Such turbulence is not a real danger to hang-gliders. *b,* If winds are strong, the turbulence is also strong and the effects are really dangerous to hang-gliders flying in the leeward side, although the windward area may still be safe.

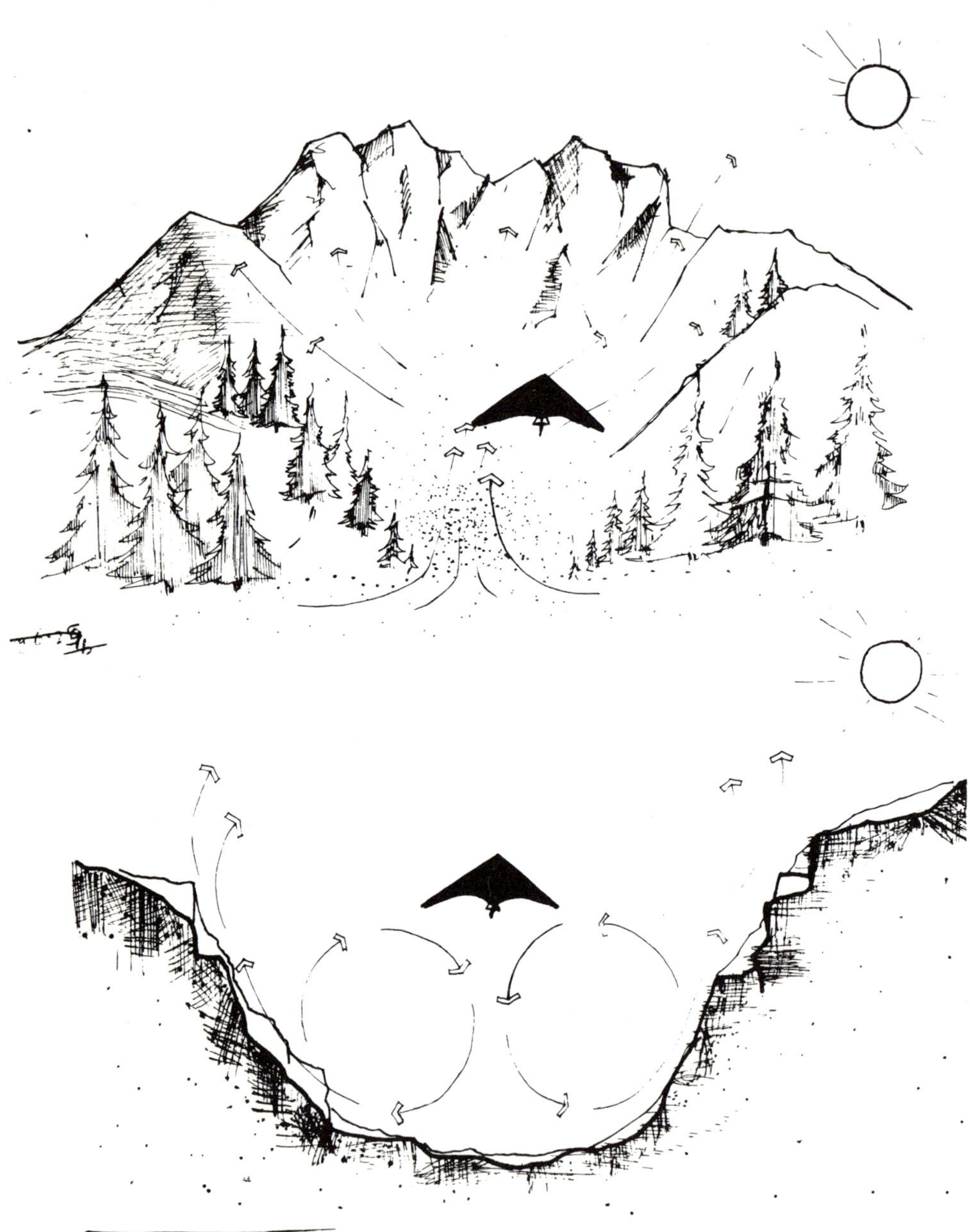

Figure 6-24. *a,* Mountain winds are originated during the day by the radiation of the sun reflected by the mountain slopes, which warms up the air, making it rise over the slopes. Cold air is drawn in the valley, creating the valley wind that flows from the valley toward the mountains. *b,* General circulation in mountain regions during the day.

Figure 6-25. *a,* In late afternoon or at night, the mountain slopes become cold; thus the adjacent layers of air are cooled by the slopes. Cool air runs down the slopes, creating a cold wind that goes from the mountains into the valley. This is a mountain wind that comes from the mountains. *b,* General circulation of winds at night in mountain regions.

REFERENCES

Amore, J. E. 1952. *Perfum, Essent. Oil Rec.* **43**: 321.

Amore, J. E. 1963. Stereochemical theory of olfaction. *Nature* 912–913.

ASTM. 1968. *Manual on Sensory Testing Methods.*

Crocker, and Henderson. 1927.

Davies, J. T. 1965. A theory of the quality of odours. *J. Theoret. Biol.* **8**: 1.

Gandemer, J. 1978. Building Research Translation: Discomfort Due to Wind Near Buildings: Aerodynamic Concepts. NBS Technical Note 710–9. U.S. Department of Commerce, pp. 32–33.

Henning, H. 1916.

Horonjeff, R. 1962. *Planning and Design of Airports.* San Francisco: McGraw-Hill Book Co.

Jackson, Hughlings. 1899.

Moncrieff, R. W. 1966. *The Chemical Senses.* London: Leonard Hill, p. 44.

Moses, and Carson. 1969.

Wise, A. F. E.

Zwaardemaker, H. 1895.

7
Wind Power

As the search for energy is now probing into the realm of the wind and the sun, what will we see in our rural and urban landscapes in the future? Will huge wind turbines clutter the oceanscapes for large-scale energy generation? Will city rooftops be full of wind turbines competing with each other to capture the winds? Will wind-produced forms, which originate as a response to wind forces, improve the architecture of the future? And, if so, to what extent?

In using energy to produce mechanical work, humans, since the beginning of civilization, have focused their attention on those forces found explicitly in nature. Animal, hydraulic, and wind energy did not require transformation into other energy types. Instead, the use of such energy types necessitated the invention of mechanisms whose complexity has constantly increased with the progress of technology. In particular, since ancient times, the kinetic energy of the wind has been used to propel ships and machines, such as windmills, that preceded the industrial age.

Wind power has been used for the production of electricity itself, although only sporadically and in small scale. However, because of the worldwide energy crisis, wind energy once again is being studied for large-scale application, along with other sources heretofore neglected. In the utilization of the wind or other natural energy forms, the unlimited potential of the source must be recognized in contraposition to the limited supply of fossil fuels presently used.

Solar energy, which generates the wind, is considered unlimited. The relationship between the dynamic forces in the wind and solar radiation is direct. Air masses of different temperature and density are subjected to different gravity forces that generate their motion. The differences in density are caused by the thermal differences in the atmosphere due to the uneven distribution of the radiation coming from the sun or reflected by the earth.

Another major characteristic of wind energy is the absence of any form of pollution, either in its direct utilization for the production of mechanical work or in its transformation into electrical energy. Yet, despite such positive factors, wind energy was neglected until a few years ago. With the harnessing of nuclear energy and the consequential opening of new horizons, and with the low cost of fossil fuels, wind and other energy sources that could not be used directly on a large scale were drastically neglected until the current energy crisis.

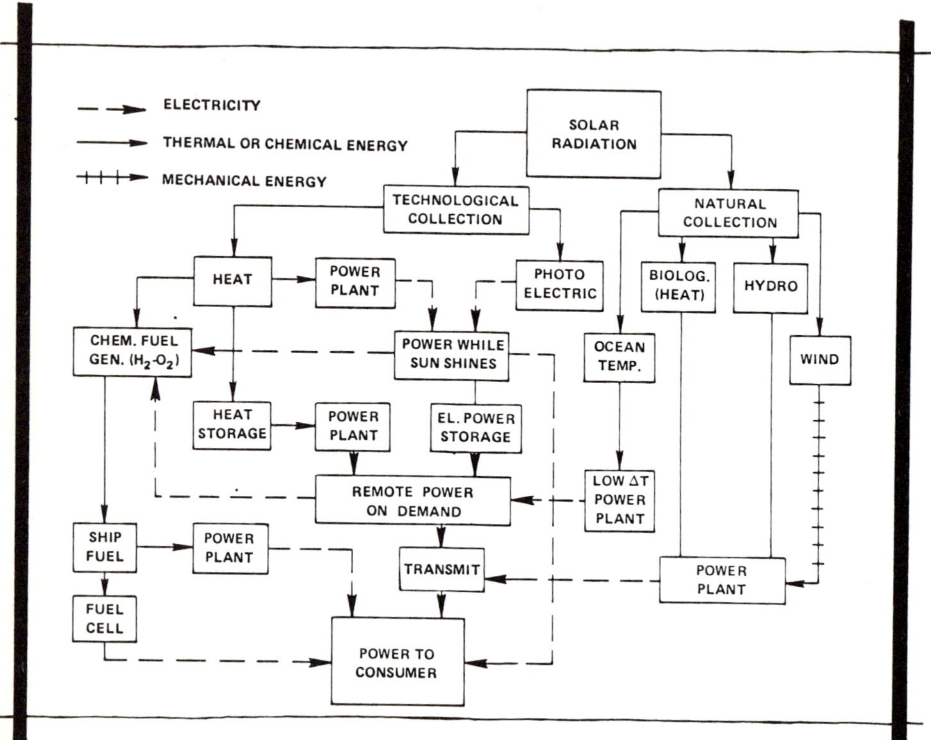

Figure 7-1. Conversion of solar energy to electricity. Note the role of the wind in the overall program. (From *Solar Energy as a National Energy Resource.* NSF/NASA, Solar Energy Panel)

In the wake of the energy crisis, general research in the United States has converged on the applied science aspects, which are divided into the following:

- *private sector,* supported by industries, pursuing short-term applications with low financial risk
- *government sector,* financed by public funds and concerned with long-term projects usually disregarded by the private sector due to their risk as financial investments

Wind and solar energy (Figure 7-1) in general are part of the government sector in the United States. Many other countries also participate in this type of research, and international cooperation is widespread. For instance, the International Energy Agency (IEA), founded in 1974 and based in Paris, includes Austria, Belgium, Italy, Japan, Luxembourg, Netherlands, New Zealand, Norway, Spain, Sweden, Switzerland, Turkey, the United Kingdom, and the United States. Its major goal is to pursue research-and-development programs for developing new energy sources that include wind energy transformation. The role of wind energy in the overall program that IEA is pursuing is on the same level with that of other major energy sources* as indicated below:

Energy conservation
- Urban planning
- Building energy loads
- Heat pumps
- Combustion
- Cascading
- Heat transfer/exchangers

Coal technology
- Technical information service
- Economic assessment service
- World coal-resources data bank
- Mining technology clearinghouse
- Fluidized bed combustion
- Low Btu coal gasification
- Coal pyrolysis
- Treatment of coal effluents

Nuclear power
- Reactor safety experiments

Fusion energy
- Superconducting magnets
- Plasma/wall interactions
- Intense neutron source

* From U.S. Department of Energy. *Solar, Geothermal, Electric and Storage Systems Program Summary Document.* DOE/ET-0102, July 1979, p. XII-4.

Geothermal energy
  Man-made energy systems
Solar energy
  Heating system performance
  Cooling system performance
  Development of heating and cooling components
  Solar collector testing
  Instrumentation package
  Meteorological data
  Small solar demonstration
Biomass conversion
  Technical information service
Ocean energy
  Wave power
Wind energy
  Large-scale wind demonstration
  Technology assessment
Hydrogen
  Hydrogen production from water

In the United States, international programs of research, development, and demonstration on wind energy and other new sources are conducted by the Office of Solar, Geothermal, Electric and Storage System (ETS) under the U.S. Department of Energy. The major goal of these programs is sharing information with other industrialized countries. They also include helping developing countries to use their local energy resources. In this respect, of course, wind is of major importance since it is available in practically any country.

The ETS programs that the United States shares with other countries include: solar electric applications; fuels from biomass; geothermal energy development; energy storage systems; electric energy systems. Under the solar electric applications, the wind programs are as follows:

ETS DIVISION	SPONSORING ORGANIZATION OR COOPERATING COUNTRY	STATUS
Wind energy conversion	IEA	Ongoing
Large-scale wind energy conversion	IEA	Ongoing
Wind energy	Spain	Ongoing
Wind energy	Saudi Arabia	Ongoing
Large wind—200 kW	IEA	Under negotiation

Multilateral agreements between the United States and other countries through IEA on wind energy transformation systems cover a time period that started in October 1977 and ended in October 1980, with provisions for automatic extension.

The program* on wind includes the following four basic points:

- environmental and meteorological aspects
- evaluation siting models
- integration into electricity supply systems
- rotor stressing and operation of large-scale systems

Part of the government-sponsored research is conducted in federal laboratories and part is contracted to private institutions. For example, the Lockheed-California Co. and the General Electric Co. Space Division received a $550,000 contract in 1975 for the study of wind-power conversion. The Kaman Aerospace Corporation in Bloomfield, Connecticut, had a NASA contract for the design of 1-MW and 100-kW wind-power plants. Honeywell Information Systems, Inc., Minneapolis, Minnesota, had a contract with Minnesota Power & Light Co. (*Energy Users Report,* 1975). The municipality of Hart, Michigan, received a $93,400 grant from the National Science Foundation (RANN) for a feasibility study on a wind-power plant conducted at the University of Michigan.

Much more significant was the communication from ERDA announcing the erection of experimental wind generators on the order of megawatts.* The first has been com-

* Op. cit., p. XII-10.

* Public interview by L. Divone. See *Energy Users Report,* May 29, 1975.

pleted and is currently operating in Boone, North Carolina, under testing conditions and some modifications. But most important was the work done at General Electric and Kaman Aerospace on wind power conversion.

The technical feasibility of producing and storing electrical energy from wind energy was established years ago. The economic feasibility of the process, however, is the final obstacle to large-scale application of wind utilization. The rising costs of energy currently produced and the diminishing number of energy sources will make wind energy more and more economical in the future. According to Vargo (1974), however, wind energy is already economically competitive with conventional electric energy.

Wind energy is a scientific reality. Although the economic feasibility of wind power in the United States could discourage private application, for energy-poor countries where the cost of conventional energy is high, wind power may already be economically feasible.

HISTORY

Although wind-powered mechanisms have been used since ancient times, the first historical notes on wind utilization refer to Persia and Asia Minor in the seventh century A.D. Data on European applications refer to France in 1180 and England in 1191. The evolution has been slow but continuous, with two major phases: the first in which wind energy was directly used for mechanical work, as in windmills, sawmills, water lifting, etc., and a second phase in which mechanical energy was changed into electrical energy.

Since their first appearance, windmills have undergone many changes and improvements in their major components. For instance, the major structure varied from wood frames to more stable masonry works and subsequently to steel frames. Other changes include:

1. The rotational axis, which was vertical, became horizontal.
2. The rotation of the wind machine according to the direction of the wind.
3. The shape of the propeller blades according to aerodynamic optimization and structural requirements.
4. Variability of the angulation of the blades to change the amount of wind intercepted, especially for high-velocity winds that could be destructive; feathering, etc.

Many names in various countries are associated with the development of wind machines through the centuries. Notable are:

- In Holland: Cornelis Cornelisz constructed the first sawmill, 1592; A. J. Dekker worked on improving the aerodynamic shape of the blades, 1927.
- In Great Britain: Edmund Lee worked on orientation controls, 1745; John Smeaton initiated aerodynamic studies on the blades, 1759; Andrew Meikle studied the mechanical angulation of the blades, 1772; Thomas Mead worked to perfect the centrifugal regulator governing the grinding fineness, 1787; Stephen Hooper, mechanical orientation of the blades, 1789; William Cubitt, automatic control of the blades, 1807; R. Catchpole, air-brake mechanism acting on the blades, 1860; E. L. Burne, improvements in the shape of the blade.
- In Germany: K. Bilau, variable orientation of the blades, about 1920.
- In the United States: Daniel Halladay, application of wind motor to water pumps, 1854.

Among the major contributors in the development of conversion of wind energy into electrical energy was P. La Cour in Denmark (1890). There, in fact, electric energy from the wind was produced on a commercial scale until 1931, when it became more economical to import electricity from Sweden.

Another important application in electrical energy conversion is the power plant built in Crimea in 1931 that has been very well documented in technical literature. Similarly, in the United States, the Grandpa's Knob wind power station built in 1941 in Vermont demonstrated the technical feasibility of the process for large-scale applications (Putnam 1948). This 1250-kW power plant equipped with a Smith-Putnam turbine operated for several months furnishing electrical power that was directly connected to the power lines. However, a mechanical breakage stopped the operation that could only be restored in 1945 after the war. Then, after a few months, the structural failure of one of the blades weighing over 2 tons put the plant out of operation.

In England, in 1950, another electrical wind-power plant was built at Costa Head, Orkney Islands (*Engineering* 1955). Several other plants of smaller capacity were built sporadically in various locations. Among them were the island of Sylt, Germany, in the North Sea, and the island of Texel, Holland. Many experiments in Holland included the conversion of old windmills to the production of electric power. Another similar application is the electric wind-power plant "Nordwind" (with a turbine diameter 15 m long). It is built on the island of Neuwark, Germany, in the North Sea, which provides electricity locally (Mackenthun 1951).

In addition to the numberless applications for private water pumping or small-size electric generators in rural areas all over the world, countries such as Portugal used wind power on a large scale. In 1965, over 1000 windmills were still in use for practical commercial operation. This, of course, is justified by local economic conditions, where energy cost is particularly high due to scarce energy resources.

The enthusiasm for wind-power applications was stopped by World War II. After the war, however, interest was resumed. Data on research and development in this period are available for several countries, including Europe in general, Germany, England, France, Israel, Italy, Holland, and the Soviet Union.* The United Nations conference on New Sources of Energy (1961) assessed the state of the art and made predictions for the future. Wind power was included among the energy sources considered.

The economic conditions at the end of the 1950s and during the 1960s discouraged further wind-power research. With the 1970s, however, the energy crisis revived this interest. Federal and private research and development programs in the United States have accelerated the progress. Major accomplishments in recent years include the 100-kW experimental station designed by the Lewis-Research Center (NASA) and built at Sandusky, Ohio (Thomas and Sholes 1975), and the larger ones that followed.

WIND CONVERSION (THEORETICAL VALUES)

A column of air in motion that passes through the circular area swept by the blades of a windmill possesses a certain amount of kinetic energy that can be calculated as follows. Assuming:

*See *Arch. Energiewirtsch, Sonerdienst* 1950; Asta 1953; Carrer 1949; Fateev 1952; Glausnizer 1965; Gross 1959; Huetter 1954a, 1954b; Haldane and Holding 1952; Levy 1968; *Revue de l'Aluminium* 1958; *Rivista di Meteorologia* 1942; Schmidt 1955; Serra 1953.

$\frac{\pi D^2}{4}$ = area swept by the wind turbine whose diameter is D

ρ = air density

L = length of the air column

t = unit of time

V = wind velocity

The total kinetic energy of the column per unit of time is:

$$\frac{MV^2}{2t}$$

But the mass of the air column is:

$$M = \frac{\pi D^2}{4} L\rho$$

Substituting in the previous expression, we obtain:

$$\frac{\pi D^2}{4} L\rho \frac{V^2}{2t}$$

But:

$$\frac{L}{t} = V$$

and substituting, we obtain:

$$\frac{\pi D^2 \rho V^3}{8}$$

This is the total energy per unit of time possessed by a column of air passing at a speed V through a wind turbine whose diameter is D.

Only a part of the total kinetic energy can be taken from the air column and transformed. A certain amount of energy must be retained by the air so that it can continue to move after having passed through the turbine. If all the kinetic energy were converted, the air column would come to a stop, and the process could not continue.

The maximum amount of energy that can be converted is 59.7%. However, this is a theoretical value based on ideal conditions of maximum efficiency. Such theoretical value is:

$$\frac{59.7}{100}\left(\frac{\pi D^2 \rho V^3}{8}\right)$$

In reality, the mechanical efficiency of the wind turbine reduces substantially the theoretical value indicated above.

POWER AUGMENTATION

The column of air passing through the turbine has a cross-sectional area equal to that swept by the blades of the turbine. However, if a funnel with a larger cross-sectional area is placed in front of the turbine, the column of air forced to pass through the turbine is increased in size (see Figure 7-2). A larger column of air has more energy; thus, for the same turbine, more energy can actually be captured and converted. The funnel, usually referred to as the *diffuser,* can even double the power output of the turbine.

Downstream from the turbine, the diffuser is gradually reduced in section, causing an increase of the velocity and pressure of the air. With respect to the undisturbed conditions of the wind, the air velocity can be raised between 20% and 60%.

TURBINES

A complete presentation of all the different types of wind turbines would be quite a task. Only the significant examples of vertical and horizontal types, including primitive

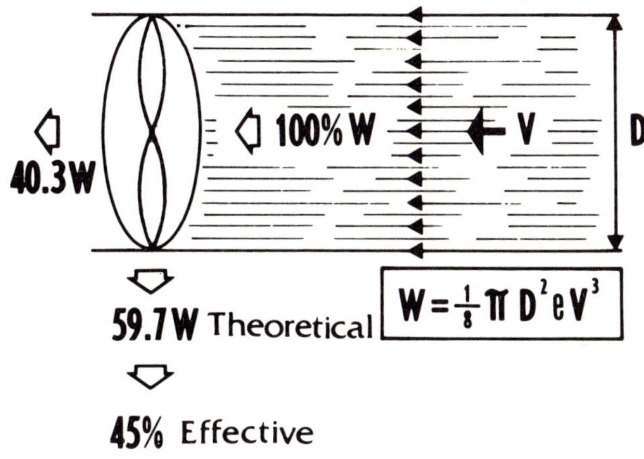

a

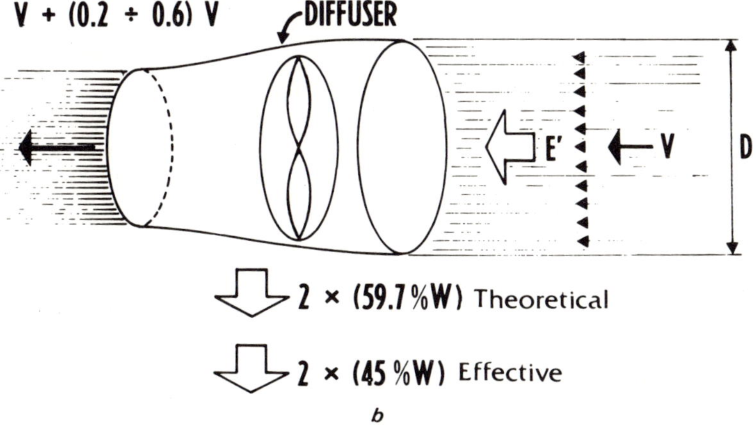

b

Figure 7-2. *a,* Theoretical and effective percentage of the kinetic energy that can be extracted from the wind. *b,* Augmentation of the percentage of energy that can be extracted from the wind by means of enlarged shrouds.

and present ones, will be mentioned here (Bathe 1948; Caldwell 1934; Darrieus 1931; McCoull 1973; Ormiston 1971; Savonius n.d., 1925a, 1925b, 1946; Sweeney and Nixon 1973; Vadot 1954).

Vertical axis turbines, which featured early prototypes in ancient times, have recently made a comeback with the Savonius, the Darrieus, and, the latest, the eggbeater, which derives its name from its resemblance to the household applicance (Rangi et al. 1974; Vance 1973). (See Figures 7-3 and 7-4.)

WIND INTERMITTENCE AND STORAGE SYSTEMS

The wind is quite erratic. Its velocity, for instance, varies considerably in direction and magnitude from one instant to another. If we observe the wind over long periods, its changes are less erratic and can be predicted. Over a year, average conditions are relatively dependable. On the basis of a seasonal periodicity proved by meteorological observations, wind energy can be included among other feasible energy sources. However, balancing systems must be added in order to stabilize the intermittence of the wind by means of storage elements that provide energy when the wind ceases.

If wind energy is used in combination with other energy types that can be controlled, the intermittence of the wind does not pose problems because the two types form a single balance system. In fact, the connection of a wind power plant with a network of other types of electric power plants should not cause a major problem. The variability of power being produced by a wind turbine can be compared to the variability of the demand that exists on the distribution lines. Consequently, as it is possible to efficiently regulate the power output in relationship to a varying demand, it will be equally possible to regulate the output with regard to the intermittence of the wind as an auxiliary power source. Testing on Mod-0 (100 kW) experimental wind stations by NASA has proved the feasibility of combining the wind turbines with conventional power systems.

Considering, on the other hand, self-sufficient wind-power plants, it is possible to provide a continuous output only with storage systems. Of these, there are several that are technologically feasible, including the following:

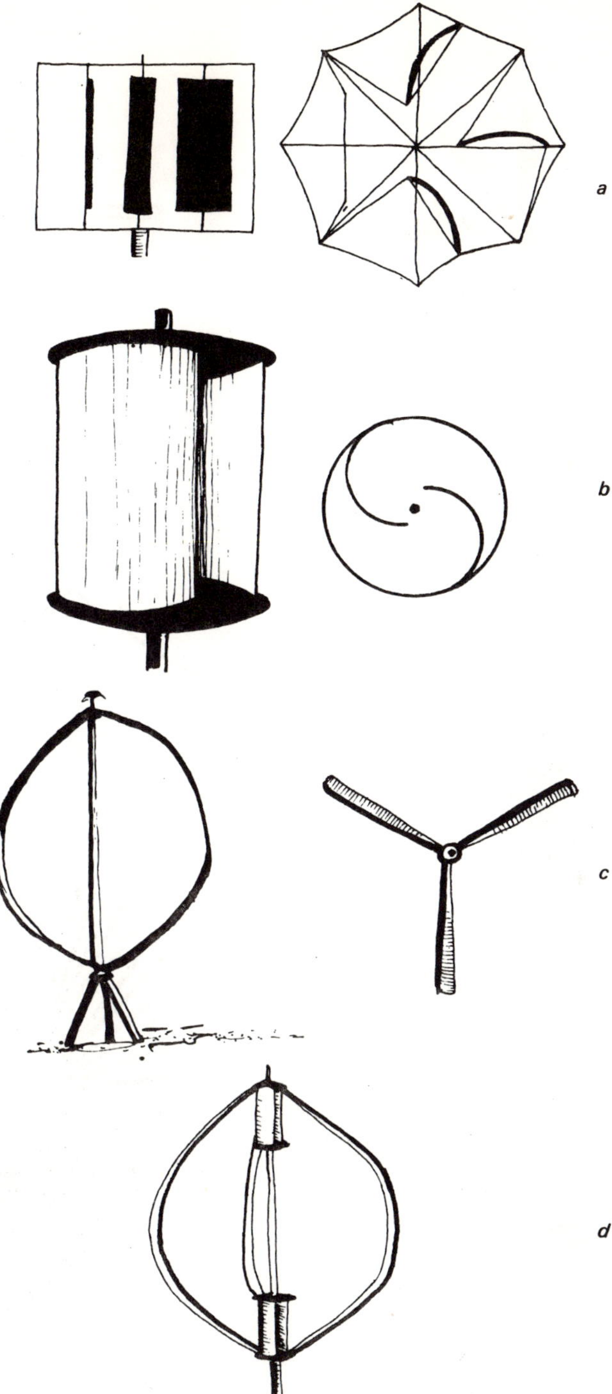

Figure 7-3. *a,* Windmills with vertical axis. Prototype of historical significance; *b,* the Savonius rotor (1920), S. J. Savonius; *c,* the Darrieus rotor (1925), G. J. M. Darrieus; *d,* eggbeater, combination (Savonius and Darrieus) designed by Sandia Laboratories and experimented with by NASA–Langley Laboratories.

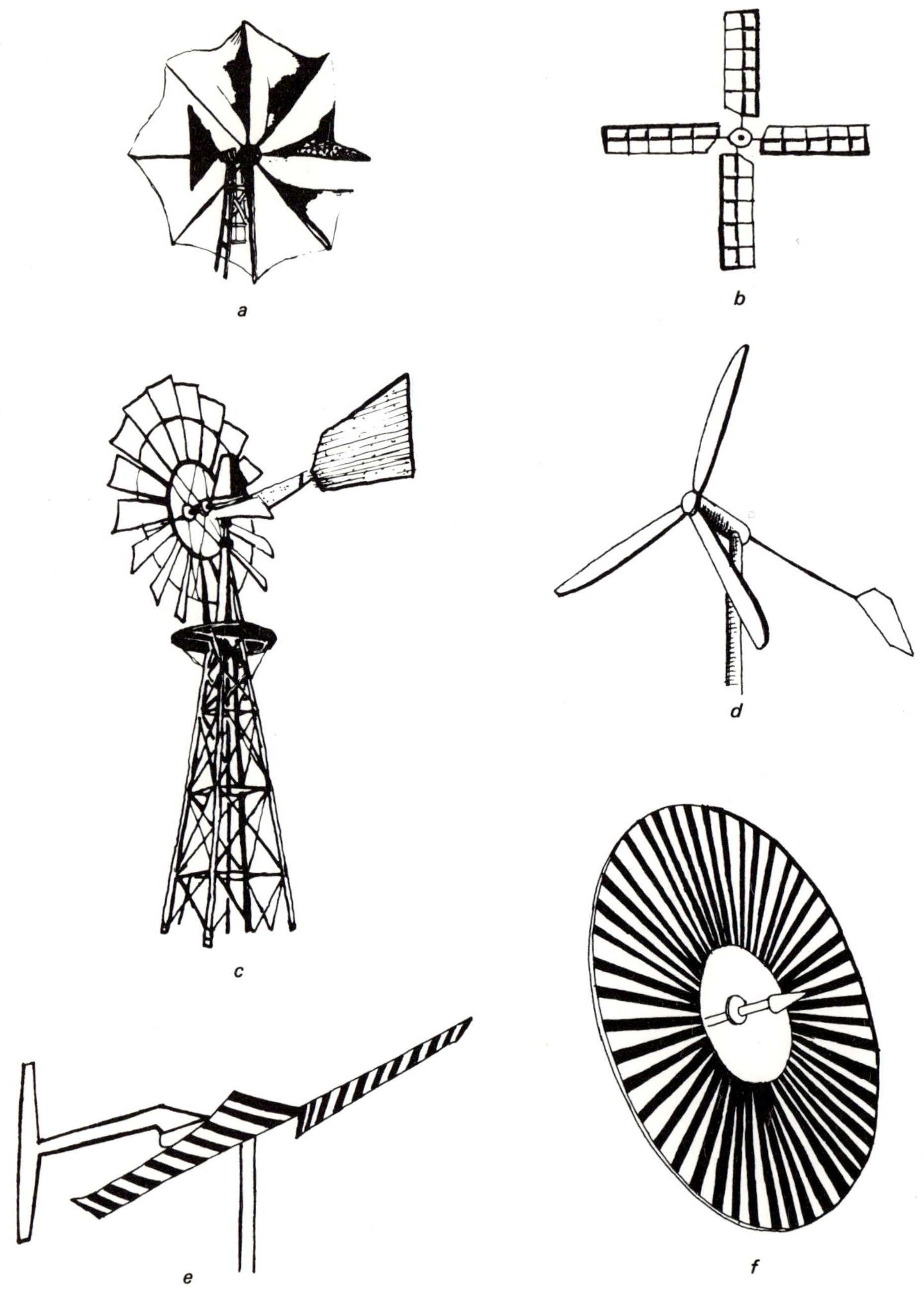

Figure 7-4. Windmills with horizontal axis. Prototype of historical significance. *a,* English type rotor with canvas sails; *b,* Dutch type rotor with wood blades; *c,* windmill typical for pumping water; *d,* modern windmill with two or three blades used for electric power generation. *e, Princeton Sailwing,* designed by T. Sweeney at Princeton University, Gruman patent. *f, The Bicycle Wheel,* designed by T. Chalk at Oklahoma State University.

- Electrical batteries for small-scale home use
- Flywheels*
- Storage of potential energy by pumping water in a basin at higher elevation
- Storage of potential energy by pressurizing fluids, which will give it back as they are allowed to expand
- Production of hydrogen from water through electrolysis

The scheme in Table 7-1 synthesizes the transformation process from an intermittent wind source to a continuous and controlled electrical output. The process is basically divided into the following intermediate steps:

1. Transformation of kinetic wind energy into electrical energy through a wind turbine and an alternator. The system is probably mounted on a floating structure on the ocean in optimum wind conditions.
2. Through submarine cables, electricity is carried to other floating stations where seawater is transformed into oxygen and hydrogen. The hydrogen is the fuel to be used for the next transformation, while the oxygen is just a by-product.
3. Storage of the hydrogen in submerged tanks under pressure; at this stage, the controlled process begins as the hydrogen efflux is regulated according to power demand.
4. Production of electricity and water from the reaction between hydrogen and atmospheric oxygen. The process occurs in land-based plants where the hydrogen is conveyed through submarine piping.

* Flywheels have been considered poor storage systems. Their storage capacity is short in time and very small in comparison to other systems (about 10% of that of conventional electric batteries). The installations are also potentially dangerous. Researchers such as Rabenhorst (1973), however, have completely changed such evaluation.

Table 7-1. Process of transformation from an intermittent energy source of continuous and controlled electrical energy.

STAGES	PROCESSES
1	Wind
2	Wind turbine and alternator (ocean-based plants)
3	Electric power transmission (submarine cables)
4	Electrolytic process
5a	Hydrogen gas, main product (submarine conduit)
5b	Oxygen by-product
6	Pressurized submarine tanks
7	Hydrogen (submarine conduit, controlled flow)
8	Power plant (combination atmospheric oxygen and hydrogen)
9a	Electric power
9b	Water by-product

Other theoretical schemes are indicated in Figure 7-5. The comparative efficiency of each system is included, which varies from a minimum of 5.5% to a maximum of 29%.

CURRENT RESEARCH

Scientists at the NASA Lewis Research Center in Cleveland, Ohio, designed and built an experimental station at Plum Brook Station near Sandusky, Ohio, which became operational in August 1975.

The 100-kW wind turbine is mounted on a horizontal axis and includes a rotor with a two-bladed propeller, 125 ft in diameter, that drives a synchronous alternator through a gearbox. The rotor rotates at a constant speed of 40 rpm and starts up at a minimum wind speed of 9 mph. At 9.5 mph, the alternator is already producing power. The blade angle naturally changes with the wind velocity in order to keep

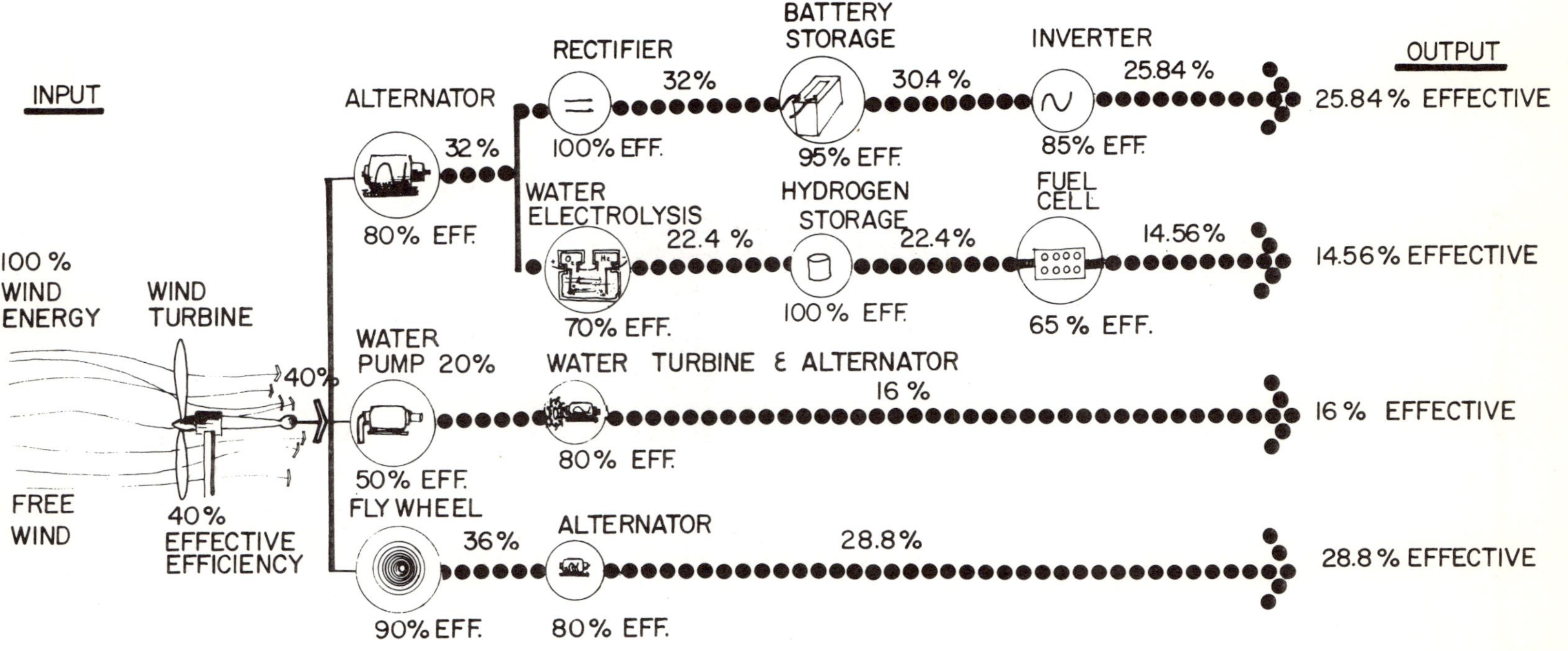

Figure 7-5. Power conversion.

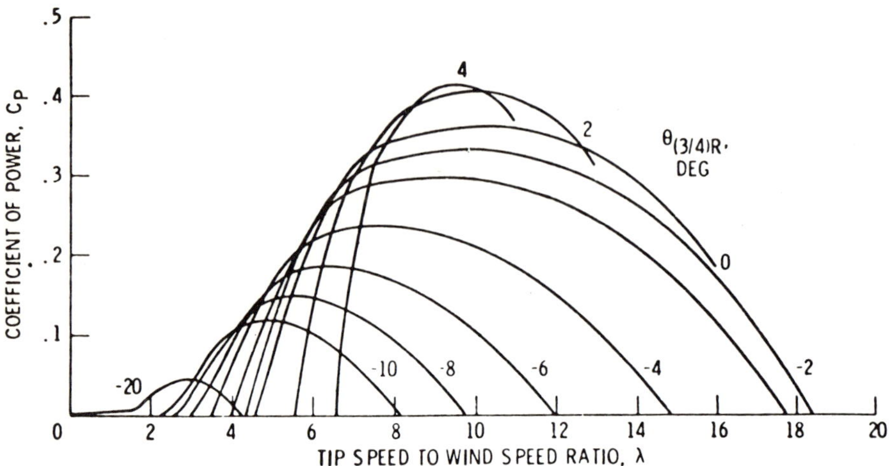

Figure 7-6. Curves showing the variation of the power coefficient (C_p) as a function of the rotor-tip speed to wind-speed ratio (λ). Each curve was determined for a given blade pitch angle (θ) expressed in $(3/4)R$ degrees. Notice that C_p is defined as the ratio of the extracted power to the power of the wind in the rotor disk area. (Courtesy of NASA)

the rotor speed constant. Energy output increases as the wind speed increases. A maximum energy of 100 kW is reached at 18-mph wind speed. For higher wind speeds, the blade angle is such that the additional wind is spilled off to avoid structural overstressing. The coefficient of power C_p is 0.375 at a wind speed of 18 mph, where

$$C_p = \text{power output/theoretical wind power in the area swept by blades}$$

Thus, the output is 37.5% of the total power contained in the wind. (See Figure 7-6.)

At maximum efficiency (18-mph wind speed), the *tip speed ratio,* λ, is 10, where

$$\lambda = \text{speed of the tip of the blade/speed of undisturbed wind}$$

Cost analysis of the actual construction dissipates any previous doubts regarding economic feasibility, when estimates were based on theoretical speculation. Actual cost was $550,000 or $5500 per kilowatt. With some modifications, it was estimated that other 200-kW plants could be built for a reduced unit cost of $2340 per kilowatt.

This experiment was the first realization of large-scale plants since the experiment in Vermont in 1945. It marked the beginning of a new era for wind power.

The Federal Wind Energy Program was initiated by the National Science Foundation (NSF) in 1973. In 1974, a 5-year program was established. The construction of the MOD-O 100-kW Plum Brook Station was just the beginning and was followed by other wind turbines of larger capacity. NASA's Lewis Research Center was charged with the task of designing, building, and testing the MOD-O. In January 1975, ERDA replaced the NSF in managing the program. (See Figure 7-7.)

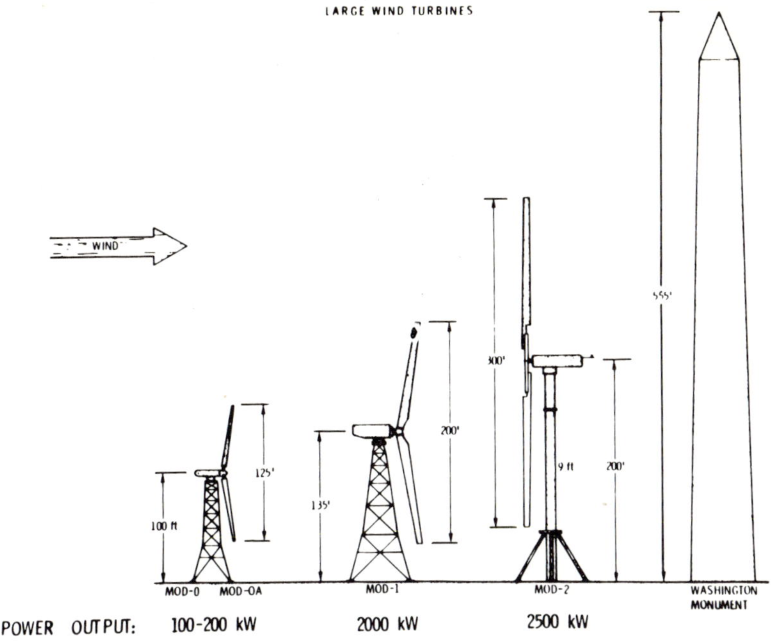

Figure 7-7. Overall federal wind program showing various types of wind turbines, including the identification code number, the nominal power output, and physical dimensions. (Courtesy of NASA)

Two studies of great interest were completed in 1975 by two private research groups: General Electric Co. and Kaman Aerospace Corporation under $0.5-million contracts from ERDA. The basic objective was the optimization of design systems and cost analysis of energy production. Each of the two studies was conducted independently from the other to insure a stricter dependability of the results and to allow a comparison of the two.

Both studies were in agreement, and the basic conclusions included the following recommendations and general findings:

- turbine: two blades; regulation of the blades; constant velocity; location downwind with respect to tower
- alternator: synchronous
- transmission: gearbox
- nominal power: 500- and 1500-kW optimum power outputs for respective wind velocities of 12 and 18 mph
- Unit cost per kWh for a 12- and 18-mph wind of, respectively, $0.03 to $0.07 and $0.005 to $0.15

Note that such costs were competitive with the current electric energy costs that at the time varied between $0.01 and $0.05 per kWh.

In the 1970s, the U.S. Department of Energy replaced ERDA; however, NASA's Lewis Research Center is still continuing the program.

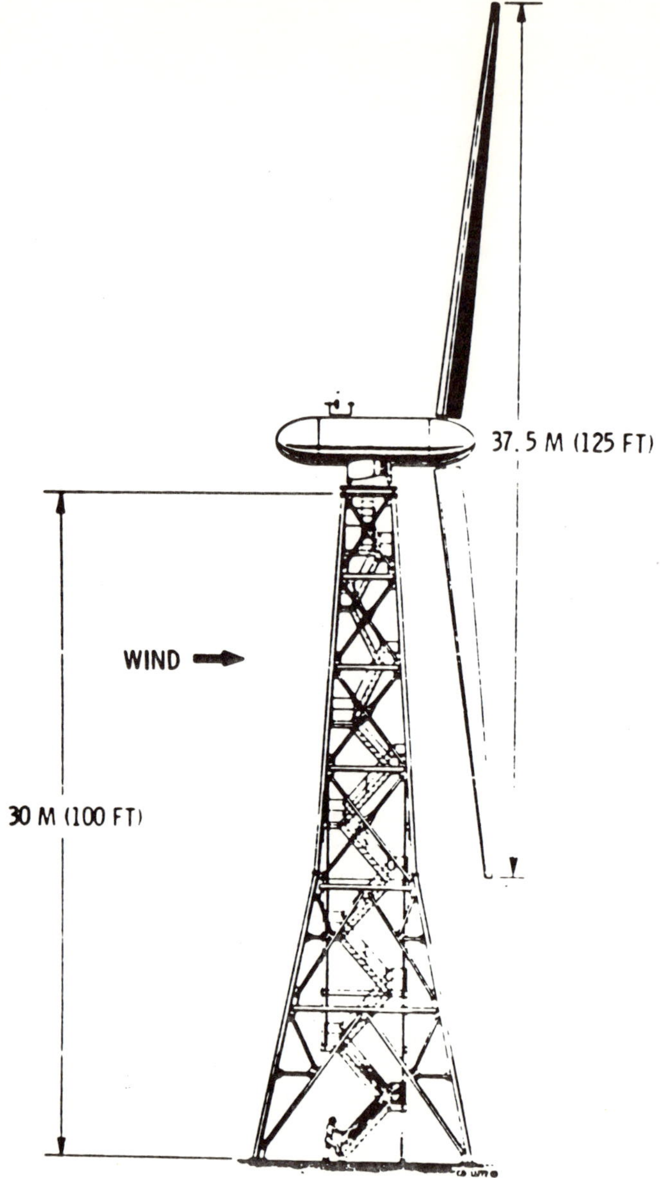

Figure 7-8. MOD-O before modifications were made. These included removal of staircase that cast a wind shadow on the blades, and the stiffening of the yaw mechanism. (Courtesy of NASA)

MOD-O was actually producing energy at 100 kW of power starting in December 1975. With a few modifications, such as the removal of the staircase in the tower to reduce its wind blockage, and stiffening of the yaw drive system, the station gave exceptionally good results. The station was connected to the Ohio Edison Co. utility system with satisfactory performance. (See Figures 7-8 and 7-9.)

Figure 7-9. Model of MOD-O wind turbine. Notice the absence of the staircase that eliminates the wind shadow on the rotor. (Courtesy of NASA)

Following MOD-O (100 kW) are: MOD-OA (200 kW), MOD-1 (2000 kW), and MOD-2 (2500 kW). A MOD-OA was installed at Clayton, New Mexico, in November 1977. Another was installed in June 1978 at Culebra, Puerto Rico. Another MOD-OA was scheduled to be installed at Block Island, Rhode Island, in April 1979 (see Figure 7-10). Design specifications for MOD-OA are shown in Table 7-2.

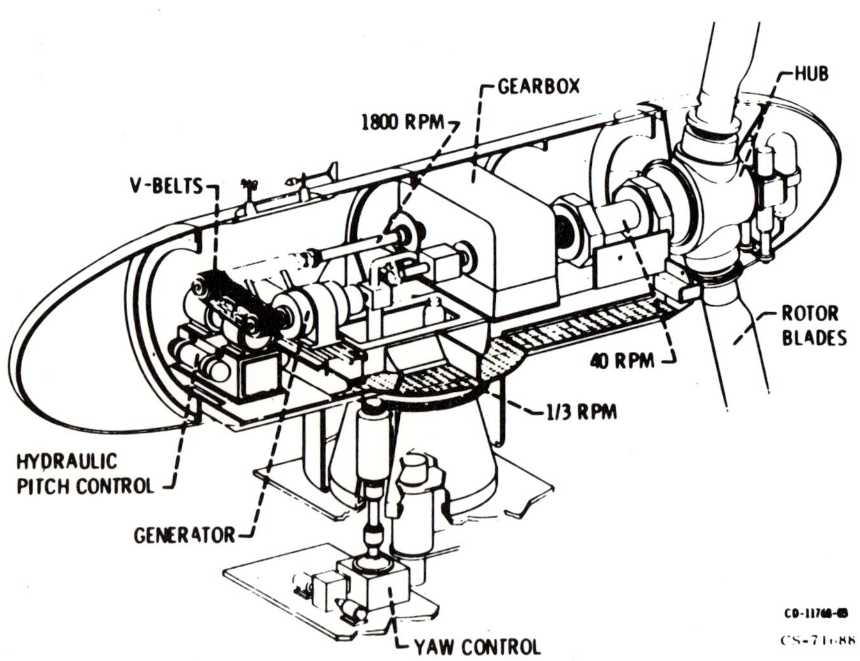

Figure 7-10. Drive train assembly and yaw system of MOD-O wind turbine. (Courtesy of NASA)

A MOD-1 was installed at Boone, North Carolina, in November 1978.

Four MOD-2's were scheduled for 1980. Several sites with adequate wind conditions have already been selected on a preliminary basis, and a few have been definitely chosen (see Figure 7-11).

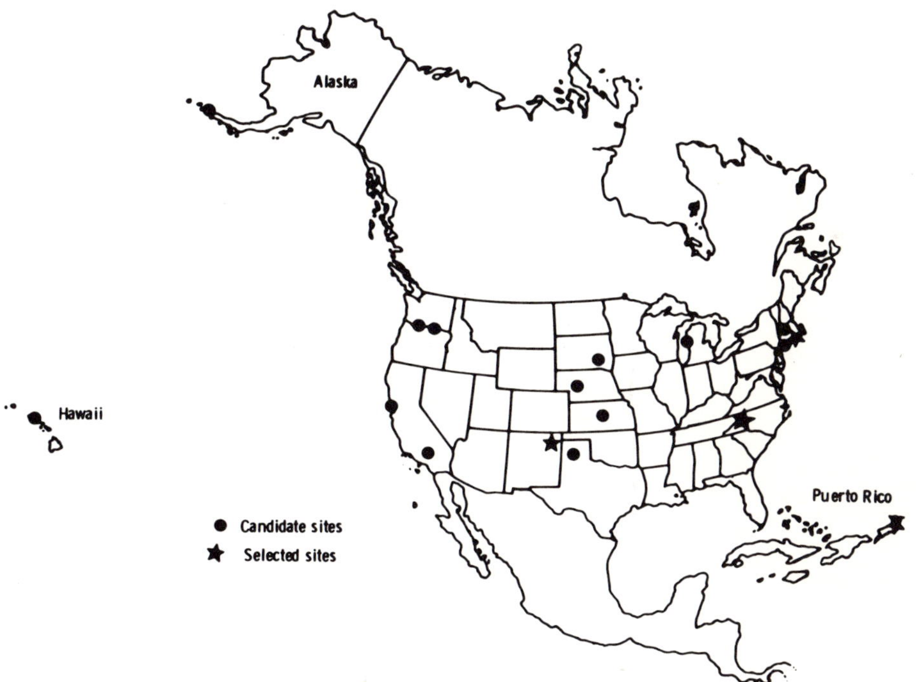

Figure 7-11. Candidate and selected sites in the United States for the installation of experimental wind power stations of various sizes (from 100 kW to 2500 kW), within 1980. (Courtesy of NASA)

Table 7-2. Design specifications for 200-kW wind turbine.

Rotor	
Number of blades	2
Diameter, ft	125
Speed, rpm	40
Direction of rotation	Counterclockwise (looking upwind)
Location relative to tower	Downwind
Type of hub	Rigid
Method of power regulation	Variable pitch
Cone angle, deg	7
Tilt angle, deg	0
Blade	
Length, ft	59.9
Material	Aluminum
Weight, lb/blade	2300
Airfoil	NACA 23000
Twist, deg	26.5
Solidity, percent	3
Tip chord, ft	1.5
Root chord, ft	4
Chord taper	Linear
Tower	
Type	Pipe truss
Height, ft	93
Ground clearance, ft	37
Hub height, ft	100
Access	Hoist
Transmission	
Type	Three-stage conventional
Ratio	45:1
Rating, hp	460
Generator	
Type	Synchronous ac
Rating, kVA	250
Power factor	0.8
Voltage, V	480 (three phase)
Speed, rpm	1800
Frequency, Hz	60
Orientation drive	
Type	Ring gear
Yaw rate, rpm	1/6
Yaw drive	Electric motors
Control system	
Supervisory	Microprocessor
Pitch actuator	Hydraulic
Performance	
Rated power, kW	200
Wind speed at 30 ft, mph (at hub):	
Cut-in	6.9 (9.5)
Rated	18.3 (22.4)
Cut-out	34.2 (40)
Maximum design	125 (150)
Weight (klb)	
Rotor (including blades)	12.2
Above tower	44.9
Tower	44.0
Total	88.9
System life	
All components, yr	30

Work in wind conversion technology is also being conducted outside the United States. In the Federal Republic of Germany, new research and development programs on wind power are presently underway. Government-sponsored projects include several wind power stations. When completed, the first of these, GROWIAN I, will have a power output between 2000 and 3000 kW. Its propeller will sweep an area 328 ft (100 m) in diameter. The proposed site is near the estuary of the Elbe River on the North Sea. Another power station of larger capacity, GROWIAN II, is also under construction. It will have a power output of 5000 kW. Responsible for this second project is Messerschmitt-Bölkow-Blohm in Ottobeuren, Germany.

WIND-WAVE ENERGY

As the wind blows over the oceans or other bodies of water, part of its kinetic energy is transferred to the water surface through friction. The water surface is therefore set in motion, creating a typical wave, which possesses the transformed wind energy. Such energy can be an invaluable source for human needs if an efficient technical methodology can be used to harness it. Ready and available along the coastlines, extending offshore as necessary over the immense ocean surface, this energy is there, free and inexhaustible. Like the direct wind-energy conversion and similar to hydraulic energy, the kinetic wind-wave energy is clean, practically free, dependable, and, as we said, inexhaustible. Therefore, exploring the potential of this source is an inviting prospect. Current research seems to indicate that energy conversion from wind-waves to electric energy is not only theoretically possible, but also practically obtainable through existing technical methodologies. Whether in shallow or in deep water, the energy extraction seems feasible. Of course, it is asked whether the tapping of this source can have environmental repercussions on the littoral regions, and science gives an optimistic reply. The economics of this energy source, however, is still not totally clear in terms of values. However, as mentioned, in the case of direct wind-energy conversion, the economics of a system is an artificial condition that could be changed at will, even in a capitalist system in which a regulated market could replace a free market.

As discussed in Chapter 5, waves generated by the wind are a function of several parameters whose contribution can be better understood on the basis of the following fundamental principles. We have seen that the fetch F is the length of the path covered by the wind over water. The height H of the wave increases proportionally to the fetch, up to a certain value of the fetch, F_{min}. Similarly, the wave height increases proportionally to the duration of the wind up to a certain value of the wind duration. Beyond such values of fetch and duration, the wave height remains constant. At this point, we say that the sea is "fully developed." Also, we say that the wind is transferring the maximum possible energy to the water so that wave height, wave period, and energy transfer remain constant. All this is valid for a given value of the wind velocity; in other words, for each wind velocity, there is a fixed condition of fully developed sea with its characteristic wave height, period, length, and energy transfer.

If the fetch is exactly F_{min}, there will be a reduction of energy and consequential reduction of wave height as the wave energy is extracted from the water by some sort of energy converter. On the other hand, if the fetch length is larger than F_{min}, the water surface can absorb more energy from the wind while energy is extracted from the water itself. On this principle, Isaacs et al. (1976) proposed that energy extraction from the ocean waves could be done under such conditions so that the wave action along the shores would not be altered at all.

The seasonal characteristics of the wind are also reflected in the characteristics of the ocean waves so that we notice that maximum wave action is observed in the winter months, decreasing constantly with a minimum value in the summer. McCormick (1979) calculated the potential wave energy available along the coasts of the United States. The highest values he found were along the Oregon-Washington coastline, where in the month of March the average power per crest length was over 25

MW/mi and the minimum, in August, was about 8 MW/mi. Similarly, for the North Atlantic coastline, the average power per crestline was 3 MW/mi in the winter and 2 MW/mi in the summer.

Several systems for wind-wave energy conversion have been studied. In 1920, the French inventor Bouchaux Praceique designed and built a turbine driven by wave motion and used it to provide power to his house by the sea. The rising and falling of the waves alternately increased the water level in an underground tank that was connected by a pipe to the sea. As the water level rose in the tank, the air above was compressed. Such compressed air in turn powered the wind turbine, which in turn drove the generator. As the wave trough lowered the water level in the tank, the reduced air pressure in the tank opened an exterior, bypass valve that allowed more air to enter the tank at atmospheric pressure, so that the cycle could be repeated giving a pulsating pressure to the turbine. Whether practical or not for wide-range use, it is important to note that the system actually worked.

In 1976, Isaacs et al. developed a deepwater wind-energy converter of practical application. The device operates on the concept of raising water above the sea level and compressing air at the same time that a float moves vertically due to the wave action. The water head combined with the air pressure drives water through a turbine, which is connected to an electrical generator. More precisely, the apparatus consists of a vertical pipe connected at the top to a float, which lifts and drops the pipe as the waves move the float vertically. A one-way valve at the bottom of the pipe opens when the pipe plunges down and closes as the pipe is lifted, so that the additional water that enters the pipe each time the pipe plunges is lifted above the level of the sea.

Two parallel and independent research works, conducted in 1971 by Masuda and Rodriguez on wave-energy conversion systems, were based on the phenomenon of resonance. Equally successful, the devices have well proved their efficiency and are commercially available. In fact, the Masuda type is popular in Japan for its application in lighthouses. The device consists of a vertical pipe connected to a float. It is partially filled with water in the lower part and has an air chamber on top. As the pipe and the float move vertically according to the wave motion, the air inside is partially compressed periodically with the same period of the wave's motion. The resultant resonance between the rhythmical motion of the waves and the pressure change of the air makes the pressure of the air increase in magnitude according to a power flow.

A converter designed by J. S. DeMaree of Suppliers Inc., Lexington, Kentucky, is based on the principle of inducing the waves to break at a specific point, where a series of devices rotating around a horizontal axis are set in motion by the passage of the wave underneath them. The mechanism is rigidly supported on a structure bedding on the ocean bottom rather than floating like most of the other converters.

McCormick (1979) described a wave-energy converter system of unique design that has been proposed by the Bolding-Alexander Corporation of Rialto, California. The apparatus consists of a fixed structure bedding on the bottom, on the surf area, such that the surge will pass through it and in so doing compress the air contained in a chamber. The compressed air is then used as the energy source for driving a generator. It is interesting that during the backrush of the waves, a system of values close the entry passages and open a relief valve, allowing more air to enter, thus preventing the wave suction. Obviously, this device can only be used in shallow water.

Another wind-wave energy converter, which operates only in deep waters, is the one designed by Salter et al. (1974).

As mentioned, the detrimental effects from capturing wind-wave energy is minimal. For the sake of clarity, it should be mentioned that a large number of energy converters placed offshore can alter the natural process of silting and erosion that takes place along the shores, changing somewhat the coastline profile. On the other hand, those converters that utilize the energy of broken waves are not really causing any change in the existing natural conditions, because they are just recouping that part of energy that would be lost through friction with the bottom and/or percolation if not tapped.

Will the future bring any new breakthrough in this direction? Since the technical feasibility of converting wind-wave energy into a usable form has already been proved, it is a matter of economic decision whether this alternative will be exploited or not. It is encouraging to see that government efforts in this area of research are presently underway. A list and description of some of these projects, as well as proposed projects, are presented in the following pages.

OCEAN WAVES ENERGY CONVERSION TECHNIQUES

The following ocean waves energy conversion techniques were prepared by Dr. Michael McCormick (1979) under contract with the U.S. Department of Energy.

WAVE ENERGY CONVERSION TECHNIQUES

TECHNIQUE	COMMON DEVICES	DOE FUNDED RESEARCH/ DEVELOPMENT (CONTRACTOR)
HEAVING BODIES	CIRCULAR CYLINDRICAL SPARS	SCRIPPS INST. (PROF. J. ISAACS) PAST FUNDED.

BASIC PRINCIPLE

A BODY IN PURE HEAVE RESONATES WITH THE WAVE PRODUCING AMPLIFIED BODY MOTIONS.

EXTRACTION EFFICIENCY

(A) UP TO 50%, DEPENDING ON THE SYSTEM DAMPING AND SUBSYSTEM.

(B) RADIATION WAVE FOCUSING MAY INCREASE THE EFFICIENCY, (ANTENNA EFFECT).

SCHEME 1

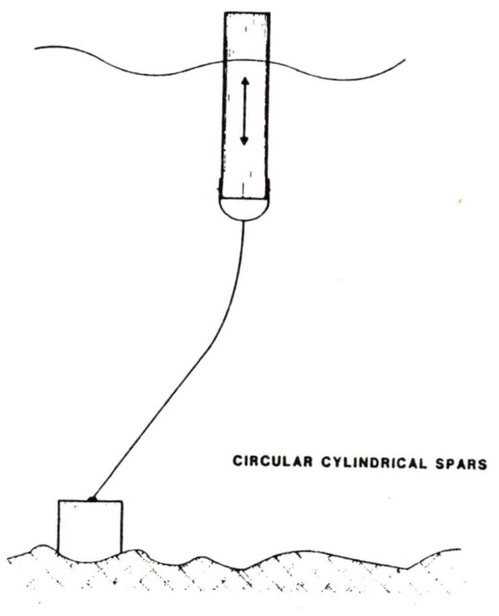

CIRCULAR CYLINDRICAL SPARS

HEAVING BODIES

SUBSYSTEM	METHOD	COMMENTS
ENERGY CONVERSION	INDUCTANCE TYPE	SOMEWHAT INEFFICIENT IN MECHANICAL TO ELECTRICAL ENERGY CONVERSION.
ENERGY CONVERSION	COMPRESSED AIR IN AN ACCUMULATOR	REQUIRES A HYDRO - OR - AIR-TURBINE AND A ONE-WAVE CHECK-VALVE.
MOORING	SLACK-SINGLE POINT	EASE IN MOORING IS HELPED BY THE SWIVEL.
ANCHOR	CLUMP	INEXPENSIVE

DEVELOPMENT STATUS

(A) NORWEGIANS (FALUES AND BUDAL) ARE DEVELOPING THE FOCUSING ASPECT.

(B) JOHN ISACCS (SCRIPPS) HAS DEMONSTRATED THE EFFECTIVENESS OF THE ACCUMULATOR.

CRITICAL PROBLEM AREAS EFFICIENCY IS LOW IN MECHANICAL TO ELECTRICAL ENERGY CONVERSION.

SCHEME 1

WAVE ENERGY CONVERSION TECHNIQUES

TECHNIQUE	COMMON DEVICES	DOE FUNDED RESEARCH/ DEVELOPMENT (CONTRACTOR)
PITCHING OR ROLLING BODIES	HEMISPHERICAL FLOATS	

BASIC PRINCIPLE

THE WAVE INDUCED ROTATIONAL MOTIONS (ABOUT THE MOORING SWIVEL AXIS) IS CONVERTED INTO ELECTRICAL ENERGY - DEVICE HAS A RESONANT ROLL OR PITCH.

EXTRACTION EFFICIENCY

(A) UP TO 50% DEPENDING ON THE SYSTEM DAMPING AND SUBSYSTEM.

(B) RADIATION WAVE FOCUSING MAY INCREASE THE EFFICIENCY (ANTENNA EFFECT).

SCHEME 2

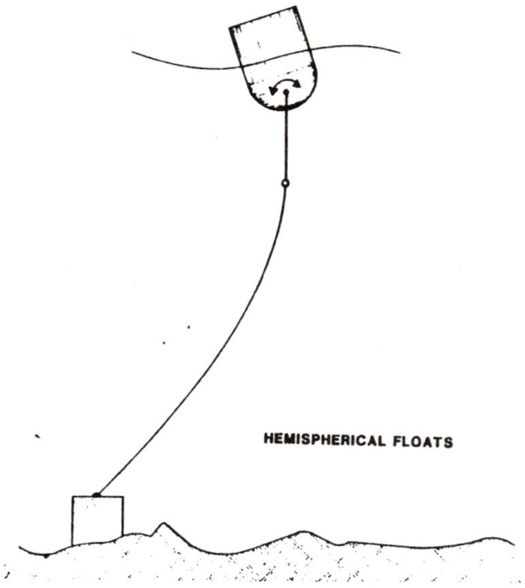

HEMISPHERICAL FLOATS

PITCHING or ROLLING BODIES

SUBSYSTEM	METHOD	COMMENTS
ENERGY CONVERSION	UNBALANCED ROTOR	WATANABE'S (U. OF HAWAII) TESTS SHOW PROMISE.
MOORING	SLACK-SINGLE POINT	(A) SWIVEL AXIS SHOULD BE THROUGH THE C.G. (B) EASE IN DEPLOYMENT.
ANCHORING	CLUMP	INEXPENSIVE AND EASE IN POSITIONING

DEVELOPMENT STATUS

WATANABE (U. OF HAWAII) HAS PERFORMED EXTENSIVE TESTS.

CRITICAL PROBLEM AREAS

(A) ROTOR MUST BE IN WATER TIGHT CHAMBER.

(B) BALANCE MAY BE A PROBLEM IN ROTOR GENERATOR ASSEMBLY.

SCHEME 2

WAVE ENERGY CONVERSION TECHNIQUES

TECHNIQUE	COMMON DEVICES	DOE FUNDED RESEARCH/ DEVELOPMENT (CONTRACTOR)
CAVITY RESONATOR (PNEUMATIC TYPE)	(A) CIRCULAR VERTICAL CENTER PIPE (B) DOUBLE - ACTING TURBINE	IEA JAPANESE STUDY

BASIC PRINCIPLE

THE WAVE CAUSES A RESONANCE OF THE INTERNAL WATER COLUMN WHICH IN TURN EXCITES THE AIR IN THE CHAMBER. THIS AIR EXCITES THE TURBO-GENERATOR SYSTEM.

EXTRACTION EFFICIENCY

(A) UP TO 50% IF TUNED TO THE WAVE.

(B) ANTENNA EFFECT MAY OCCUR INCREASING THE EFFICIENCY.

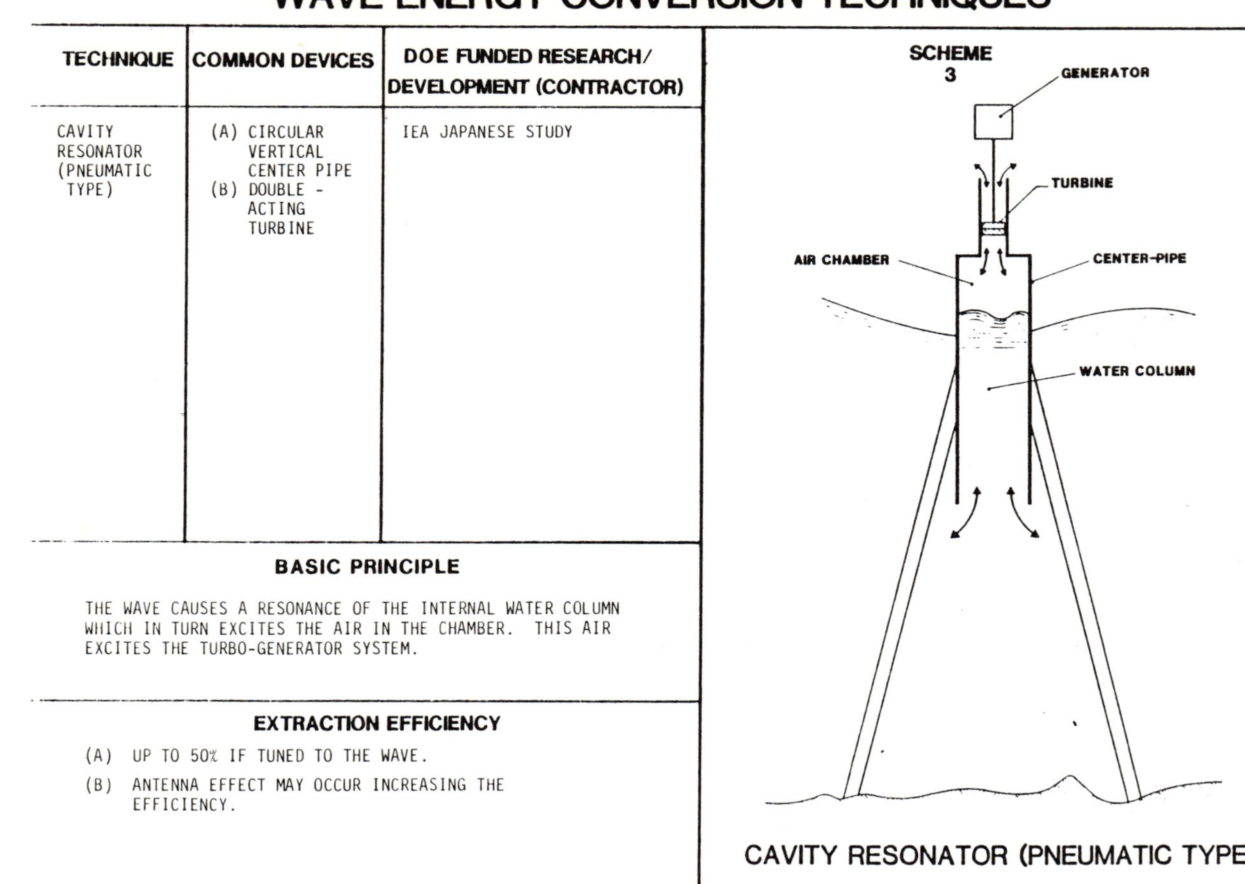

CAVITY RESONATOR (PNEUMATIC TYPE)

SUBSYSTEM	METHOD	COMMENTS
ENERGY CONVERSION	TURBO-GENERATOR	THE MOST EFFICIENT DESIGN WILL BE IDENTIFIED IN THE IEA JAPANESE STUDY.
MOORING	(A) SEA LEGS (B) SINGLE-POINT SLACK	(A) IN RELATIVELY SHALLOW WATER SEA LEGS ARE BEST USED. (B) IN DEEP WATER A SINGLE-POINT MOOR TO A FLOTATION COLLAR IS NECESSARY.
ANCHORING	IMBEDMENT	DUE TO THE RELATIVELY LARGE SIZE OF THE SYSTEM EMBEDMENT IS REQUIRED FOR EITHER THE SEA LEGS OR SINGLE-POINT MOOR.

DEVELOPMENT STATUS

THE DEVICE WAS DEVELOPED INDEPENDENTLY BY MASUDA (JAPAN) AND RICKAFRANCA OF THE PHILIPPINES. SMALLER DEVICES ARE COMMERCIALLY AVAILABLE, E.G., NAVIGATION BUOYS. LARGE SCALE DEVELOPMENT WILL DEPEND ON THE IEA JAPANESE STUDY.

CRITICAL PROBLEM AREAS

TURBINE, EFFICIENCY IS NORMALLY LOW FOR LOW AIR VELOCITY.

SCHEME 3

WAVE ENERGY CONVERSION TECHNIQUES

TECHNIQUE	COMMON DEVICES	DOE FUNDED RESEARCH/ DEVELOPMENT (CONTRACTOR)	SCHEME 4
WAVE FOCUSING	(A) RESONATING BODIES (B) SURGE FUNNELING (C) FRESNEL DIFFRACTION (D) REFRACTION		SURGE FUNNELING FRESNEL DIFFRACTION REFRACTION WAVE FOCUSING

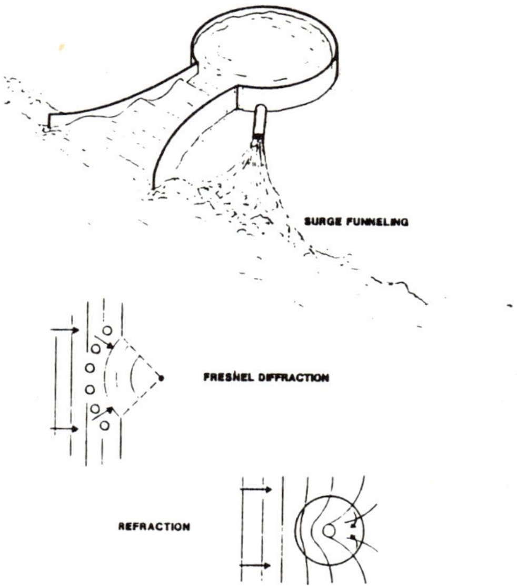

BASIC PRINCIPLE

USING ANY OF THE FOCUSING TECHNIQUES CONCENTRATES WAVE ENERGY FROM A LARGE CREST LENGTH ON A RELATIVELY SMALL REGION WITH LITTLE WAVE ENERGY LOSS.

EXTRACTION EFFICIENCY

THIS DEPENDS ON THE MECHANICAL TO ELECTRICAL ENERGY CONVERSION SYSTEM. MOST INVOLVE EITHER INDUCTANCE DEVICES (A) AND (C) OR TURBO-GENERATORS (B) AND (D).

SUBSYSTEM	METHOD	COMMENTS
ENERGY CONVERSION	(A) INDUCTANCE (B) TURBO-GENERATOR	SEE SCHEMES 1 AND 3
MOORING		(A) SEE SCHEME 1 (B) PILES USED TO FOCUS (D) MULTI-POINT MOORINGS REQUIRED

DEVELOPMENT STATUS

(A) SEE SCHEME 1

(B) SUGGESTED BY PROF. ISAACS (SCRIPPS) AND NORWEGIANS NO SIGNIFICANT R&D TO DATE.

(C) SUGGESTED BY DESERT RESEARCH INSTITUTE AND NORWEGIANS. EXPERIMENTAL STUDY IS UNDERWAY AT THE NAVAL ACADEMY.

(D) LOCKHEED CALIFORNIA IS CONDUCTING A TEST PROGRAM.

CRITICAL PROBLEM AREAS

(A) SEE SCHEME 1

(B) CAPITAL COSTS WOULD BE HIGH. ALSO, THIS TECHNIQUE COULD BE ENVIRONMENTALLY UNACCEPTABLE.

(C) FROM PRELIMINARY EXPERIMENTAL RESULTS, THE TECHNIQUE IS OF QUESTIONABLE VALUE.

(D) THE MOST PROMISING

SCHEME 4

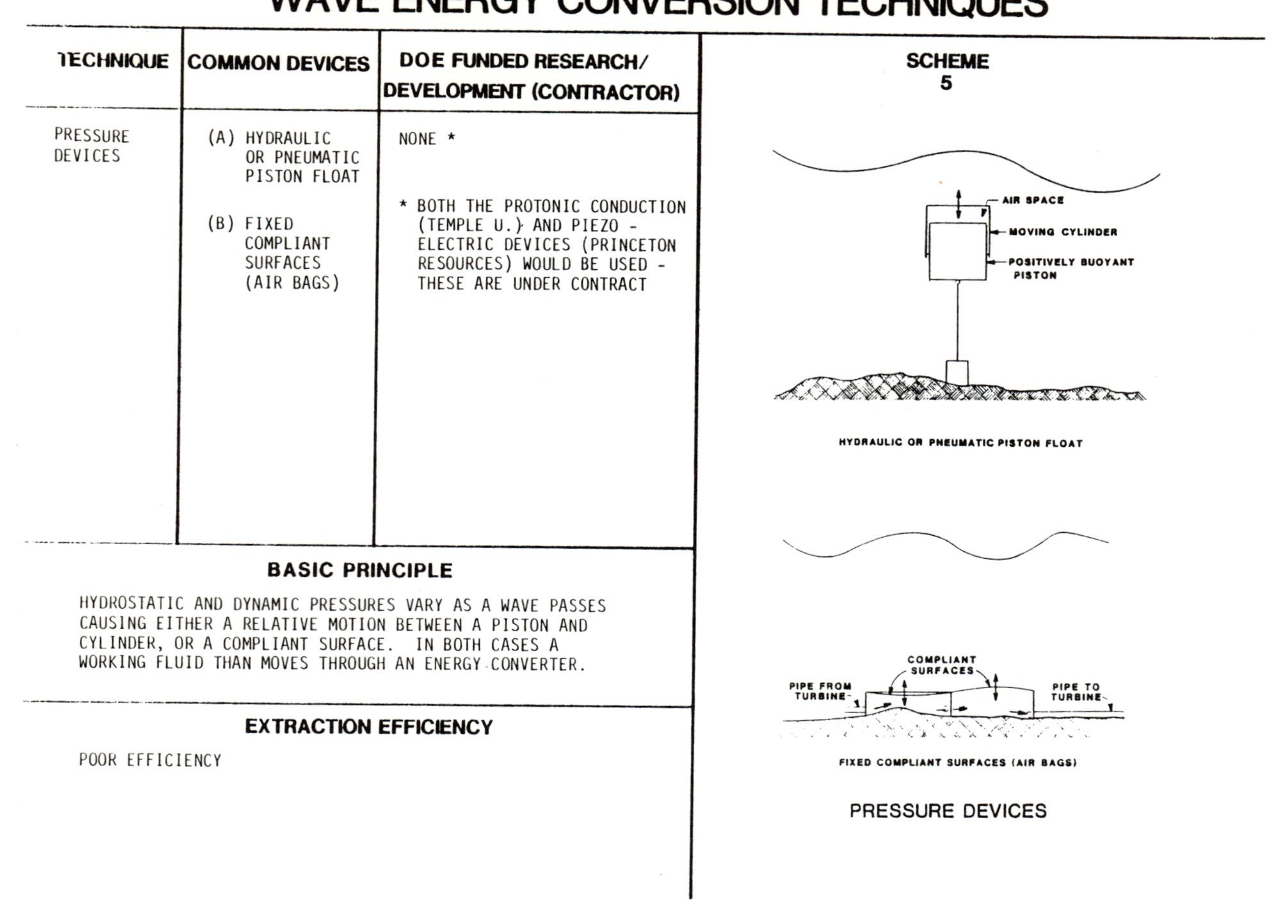

WAVE ENERGY CONVERSION TECHNIQUES

TECHNIQUE	COMMON DEVICES	DOE FUNDED RESEARCH/ DEVELOPMENT (CONTRACTOR)
PRESSURE DEVICES	(A) HYDRAULIC OR PNEUMATIC PISTON FLOAT (B) FIXED COMPLIANT SURFACES (AIR BAGS)	NONE * * BOTH THE PROTONIC CONDUCTION (TEMPLE U.) AND PIEZO - ELECTRIC DEVICES (PRINCETON RESOURCES) WOULD BE USED - THESE ARE UNDER CONTRACT

BASIC PRINCIPLE

HYDROSTATIC AND DYNAMIC PRESSURES VARY AS A WAVE PASSES CAUSING EITHER A RELATIVE MOTION BETWEEN A PISTON AND CYLINDER, OR A COMPLIANT SURFACE. IN BOTH CASES A WORKING FLUID THAN MOVES THROUGH AN ENERGY CONVERTER.

EXTRACTION EFFICIENCY

POOR EFFICIENCY

SCHEME 5

HYDRAULIC OR PNEUMATIC PISTON FLOAT

FIXED COMPLIANT SURFACES (AIR BAGS)

PRESSURE DEVICES

SUBSYSTEM	METHOD	COMMENTS
ENERGY CONVERSION	PNEUMATIC	COMPRESSED AIR DRIVES A TURBINE
ENERGY CONVERSION	HYDRAULIC	HYDRAULIC FLUID DRIVES A TURBINE
MOORING	(A) SINGLE POINT (B) EMBEDMENT	(A) USED WITH A CLUMP ANCHOR (B) WORKING FLUID IS PUMPED VIA PIPE TO GENERATOR

DEVELOPMENT STATUS

(A) AVCO BUILT A SMALL MODEL AND TESTED AT SEA IN THE EARLY 1960's. PROJECT NOT CONTINUED.

(B) HUMBOLT STATE UNIVERSITY (CALIFORNIA) TESTED AT SEA. PROJECT NOT CONTINUED.

CRITICAL PROBLEM AREAS

BOTH SYSTEMS ARE DIFFICULT TO CONTROL OR REGULATE. ALSO, BOTH SUFFER FROM VERY LOW EFFICIENCY.

SCHEME 5

WAVE ENERGY CONVERSION TECHNIQUES

TECHNIQUE	COMMON DEVICES	DOE FUNDED RESEARCH/ DEVELOPMENT (CONTRACTOR)
SURGING DEVICES	(A) ARTIFICIAL BEACH AND CONVEYOR BELT (B) COMPRESSED AIR SYSTEM	NONE

BASIC PRINCIPLE

BROKEN WAVES IN THE SURF ZONE SURGE ONTO THE SHORE. THE SURGE EXCITES A WHEEL OR BELT AS IN (A), OR COMPRESSES AIR AS IN (B).

EXTRACTION EFFICIENCY

POOR EFFICIENCY

SCHEME 6

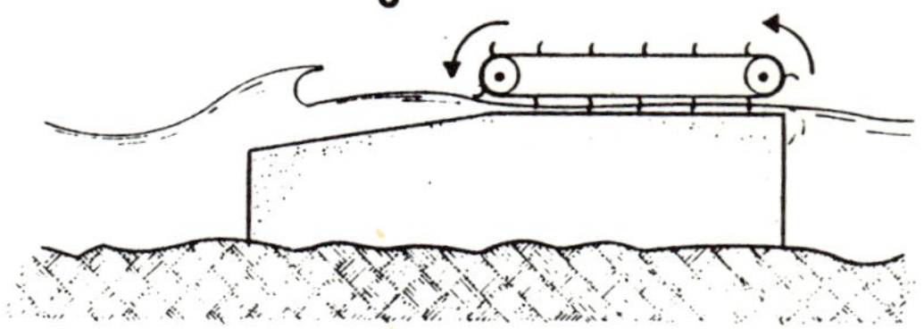

ARTIFICIAL BEACH AND CONVEYOR BELT

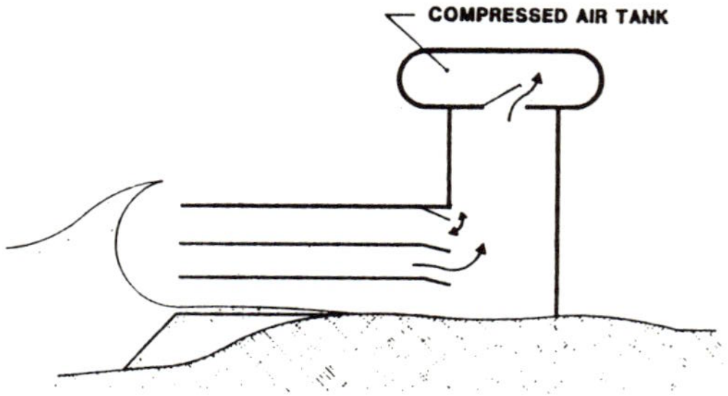

COMPRESSED AIR SYSTEM

SURGING DEVICES

SUBSYSTEM	METHOD	COMMENTS
ENERGY CONVERSION	(A) WHEEL OR BELT DRIVEN GENERATOR	NONE
ENERGY	(B) COMPRESSED AIR DRIVEN GENERATOR	NONE

DEVELOPMENT STATUS

(A) DeMAREE CO. BUILT A MODEL THAT DID NOT WORK.

(B) BOLDING-ALEXANDER DEVELOPED THE IDEA. RECENTLY REVISED BY MR. BOLDING.

CRITICAL PROBLEM AREAS

(A) INEFFICIENT AND TOO MANY MOVING PARTS EXPOSED TO SALT WATER.

(B) INEFFICIENT COMPRESSION.

NOTE: THE COSTS OF BOTH ARE ENORMOUS. FURTHERMORE, THE POWER IN A BROKEN WAVE IS FAR LESS THAN THE VALUE IN DEEP WATER.

SCHEME 6

WAVE ENERGY CONVERSION TECHNIQUES

TECHNIQUE	COMMON DEVICES	DOE FUNDED RESEARCH/ DEVELOPMENT (CONTRACTOR)
(A) FLAPPING BODIES	(A) HINGED BOARD WITH WEIGHT OR SPRING	(A) NONE
(B) PADDLES	(B) PADDLE WHEELS	(B) NONE

BASIC PRINCIPLE

(A) FLAP IS EXCITED BY A BROKEN WAVE, THEN RETURNS TO ORIGINAL POSITION. ROTATION EXCITES A GENERATOR.

(B) PADDLE ROTATION EXCITES A GENERATOR.

EXTRACTION EFFICIENCY

POOR FOR BOTH

SCHEME 7

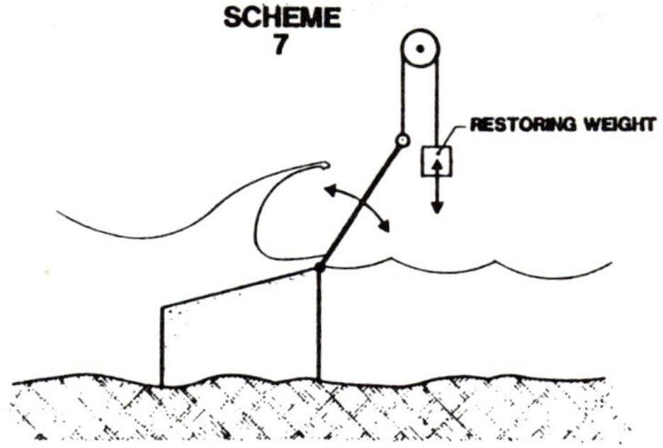

RESTORING WEIGHT

HINGED BOARD WITH WEIGHT OR SPRING

WAVE DIRECTION

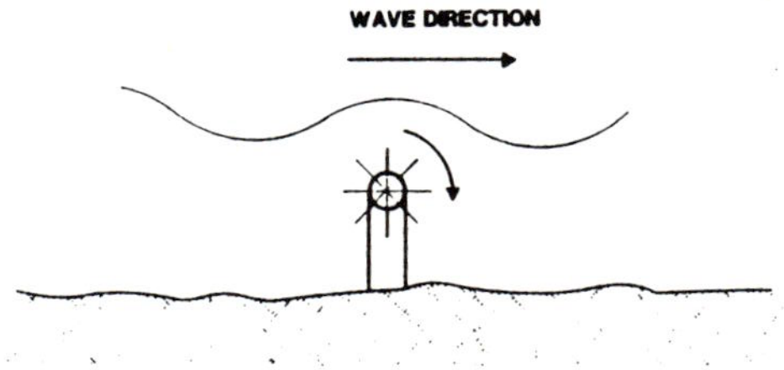

PADDLE WHEELS

FLAPPING BODIES & PADDLES

(A) RESTORING MECHANISM	(A) WEIGHT (B) SPRING	(A) WEIGHT AT THE END OF A PULLEY-ROPE. (B) SPRING IS COILED AT HINGE.

DEVELOPMENT STATUS

NONE

CRITICAL PROBLEM AREAS

INEFFICIENCY AND FOULING OF ROTATING PARTS.

SCHEME 7

WAVE ENERGY CONVERSION TECHNIQUES

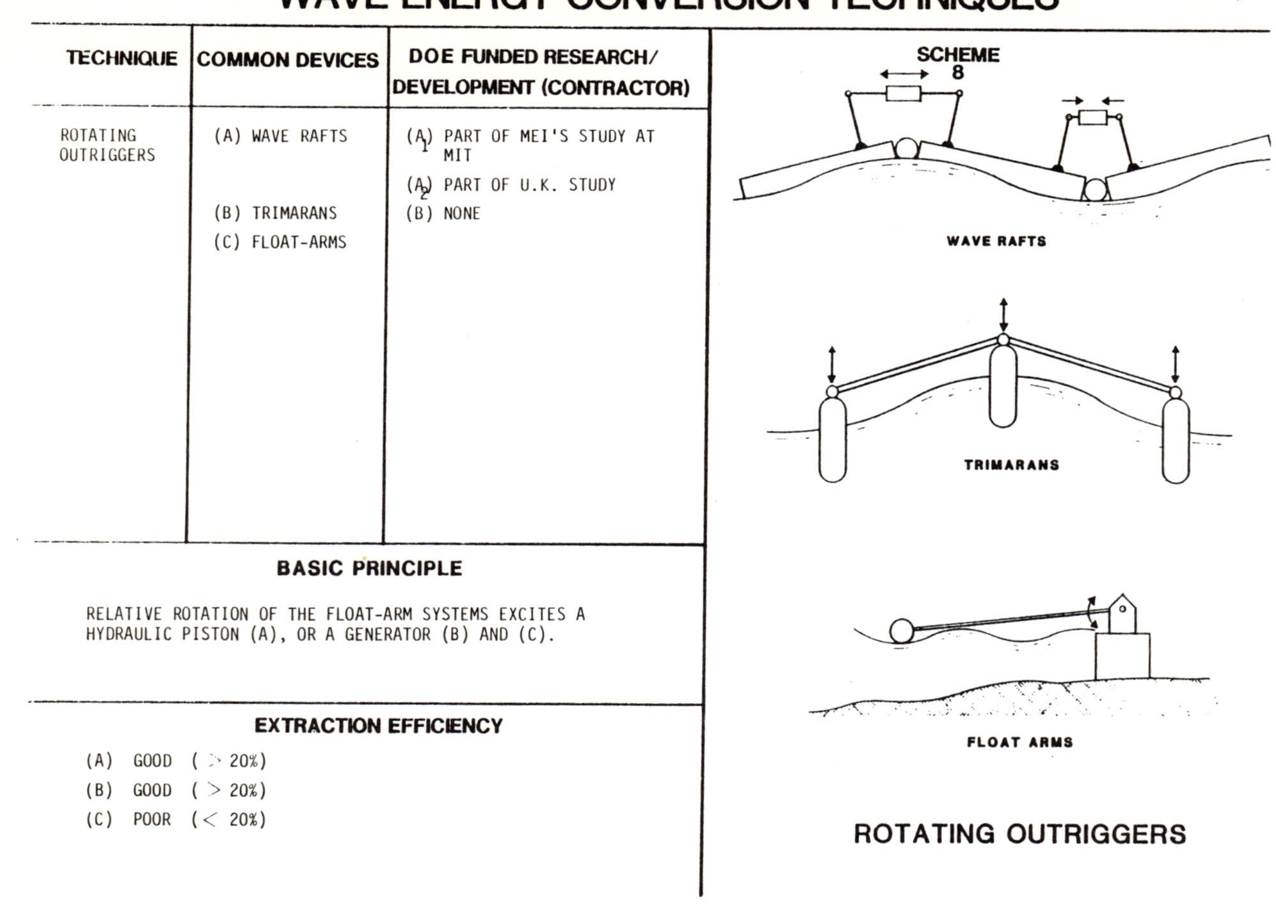

TECHNIQUE	COMMON DEVICES	DOE FUNDED RESEARCH/ DEVELOPMENT (CONTRACTOR)
ROTATING OUTRIGGERS	(A) WAVE RAFTS (B) TRIMARANS (C) FLOAT-ARMS	(A_1) PART OF MEI'S STUDY AT MIT (A_2) PART OF U.K. STUDY (B) NONE

BASIC PRINCIPLE

RELATIVE ROTATION OF THE FLOAT-ARM SYSTEMS EXCITES A HYDRAULIC PISTON (A), OR A GENERATOR (B) AND (C).

EXTRACTION EFFICIENCY

(A) GOOD ($>$ 20%)
(B) GOOD ($>$ 20%)
(C) POOR ($<$ 20%)

SUBSYSTEM	METHOD	COMMENTS
ENERGY CONVERSION	(A) HYDRAULIC PISTONS	DOUBLE PISTONS MOVE A HYDRAULIC FLUID THROUGH A TURBINE.
ENERGY CONVERSION	(B) AND (C) GENERATOR	GENERATOR IS LOCATED AT THE HINGE.

DEVELOPMENT STATUS

(A) COCKERELL'S RAFTS BEING DEVELOPED IN U.K. HAGAN RAFTS WERE BEING DEVELOPED IN U.S. BY THE WILLIAMS CORP. OF NEW ORLEANS: PROJECT DISCONTINUED. U.K. HAS HAD "SECOND THOUGHTS" DUE TO MOORING DIFFICULTIES AND EXPENSE.

CRITICAL PROBLEM AREAS

(A) AND (B): MOORING IS EXTREMELY DIFFICULT.

(B) AND (C): ROTATING PARTS FOUL IN SALT WATER .

(C) VERY POOR EFFICIENCY.

SCHEME 8

WAVE ENERGY CONVERSION TECHNIQUES

TECHNIQUE	COMMON DEVICES	DOE FUNDED RESEARCH/ DEVELOPMENT (CONTRACTOR)
COMBINATION DEVICES	HEAVING AND PITCHING FLOATS FORCED TO ROTATE	(A) PART OF MEI'S STUDY AT MIT. (B) PART OF U.K. STUDY BY SALTER AT U. OF EDINBURGH

BASIC PRINCIPLE

BODY ROTATES DUE TO DYNAMIC PRESSURE ON "BEAK", AND ALSO DUE BEAK'S BUOYANCY.

EXTRACTION EFFICIENCY

GREATER THAN 90% UNDER CERTAIN WAVE CONDITIONS.

SCHEME 9

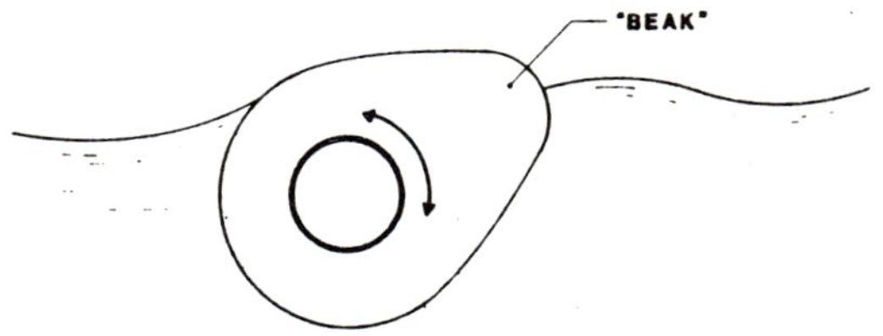

HEAVING & PITCHING

FLOATS FORCED TO ROTATE

COMBINATION DEVICES

SUBSYSTEM	METHOD	COMMENTS
ENERGY CONVERSION	HYDRAULIC	HYDRAULIC FLUID SLOWLY PULSATES THROUGH LINES TO A TURBINE.
MOORING	UNKNOWN	MOORING IS DIFFICULT SINCE ALIGNMENT OF ROTATING PARTS MUST BE MAINTAINED.

DEVELOPMENT STATUS

THE U.K. HAS BEEN DEVELOPING THE SALTER "DUCKS". MOORING DIFFICULTIES AND EXPENSE HAVE CAUSED A PROGRAM REVISION.

CRITICAL PROBLEM AREAS

(A) MOORING IS DIFFICULT AND EXPENSIVE - MAIN PROBLEM.

(B) MECHANICAL TO ELECTRICAL ENERGY CONVERSION IS POOR.

SCHEME 9

FUTURE DEVELOPMENTS

The application of a wind power conversion system in combination with solar energy is a noteworthy concept still to be explored. Although it seems theoretically possible, its economic feasibility has yet to be proved. The basic criterion consists of creating an air current that would be accelerated to a designated high speed as it passed through a wind turbine connected to an electric generator. The artificial wind would be generated by energizing large masses of air through solar heating. A large tensile structure, consisting of a transparent plastic membrane supported by a cable, would act as a greenhouse. The air under the structure would be heated by the sun's rays going through the plastic membrane, in accordance with the greenhouse effect. The tensile structure would be pitched upward toward a central point so that the hot air, which tends to rise, would be funneled and gradually accelerated toward that point; there it would be conveyed to a turbine. A major advantage of such an installation is that the primary energy—solar radiation—is more predictable than the natural wind in certain geographical areas. The insolation factor for the site, the size of the structure, the dimensions of the turbine, and the power output of the generator are all mutually related parameters whose coordination produces a large spectrum of possibilities. Potential sites for such a project would be at low latitudes because, as one gets near the equator, the insolation factor and the energy output increase while other parameters remain constant.

Much more farfetched is the idea of exploiting the jet stream, of which the velocity—up to 200 mph—represents a large and dependable energy source. To accomplish such a task, both wind turbine and generator would have to be kept aloft at 30,000-ft altitudes. If the apparatuses could be built within certain weight limits, they could be raised and kept aloft by the combination of two systems: balloons and kites. Large balloons, and smaller ones, shaped as Rogallo kites and anchored to the ground, could be interconnected so that they could lift turbines and generators and keep them in place. The continuity of the jet stream could guarantee a constant lifting force on the kites and a steady electric energy output. To conduct the electricity to the ground, it is assumed that the large-size anchoring cables would act simultaneously as anchor lines and as conductors. This idea, which probably seems to border on science fiction, is only at the conceptual stage, but has great potential for future development.

REFERENCES

Asta, A. 1953. Esperienza zull'utilizzazione dell'energia del vento per produzione di energia elettrica. *La Ricerca Scientifica* 23: 537–558.

Bathe, G. 1948. Horizontal Windmills, Draft Mills and Similar Airflow Engines. Allen, Lane and Scott, Philadelphia, Pennsylvania.

Caldwell, F. W. 1934. Aircraft-propeller development and testing summarized, *SAE Journal* 35(3).

Carrer, A. 1949. Generatori a corrente continua per l' utilizzazione dell'energia dell vento con schema tipo Ward-Leonard a funzionamento invertito. *Elettrotecnica* 36: 376–382.

Darrieus, G. J. M. 1931. Turbine Having Its Rotating Shaft Transverse to the Flow of the Current. U.S. Patent No. 1835018.

Energy Users Report. 1975. D-4. No. 96. June 1975.

Fateev, E. M. and I. V. Rozhdestvenskii. 1952. Achievements of Soviet wind power engineering. *Vestnik Mashinostroeniya* 9: 24–27.

Frenkiel, J. 1956. Wind-power research in Israel. *Wind and Solar Energy, Proceedings of the New Delphi Symposium.* UNESCO, Paris.

———. 1956. Exploitation of wind energy in Israel. *Journal of the Association of English Architects in Israel* 14(5): 50–52.

Glausnizer, R. 1965. The importance of wind-power to the German energy economy. *Elektrizitätswirtschaft,* 7.

Gross, A. T. n.d. Wind power usage in Europe? *Ergeb der Jahrrestagung der Studienges. Windkraft eV Stuttgart* DK 621, 311, 24, 003, 1, No. 4, pp. 1–6.

———. 1959. Windkraftnutzung in Europa. *Brennstoff-Kraft* 11: 414–419.

Haldane, T. G. N. and E. W. Holding. 1952. Recent developments in large-scale wind-power generation in Great Britain. *Transactions of the Fourth World Power Conference, Lund, Humphries, London* 4: 2501–2510.

Huetter, U. 1954a. The development of wind power installations for electrical power generation in Germany, *Z. Brennest-Waerme-Kratt* (*Düsseldorf*) 6: 270–278.

———. 1954b. The Use of Wind Energy for Generating Electric Current in Western Germany. World Power Conference, Rio de Janeiro.

Isaacs, J. D., G. L. Wick, and Schmitt. 1976. Utilization of the energy from ocean waves. Wave and Salinity Gradient Energy Conversion Workshop Proceedings, Newark, Delaware, May 24, 1976, pp. F.1–F.36.

Les aérogénérateurs à hélices en alliage léger de l'électricité de France. *Revue de l'Aluminium* 35: 1229–36, 1958.

Levy, N. 1968. Current state of wind power research in the Soviet Union. Brace Research Institute.

Mackenthun. 1951. Windkraftanlage Neuwork. *Elektrizitäts.*

Masuda and R. Rodriguez. 1971.

McCoull, J. 1973. Windmills. *Environment* 16–16.

McCormick, M. 1979. *Ocean Waves Energy Conversion Techniques.* Prepared under contract to U.S. Department of Energy.

McCormick, M. E. 1979. Salinity Gradients, Tides and Waves as Energy Sources.

Ormiston, R. A. 1971. Theoretical and experimental aerodynamics of the sailwing. *Journal of Aircraft,* 8(2).

The problem of wind utilization in Italy. *Rivista di Meteorologia* 4(1), 1942.

Putnam, P. C. 1948. *Power from the Wind.* New York: Van Nostrand Reinhold.

Rabenhorst, David W. 1973. Superflywheel. *Proceedings of the Solar Heating and Cooling from Buildings Workshop,* Washington, D.C., March 21–23, 1973.

Rangi, R. S., P. South, and R. J. Templin. 1974. Wind power and the vertical-axis wind turbine developed at the National Research Council, Canada. *Division of Mechanical Engineering and National Aeronautical Establishment, Quarterly Bulletin* 2: 1–14.

Salter, S. H., D. C. Jeffrey, and J. R. M. Taylor. 1976. Wave power nodding duck wave energy extractors. *Energy Digest* (London) 5(4): 19–21.

Savonius, S. J. 1931. The Savonius rotor and its applications. *Mechanical Engineering* 53: 333–338.

———. 1925a. The Wing Rotor in Theory and Practice. Helsingford, Finland.

———. 1925b. The wing rotor. *The Engineer* 140: 193.

———. 1946. *The Savonius Rotor.* New York University, College of Engineering.

Schmidt, K. O. 1955. Die Windkraftanlage der Deutschen Bundespost auf dem Schöneberg. *Elektrotechnik* 37 (51/52).

Serra, L. 1953. Le vent en France et ses possibilités d'utilisation. *Metéorologie* 273–292.

Sweeney, T. E. and W. B. Nixon. 1973. An Introduction to the Princeton Sailwing Windmill. NASA Lewis Res. Center Wind Energy Conversion Systems, pp. 70–72.

Thomas, R. L. and J. E. Sholes. 1975. Preliminary Results of the Large Experimental Wind Turbine Phase of the National Wind Energy Program. NASA TM-71796.

United Nations Conference on New Sources of Energy. 1961. Wind power. *Proceedings on Solar Energy, Wind Power and Geothermal Energy.* Vol. 7. Rome, August 21–31, 1961. United Nation Conference on New Sources of Energy, New York.

Vadot, L. L. 1954. Etude synoptique des différents types d'éoliennes. Houille Blanche, Grenoble.

Vance, W. 1973. Vertical axis wind rotors: status and potential. *Wind Energy Conversion Systems.* NASA, Lewis Research Center.

Vargo, D. J. 1974. Energy Developments in the 20th Century. Presented at the Fourth Annual Regulatory Information Systems Conference, St. Louis, Missouri, September 10–12, 1974.

Wind driven generator on Costa Head. *Engineering,* July 8, 1955, p. 180

Wind power production in Great Britain. 1950. *Architects Energiewirtsch, Sonderdienst* 239.

Appendix A
Windbreaks

A windbreak is a strip or belt of trees or shrubs established within or adjacent to a field. Its purpose is to reduce soil blowing; conserve soil moisture; protect crops, orchards, livestock, and wildlife; and increase the natural beauty of the landscape. Windbreaks are applicable in or along open fields where protection against wind damage to soils, crops, or livestock can be accomplished, or where strips of trees or shrubs will increase the natural beauty of an area and/or provide food and cover for wildlife.

The degree of wind erosion is dependent directly on the physical character and condition of the soil. Only dry soils are moved. Wet or damp soils are not appreciably affected. The structure of soil in an air-dry state is the index to its erodibility. There are three types of soil movement in the process of wind erosion, usually operating simultaneously:

1. *saltation:* movement of particles in short bounces on the ground
2. *suspension:* fine dust particles carried suspended in air
3. *surface creep:* movement of large particles on the ground both by wind and by bombardment by smaller particles

SPECIFICATIONS

Location of and Distance Between Windbreaks

To determine the location of and distance between windbreaks:

1. Calculate the area or distance sheltered. Determine the future height of the windbreak and its alignment to the prevailing wind. The formula of $10H$ ($10 \times$ Height of barrier) is used to determine distance sheltered. When the windbreak is aligned near to right angles (0° to 22°) to the wind, the sheltered distance decreases as angle of wind hitting the break increases.

2. Try to locate windbreaks along farm boundaries, field boundaries, public roads, private roads, or ridge tops. In open fields, plant windbreaks at right angles to prevailing wind.

Effective Height of Barrier (Ft)	*Distance Sheltered (Ft)*	
	Prevailing Wind Angle	
	0° to 22.5°	*60°*
80	800	400
65	650	325
50	500	250
35	350	175
20	200	100
10	100	35

When slopes facing the wind are over 2%, use the following correction factors to determine distance sheltered:

% Slope Facing Wind—Along Wind Direction	*Barrier Height Correction Factor*
3	0.77
4	0.72
5	0.67
6	0.62

For example, a 65-ft barrier with 60° alignment on land with a 4% slope will provide a sheltered distance of 234 ft ($325 \times 0.72 = 234$ ft).

Natural Shelterbelts

When clearing forested land for agriculture, consider leaving shelterbelts of native trees to provide wind-erosion protection and wildlife travel lanes and escape cover. Natural shelterbelts offer immediate wind protection without the time lag of waiting for trees to grow. Natural shelterbelts are effective and valuable in eastern North Carolina where light organic soils cleared for agriculture can be subject to severe wind erosion.

Shelterbelts of 33 to 330 feet in width offer effective wind protection as well as wildlife escape cover and habitat and also can produce marketable timber products. The distance sheltered by natural shelterbelts should be calculated using the methods described in Location of and Distance Between Windbreaks.

Species

1. Consider the potential growth and future economic value of plants and trees, the plant's usefulness as food and cover for wildlife, and the appearance or beauty of the windbreak in the landscape. Coniferous evergreens such as loblolly pine, eastern red cedar, and Arizona cypress give year-round protection and are excellent for windbreaks.
2. Plant species that are adapted to the soil-site conditions. Species in Table A-1 are suited generally to the sandhills, coastal plain, and flatwoods of North Carolina.
3. Tree and shrub species that are to be planted should be selected from Table A-1.

Table A-1. Recommended tree and shrub species for shelterbelts.

SPECIES	AVERAGE EFFECTIVE HEIGHT (FT)[a]
Loblolly pine (*Pinus Taeda*)	70–90
Slash pine (*Pinus elliottii*)	70–90
Eastern red cedar (*Juniperus virginiana*)	45–65
Arizona cypress (*Cupressus arizonica*)	40–60
Carolina laurelcherry (*Prunus caroliniana*)	30–40
California privet (*Ligustrum ovalifolium*)[b]	15–30
Chinese photonia (*Photinia serrulata*)	15–20
Bush honeysuckle (*Lonicera maackii*)[c]	8–12
Autumn olive (*Elaeagnus umbellata*)[c]	8–12
Shrub lespedeza (*Lespedeza bicolor*)[d]	8–12

[a] Trees at 50 years of age; shrubs at maturity.
[b] There are several suitable varieties of ligustrum; estimate height for variety used.
[c] Not evergreen, but normally in full leaf by critical wind erosion period.
[d] Not evergreen; should be planted in combination with evergreens.

Spacing and Arrangment

1. Plant two or three rows of trees with rows 6 ft apart.
2. Pines should be no more than 8 ft apart in the row; red cedar and Arizona cypress should be no more than 5 to 6 ft apart.
3. Stagger each tree in the row between the trees of adjacent rows.
4. Where shrubs are used, plant a belt of them no less than 4 ft wide on the windward side of the field to be protected.
5. Plant privet at least 4 ft apart in the row.
6. Plant shrub lespedezas in rows 2 ft apart with individual plants 18 in. apart in the row.

Site Preparation

Prepare a suitable planting site by eliminating competing vegetation. Sod land should be broken, worked thoroughly, and fallowed if necessary.

Planting Methods

1. Mechanical tree planters may be used where soil, erosion conditions, and slope permit.
2. Hand planting: Use a planting bar (dibble) or a mattock where it is impractical to use a machine planter. Planting bars work very well in lighter sandy soils.

Culture

Reducing competition from weeds, grass, or brush is important for the survival and growth of the windbreak seedlings. Competition can be reduced by:

1. Mowing between rows annually for at least 2 yrs after establishment. Plan spacing between rows to permit passage of equipment.
2. Spraying vegetative competition with approved herbicide.
3. Mulching.

Protection and Maintenance Requirements

1. Protect from fire and grazing damage by domestic livestock.
2. Replace dead trees and shrubs with the same species the first year following planting.

Appendix B
Local Climatological Data for A Region of the United States

Local Climatological Data

Annual Summary With Comparative Data

1977

BOSTON, MASSACHUSETTS

Narrative Climatological Summary

Climate is the composite of numerous weather elements. Three important influences are responsible for the main features of Boston's climate. First, the latitude (42° N) places the city in the zone of prevailing west to east atmospheric flow in which are encompassed the northward and southward movements of large bodies of air from tropical and polar regions. This results in variety and changeability of the weather elements. Secondly, Boston is situated on or near several tracks frequently followed by systems of low air pressure. The consequent fluctuations from fair to cloudy or stormy conditions reinforce the influence of the first factor, while also assuring a rather dependable precipitation supply. The third factor, Boston's east-coast location, is a moderating factor affecting temperature extremes of winter and summer.

Hot summer afternoons are frequently relieved by the locally celebrated "sea-breeze," as air flows inland from the cool water surface to displace the warm westerly current. This refreshing east wind is more commonly experienced along the shore than in the interior of the city or the western suburbs. In winter, under appropriate conditions, the severity of cold waves is reduced by the nearness of the then relatively warm water. The average date of the last occurrence of freezing temperature in spring is April 8; the latest is May 3, 1874 and 1882. The average date of the first occurrence of freezing temperature in autumn is November 7; the earliest on record is October 5, 1881. In suburban areas, especially away from the coast, these dates are later in spring and earlier in autumn by up to one month in the more susceptible localities.

Boston has no dry season. For most years the longest run of days with no measurable precipitation does not extend much more than two weeks. This may occur at any time of year. Most growing seasons have several shorter dry spells during which irrigation for high-value crops may be useful.

Much of the rainfall from June to September comes from showers and thunderstorms. During the rest of the year, low pressure systems pass more or less regularly and produce precipitation on an average of roughly one day in three. Coastal storms, or "northeasters," are prolific producers of rain and snow. The main snow season extends from December through March. The average number of days with four inches or more of snowfall is four per season, and days with seven inches or more come about twice per season. Periods when the ground is bare or nearly bare of snow may occur at any time in the winter.

Relative humidity has been known to fall as low as 5% (May 10, 1962), but such desert dryness is very rare. Heavy fog occurs on an average of about two days per month with its prevalence increasing eastward from the interior of Boston Bay to the open waters beyond.

The greatest number of hours of sunshine recorded in any month was 390, or 86% of possible, in June 1912, while the least was 60 hours, or 21%, in December 1972.

Although winds of 32 m.p.h. or higher may be expected on at least one day in every month of the year, gales are both more common and more severe in winter.

NATIONAL OCEANIC AND ATMOSPHERIC ADMINISTRATION / ENVIRONMENTAL DATA SERVICE / NATIONAL CLIMATIC CENTER ASHEVILLE, N.C.

Meteorological Data For The Current Year

Station: BOSTON, MASSACHUSETTS # 14739 GEN LOGAN INTERNATIONAL AP Standard time used: EASTERN Latitude: 42° 22' N Longitude: 71° 02' W Elevation (ground): 15 feet Year: 1977

Month	Temperature °F: Averages: Daily maximum	Daily minimum	Monthly	Extremes: Highest	Date	Lowest	Date	Degree days Base 65 °F: Heating	Cooling	Precipitation in inches: Water equivalent: Total	Greatest in 24 hrs.	Date	Snow, Ice pellets: Total	Greatest in 24 hrs.	Date	Relative humidity, pct.: Hour 01 (Local time)	Hour 07	Hour 13	Hour 19	Wind: Resultant: Direction	Speed m.p.h.	Average speed m.p.h.	Fastest mile: Speed m.p.h.	Direction	Date	Percent of possible sunshine	Average sky cover, tenths, sunrise to sunset	Number of days: Sunrise to sunset: Clear	Partly cloudy	Cloudy	Precipitation .01 inch or more	Snow, Ice pellets 1.0 inch or more	Thunderstorms	Heavy fog, visibility ¼ mile or less	Temperature °F: Maximum: (b) 90° and above	32° and below	Minimum: 32° and below	0° and below	Average station pressure mb: Elev. 29 feet m.s.l.
JAN	29.8	16.7	23.3	47	10	3	18	1290	0	4.41	1.74	10-11	23.2	13.8	7	61	63	55	56	28	9.4	14.1	39	W	11	67	5.1	14	7	10	11	4	1	1	0	17	31	0	1007.5
FEB	36.6	24.8	30.7	48	25	9	17	956	0	2.40	1.05	24-25	5.9	3.8	20-21	60	61	56	56	28	7.1	13.1	38	W	25	57	6.0	9	6	13	12	2	0	4	0	7	22	0	1010.8
MAR	52.5	36.8	44.7	81	30	27	2	623	1	4.76	1.60	22-23	10.7	9.6	18-19	69	69	55	61	30	5.5	13.1	60	NE	22	61	5.9	9	8	14	13	2	2	3	0	0	10	0	1012.2
APR	61.2	41.4	51.3	88	22	26	9	414	13	4.07	1.53	23-24	T	T	8	67	67	50	57	28	4.3	14.0	39	NW	3	71	4.8	15	6	9	9	0	0	0	0	0	4	0	1015.9
MAY	73.0	52.2	62.6	92	17	35	10	158	92	3.52	3.09	9-10	0.5	0.5	10	65	63	51	57	27	3.4	12.4	44	NE	9	79	5.1	11	10	10	6	0	2	0	2	0	0	0	1013.2
JUN	75.8	59.0	67.4	90	28	50	8	43	124	2.49	0.94	10-11	0.0	0.0		82	77	63	72	25	3.0	10.7	35	SW	17	67	6.9	3	12	15	10	0	4	1	1	0	0	0	1010.2
JUL	84.1	65.7	74.9	102	21	57	27	0	314	2.21	1.19	25	0.0	0.0		74	70	52	62	26	4.9	10.6	46	N	18	77	5.1	14	8	9	7	0	4	1	9	0	0	0	1015.2
AUG	82.3	64.5	73.4	95	29	56	26	4	272	2.91	0.71	1-2	0.0	0.0		86	81	63	77	24	4.5	9.4	38	SW	5	68	6.4	6	12	13	12	0	4	3	5	0	0	0	1016.3
SEP	71.1	57.7	64.4	92	2	47	22	85	75	4.03	1.55	20-21	0.0	0.0		86	86	70	80	27	0.5	10.1	38	SW	26	53	6.1	8	10	12	13	0	1	1	1	0	0	0	1015.9
OCT	62.1	48.5	55.3	77	2	37	24	304	6	4.63	1.10	1-2	0.0	0.0		80	81	65	75	28	3.0	11.3	36	NE	14	52	6.1	9	8	14	12	0	1	2	0	0	0	0	1014.9
NOV	53.6	42.6	48.1	67	16	29	28	498	0	2.54	1.45	8-9	0.7	0.7	28	77	80	69	73	27	1.9	12.0	37	W	27	30	7.7	5	4	21	10	0	3	3	0	0	3	0	1019.3
DEC	40.7	27.7	34.2	55	21	5	12	948	0	6.20	1.52	21-22	5.2	1.8	5-6	76	79	65	70	30	5.4	14.2	39	NE	6	46	5.8	12	4	15	16	3	0	3	0	8	19	0	1014.6
YEAR	60.2	44.8	52.5	102	JUL 21	3	JAN 18	5323	897	44.17	3.09	MAY 9-10	46.2	13.8	JAN 7	74	73	60	66	28	4.2	12.1	60	NE	MAR 22	62	5.9	115	95	155	131	11	22	22	18	32	89	0	1013.8

Normals, Means, And Extremes

Month	Temperatures °F: Normal: Daily maximum	Daily minimum	Monthly	Extremes: Record highest	Year	Record lowest	Year	Normal Degree days Base 65 °F: Heating	Cooling	Precipitation in inches: Water equivalent: Normal	Maximum monthly	Year	Minimum monthly	Year	Maximum in 24 hrs.	Year	Snow, Ice pellets: Maximum monthly	Year	Maximum in 24 hrs.	Year	Relative humidity pct.: Hour 01 (Local time)	Hour 07	Hour 13	Hour 19	Wind: Mean speed m.p.h.	Prevailing direction	Fastest mile: Speed m.p.h. *	Direction	Year	Pct. of possible sunshine	Mean sky cover, tenths, sunrise to sunset	Mean number of days: Sunrise to sunset: Clear	Partly cloudy	Cloudy	Precipitation .01 inch or more	Snow, Ice pellets 1.0 inch or more	Thunderstorms	Heavy fog, visibility ¼ mile or less	Temperatures °F: Max.: (b) 90° and above	32° and below	Min.: 32° and below	0° and below	Average station pressure mb.: Elev. 29 feet m.s.l.
(a)				26		26					26		26		26		42		42		13	13	13	13	20	15	19	19		42	42	42	42	42	26	42	42	42	13	13	13	13	5
J	35.9	22.5	29.2	63	1974	-12	1957	1110	0	3.69	9.54	1958	0.89	1970	2.07	1958	32.5	1948	13.8	1977	65	66	57	61	14.2	NW	61	NW	1974	54	6.2	9	7	15	12	3	*	2	0	11	26	1	1014.8
F	37.5	23.3	30.4	68	1957	-4	1961	969	0	3.54	7.08	1969	1.15	1968	2.68	1969	41.3	1969	19.4	1958	64	67	57	60	14.1	WNW	57	S	1970	56	6.2	8	7	13	11	3	*	2	0	8	24	*	1014.0
M	44.6	31.5	38.1	81	1977	6	1967	834	0	4.01	11.00	1953	1.48	1962	4.13	1968	31.2	1956	17.7	1960	67	68	57	62	14.0	NW	60	NE	1977	57	6.4	8	8	15	12	2	1	2	0	2	17	0	1013.9
A	56.3	40.8	48.6	94	1976	17	1954	492	0	3.49	7.82	1958	1.24	1966	2.31	1973	3.3	1967	3.1	1956	67	66	53	59	13.4	WNW	52	NW	1963	57	6.4	7	9	14	11	*	1	2	*	0	3	0	1012.4
M	67.1	50.1	58.6	93	1964	34	1956	218	20	3.47	13.38	1954	0.53	1964	5.74	1954	0.5	1977	0.5	1977	72	70	57	64	12.2	SW	50	NE	1967	59	6.5	6	11	14	11	0	2	3	*	0	0	0	1013.1
J	76.6	59.3	68.0	100	1952	46	1965	27	117	3.19	8.63	1959	0.48	1953	2.46	1960	0.0		0.0		78	74	60	68	11.3	SW	40	NW	1959	63	6.3	7	10	13	11	0	4	2	3	0	0	0	1014.1
J	81.4	65.1	73.3	102	1977	54	1965	0	260	2.74	8.12	1959	0.52	1952	2.42	1959	0.0		0.0		76	73	56	66	10.8	SW	46	N	1977	66	6.1	7	12	12	9	0	5	2	5	0	0	0	1013.7
A	79.3	63.3	71.3	102	1975	47	1965	8	203	3.46	17.09	1955	0.83	1972	8.40	1955	0.0		0.0		78	75	57	68	10.8	SW	45	SW	1971	67	5.6	9	11	11	10	0	4	2	3	0	0	0	1016.3
S	72.2	56.7	64.5	100	1953	38	1965	76	61	3.16	8.31	1954	0.35	1957	5.64	1954	0.0		0.0		80	79	61	72	11.2	SW	57	S	1960	63	5.5	10	8	12	9	0	2	2	1	0	0	0	1016.5
O	63.2	47.5	55.4	90	1963	28	1976	301	0	3.02	8.68	1962	0.96	1967	4.26	1962	T	1975	T	1975	75	76	57	67	12.1	SW	45	NW	1963	61	5.5	11	8	12	9	0	1	2	0	0	1	0	1017.1
N	51.7	38.7	45.2	77	1974	17	1972	594	0	4.51	8.18	1969	0.64	1976	3.33	1955	10.0	1938	8.0	1940	72	74	61	67	13.0	SW	54	NE	1968	51	6.4	8	7	15	11	*	*	2	0	*	6	0	1014.9
D	39.3	26.6	33.0	70	1966	-4	1962	992	0	4.24	9.74	1969	1.03	1955	4.17	1969	27.9	1970	13.0	1960	69	72	61	66	13.8	WNW	49	NW	1962	52	6.2	9	7	15	12	2	*	1	0	6	21	*	1015.4
YR	58.7	43.8	51.3	102	JUL 1977	-12	JAN 1957	5621	661	42.52	17.09	AUG 1955	0.35	SEP 1957	8.40	AUG 1955	41.3	FEB 1969	19.4	FEB 1958	72	72	58	65	12.6	SW	61	NW	JAN 1974	60	6.1	99	105	161	128	11	19	23	12	27	98	1	1014.7

Means and extremes above are from existing and comparable exposures. Annual extremes have been exceeded at other sites in the locality as follows: Highest temperature 104 in July 1911; lowest temperature -18 in February 1934; minimum monthly precipitation T in March 1915; fastest mile wind 87 S in September 1938.

(a) Length of record, years, through the current year unless otherwise noted, based on January data.
(b) 70° and above at Alaskan stations.
* Less than one half.
T Trace.

NORMALS - Based on record for the 1941-1970 period.
DATE OF AN EXTREME - The most recent in cases of multiple occurrence.
PREVAILING WIND DIRECTION - Record through 1963.
WIND DIRECTION - Numerals indicate tens of degrees clockwise from true north. 00 indicates calm.
FASTEST MILE WIND - Speed is fastest observed 1-minute value when the direction is in tens of degrees.

Average Temperature

Year	Jan	Feb	Mar	Apr	May	June	July	Aug	Sept	Oct	Nov	Dec	Annual
1938	28.0	30.6	39.1	48.7	55.8	67.2	71.8	73.6	62.2	56.0	46.4	34.0	51.1
1939	27.8	32.2	33.2	43.6	56.4	65.6	72.3	73.8	64.5	54.2	40.8	33.4	49.8
1940	23.0	29.6	33.1	43.6	56.2	65.1	71.6	68.4	63.3	50.6	42.9	34.3	48.5
1941	25.2	29.4	33.4	51.6	59.8	68.2	71.4	70.6	66.0	55.9	48.0	35.2	51.2
1942	28.6	27.0	40.8	49.7	60.8	67.4	71.1	70.7	64.6	55.8	43.5	28.6	50.7
1943	25.5	30.6	36.0	43.4	57.6	71.2	74.1	71.0	63.0	53.8	43.3	29.6	49.9
1944	31.0	29.6	34.1	44.6	63.2	67.0	73.8	74.7	65.0	53.8	43.2	30.8	50.9
1945	23.8	30.5	46.3	52.5	55.2	66.6	71.9	70.8	67.1	53.1	45.5	28.5	51.0
1946	28.4	27.6	47.2	46.2	58.0	67.5	70.8	67.8	65.8	58.4	47.8	34.8	51.7
1947	32.6	29.6	37.7	47.2	56.9	65.4	74.4	73.2	64.8	61.6	41.2	30.4	51.2
1948	23.4	26.6	38.0	48.0	55.0	63.6	74.5	73.7	65.7	54.2	49.6	36.3	50.7
1949	34.6	34.4	39.2	50.8	60.4	71.6	76.2	74.4	63.2	58.4	43.5	36.8	53.6
1950	36.2	28.0	33.7	46.8	55.8	69.1	73.5	70.9	61.5	56.3	47.8	35.5	51.3
#1951	34.0	34.7	39.1	51.0	59.1	66.0	74.0	70.7	65.6	54.8	42.6	35.0	52.2
1952	32.6	32.5	37.2	50.5	57.2	70.7	77.5	72.3	66.3	53.0	44.9	35.7	52.6
1953	34.7	35.0	39.1	48.9	58.4	70.5	73.2	72.0	66.3	56.2	48.6	40.2	53.6
1954	26.0	36.4	38.8	50.1	56.3	66.6	72.2	70.1	63.2	58.6	44.4	34.4	51.4
1955	28.5	32.0	37.6	49.2	62.8	66.8	77.2	74.5	64.5	55.1	41.9	26.6	51.4
1956	30.6	32.5	33.6	45.6	55.4	68.9	71.7	71.9	61.1	54.4	46.1	36.0	50.7
1957	23.4	34.7	39.1	49.4	59.9	71.3	74.1	69.3	67.3	54.6	47.2	40.0	52.5
1958	31.0	25.5	39.1	48.8	56.6	63.9	72.4	72.4	64.6	52.6	46.6	26.4	50.0
1959	28.7	26.7	37.0	49.0	62.6	64.6	74.7	74.1	68.1	55.1	44.4	36.3	51.8
1960	30.9	35.3	32.7	48.3	59.7	69.6	73.1	72.1	63.7	53.9	48.0	29.5	51.4
1961	25.0	31.6	36.8	45.3	56.3	68.9	72.1	72.5	69.0	57.3	44.6	32.8	51.0
1962	28.7	26.7	38.5	49.4	57.2	68.4	70.4	70.0	62.9	54.1	41.7	30.0	49.8
#1963	29.5	25.9	39.1	48.9	59.4	69.5	74.7	70.4	60.8	60.0	48.3	25.9	51.0
#1964	31.7	29.1	38.7	46.1	60.3	67.1	71.5	66.4	62.0	52.5	44.1	32.4	50.2
1965	25.4	28.0	35.8	44.2	59.5	67.4	71.0	70.5	62.5	52.8	42.1	36.1	49.6
1966	28.8	31.3	39.8	45.9	57.3	69.4	74.9	71.3	63.5	54.5	46.9	34.2	51.5
1967	35.1	26.4	33.2	44.9	51.7	67.2	73.0	70.9	62.7	53.8	40.1	35.0	49.5
1968	25.6	26.1	39.1	49.6	56.1	64.9	75.2	70.7	65.0	57.9	43.8	30.9	50.4
1969	29.3	29.5	35.4	50.6	58.5	69.3	71.0	74.3	63.7	54.3	44.9	33.4	51.2
1970	23.0	32.3	37.4	49.0	59.6	67.0	74.3	73.6	65.6	54.9	44.8	28.9	50.9
1971	23.8	30.5	36.7	45.1	55.7	69.1	73.4	73.4	68.0	59.8	43.1	36.3	51.2
1972	33.0	29.6	36.3	44.9	57.6	65.4	73.8	71.5	65.7	51.8	42.3	33.0	50.4
1973	31.4	30.1	43.3	49.9	57.0	70.0	74.3	74.8	64.4	55.6	45.8	39.6	53.0
1974	31.7	29.1	38.7	50.9	54.7	64.8	72.4	72.0	63.7	50.1	45.3	37.8	50.9
1975	34.9	32.1	36.9	45.1	61.5	67.5	75.9	72.9	63.9	57.3	51.8	34.4	52.8
1976	26.1	37.3	41.2	55.1	60.2	73.4	72.9	72.0	64.9	52.3	41.9	29.0	52.2
1977	23.3	30.7	44.7	51.3	62.6	67.4	74.9	73.4	64.4	55.3	48.1	34.2	52.5
RECORD													
MEAN	28.8	29.2	37.0	47.0	57.8	67.0	72.6	70.8	64.0	54.2	43.5	32.6	50.4
MAX	36.3	36.9	44.5	55.1	66.4	75.8	80.9	78.8	72.0	62.1	50.6	39.6	58.3
MIN	21.2	21.5	29.5	38.9	49.1	58.2	64.2	62.7	56.0	46.2	36.3	25.5	42.4

Heating Degree Days

BOSTON, MA

Season	July	Aug	Sept	Oct	Nov	Dec	Jan	Feb	Mar	Apr	May	June	Total
1957-58	0	7	57	317	529	766	1046	1102	796	481	262	85	5448
1958-59	4	2	84	386	547	1190	1118	1065	862	476	142	92	5968
1959-60	1	7	79	319	611	885	1048	855	992	493	166	33	5489
1960-61	0	5	103	335	503	1094	1231	928	865	587	287	22	5960
1961-62	6	3	51	246	604	991	1118	1066	814	467	271	35	5672
1962-63	6	13	105	330	691	1078	1094	1087	798	477	196	38	5913
#1963-64	1	3	160	198	495	1207	1026	1033	808	559	187	57	5734
1964-65	14	26	140	380	620	1004	1220	1032	900	617	195	80	6228
1965-66	2	37	136	371	680	888	1115	936	776	566	258	46	5811
1966-67	0	1	88	322	535	950	921	1075	977	596	403	58	5926
1967-68	0	4	110	347	739	923	1214	1122	797	454	270	76	6056
1968-69	1	9	46	247	630	1050	1099	987	911	430	208	21	5639
1969-70	2	3	107	326	595	973	1295	909	846	473	184	52	5765
1970-71	0	0	68	314	598	1113	1269	962	868	586	287	25	6090
1971-72	0	2	37	169	651	882	985	1021	883	598	250	54	5532
1972-73	3	4	51	405	673	985	1033	971	666	450	258	24	5523
1973-74	0	2	94	289	570	782	1023	1000	809	429	335	77	5410
1974-75	0	2	102	458	587	836	925	918	866	590	162	59	5505
1975-76	0	8	70	239	395	941	1198	800	733	331	166	16	4897
1976-77	1	10	55	393	688	1108	1290	956	623	414	158	43	5739
1977-78	0	4	85	304	498	948							

Cooling Degree Days

Year	Jan	Feb	Mar	Apr	May	June	July	Aug	Sept	Oct	Nov	Dec	Total
1969	0	0	0	9	13	156	196	297	74	1	0	0	746
1970	0	0	0	0	25	118	294	273	91	9	0	0	810
1971	0	0	0	0	6	155	269	271	132	15	1	0	849
1972	0	0	0	0	26	74	279	213	79	0	0	0	671
1973	0	0	0	7	18	180	296	316	84	3	0	0	904
1974	0	0	0	10	22	81	235	226	68	1	3	0	646
1975	0	0	0	0	60	139	345	261	44	9	4	0	862
1976	0	0	0	43	25	276	251	231	61	8	0	0	895
1977	0	0	1	13	92	124	314	272	75	6	0	0	897

Precipitation

Year	Jan	Feb	Mar	Apr	May	June	July	Aug	Sept	Oct	Nov	Dec	Annual
1938	4.91	2.38	2.42	3.22	4.42	6.30	9.46	3.31	6.00	2.43	2.89	2.80	50.54
1939	2.18	3.79	5.23	4.54	1.29	2.70	0.75	2.14	1.01	4.77	1.14	2.91	32.45
1940	1.68	4.78	3.83	4.58	3.28	1.80	3.17	0.85	2.32	0.76	6.24	2.76	36.05
1941	4.21	1.70	3.40	1.70	2.43	4.29	2.90	1.55	1.18	1.92	2.40	3.19	30.87
1942	3.69	3.45	7.01	1.59	2.11	4.24	4.14	2.09	1.96	2.78	4.69	4.72	42.47
1943	3.74	1.23	4.02	2.64	4.56	1.49	3.91	1.28	1.41	4.82	2.16	0.99	32.25
1944	2.03	2.15	3.92	3.52	0.25	5.35	1.61	1.79	5.36	2.58	5.68	2.83	37.07
1945	3.67	4.09	1.90	2.02	4.47	6.44	2.12	4.27	1.81	2.23	6.86	7.42	47.30
1946	4.18	3.00	1.50	2.62	4.91	2.76	2.22	9.92	2.04	0.34	0.98	3.60	38.07
1947	2.45	1.44	2.30	4.15	4.36	2.88	3.98	2.19	3.95	1.13	5.13	3.95	37.91
1948	5.11	2.08	3.14	2.62	5.37	4.50	4.53	1.24	0.67	4.84	5.16	1.25	40.51
1949	3.21	3.25	1.66	3.23	2.53	0.93	1.10	2.12	6.47	1.60	3.71	1.64	31.45
1950	3.86	3.81	2.99	2.38	1.55	1.10	1.45	3.14	0.89	1.99	6.17	3.37	32.70
#1951	4.04	3.71	4.41	3.06	4.81	4.31	2.13	3.23	2.00	3.98	6.60	4.69	46.97
1952	4.31	4.71	4.41	4.41	3.57	3.26	0.52	6.86	1.13	1.61	1.72	4.09	40.60
1953	6.28	4.14	11.00	6.04	5.06	0.48	2.76	1.81	2.50	4.91	7.66	5.09	57.73
1954	3.26	3.37	3.33	5.25	13.38	2.78	2.50	5.64	8.31	3.58	5.52	5.40	62.32
1955	0.92	4.11	5.42	4.12	0.99	3.52	4.28	17.09	2.40	6.94	5.68	1.03	56.50
1956	6.99	4.36	5.39	2.94	1.85	2.03	3.32	1.46	5.07	4.39	3.46	6.13	47.39
1957	2.47	1.34	3.38	3.78	3.63	1.62	0.64	1.71	0.35	2.67	5.75	6.58	33.92
1958	9.54	5.87	4.48	7.82	4.45	2.96	3.91	5.37	7.50	4.62	3.35	1.78	61.65
1959	2.72	3.45	5.81	4.44	1.24	8.63	8.12	2.93	0.63	4.60	4.20	4.64	51.41
1960	3.04	4.84	3.23	3.51	3.80	3.46	5.18	1.64	5.97	2.48	2.49	4.82	44.46
1961	2.92	4.94	4.71	6.59	4.51	1.67	3.29	3.17	7.04	2.46	3.18	3.36	47.84
1962	3.11	4.16	1.48	3.85	1.86	2.33	1.61	3.72	4.10	8.68	3.80	4.53	43.23
1963	3.13	2.60	4.39	1.48	2.86	1.92	1.72	1.67	3.05	1.25	7.74	3.03	34.84
1964	4.56	4.67	3.48	3.69	0.53	1.91	3.12	1.78	2.65	2.82	2.18	5.08	36.47
1965	2.64	3.17	2.22	2.32	0.93	2.99	0.55	1.48	2.01	1.59	2.08	1.73	23.71
1966	5.29	3.48	1.98	1.24	2.66	3.40	3.21	1.25	3.42	2.62	4.43	3.03	36.01
1967	2.28	4.05	4.67	4.83	7.32	3.48	2.47	5.74	2.00	0.96	3.38	6.42	47.60
1968	3.85	1.15	7.86	1.72	3.26	5.65	0.55	1.63	1.79	1.85	6.74	6.23	42.28
1969	2.26	7.08	2.63	4.37	1.96	0.63	2.98	1.89	4.42	1.64	8.18	9.74	47.78
1970	0.89	4.65	4.32	2.79	3.01	4.62	1.27	4.12	2.60	2.63	4.09	6.92	41.91
1971	1.88	5.05	3.08	2.92	3.72	1.74	2.84	1.59	1.55	2.16	6.74	2.40	35.67
1972	2.05	5.29	5.37	3.34	5.26	6.76	2.19	0.83	5.94	2.98	7.02	6.08	53.11
1973	3.12	2.13	2.20	5.65	3.76	4.68	4.83	2.78	1.95	2.71	1.74	7.20	42.75
1974	3.22	3.24	4.01	3.86	2.87	2.29	1.54	3.41	7.03	3.12	1.73	3.92	40.24
1975	5.70	3.37	2.74	2.40	1.78	2.10	2.35	5.52	5.49	4.41	5.13	4.80	45.79
1976	5.29	2.45	2.42	2.00	1.98	0.58	4.30	7.99	1.56	4.16	0.64	3.35	36.72
1977	4.41	2.40	4.76	4.07	3.52	2.49	2.21	2.91	4.03	4.63	2.54	6.20	44.17
RECORD													
MEAN	3.60	3.38	3.84	3.56	3.24	3.13	3.15	3.60	3.24	3.24	3.85	3.69	41.52

Snowfall

Season	July	Aug	Sept	Oct	Nov	Dec	Jan	Feb	Mar	Apr	May	June	Total
1938-39	0.0	0.0	0.0	0.0	10.0	1.3	7.5	4.7	16.5	0.2	0.0	0.0	40.2
1939-40	0.0	0.0	0.0	0.0	1.0	6.5	4.6	23.8	0.3	1.5	0.0	0.0	37.7
1940-41	0.0	0.0	0.0	T	8.5	4.5	20.4	1.3	13.1	0.0	T	0.0	47.8
1941-42	0.0	0.0	0.0	T	T	0.2	8.2	6.2	7.9	1.5	0.0	0.0	24.0
1942-43	0.0	0.0	0.0	0.0	T	6.8	26.4	4.5	8.0	T	0.0	0.0	45.7
1943-44	0.0	0.0	0.0	0.0	T	0.3	5.1	9.6	12.7	T	0.0	0.0	27.7
1944-45	0.0	0.0	0.0	0.0	1.2	6.9	24.3	26.3	0.5	0.0	T	0.0	59.2
1945-46	0.0	0.0	0.0	0.0	4.9	24.6	9.8	9.8	0.2	1.5	0.0	0.0	50.8
1946-47	0.0	0.0	0.0	0.0	T	4.9	4.0	9.0	0.9	0.6	0.0	0.0	19.4
1947-48	0.0	0.0	0.0	0.0	1.1	26.8	32.5	17.0	11.8	0.0	0.0	0.0	89.2
1948-49	0.0	0.0	0.0	0.0	1.1	5.5	13.7	6.9	9.9	0.0	0.0	0.0	37.1
1949-50	0.0	0.0	0.0	0.0	1.4	2.3	7.9	15.2	5.1	T	0.0	0.0	31.9
1950-51	0.0	0.0	0.0	0.0	T	2.7	13.9	9.2	3.9	0.0	0.0	0.0	29.7
#1951-52	0.0	0.0	0.0	0.0	T	8.4	10.7	19.1	1.4	0.0	0.0	0.0	39.6
1952-53	0.0	0.0	0.0	T	T	2.4	13.6	11.4	0.2	2.2	T	0.0	29.8
1953-54	0.0	0.0	0.0	0.0	T	T	19.2	1.9	0.4	2.1	0.0	0.0	23.6
1954-55	0.0	0.0	0.0	0.0	T	10.3	0.9	6.5	7.0	0.4	0.0	0.0	25.1
1955-56	0.0	0.0	0.0	0.0	2.5	1.8	7.7	14.5	31.2	3.2	0.0	0.0	60.9
1956-57	0.0	0.0	0.0	0.0	0.6	15.4	20.6	2.8	11.5	1.1	0.0	0.0	52.0
1957-58	0.0	0.0	0.0	T	0.1	T	6.6	23.9	12.0	2.1	T	0.0	44.7
1958-59	0.0	0.0	0.0	0.0	0.1	4.6	4.1	10.7	14.6	T	0.0	0.0	34.1
1959-60	0.0	0.0	0.0	0.0	0.6	5.5	10.2	2.3	22.3	T	0.0	0.0	40.9
1960-61	0.0	0.0	0.0	T	T	16.9	18.7	14.9	9.0	2.0	0.0	0.0	61.5
1961-62	0.0	0.0	0.0	T	0.9	11.4	2.5	28.7	1.1	0.1	0.0	0.0	44.7
1962-63	0.0	0.0	0.0	T	0.9	5.3	6.5	4.6	13.6	T	0.0	0.0	30.9
1963-64	0.0	0.0	0.0	T	0.0	17.7	14.4	23.2	7.7	T	0.0	0.0	63.0
1964-65	0.0	0.0	0.0	T	T	12.2	22.2	4.7	9.7	1.6	0.0	0.0	50.4
1965-66	0.0	0.0	0.0	0.0	T	2.3	26.4	12.1	3.3	T	T	0.0	44.1
1966-67	0.0	0.0	0.0	0.0	0.0	9.9	0.5	23.5	22.9	3.3	T	0.0	60.1
1967-68	0.0	0.0	0.0	0.0	2.2	14.7	17.7	3.4	6.8	0.0	0.0	0.0	44.8
1968-69	0.0	0.0	0.0	0.0	0.4	5.1	0.9	41.3	6.1	T	0.0	0.0	53.8
1969-70	0.0	0.0	0.0	T	T	12.6	7.4	10.5	18.2	0.1	0.0	0.0	48.8
1970-71	0.0	0.0	0.0	T	T	27.9	12.0	8.1	7.4	1.9	0.0	0.0	57.3
1971-72	0.0	0.0	0.0	0.0	2.8	7.9	7.8	16.5	12.1	0.4	0.0	0.0	47.5
1972-73	0.0	0.0	0.0	T	0.6	3.3	3.6	2.5	0.3	T	0.0	0.0	10.3
1973-74	0.0	0.0	0.0	0.0	0.0	T	16.0	17.8	0.1	3.0	0.0	0.0	36.9
1974-75	0.0	0.0	0.0	0.0	2.0	3.6	2.2	17.0	1.8	1.0	0.0	0.0	27.6
1975-76	0.0	0.0	0.0	T	0.1	19.3	15.0	1.4	10.8	T	0.0	0.0	46.6
1976-77	0.0	0.0	0.0	0.0	1.0	17.2	23.2	5.9	10.7	T	0.5	0.0	58.5
1977-78	0.0	0.0	0.0	0.0	0.7	5.2							
RECORD													
MEAN	0.0	0.0	0.0	T	1.2	8.1	12.2	11.8	8.1	0.7	T	0.0	42.1

# Indicates a station move or relocation of instruments. See Station Location table.

Record mean values above are means through the current year for the period beginning in 1872 for temperature, 1871 for precipitation and 1936 for snowfall. Data are from City Office locations through 1935 and from Airport locations thereafter.

14739 Boston, MA

STATION LOCATION

BOSTON, MASSACHUSETTS

Location	Occupied from	Occupied to	Airline distance and direction from previous location	Latitude North	Longitude West	Elevation above: Sea level	Elevation above: Ground								Elevation above: Sea level	Remarks
						Ground at temperature site	Wind instruments	Extreme thermometers	Psychrometer	Telepsychrometer	Tipping bucket rain gage	Weighing rain gage	8" rain gage	Hygrothermometer	Pyranometer	
CITY																
Old State House, corner State & Devonshire Sts.	10/20/70	1/09/71		42° 21'	71° 04'	16										Ground elevation approximate.
103 Court Street	1/10/71	8/12/75	600 ft. NW	42° 21'	71° 04'	40										Ground elevation approximate.
Equitable Building Corner Milk & Devonshire Streets	8/12/75	10/01/84	1200 ft. SE	42° 21'	71° 04'	12	172	156	156				162			
Old U. S. Post Office and Courthouse Milk, Devonshire, Congress & Water Streets East Tower	10/01/84	6/07/29	300 ft. NE	42° 21'	71° 04'	17	188	115	115			154	174 154			8 inch rain gage moved from bad exposure atop east tower to west tower, 154 feet above ground on 7/1/91. Marvin Weighing Rain and Snow Gage installed 1/1/98.
Young's Hotel Building Corner City Hall Avenue and Court Street	6/07/29	9/29/33	700 ft. NW	42° 21'	71° 04'	40	165	106	106		96		96			Anemometer atop City Hall Annex, across City Hall Avenue.
New U. S. Post Office and Courthouse Same site as old	9/29/33	6/06/64	700 ft. SE	42° 21'	71° 04'	20	360	337	336		329		328		a335	Observation Program transferred to Airport 1/1/36. a - Added 1/14/44.
U. S. Custom House India and State Streets	6/06/64	Present	1140 ft. NE	42° 22'	71° 03'	12									b157	b - Located atop Boston Building, 1/3 mile West of Custom House. Decommissioned 11/13/68.
AIRPORT																
U. S. Army Hangar No. 1 Boston Airport East Boston	10/15/26	4/01/27		42° 22'	71° 02'	3										Pibal only.
Section F, Army Base South Boston	4/01/27	11/01/27	1-3/4 mi. S	42° 21'	71° 02'		143									Pibal only.
Shack 25 feet South of Commercial Hangar Boston AP, East Boston	11/01/27	7/01/29	1-3/4 mi. N	42° 21'	71° 02'	2	22		4							Pibal only.
Shack 200 feet SW of East Coast Hangar Boston AP, East Boston	7/01/29	5/01/30	1/8 mi. SW	42° 22'	71° 02'	12	24		4							Pibal only to 2/16/30.
Administration Building Boston Municipal Airport East Boston	5/01/30	11/01/45	1/8 mi. NW	42° 22'	71° 02'	12	50	31	31		a3	b3	3			a - Added 1/1/36. b - Added 2/1/38. Official synoptic records began 1/1/36.
Administration Building Boston Municipal Airport East Boston	11/01/45	11/22/51	Same	42° 22'	71° 02'	12	*62	*33	*33		#32	#32	#32			* - Installed on 30 foot instrument tower on roof 9/17/37. # - Gages moved to roof 3/10/44.
Gate No.11, Boutwell Building, Logan Int'l. Airport, East Boston	11/22/51	12/05/63	5/8 mi. E	42° 22'	71° 01'	15	X33	20	20		19	19	18			X - 34 feet to 7/20/54 and 75 feet to 8/23/57.
General Aviation Adm. Building, West Wing, Logan International AP	12/05/63	Present	5/8 mi. W	42° 22'	71° 02'	d15	22	e33 f6	e33 f5		33 f5	33 f5	33 f5	c4		Instrument relocations completed 12/11/63. c - Commissioned on field site 4/1/64. d - 12 feet to 4/1/64. e - Standby status after 4/1/64. f - Effective 8/5/71.

Appendix C
Decennial Census of United States Climate Summary of Hourly Observations Logan International Airport

U. S. DEPARTMENT OF COMMERCE
LUTHER H. HODGES, Secretary
WEATHER BUREAU
F. W. REICHELDERFER, Chief

CLIMATOGRAPHY OF THE UNITED STATES NO. 82 – 19

DECENNIAL CENSUS OF UNITED STATES CLIMATE–

SUMMARY OF HOURLY OBSERVATIONS

75th Meridian Time Zone

BOSTON, MASSACHUSETTS

Logan International Airport

1951 - 1960

Washington, D.C.: 1962
Reprinted February 1974

For sale by the Superintendent of Documents, U.S. Government Printing Office, Washington 25, D.C. - Price 10 cents

PREFACE AND EXPLANATION OF TABLES

This summary of surface weather data is one of a series prepared under the Decennial Census of United States Climate, 1960 program. Similar summaries are being published for most Weather Bureau stations. All data are based on monthly data published in Local Climatological Data Supplements for all or part of the period 1951 - 1960. Where the full 10-year period is not covered by the monthly data, summaries are based on the period 1956 - 1960.

This series supersedes the series entitled "Climatography of the United States No. 30-Summary of Hourly Observations", which was published in 1956. It differs from this earlier information in that a longer, more representative period is used for summarization wherever possible. Comparability between stations is improved in the Decennial Census series by the use of identical 10-year and 5-year periods.

The tables in this pamphlet are similar to Tables A through E in the monthly Local Climatological Data Supplements except that Tables B, D, and E give percentage frequencies instead of total occurrences. In these tables all hourly observations are represented unless otherwise specified. The total number of observations used is indicated on each page under the month name. In the percentage tables "+" indicates more than 0 but less than 0.5. Values are to the nearest whole percent, but not adjusted to make their sums exactly equal to column or row totals.

STATION LOCATION

Boston, Massachusetts
Logan International Airport

Location	Occupied from	Occupied to	Airline distance and direction from previous location	Latitude	Longitude	Elevation above (feet)											Remarks
						Sea level		Ground									
						Ground	Actual barometer elevation (H_2)	Wind instruments	Extreme thermometers	Psychrometer	Telepsychrometer	Tipping bucket rain gage	Weighing rain gage	8" rain gage			
Administration Bldg., 3rd Floor, Boston Municipal Airport, E. Boston	11- 1-45	11-22-51	1/8 mi. NW	42° 22'N	71° 02' W	12	42	62	33	33		32	32	32			Due to aircraft interference wind instruments moved 1/8 mi.E and relocated on top of temporary Airport control tower.
Gate No. 11, Boutwell Bldg., Logan Int. Airport, E. Boston.	11-22-51	Present	5/8 mi. E	42° 22'N	71° 01' W	15	34	33	20	20		19	19	18			Wind system relocated between runways 4R and 4L spot .55 mi. SSE of station and 1140 ft. northwest of touchdown on 4R.

LOCATION AND TOPOGRAPHY:

Logan Airport, 1.9 miles east-northeast of Boston Post Office building, is located on a filled extension of East Boston, and is virtually surrounded by water. The Blue Hills, 8 to 10 miles south-southwest, are the highest terrain in the immediate vicinity (500-600 feet MSL). To the west are highlands 1,500 to 2,300 feet MSL, while northwest and north are the Green Mountains and the Presidential Range with many peaks above 5,000 feet MSL.

SMOKE SOURCES:

Smoke pollution may occur with winds north through west to south. The major source lies between west-northwest and west-southwest. On the average, the time of minimum visibility due to smoke varies from 1-1/4 hours after sunrise in winter to 2-1/4 hours in July. Minimum visibility in summer due to smoke is about double that in winter.

A TEMPERATURE AND WIND SPEED–RELATIVE HUMIDITY OCCURRENCES:

WIND	0-4 M.P.H.						5-14 M.P.H.						15-24 M.P.H.						25 M.P.H. AND OVER						TOTAL OBS.
REL. HUMID / TEMP. (°F)	UNDER 30	30-49%	50-69%	70-79%	80-89%	90-100%	UNDER 30	30-49%	50-69%	70-79%	80-89%	90-100%	UNDER 30	30-49%	50-69%	70-79%	80-89%	90-100%	UNDER 30	30-49%	50-69%	70-79%	80-89%	90-100%	
64/ 60								2							2	1		1				3	3	2	14
59/ 55								1	5	1	1				7	10	5	7			1	3	6	11	58
54/ 50								6	15	11	25	24		2	17	11	38	33		2	1	2	7	11	205
49/ 45			1	1		1		5	15	18	28	26		8	15	16	28	26		4	3	2	2	2	201
44/ 40		4	13	6	3	12	2	36	84	37	44	81	2	34	73	24	29	69		1	12	1	1	74	642
39/ 35		3	17	8	13	46	2	84	156	137	130	136	3	93	173	60	72	118		15	14	4	3	34	1321
34/ 30	1	12	33	20	29	28	5	140	350	134	107	191	1	133	192	58	65	101	3	24	35	4	5	35	1706
29/ 25	2	16	21	11	10	7	17	151	275	79	56	36	13	138	274	49	34	24	6	37	50	9	11	3	1329
24/ 20		8	6	4	7	1	11	119	141	44	36	6	11	95	240	29	20	5	4	22	32	8	2		851
19/ 15	1	9	7	2	1		7	94	71	20	19	2	20	116	117	22	10	1	3	48	19	6			595
14/ 10		1	4					32	25	5	20	1	9	68	69	25	4		7	13	15	2			300
09/ 05								15	20		1		2	41	47	6	2		7	1	6				148
04/ 00								8	11	1				6	7	10	2		3	3					51
-01/-05									2					1	6					2					11
-06/-10														1	1					3					5
-11/-15															3										3
TOTAL	4	53	102	52	63	95	44	693	1170	487	467	503	61	736	1243	321	309	385	33	175	188	44	40	172	7440

B PERCENTAGE FREQUENCIES OF WIND DIRECTION AND SPEED:

DIRECTION	0-3	4-7	8-12	13-18	19-24	25-31	32-38	39-46	47 OVER	TOTAL	AV SPEED
	HOURLY OBSERVATIONS OF WIND SPEED IN MILES PER HOUR										
N	+	1	2	3	1	+	+	+		7	12.3
NNE	+	+	1	1	1	+	+			3	15.8
NE	+	1	1	+	+	+	+	+	+	3	17.4
ENE	+	+	+	+	+	+	+			2	16.8
E	+	+	+	+	+	+	+			2	12.9
ESE	+	1	+	+	+					1	8.2
SE	+	1	+	1	+					2	10.0
SSE	+	1	1	1	+	+		+		2	11.5
S	+	1	1	1	+	+				3	11.3
SSW	+	1	1	2	1	+	+	+		5	14.5
SW	+	1	3	3	1	+	+			8	14.3
WSW	+	1	2	3	1	+	+			7	12.6
W	+	1	2	3	2	+	+			8	15.1
WNW	+	1	3	5	3	1	+	+		14	15.3
NW	+	1	3	7	5	2	1	+	+	20	17.6
NNW	+	1	3	5	3	+	+			13	14.8
CALM	1									1	
TOTAL	3	10	25	35	18	7	2	+	+	100	14.7

C OCCURRENCES OF PRECIPITATION AMOUNTS:

INTENSITIES	A.M. HOUR ENDING AT 1	2	3	4	5	6	7	8	9	10	11	NOON	P.M. HOUR ENDING AT 1	2	3	4	5	6	7	8	9	10	11	MID	NO OF DAYS WITH
	FREQUENCY OF OCCURRENCE FOR EACH HOUR OF THE DAY																								
TRACE	30	30	38	48	48	39	34	46	49	47	52	37	30	31	45	36	30	27	27	30	35	30	30	28	53
01 IN	6	15	15	12	12	14	15	9	14	16	16	21	14	19	10	16	16	14	14	12	5	15	12	7	14
02 TO 09 IN	28	22	21	20	19	23	26	26	24	26	23	19	29	27	25	19	18	19	22	23	22	18	19	26	41
10 TO 24 IN	1	5	6	4	6	6	5	5	2	4	4	8	5	2	1	5	3	5	4	3	4	5	3	5	16
25 TO 49 IN	2			2	1	1	1		1		1		1		1										20
50 TO 99 IN																									27
1 00 TO 1 99 IN																									9
2 00 IN AND OVER																									
TOTAL	67	72	80	86	86	83	81	86	90	93	96	85	79	79	82	76	67	65	67	68	66	68	64	66	180

D PERCENTAGE FREQUENCIES OF CEILING–VISIBILITY:

VISIBILITY (MILES)	CEILING (FEET) 0	100-200	300-400	500-900	1000-1900	2000-2900	3000-4900	5000-9500	OVER 9500	TOT
0 TO 1/8	+	+	+	+					+	1
3/16 TO 3/8	+	+	+	+						+
1/2 TO 3/4		+	1	1	+	+				2
1 TO 2 1/2		+	1	3	2	1	+	+	1	8
3 TO 6		+	1	3	3	1	2	2	8	20
7 TO 15			+	1	3	4	6	6	50	69
20 TO 30										
35 OR MORE										
TOTAL	+	1	3	8	8	6	8	9	59	100

E PERCENTAGE FREQUENCIES OF SKY COVER, WIND, AND RELATIVE HUMIDITY:

HOUR OF DAY	CLOUDS SCALE 0-10: 0-3	4-7	8-10	WIND SPEED (M.P.H.): 0-3	4-12	13-24	25-& OVER	RELATIVE HUMIDITY (%): 0-29	30-49	50-69	70-79	80-89	90-100
00	40	4	56	2	35	55	7	+	17	39	14	14	16
01	39	5	56	3	38	52	7	+	16	40	15	14	15
02	39	6	55	3	36	51	9	1	14	43	13	13	17
03	39	5	56	2	37	52	9	+	15	39	15	15	16
04	39	7	55	2	35	55	8	+	15	39	15	14	16
05	38	6	56	2	36	54	8	+	14	39	15	15	16
06	35	6	60	3	35	55	7	+	14	38	17	13	17
07	33	5	61	4	35	53	8	1	13	40	14	14	18
08	33	6	61	3	34	54	10	1	15	40	16	12	17
09	35	9	56	5	34	51	11	1	19	39	14	10	17
10	36	6	57	5	33	51	11	1	24	36	13	12	15
11	34	11	55	5	33	52	11	2	28	35	8	12	15
12	33	11	56	3	34	52	12	3	32	33	9	8	15
13	33	9	57	3	34	52	12	6	35	27	9	9	13
14	34	10	57	3	33	54	11	6	33	30	8	9	14
15	32	13	55	2	35	54	9	5	35	29	7	11	13
16	35	7	58	3	35	54	8	5	35	28	9	10	14
17	37	6	57	2	35	55	7	5	28	32	12	10	14
18	38	10	52	2	36	54	8	3	28	33	10	11	15
19	45	5	49	4	37	52	8	2	25	34	11	12	15
20	42	9	49	3	36	53	8	2	21	39	11	12	15
21	40	9	52	4	34	54	8	1	21	40	12	11	16
22	40	5	55	3	37	53	7	1	18	41	12	11	16
23	41	6	54	2	36	54	7	+	19	39	12	14	16
AVG	37	7	56	3	35	53	9	2	22	36	12	12	16

BOSTON, MASS.
Logan Int.AP

FEBRUARY
6792 Obs.

A TEMPERATURE AND WIND SPEED–RELATIVE HUMIDITY OCCURRENCES:

WIND	0-4 M.P.H.						5-14 M.P.H.						15-24 M.P.H.						25 M.P.H. AND OVER						TOTAL OBS.
REL. HUMID. / TEMP. (°F)	UNDER 30	30-49%	50-69%	70-79%	80-89%	90-100%	UNDER 30	30-49%	50-69%	70-79%	80-89%	90-100%	UNDER 30	30-49%	50-69%	70-79%	80-89%	90-100%	UNDER 30	30-49%	50-69%	70-79%	80-89%	90-100%	
69/ 65								2							5										7
64/ 60								2	2						5	1				1			2		13
59/ 55								8	9	4				2	5	4	3	9		5	1		7	4	61
54/ 50			1	2		1	1	13	14	11	9	21	3	17	8	14	26	24		1	3	4	8	19	200
49/ 45		1	6	5		2	3	30	50	27	15	31	2	37	35	10	13	19		4	8	1	2	1	302
44/ 40		1	13	4	11	12	5	57	98	100	82	104		58	105	41	31	54	1	16	26	11	3	9	842
39/ 35		8	16	13	17	19	4	81	221	142	147	183	14	118	199	68	81	137	5	37	34	5	17	34	1600
34/ 30		5	22	9	15	33	4	97	308	135	103	122	12	134	214	44	38	68	4	45	34	10	7	29	1492
29/ 25	2	4	12	14	6	6	9	70	145	100	87	49	8	133	144	32	30	20	5	58	12	9	3	19	977
24/ 20		6	7		4	1	12	52	114	23	47	6	10	77	109	11	24	8	8	40	17	1	5	2	584
19/ 15		2	7	3			10	60	42	1	11	1	10	81	69	9	10		1	25	9		1		352
14/ 10							2	26	22	4	3	2	2	62	35	10	1		4	24	4	1			202
09/ 05								11	14		4		5	37	21	6	2			6					106
04/ 00								7	4				1	15	14					1					42
-01/-05														4	8										12
TOTAL	2	27	84	50	53	74	50	516	1043	547	508	519	67	775	976	250	259	339	28	263	148	42	55	117	6792

B PERCENTAGE FREQUENCIES OF WIND DIRECTION AND SPEED:

DIRECTION	HOURLY OBSERVATIONS OF WIND SPEED IN MILES PER HOUR 0-3	4-7	8-12	13-18	19-24	25-31	32-38	39-46	47 OVER	TOTAL	AV SPEED
N	+	1	2	2	+	+		+		6	12.7
NNE	+	+	1	1	+	+	+	+		3	13.1
NE	+	1	1	1	1	+	+	+	+	4	15.0
ENE	+	1	1	+	+	+		+		2	12.9
E	+	1	1	1	+	+	+	+		4	14.2
ESE	+	1	1	1	+	+				4	12.4
SE	+	1	1	1	+	+				3	10.4
SSE	+	1	1	1	+	+				3	10.7
S	+	1	2	2	1	+	+	+		5	13.0
SSW	+	1	1	1	1	+	+	+		5	14.3
SW	+	+	2	3	1	+	+	+		7	14.3
WSW	+	+	2	3	1	+	+			7	14.6
W	+	1	2	4	2	1	+	+		9	17.3
WNW	+	1	3	5	4	2	+	+		15	17.5
NW	+	1	3	5	3	2	+	+		14	17.1
NNW	+	1	3	4	2	1	+			11	15.2
CALM	1									1	
TOTAL	2	11	25	35	17	8	2	+	+	100	15.0

C OCCURRENCES OF PRECIPITATION AMOUNTS:

INTENSITIES	FREQUENCY OF OCCURRENCE FOR EACH HOUR OF THE DAY — A M HOUR ENDING AT 1	2	3	4	5	6	7	8	9	10	11	NOON	P M HOUR ENDING AT 1	2	3	4	5	6	7	8	9	10	11	MID	NO OF DAYS WITH
TRACE	27	36	41	45	33	36	27	35	36	37	41	36	32	36	36	38	32	23	29	22	29	37	28	20	43
01 IN	16	9	13	15	19	12	17	11	16	18	14	12	12	10	8	9	11	10	6	10	11	5	8	15	15
02 TO 09 IN	10	10	15	14	14	17	22	16	17	17	13	14	12	15	11	14	14	22	16	16	12	14	15	13	43
10 TO 24 IN	9	10	6	2	3	5	4	6	4	5	6	3	6	4	4	5	6	2	4	5	4	4	4	5	15
25 TO 49 IN	1	1		1	1			2	2	2	1	1			1			1	1		1	1			20
50 TO 99 IN			1																						23
1 00 TO 1 99 IN																									5
2 00 IN AND OVER																									2
TOTAL	63	66	76	77	70	70	70	70	75	79	75	66	62	65	60	66	63	58	56	53	57	61	55	53	166

D PERCENTAGE FREQUENCIES OF CEILING–VISIBILITY:

VISIBILITY (MILES)	CEILING (FEET) 0	100-200	300-400	500-900	1000-1900	2000-2900	3000-4900	5000-9500	OVER 9500	TOT
0 TO 1/8	+	+	+	+					+	+
3/16 TO 3/8	+	+	+	+					+	1
1/2 TO 3/4		+	1	1	+	+	+	+	+	2
1 TO 2 1/2		+	1	2	1	+	+	+	1	7
3 TO 6			1	3	2	1	1	1	7	17
7 TO 15	+		+	1	3	4	6	7	53	74
20 TO 30										
35 OR MORE										
TOTAL	+	1	3	7	7	5	8	9	60	100

E PERCENTAGE FREQUENCIES OF SKY COVER, WIND, AND RELATIVE HUMIDITY:

HOUR OF DAY	CLOUDS SCALE 0-10: 0-3	4-7	8-10	WIND SPEED (M.P.H.): 0-3	4-12	13-24	25-& OVER	RELATIVE HUMIDITY (%): 0-29	30-49	50-69	70-79	80-89	90-100
00	44	6	50	3	39	52	6		18	38	13	16	16
01	41	6	53	3	39	52	7		15	39	16	16	15
02	43	5	52	2	37	53	8		15	38	14	17	16
03	44	4	53	2	39	50	8		13	39	14	17	16
04	40	8	53	2	43	48	7		13	38	15	16	18
05	42	6	51	2	42	50	6		13	36	16	17	18
06	37	9	54	2	40	51	7		13	37	14	18	18
07	32	12	56	2	39	53	7		13	37	15	17	19
08	35	10	55	3	39	48	10		18	35	13	16	18
09	36	8	56	5	31	52	12	1	23	33	15	13	16
10	33	12	55	4	33	49	14	2	27	34	11	11	15
11	31	11	58	2	34	51	14	2	33	34	8	8	15
12	31	12	57	3	31	50	16	5	36	29	9	9	13
13	33	10	57	2	33	51	14	5	36	26	12	9	12
14	36	9	55	2	29	54	14	7	36	24	13	8	11
15	31	12	56	2	30	54	15	6	36	23	13	8	13
16	37	8	55	1	29	57	13	6	35	24	14	9	12
17	39	9	53	2	35	54	9	5	30	27	12	13	13
18	42	8	50	2	39	53	6	4	25	31	13	12	15
19	41	11	48	2	37	52	8	4	25	31	13	11	16
20	41	9	51	4	37	52	7	2	23	36	11	11	17
21	40	10	50	3	35	55	7	1	23	34	14	12	15
22	37	11	52	2	33	58	7	+	20	37	14	11	17
23	42	9	49	4	39	51	6	1	20	35	12	15	16
AVG	38	9	53	2	36	52	10	2	23	33	13	13	15

BOSTON, MASS.
Logan Int.AP

MARCH
7440 Obs.

A TEMPERATURE AND WIND SPEED–RELATIVE HUMIDITY OCCURRENCES:

WIND	0-4 M.P.H.						5-14 M.P.H.						15-24 M.P.H.						25 M.P.H. AND OVER						TOTAL OBS.
R.H. HUMID. / TEMP. (°F)	UNDER 30	30-49%	50-69%	70-79%	80-89%	90-100%	UNDER 30	30-49%	50-69%	70-79%	80-89%	90-100%	UNDER 30	30-49%	50-69%	70-79%	80-89%	90-100%	UNDER 30	30-49%	50-69%	70-79%	80-89%	90-100%	
74/ 70													1												1
69/ 65								4	1				1	11	3						1				21
64/ 60							3	9	5				6	12	12					1	4				52
59/ 55		1					2	22	4	2	5	1	14	16	6	3	19	11	1	1	3		3	3	117
54/ 50		1	1			4	10	31	28	12	20	20	27	53	10	9	12	16	5	5	1	2	2	5	274
49/ 45	1		1	2	1	3	25	91	81	40	30	45	31	111	50	27	20	54	6	14	7	1	3	15	659
44/ 40	1	1	8	18	17	12	16	120	242	170	131	132	35	146	168	82	58	73	7	25	21	10	2	14	1509
39/ 35	1	6	12	11	35	29	19	102	292	165	230	230	24	145	259	111	134	165	7	48	23	7	19	53	2127
34/ 30		5	9	7	4	17	10	86	228	88	67	98	25	199	195	56	49	111	7	32	27	4	6	64	1394
29/ 25		8	9	1	2	3	7	77	142	19	20	13	12	160	128	27	23	21	5	44	24	7	12	44	808
24/ 20			1				5	33	39	5	11	6	11	97	54	6	17	1	4	11	3		3		307
19/ 15	1	1					4	15	9				6	55	22		10			6		3	4		136
14/ 10								12	1					16	5					1					35
TOTAL	4	23	41	39	59	68	101	602	1072	501	514	545	193	1021	912	321	342	452	42	188	114	34	54	198	7440

B PERCENTAGE FREQUENCIES OF WIND DIRECTION AND SPEED:

DIRECTION	HOURLY OBSERVATIONS OF WIND SPEED IN MILES PER HOUR 0-3	4-7	8-12	13-18	19-24	25-31	32-38	39-46	47 OVER	TOTAL	AV SPEED
N	+	1	2	2	1	+	+	+		6	14.3
NNE	+	+	1	1	1	+	+	+	+	4	14.8
NE	+	1	1	2	1	+	+	+	+	5	16.8
ENE	+	+	1	1	1	1	+	+		5	17.6
E	+	+	1	2	1	+	+	+		6	16.0
ESE	+	1	2	2	1	+	+			5	14.1
SE	+	1	2	1	+	+				5	12.2
SSE	+	1	1	1	+	+	+			3	11.9
S	+	1	1	1	+	+				3	10.8
SSW	+	1	1	1	+	+	+			3	13.7
SW	+	1	2	2	1	+				6	13.6
WSW	+	+	1	2	1	+	+			5	15.2
W	+	+	1	4	2	1	+	+		8	17.9
WNW	+	+	2	5	3	1	+			11	17.1
NW	+	+	3	6	3	1	+	+		14	17.2
NNW	+	1	3	5	2	+	+			10	14.6
CALM	+									+	
TOTAL	2	8	25	38	19	6	2	+	+	100	15.4

C OCCURRENCES OF PRECIPITATION AMOUNTS:

INTENSITIES	FREQUENCY OF OCCURRENCE FOR EACH HOUR OF THE DAY — A.M. HOUR ENDING AT 1	2	3	4	5	6	7	8	9	10	11	NOON	P.M. HOUR ENDING AT 1	2	3	4	5	6	7	8	9	10	11	MID	NO OF DAYS WITH
TRACE	37	40	41	36	39	41	35	38	36	42	46	38	42	38	39	37	35	40	34	28	39	37	33	42	60
.01 IN	10	15	11	14	12	9	12	14	14	10	10	13	6	11	11	20	14	13	12	16	13	20	20	17	15
.02 TO .09 IN	15	12	16	17	20	20	24	19	24	22	20	21	22	22	22	14	23	25	27	23	17	14	14	13	29
.10 TO .24 IN	4	5	4	6	7	6	6	9	5	6	6	6	8	8	7	12	11	8	6	3	3	2	6	5	21
.25 TO .49 IN	1	2	3	2	1	1	2									1				1	1				16
.50 TO .99 IN																									28
1.00 TO 1.99 IN																									13
2.00 IN AND OVER																									1
TOTAL	67	74	75	75	79	77	79	80	79	80	82	78	78	79	79	84	83	86	79	71	73	73	73	77	183

D PERCENTAGE FREQUENCIES OF CEILING–VISIBILITY:

VISIBILITY (MILES)	CEILING (FEET) 0	100-200	300-400	500-900	1000-1900	2000-2900	3000-4900	5000-9500	OVER 9500	TOT.
0 TO 1/8	+	+		+	+				+	+
3/16 TO 3/8	+	+	+	+	+	+				1
1/2 TO 3/4		+	+	1	+	+			+	2
1 TO 2 1/2		+	1	2	1	+	+	+	+	7
3 TO 6		+	1	4	2	1	1	1	3	13
7 TO 15			+	1	4	4	7	10	51	78
20 TO 30										
35 OR MORE										
TOTAL	+	1	3	9	7	4	9	12	55	100

E PERCENTAGE FREQUENCIES OF SKY COVER, WIND, AND RELATIVE HUMIDITY:

HOUR OF DAY	CLOUDS SCALE 0-10: 0-3	4-7	8-10	WIND SPEED (M.P.H.): 0-3	4-12	13-24	25-& OVER	RELATIVE HUMIDITY (%): 0-29	30-49	50-69	70-79	80-89	90-100
00	40	5	55	3	37	55	5		20	35	12	17	15
01	40	6	54	4	38	54	5	1	17	38	11	14	19
02	39	6	55	4	36	55	5	1	16	37	12	14	20
03	39	6	55	2	41	53	5	1	13	38	13	16	19
04	40	5	55	3	41	51	6		15	36	12	16	20
05	39	7	54	2	40	52	6	+	16	32	16	16	20
06	34	10	56	1	42	52	5	+	14	34	16	15	20
07	36	7	57	1	39	55	6	1	19	31	13	16	20
08	36	8	56	2	31	56	11	1	25	32	11	13	18
09	35	5	59	2	27	62	10	3	28	31	11	11	16
10	33	9	58	2	28	59	11	4	37	24	9	12	14
11	29	13	58	2	23	64	11	7	34	27	9	8	14
12	28	13	59	1	25	61	13	9	35	24	10	9	13
13	28	13	59	+	22	64	14	13	33	23	9	9	13
14	27	14	59	+	21	65	13	14	33	21	9	10	14
15	26	15	59	+	20	68	12	15	30	22	10	10	13
16	29	14	57	+	25	62	13	14	27	23	11	11	15
17	30	14	56		29	60	11	10	30	21	12	12	15
18	36	10	55	1	36	56	7	7	27	21	16	10	19
19	36	10	55	1	36	56	7	4	27	24	14	14	17
20	36	9	55	+	40	52	7	3	26	25	13	15	18
21	38	6	55	2	40	51	7	1	25	29	11	14	19
22	38	6	56	2	38	54	6	1	23	30	15	14	18
23	38	8	54	3	38	53	6	1	20	35	13	15	17
AVG	35	9	56	2	33	57	8	5	25	29	12	13	17

BOSTON, MASS.
Logan Int.AP

APRIL
7200 Obs.

A TEMPERATURE AND WIND SPEED—RELATIVE HUMIDITY OCCURRENCES:

WIND	0-4 M.P.H.						5-14 M.P.H.						15-24 M.P.H.						25 M.P.H. AND OVER						TOTAL OBS.
REL. HUMID. / TEMP. (°F)	UNDER 30	30-49%	50-69%	70-79%	80-89%	90-100%	UNDER 30	30-49%	50-69%	70-79%	80-89%	90-100%	UNDER 30	30-49%	50-69%	70-79%	80-89%	90-100%	UNDER 30	30-49%	50-69%	70-79%	80-89%	90-100%	
89/ 85													3						1						4
84/ 80								2					4	11					2	3					22
79/ 75							3	2					10	17	3				7	6					48
74/ 70							6	6	6				11	32	13				6	3	2				85
69/ 65							17	31	29	3	1		20	42	30	6			2	11	6				198
64/ 60		2	2				18	65	44	25	22	7	49	71	37	18	17	19	13	13	7		5	1	435
59/ 55	1	4		9	4	3	43	124	88	32	35	40	49	122	82	15	35	20	11	21	4	3	5	13	763
54/ 50		10	25	3	16	13	27	147	246	95	97	73	37	162	132	31	35	20	13	28	11	3	2	1	1227
49/ 45		8	19	9	32	48	14	144	270	139	200	245	25	134	183	83	72	84	9	23	32	3	10	27	1813
44/ 40		6	12	15	12	27	3	88	199	158	224	250	13	76	188	104	119	156	5	14	15	9	14	59	1766
39/ 35		1	7	7	2		4	26	83	56	39	52	1	57	94	25	32	51		8	15	10	10	50	630
34/ 30		2	1			3		19	19	5	7	11	1	22	18	9	7	17		3	2		5	27	178
29/ 25								1	4					7						5					17
24/ 20								6					1	5											12
19/ 15															2										2
TOTAL	1	33	66	43	66	94	135	661	988	513	625	678	224	758	782	291	317	367	69	138	94	28	51	178	7200

B PERCENTAGE FREQUENCIES OF WIND DIRECTION AND SPEED:

DIRECTION	HOURLY OBSERVATIONS OF WIND SPEED (IN MILES PER HOUR) 0-3	4-7	8-12	13-18	19-24	25-31	32-38	39-46	47 OVER	TOTAL	AV SPEED
N	+	1	1	1	+	+	+			4	13.1
NNE	+	+	1	1	1	1	+	+		4	16.3
NE	+	1	1	2	1	1	+	+		6	16.5
ENE	+	1	1	2	1	1	+	+	+	6	15.4
E	+	1	2	2	1	+	+			6	12.4
ESE	+	1	2	3	1	+				7	13.3
SE	+	1	2	2	+	+				5	12.0
SSE	+	1	1	1	+	+				3	10.2
S	+	1	2	1	+	+	+			5	12.0
SSW	+	1	2	2	2	1	+			8	15.2
SW	+	1	3	4	2	1	+			10	14.9
WSW	+	1	2	3	1	+		+		6	14.4
W	+	+	2	3	1	1	+			6	16.0
WNW	+	1	2	4	2	1	+	+		9	16.0
NW	+	+	2	3	2	1	+			8	16.1
NNW	+	1	2	2	1	+	+			6	13.4
CALM	+									+	
TOTAL	2	10	28	37	15	6	1	+	+	100	14.5

C OCCURRENCES OF PRECIPITATION AMOUNTS:

INTENSITIES	FREQUENCY OF OCCURRENCE FOR EACH HOUR OF THE DAY — A.M. HOUR ENDING AT 1	2	3	4	5	6	7	8	9	10	11	NOON	P.M. HOUR ENDING AT 1	2	3	4	5	6	7	8	9	10	11	MID	NO OF DAYS WITH
TRACE	39	43	28	34	36	49	39	44	50	42	47	48	45	52	42	37	28	28	30	37	32	37	35	35	55
.01 IN	8	14	19	12	21	10	9	10	11	8	11	9	14	8	13	11	10	13	19	10	18	18	8	9	7
.02 TO .09 IN	23	22	18	25	12	15	25	25	14	17	12	15	17	15	14	18	25	23	20	26	18	16	24	19	35
.10 TO .24 IN	6	7	8	4	5	5	5	2	10	3	3	4	4	3	6	8	3	4	6	4	7	3	4	8	25
.25 TO .49 IN	2		1		1		1				1		1		1	1		1	1	2		1	2	1	30
.50 TO .99 IN																					1				19
1.00 TO 1.99 IN																									9
2.00 IN AND OVER																									1
TOTAL	78	86	74	75	75	79	79	81	85	70	74	76	81	78	76	75	66	69	76	79	76	75	73	72	181

D PERCENTAGE FREQUENCIES OF CEILING—VISIBILITY:

VISIBILITY (MILES)	CEILING (FEET) 0	100-200	300-400	500-900	1000-1900	2000-2900	3000-4900	5000-9500	OVER 9500	TOT.
0 TO 1/8	+	+	+							+
3/16 TO 3/8		+	+		+					+
1/2 TO 3/4		1	1	+	+	+	+		+	1
1 TO 2 1/2		+	1	2	1	+	+	+	+	5
3 TO 6		+	1	4	2	1	1	1	4	14
7 TO 15			+	3	5	2	4	12	53	79
20 TO 30									+	+
35 OR MORE										
TOTAL	+	1	3	9	8	3	5	14	57	100

E PERCENTAGE FREQUENCIES OF SKY COVER, WIND, AND RELATIVE HUMIDITY:

HOUR OF DAY	CLOUDS SCALE 0-10: 0-3	4-7	8-10	WIND SPEED (M.P.H.): 0-3	4-12	13-24	25-& OVER	RELATIVE HUMIDITY (%): 0-29	30-49	50-69	70-79	80-89	90-100
00	34	7	58	3	49	44	3		15	31	14	18	21
01	35	6	59	3	50	43	4		13	30	15	18	23
02	36	8	56	2	52	43	3		12	31	16	17	25
03	37	10	53	3	52	40	5		8	31	17	18	26
04	33	9	58	3	51	42	4		9	28	17	20	26
05	32	8	60	3	48	44	5		7	30	18	21	25
06	32	9	59	3	45	46	6		9	32	19	16	25
07	32	8	60	4	40	50	6		16	34	17	13	21
08	32	9	59	6	34	54	6	1	27	31	13	12	17
09	31	8	61	2	38	50	10	4	37	22	8	12	17
10	29	11	60	1	33	56	9	9	36	22	9	9	15
11	28	13	59	2	28	60	11	11	36	21	8	11	13
12	29	13	58	1	24	61	13	16	29	24	8	12	11
13	27	15	58	+	23	64	12	19	29	22	8	12	10
14	25	17	58	+	20	65	15	18	30	22	6	14	11
15	25	16	59	1	23	66	11	16	33	18	10	14	10
16	26	13	61	1	23	63	13	16	30	20	11	12	12
17	29	13	58	1	28	59	13	13	27	24	10	12	13
18	26	15	59	1	33	56	10	9	27	25	11	10	18
19	29	15	56	3	38	52	7	6	23	28	11	12	20
20	35	7	58	3	41	50	6	3	19	31	10	14	21
21	34	6	60	3	47	44	6	1	20	31	11	16	21
22	35	8	56	4	46	45	4	1	19	29	13	19	20
23	33	10	57	3	49	43	5	1	18	27	14	19	21
AVG	31	11	58	2	38	52	8	6	22	27	12	15	18

BOSTON, MASS.
Logan Int.AP

MAY
7440 Obs.

A TEMPERATURE AND WIND SPEED–RELATIVE HUMIDITY OCCURRENCES:

WIND	0-4 M.P.H.						5-14 M.P.H.						15-24 M.P.H.						25 M.P.H. AND OVER						TOTAL OBS.
REL. HUMID. / TEMP. (°F)	UNDER 30	30-49%	50-69%	70-79%	80-89%	90-100%	UNDER 30	30-49%	50-69%	70-79%	80-89%	90-100%	UNDER 30	30-49%	50-69%	70-79%	80-89%	90-100%	UNDER 30	30-49%	50-69%	70-79%	80-89%	90-100%	
94/ 90							1	3					1	6											11
89/ 85		1					8	21	4				7	15	5				6	2					69
84/ 80			1				15	14	14				27	26	18				2	3	3				123
79/ 75		1		2			18	38	30	8			33	58	44				7	11	2				252
74/ 70		2	4	4	3	3	33	85	57	45	15	1	51	85	41	22	5		9	8	3				476
69/ 65	1	3	5	10	7	2	30	103	163	44	78	22	33	142	59	18	18	14	9	12	5		3		781
64/ 60		5	11	19	7	12	28	131	191	143	104	81	37	150	116	34	19	9	2	18	14	1	2		1134
59/ 55	1	5	13	22	37	49	23	131	280	136	241	169	26	136	198	69	49	26	8	18	8	1	6	4	1656
54/ 50			18	23	19	41	10	82	263	141	196	295	18	78	168	85	46	78	3	11	9	5	7	18	1614
49/ 45		1	12	6	10	16	2	48	165	94	133	163	5	28	87	32	78	91	1	6	3	1	7	19	1008
44/ 40			2	2	2	6		11	41	11	23	35	1	15	48	5	20	58		5	4		5	10	304
39/ 35														5	3						2				10
34/ 30														1	1										2
TOTAL	2	18	66	88	85	129	168	667	1208	622	790	766	239	745	788	265	235	276	47	94	53	8	30	51	7440

B PERCENTAGE FREQUENCIES OF WIND DIRECTION AND SPEED:

DIRECTION	HOURLY OBSERVATIONS OF WIND SPEED (IN MILES PER HOUR)										AV SPEED
	0 - 3	4 - 7	8 - 12	13 - 18	19 - 24	25 - 31	32 - 38	39 - 46	47 OVER	TOTAL	
N	+	1	1	1	+	+	+			4	12.0
NNE	+	+	1	1	+	+	+			3	13.4
NE	+	1	2	1	1	+	+			5	13.7
ENE	+	1	2	3	1	+	+			6	13.4
E	+	1	2	3	1	+				6	12.8
ESE	+	1	2	3	1	+				7	14.3
SE	+	1	2	2	+	+				6	11.1
SSE	+	1	1	1	+					3	9.3
S	+	1	2	1	1	+				5	11.5
SSW	+	1	3	3	1	+	+	+		9	13.8
SW	+	1	4	5	2	+	+			12	14.1
WSW	+	1	3	2	1	+				6	12.6
W	+	1	2	2	1	+	+			6	13.6
WNW	+	1	3	3	1	+				8	13.9
NW	+	1	2	3	1	1	+			8	15.3
NNW	+	+	2	2	1	+	+	+		6	14.4
CALM	+									+	
TOTAL	3	12	32	37	12	3	+	+		100	13.3

C OCCURRENCES OF PRECIPITATION AMOUNTS:

INTENSITIES	FREQUENCY OF OCCURRENCE FOR EACH HOUR OF THE DAY																								NO OF DAYS WITH
	A.M. HOUR ENDING AT												P.M. HOUR ENDING AT												
	1	2	3	4	5	6	7	8	9	10	11	NOON	1	2	3	4	5	6	7	8	9	10	11	MID	
TRACE	32	37	38	41	46	41	40	35	32	32	32	29	31	38	45	44	33	44	45	35	42	40	40	43	48
01 IN	11	12	11	12	12	10	11	12	9	9	11	9	6	14	8	6	10	7	9	17	8	8	8	10	11
02 TO 09 IN	13	11	10	12	7	15	13	13	17	15	16	19	16	11	8	15	11	23	12	12	16	14	15	12	43
10 TO 24 IN	6	8	7	4	3	4	4	5	4	3	2	1	2	1	8	5	7	3	4	2	3	6	6	4	29
25 TO 49 IN	3	1	2		2	1	5	2			1							1		2	3		1	2	21
50 TO 99 IN	1	1		1	1	1																			18
1 00 TO 1 99 IN																									7
2 00 IN AND OVER																									2
TOTAL	66	70	68	70	71	72	73	67	62	59	62	58	55	64	69	70	61	78	70	68	72	68	70	71	179

D PERCENTAGE FREQUENCIES OF CEILING–VISIBILITY:

VISIBILITY (MILES)	CEILING (FEET)									
	0	100-200	300-400	500-900	1000-1900	2000-2900	3000-4900	5000-9500	OVER 9500	TOT.
0 TO 1/8		+								+
3/16 TO 3/8		1	+		+	+			+	1
1/2 TO 3/4		1	+	+		+		+	+	1
1 TO 21/2		+	1	1	+	+	+	+	1	4
3 TO 6		+	1	4	2	+	1	1	5	14
7 TO 15			+	2	4	1	4	11	57	79
20 TO 30										
35 OR MORE										
TOTAL		2	3	7	6	2	4	12	63	100

E PERCENTAGE FREQUENCIES OF SKY COVER, WIND, AND RELATIVE HUMIDITY:

HOUR OF DAY	CLOUDS SCALE 0-10			WIND SPEED (M. P. H.)				RELATIVE HUMIDITY (%)					
	0-3	4-7	8-10	0-3	4-12	13-24	25-& OVER	0-29	30-49	50-69	70-79	80-89	90-100
00	40	10	50	6	53	40	1	+	13	30	15	19	23
01	38	13	49	4	59	35	1		12	30	15	21	23
02	41	12	47	6	56	37	1		9	30	15	23	24
03	36	11	53	8	58	33	2		7	30	17	20	25
04	32	12	56	6	61	32	1		5	31	18	21	26
05	28	14	57	7	58	33	3		6	29	21	20	25
06	34	9	57	6	54	38	3		12	32	17	19	20
07	31	13	56	4	50	43	4	1	19	32	17	14	17
08	35	11	54	2	47	47	4	3	29	30	14	11	14
09	33	14	54	1	44	50	5	5	34	29	9	10	12
10	33	14	53	1	41	53	5	10	34	25	11	9	10
11	32	13	55	1	35	58	5	13	33	26	9	8	11
12	29	16	55	1	31	62	6	16	31	25	11	9	9
13	25	22	54	1	26	67	7	18	29	25	9	11	8
14	26	18	55		27	65	8	17	29	24	13	9	9
15	25	19	57	+	25	69	6	15	29	25	10	10	10
16	26	15	60		27	67	6	16	29	24	8	13	10
17	27	14	59	1	31	61	7	13	26	25	11	14	11
18	28	14	58	2	37	57	5	10	26	26	12	12	15
19	30	11	59	3	42	51	4	4	23	29	12	16	15
20	32	13	55	5	46	45	4	3	17	32	14	17	18
21	37	11	52	5	51	42	3	1	15	33	14	18	20
22	36	11	53	4	50	44	2	+	14	33	13	22	19
23	39	9	52	6	52	40	2	+	12	32	13	22	21
AVG	32	13	55	3	44	49	4	6	20	28	13	15	16

BOSTON, MASS.
Logan Int.AP

JUNE
7200 Obs.

A TEMPERATURE AND WIND SPEED—RELATIVE HUMIDITY OCCURRENCES:

WIND	0-4 M.P.H.						5-14 M.P.H.						15-24 M.P.H.						25 M.P.H. AND OVER						TOTAL OBS.
REL. HUMID. / TEMP. (°F)	UNDER 30%	30-49%	50-69%	70-79%	80-89%	90-100%	UNDER 30%	30-49%	50-69%	70-79%	80-89%	90-100%	UNDER 30%	30-49%	50-69%	70-79%	80-89%	90-100%	UNDER 30%	30-49%	50-69%	70-79%	80-89%	90-100%	
104/100														1											1
99/ 95								8	1					12											21
94/ 90							12	22	1				5	42	4										86
89/ 85							7	77	27				5	100	33					3	1				253
84/ 80	2	3	3				30	109	118	9	1		14	97	70	7				14	7				484
79/ 75		5	7	5	1		10	138	204	64	14		17	92	101	23	9		2	7	5	2	2		708
74/ 70		7	13	13	9	1	19	131	298	134	145	32	31	98	126	72	44	7	6	8	6			1	1201
69/ 65		2	22	29	18	17	2	77	380	199	198	124	11	77	135	40	53	16	1	14	1	4	1		1421
64/ 60		3	26	18	25	49	2	64	265	188	309	209	2	58	153	33	49	15		2	3		1		1474
59/ 55		1	13	10	28	64		23	141	173	166	209		19	75	53	31	41			6	1			1054
54/ 50			3	3	8	14		4	42	53	59	110		3	16	40	30	58				2		1	446
49/ 45								1	3		7	12				1	10	17							51
TOTAL	2	21	87	78	89	145	82	654	1480	820	899	696	85	599	713	269	226	154	9	48	29	9	4	2	7200

B PERCENTAGE FREQUENCIES OF WIND DIRECTION AND SPEED:

DIRECTION	HOURLY OBSERVATIONS OF WIND SPEED (IN MILES PER HOUR)										AV SPEED
	0-3	4-7	8-12	13-18	19-24	25-31	32-38	39-46	47 OVER	TOTAL	
N	+	1	2	1	+					4	9.9
NNE	+	+	1	1	+					3	11.1
NE	+	1	2	2	1	+				5	13.1
ENE	+	1	2	2	1	+				5	12.2
E	+	1	3	3	+					7	11.7
ESE	+	1	3	3	+					7	11.2
SE	+	1	2	1	+					5	10.3
SSE	+	1	1	+	+					3	9.0
S	+	1	2	1	+	+				5	11.1
SSW	+	1	3	3	1	+	+			8	13.1
SW	+	1	6	7	2	+				16	13.4
WSW	+	1	3	3	+					7	11.7
W	+	1	2	3	1	+				7	12.7
WNW	+	1	3	3	1	+				8	13.1
NW	+	1	2	2	1	+	+			6	13.1
NNW	+	+	1	2	+	+				4	12.7
CALM	1									1	
TOTAL	3	14	37	36	8	1	+			100	12.1

C OCCURRENCES OF PRECIPITATION AMOUNTS:

INTENSITIES	FREQUENCY OF OCCURRENCE FOR EACH HOUR OF THE DAY																								NO. OF DAYS WITH
	A.M. HOUR ENDING AT												P.M. HOUR ENDING AT												
	1	2	3	4	5	6	7	8	9	10	11	NOON	1	2	3	4	5	6	7	8	9	10	11	MID.	
TRACE	23	26	32	25	30	32	38	34	28	30	30	25	24	21	17	18	23	30	27	21	23	17	25	28	58
.01 IN	8	3	2	7	9	9	6	6	4	3	5	4	7	6	11	9	12	8	7	8	5	6	9	9	11
.02 TO .09 IN	7	8	6	10	10	10	9	4	7	9	4	7	4	6	10	15	14	13	11	11	10	11	9	10	36
.10 TO .24 IN	1	2	6	4	3	5	1	3	4	2	3		4	4	2	2	3	2	2	1	1	3	3	2	14
.25 TO .49 IN	1	1			1		3	4	1					1	1	4		1	1			1			22
.50 TO .99 IN									1	1					1		1	1			1				13
1.00 TO 1.99 IN													1					1							6
2.00 IN AND OVER																									1
TOTAL	40	40	46	46	53	56	57	51	45	45	42	36	40	38	42	48	53	56	48	41	40	38	46	49	161

D PERCENTAGE FREQUENCIES OF CEILING—VISIBILITY:

VISIBILITY (MILES)	CEILING (FEET)									
	0	100-200	300-400	500-900	1000-1900	2000-2900	3000-4900	5000-9500	OVER 9500	TOT.
0 TO 1/8	+	+	+							+
3/16 TO 3/8	+	+	+	+					+	+
1/2 TO 3/4		+	+	+	+				+	1
1 TO 2 1/2		+	1	1	1	+	+	+	1	5
3 TO 6		+	+	3	2	+	1	1	8	16
7 TO 15	+		+	2	3	2	4	8	59	77
20 TO 30										
35 OR MORE									+	+
TOTAL	+	1	2	6	5	3	5	9	68	100

E PERCENTAGE FREQUENCIES OF SKY COVER, WIND, AND RELATIVE HUMIDITY:

HOUR OF DAY	CLOUDS SCALE 0-10			WIND SPEED (M.P.H.)				RELATIVE HUMIDITY (%)					
	0-3	4-7	8-10	0-3	4-12	13-24	25- & OVER	0-29	30-49	50-69	70-79	80-89	90-100
00	37	11	52	5	62	33	+		6	30	23	19	22
01	39	11	50	5	63	32			4	30	21	23	21
02	39	8	53	5	64	31			3	28	20	26	22
03	35	10	54	6	69	26			3	25	20	30	22
04	30	12	58	6	68	26			2	23	22	30	23
05	33	11	56	7	68	26			2	25	23	29	21
06	35	10	55	6	64	29	1		4	34	21	23	17
07	35	10	55	7	56	36	1		10	37	21	16	15
08	35	12	53	4	56	39	1		20	38	16	13	13
09	34	15	50	3	52	42	3	1	28	38	10	13	10
10	33	13	54	2	46	50	2	2	33	34	12	10	8
11	28	20	52	1	40	55	4	4	37	31	13	10	6
12	26	21	53	1	39	57	3	7	36	30	14	8	6
13	27	18	54	1	30	67	2	8	33	31	15	10	4
14	26	21	54	+	29	67	3	7	37	29	12	11	5
15	25	20	55	2	30	65	3	8	33	30	12	10	6
16	28	17	55	+	32	63	4	8	31	31	10	12	8
17	30	19	51	2	36	61	2	6	29	31	12	11	10
18	30	14	56	+	46	52	1	5	24	35	13	13	11
19	29	16	55	3	51	45	1	3	20	36	15	14	12
20	32	16	52	2	56	42	+	1	17	35	16	17	15
21	35	17	48	5	56	39		+	12	37	15	20	16
22	35	15	50	3	59	37	1		9	36	17	19	19
23	37	11	51	4	60	36			8	34	17	18	22
AVG	32	15	53	3	51	44	1	2	18	32	16	17	14

BOSTON, MASS.
Logan Int.AP

JULY
7440 Obs.

A TEMPERATURE AND WIND SPEED–RELATIVE HUMIDITY OCCURRENCES:

WIND	0-4 M.P.H.						5-14 M.P.H.						15-24 M.P.H.						25 M.P.H. AND OVER						TOTAL OBS.
REL. HUMID. / TEMP. (°F)	UNDER 30	30-49%	50-69%	70-79%	80-89%	90-100%	UNDER 30	30-49%	50-69%	70-79%	80-89%	90-100%	UNDER 30	30-49%	50-69%	70-79%	80-89%	90-100%	UNDER 30	30-49%	50-69%	70-79%	80-89%	90-100%	
99/ 95							2	10					3	12						1					28
94/ 90							6	66	11				4	68	19					7					181
89/ 85		2	2				12	140	102	1			12	108	66	2			3	14	2				466
84/ 80		4	9	4			5	239	269	39	4		12	145	143	21				4	3				901
79/ 75		11	20	16	4		1	217	466	257	116	9	5	125	160	67	11	6		7	1	1			1500
74/ 70		4	34	30	39	16		144	467	340	325	164	1	69	93	72	45	13		1	3	2	1		1863
69/ 65			25	27	42	112		25	377	227	357	281		12	69	31	48	37			1		1	1	1673
64/ 60			6	10	13	51			140	97	126	229		1	18	8	7	28					1	2	737
59/ 55				1		5			11	10	11	39				1		12						1	91
TOTAL		21	96	88	98	184	26	841	1843	971	939	722	37	540	568	202	111	96	3	34	10	3	3	4	7440

B PERCENTAGE FREQUENCIES OF WIND DIRECTION AND SPEED:

DIRECTION	HOURLY OBSERVATIONS OF WIND SPEED (IN MILES PER HOUR) 0-3	4-7	8-12	13-18	19-24	25-31	32-38	39-46	47 OVER	TOTAL	AV SPEED
N	+	1	1	1	+	+				3	10.1
NNE	+	+	1	+	+	+				2	10.4
NE	+	1	1	1	+					3	10.2
ENE	+	1	1	1	+					3	10.5
E	+	1	2	2	+	+				5	11.0
ESE	+	1	3	2	+	+				7	10.6
SE	+	1	2	1	+					4	10.0
SSE	+	1	2	+	+					3	9.0
S	+	1	2	1	+					5	9.5
SSW	+	2	3	3	1	+				9	11.8
SW	+	2	8	7	1	+	+			18	12.1
WSW	+	1	4	3	1	+				9	11.8
W	+	1	3	3	1	+	+			8	11.9
WNW	+	1	4	3	1	+	+			9	11.7
NW	+	1	3	2	+	+				7	11.8
NNW	+	1	2	1	+	+				4	11.2
CALM	1									1	
TOTAL	4	17	42	32	5	1	+			100	11.2

C OCCURRENCES OF PRECIPITATION AMOUNTS:

INTENSITIES	FREQUENCY OF OCCURRENCE FOR EACH HOUR OF THE DAY — A.M. HOUR ENDING AT 1	2	3	4	5	6	7	8	9	10	11	NOON	P.M. HOUR ENDING AT 1	2	3	4	5	6	7	8	9	10	11	MID	NO OF DAYS WITH
TRACE	21	22	23	28	25	28	27	31	25	25	19	26	26	25	16	30	30	26	33	22	28	21	19	20	48
.01 IN	4	3	12	5	7	4	8	6	3	2	5	2	6	4	6	4	7	7	4	8	4	6	4	6	12
.02 TO .09 IN	11	10	6	7	8	5	4	6	8	5	4	6		4	6	10	5	9	5	11	8	9	12	10	34
.10 TO .24 IN	2	6	4	1	2	2	2	1	3	3	3	2	2	4	4	2	5	1	4	3	3	3	2	3	24
.25 TO .49 IN				3		3	2	1								1	2	2				1	1		14
.50 TO .99 IN				1	1	1	2								2			1	1				1		8
1.00 TO 1.99 IN																1			1	1					9
2.00 IN AND OVER																									3
TOTAL	38	41	45	45	43	43	45	45	39	35	31	36	34	37	34	48	49	46	48	45	43	40	39	39	152

D PERCENTAGE FREQUENCIES OF CEILING–VISIBILITY:

VISIBILITY (MILES)	CEILING (FEET) 0	100-200	300-400	500-900	1000-1900	2000-2900	3000-4900	5000-9500	OVER 9500	TOT.
0 TO 1/8	+	+								+
3/16 TO 3/8		+	+				+			+
1/2 TO 3/4		+	+	+	+		+		+	1
1 TO 2 1/2		+	1	1	1	+	+	+	1	5
3 TO 6			+	2	1	+	1	2	11	17
7 TO 15			+	1	2	1	2	7	64	76
20 TO 30										
35 OR MORE										
TOTAL	+	1	2	4	4	2	3	9	76	100

E. PERCENTAGE FREQUENCIES OF SKY COVER, WIND, AND RELATIVE HUMIDITY:

HOUR OF DAY	CLOUDS SCALE 0-10: 0-3	4-7	8-10	WIND SPEED (M.P.H.): 0-3	4-12	13-24	25- & OVER	RELATIVE HUMIDITY (%): 0-29	30-49	50-69	70-79	80-89	90-100
00	44	12	45	5	71	24	+		1	36	23	20	20
01	44	12	44	5	71	25			1	33	22	23	22
02	44	12	44	6	74	20			+	27	25	21	26
03	40	13	46	5	76	18			+	24	24	24	27
04	36	15	48	6	78	16				20	28	23	29
05	35	14	51	9	77	14			+	21	31	22	26
06	38	13	49	6	74	21			1	33	26	19	22
07	36	16	47	6	66	28	+		5	46	15	20	14
08	38	16	45	7	59	34			13	48	16	14	9
09	38	16	45	5	57	37	1		28	41	14	12	6
10	37	17	46	4	55	41	1	+	38	36	11	10	5
11	31	22	47	2	54	44	+	1	44	32	11	7	5
12	27	26	47	1	42	55	2	2	45	32	10	9	3
13	25	24	51	1	35	60	4	3	45	32	9	7	4
14	26	25	49	1	34	63	3	5	44	29	11	8	4
15	25	27	48	+	36	61	2	4	42	32	11	7	5
16	27	24	49	+	44	54	3	3	39	33	12	7	6
17	33	19	48	1	46	51	1	1	36	34	13	9	7
18	33	19	49	2	49	49		1	29	35	15	12	8
19	36	15	49	3	60	37	+	1	23	36	14	16	11
20	39	16	45	5	62	33		+	14	37	17	18	14
21	40	20	40	5	62	33			10	38	15	21	15
22	46	15	39	5	63	32	+		5	39	18	21	17
23	46	13	41	5	66	28			1	41	18	22	19
AVG	36	18	46	4	59	36	1	1	19	34	17	15	14

BOSTON, MASS.
Logan Int.AP

AUGUST
7440 Obs.

A TEMPERATURE AND WIND SPEED–RELATIVE HUMIDITY OCCURRENCES:

WIND	0-4 M.P.H.						5-14 M.P.H.						15-24 M.P.H.						25 M.P.H. AND OVER						TOTAL OBS.
REL. HUMID / TEMP (°F)	UNDER 30	30-49%	50-69%	70-79%	80-89%	90-100%	UNDER 30	30-49%	50-69%	70-79%	80-89%	90-100%	UNDER 30	30-49%	50-69%	70-79%	80-89%	90-100%	UNDER 30	30-49%	50-69%	70-79%	80-89%	90-100%	
104/100								1																	1
99/ 95							4	15	1				8	8						1					37
94/ 90			1				1	27	15				3	36	19					1	2				105
89/ 85		1	6					104	79				3	67	84	1					3				348
84/ 80		6	10				1	148	222	51	8		1	75	108	14				2	2	1			649
79/ 75		3	6	7	8		13	182	351	185	115	31	7	64	94	39	26	9	1	2	4	1	3		1151
74/ 70		4	19	18	27	19	1	149	497	294	277	178		59	106	45	51	25		2	2		1	6	1780
69/ 65		2	13	28	30	76		64	402	318	354	460		27	95	43	21	34				2	2	8	1979
64/ 60			13	28	57	26		13	204	190	264	262		6	37	15	10	39		1				5	1170
59/ 55				4	8	10			72	33	34	17			19	4	2	3							206
54/ 50					1				5	6					2										14
TOTAL		16	68	85	131	131	20	703	1848	1077	1052	948	22	342	564	161	110	110	1	9	13	4	6	19	7440

B PERCENTAGE FREQUENCIES OF WIND DIRECTION AND SPEED:

DIRECTION	HOURLY OBSERVATIONS OF WIND SPEED (IN MILES PER HOUR) 0-3	4-7	8-12	13-18	19-24	25-31	32-38	39-46	47 OVER	TOTAL	AV SPEED
N	+	1	3	1	+					5	9.9
NNE	+	1	1	+	+	+	+			3	10.7
NE	+	1	2	1	+	+				4	11.0
ENE	+	1	2	1	+	+	+	+		4	11.2
E	+	1	3	2	+		+	+	+	6	10.7
ESE	+	1	3	2	+	+				6	11.0
SE	+	1	3	1					+	5	10.3
SSE	+	1	2	1	+				+	3	9.5
S	+	1	3	1	+					5	9.8
SSW	+	1	4	3	1	+	+		+	9	11.7
SW	+	2	6	5	1	+	+			14	11.6
WSW	+	1	4	3	+	+				9	11.5
W	+	1	4	2	+	+				7	11.0
WNW	+	1	4	2	+	+	+			8	11.0
NW	+	1	3	2	+	+				6	11.8
NNW	+	1	2	2	+	+				5	11.6
CALM	1									1	
TOTAL	3	16	47	29	4	+	+	+	+	100	11.0

C OCCURRENCES OF PRECIPITATION AMOUNTS:

INTENSITIES	FREQUENCY OF OCCURRENCE FOR EACH HOUR OF THE DAY — A M HOUR ENDING AT 1	2	3	4	5	6	7	8	9	10	11	NOON	P M HOUR ENDING AT 1	2	3	4	5	6	7	8	9	10	11	MID	NO OF DAYS WITH
TRACE	26	23	24	19	27	25	23	31	26	26	23	24	21	25	27	21	22	30	33	28	24	24	23	21	51
.01 IN	6	6	5	5	3	6	2	4	3	3	3	3	9	2	2	8	7	9	6	6	9	5	4	8	13
.02 TO .09 IN	8	7	11	10	10	10	13	8	6	9	11	6	8	6	2	10	13	7	11	6	3	8	8	9	23
.10 TO .24 IN	5	4	4	1	4	2		4	3	2	4	7	4	4	5	4	3	4	1	2	1	2	4	2	25
.25 TO .49 IN	1	2	1			1	2	1	3	1	3	1	2	2	2	2	3	2		1	2	1		1	17
.50 TO .99 IN				1				1	1			1			1	1	1								12
1.00 TO 1.99 IN								1	2	2	1			1	1										8
2.00 IN AND OVER																									4
TOTAL	46	42	45	36	44	44	40	50	44	43	45	42	44	40	40	46	49	52	51	43	39	40	39	41	153

D PERCENTAGE FREQUENCIES OF CEILING–VISIBILITY:

VISIBILITY (MILES)	CEILING (FEET) 0	100-200	300-400	500-900	1000-1900	2000-2900	3000-4900	5000-9500	OVER 9500	TOT
0 TO 1/8	+	+							+	+
3/16 TO 3/8	+	+	+						+	1
1/2 TO 3/4	+	+	+	+	+	+		+	+	1
1 TO 2 1/2		+	1	1	+	+	+	+	2	5
3 TO 6			1	2	1	1	1	2	11	18
7 TO 15			+	2	2	2	3	6	60	75
20 TO 30										
35 OR MORE										
TOTAL	+	1	2	5	4	3	4	8	73	100

E PERCENTAGE FREQUENCIES OF SKY COVER, WIND, AND RELATIVE HUMIDITY:

HOUR OF DAY	CLOUDS SCALE 0-10: 0-3	4-7	8-10	WIND SPEED (M. P. H.): 0-3	4-12	13-24	25- & OVER	RELATIVE HUMIDITY (%): 0-29	30-49	50-69	70-79	80-89	90-100
00	45	13	43	5	77	18			1	29	23	24	23
01	46	12	41	5	75	20	+		1	25	26	23	26
02	47	12	41	5	78	17	+		1	22	23	27	28
03	48	10	42	5	77	18			+	19	22	30	28
04	44	10	46	5	79	15				18	21	29	33
05	40	10	50	4	80	15	+			15	23	29	32
06	39	9	52	6	73	21			+	21	28	25	26
07	38	10	52	8	67	25	1		1	35	24	22	18
08	39	14	47	4	66	29	1		9	43	22	13	13
09	43	12	45	3	62	35	1		20	46	12	12	10
10	39	16	45	2	56	40	1	+	27	42	15	7	9
11	35	18	47	1	52	46	1	+	33	42	12	5	8
12	28	25	46	1	51	47	1	1	37	38	10	7	7
13	24	31	46		38	61	1	2	34	40	10	7	6
14	25	31	43	1	39	58	2	4	36	37	10	5	8
15	30	26	44	+	41	56	2	3	34	37	11	8	7
16	34	22	44	+	46	52	2	2	32	38	11	9	8
17	39	18	43	2	53	44	2	1	28	38	11	13	9
18	40	16	44	1	60	38	1		21	40	14	14	11
19	38	17	45	4	65	31	1		14	37	19	18	12
20	45	14	42	3	67	30	+		7	38	18	22	14
21	46	13	41	5	69	26	+		4	38	20	21	17
22	44	16	40	3	72	25			3	33	22	25	17
23	45	13	42	5	71	25			1	32	22	24	21
AVG	39	16	45	3	63	33	1	1	14	34	18	17	16

BOSTON, MASS.
Logan Int.AP

SEPTEMBER
7200 Obs.

A TEMPERATURE AND WIND SPEED–RELATIVE HUMIDITY OCCURRENCES:

WIND	0-4 M.P.H.						5-14 M.P.H.						15-24 M.P.H.						25 M.P.H. AND OVER						TOTAL OBS.
REL. HUMID. / TEMP. (°F)	UNDER 30	30-49%	50-69%	70-79%	80-89%	90-100%	UNDER 30	30-49%	50-69%	70-79%	80-89%	90-100%	UNDER 30	30-49%	50-69%	70-79%	80-89%	90-100%	UNDER 30	30-49%	50-69%	70-79%	80-89%	90-100%	
104/100														1											1
99/ 95														2											2
94/ 90								11	4					8	1										24
89/ 85		2					1	34	22	1				24	25										109
84/ 80			1					44	68	24	1		6	30	62	5				1	6				248
79/ 75		1	7	4	5		2	77	151	66	66	10	2	61	91	30	10			2	1	1			587
74/ 70		4	13	16	11	17	13	106	220	128	168	118	2	68	97	40	69	16		2	6		2		1116
69/ 65		5	10	11	30	31	11	127	297	172	237	165	6	53	102	47	44	15		5	6	5	4	8	1391
64/ 60		2	15	13	19	35	11	104	326	244	250	220	4	45	130	52	44	28		7	3	1		4	1557
59/ 55		6	19	19	33	27	5	42	250	218	223	241	6	25	62	48	18	56		4	4				1306
54/ 50		3	7	17	12	20		32	117	106	80	75	2	13	59	14	13	64		4	3				641
49/ 45			2	6	2			11	60	34	19	19	1	6	12	4		3							179
44/ 40								6	12	5	6	2		3		3									37
39/ 35									2																2
TOTAL		23	74	86	112	130	43	594	1529	998	1050	850	29	339	641	243	198	182		25	29	7	6	12	7200

B PERCENTAGE FREQUENCIES OF WIND DIRECTION AND SPEED:

DIRECTION	HOURLY OBSERVATIONS OF WIND SPEED (IN MILES PER HOUR)										AV SPEED
	0-3	4-7	8-12	13-18	19-24	25-31	32-38	39-46	47 OVER	TOTAL	
N	+	1	2	1	+	+				5	10.2
NNE	+	1	1	1	+	+	+			3	11.2
NE	+	1	1	2	+			+	+	4	11.7
ENE	+	+	1	2	+		+	+		4	12.6
E	+	1	2	2	+			+		5	10.8
ESE	+	1	2	1	+		+			5	10.4
SE	+	1	2	1	+		+			4	10.0
SSE	+	1	2	1	+					3	10.1
S	+	1	3	1	+				+	5	10.6
SSW	+	1	4	3	1	+				9	11.7
SW	+	1	6	6	1	+				14	12.6
WSW	+	1	3	3	+	+	+	+		8	12.2
W	+	1	3	2	+	+	+			6	11.6
WNW	+	1	4	3	+	+				8	11.7
NW	+	1	4	4	1	+	+		+	10	12.4
NNW	+	1	3	2	1	+		+		6	12.1
CALM	1									1	
TOTAL	4	13	43	33	5	1	+	+	+	100	11.5

C OCCURRENCES OF PRECIPITATION AMOUNTS:

INTENSITIES	FREQUENCY OF OCCURRENCE FOR EACH HOUR OF THE DAY — A.M. HOUR ENDING AT												P.M. HOUR ENDING AT												NO OF DAYS WITH
	1	2	3	4	5	6	7	8	9	10	11	NOON	1	2	3	4	5	6	7	8	9	10	11	MID	
TRACE	19	25	30	24	25	29	25	32	28	27	21	17	16	18	25	15	19	20	21	26	31	24	23	17	45
.01 IN	6	5	4	7	6	4	7	6	5	2	2	4	6	4	6	5	7	4	5	4	7	7	5	8	13
.02 TO .09 IN	8	9	11	13	8	11	12	6	7	8	7	11	6	10	6	8	10	12	12	10	7	9	8	9	27
.10 TO .24 IN	7	2	3	1	7	3	2	2	3	2	6	2	5	1	1	3	1	3	2	2	4	4	4	3	20
.25 TO .49 IN		1		3		1	2			2			1	2		2	1		1	1	1	1	1	1	10
.50 TO .95 IN	1				1						1		1	2	1					1					11
1.00 TO 1.99 IN												1			1										6
2.00 IN AND OVER																									4
TOTAL	36	42	48	48	47	48	48	46	43	41	37	35	35	37	40	33	38	39	41	44	50	45	41	38	136

D PERCENTAGE FREQUENCIES OF CEILING–VISIBILITY:

VISIBILITY (MILES)	CEILING (FEET) 0	100-200	200-400	500-900	1000-1900	2000-2900	3000-4900	5000-9500	OVER 9500	TOT.
0 TO 1/8	+	+		+					+	+
3/16 TO 3/8	+	+	+						+	1
1/2 TO 3/4		+	+	+	+				+	1
1 TO 2 1/2		+	1	1	+	+	+	+	1	4
3 TO 6		+	1	2	1	+	1	1	9	15
7 TO 15			+	2	3	2	4	7	61	79
20 TO 30										
35 OR MORE										
TOTAL	+	1	2	5	5	2	5	8	71	100

E PERCENTAGE FREQUENCIES OF SKY COVER, WIND, AND RELATIVE HUMIDITY:

HOUR OF DAY	CLOUDS SCALE 0-10: 0-3	4-7	8-10	WIND SPEED (M.P.H.): 0-3	4-12	13-24	25- & OVER	RELATIVE HUMIDITY (%): 0-29	30-49	50-69	70-79	80-89	90-100
00	50	10	40	6	66	28	+		1	27	22	27	23
01	53	6	41	5	67	27	+		1	22	25	27	25
02	51	10	39	6	71	23			1	19	26	25	28
03	51	8	41	7	72	21			1	16	24	30	29
04	50	9	41	7	67	25	1		1	15	24	27	33
05	46	11	43	6	69	25			+	16	23	27	34
06	42	12	46	5	69	26	+			18	22	30	30
07	45	10	46	4	65	30	1		1	29	23	27	20
08	43	11	46	4	57	37	2		7	41	21	17	14
09	41	14	46	5	49	45	1		15	47	16	12	10
10	39	15	45	3	46	49	2		25	49	9	9	8
11	41	16	43	1	45	52	2	1	32	42	10	10	6
12	38	21	42	1	41	56	2	3	38	36	11	6	7
13	38	18	44	1	33	62	4	4	36	39	9	6	6
14	39	18	43	+	37	61	2	4	35	39	8	6	7
15	39	15	46	1	41	56	1	4	32	38	13	7	6
16	41	18	41	1	42	57	1	3	28	39	13	10	6
17	40	16	45	1	49	49	1	2	24	35	18	12	8
18	40	14	46	3	60	36	1	1	18	34	20	17	10
19	44	15	40	4	59	35	2	+	13	33	18	23	13
20	46	12	42	5	59	35	1	1	8	32	21	25	14
21	43	13	44	7	58	34	1	+	4	33	20	26	16
22	46	11	43	7	63	29	+		4	29	23	26	18
23	48	10	42	5	64	30	+		1	30	24	25	20
AVG	44	13	43	4	56	39	1	1	14	32	19	19	16

BOSTON, MASS.
Logan Int.AP

OCTOBER
7440 Obs.

A TEMPERATURE AND WIND SPEED–RELATIVE HUMIDITY OCCURRENCES:

WIND	0-4 M.P.H.						5-14 M.P.H.						15-24 M.P.H.						25 M.P.H. AND OVER						TOTAL OBS.
REL. HUMID. / TEMP (°F)	UNDER 30%	30-49%	50-69%	70-79%	80-89%	90-100%	UNDER 30%	30-49%	50-69%	70-79%	80-89%	90-100%	UNDER 30%	30-49%	50-69%	70-79%	80-89%	90-100%	UNDER 30%	30-49%	50-69%	70-79%	80-89%	90-100%	
89/ 85								5	1				3	8	1										18
84/ 80								5	12					7	2										26
79/ 75	1		2	2			3	22	21	1			5	20	4	1	1								83
74/ 70	1	3	2	4	3		15	30	37	18	8	7	3	34	20	5	8	1		5			1	1	206
69/ 65		8	4	4	12	10	19	67	102	43	55	56	12	62	49	14	26	21	5	7	3	7	2	4	592
64/ 60	2	6	12	8	23	39	16	95	154	100	115	152	25	88	78	37	37	54	3	6	1	5	1	9	1066
59/ 55		5	7	10	22	71	14	146	254	133	159	420	5	95	98	49	63	71		6	5	5	14	29	1681
54/ 50	1	2	39	13	26	33	9	101	270	191	195	216	1	91	155	86	90	75		6	9	13	18	23	1663
49/ 45			11	14	14	15	4	89	224	166	156	88	2	49	150	64	58	60		11	6	5	3	12	1201
44/ 40		1	12	5	1	4		27	199	95	44	39		29	107	27	12	17		7	15				641
39/ 35			1	4	1			2	100	26	15	1		10	52	6	7	7							232
34/ 30									14	12	3				2										31
TOTAL	5	25	90	64	102	172	80	589	1388	785	750	979	56	493	718	289	302	306	8	48	39	35	39	78	7440

B PERCENTAGE FREQUENCIES OF WIND DIRECTION AND SPEED:

DIRECTION	HOURLY OBSERVATIONS OF WIND SPEED (IN MILES PER HOUR)										AV. SPEED
	0-3	4-7	8-12	13-18	19-24	25-31	32-38	39-46	47 OVER	TOTAL	
N	+	1	2	2	+	+	+			6	11.7
NNE	+	1	1	2	1	+	+			5	14.0
NE	+	1	2	2	1	+	+			5	13.7
ENE	+	1	1	2	1	+	+			5	14.9
E	+	1	2	2	1	+	+			6	14.2
ESE	+	1	2	2	+	+				5	11.2
SE	+	1	2	1	+	+				3	10.5
SSE	+	1	1	+	+	+				2	10.1
S	+	1	2	1	+	+	+			4	10.0
SSW	+	1	3	2	1	+	+			7	12.0
SW	+	1	4	4	1	+				11	12.3
WSW	+	1	2	2	+	+	+			6	12.2
W	+	1	2	2	1	+	+			6	12.6
WNW	+	1	3	3	1	+	+			9	12.6
NW	+	1	3	4	1	+	+			10	13.4
NNW	+	1	3	3	1	+				8	13.0
CALM	1									1	
TOTAL	4	13	36	34	10	3	+			100	12.5

C OCCURRENCES OF PRECIPITATION AMOUNTS:

INTENSITIES	FREQUENCY OF OCCURRENCE FOR EACH HOUR OF THE DAY																								NO OF DAYS WITH
	A.M. HOUR ENDING AT												P.M. HOUR ENDING AT												
	1	2	3	4	5	6	7	8	9	10	11	NOON	1	2	3	4	5	6	7	8	9	10	11	MID	
TRACE	24	32	33	28	32	35	34	32	31	23	29	29	31	28	30	28	20	25	22	29	33	31	34	31	34
.01 IN	10	7	5	8	10	4	7	13	7	11	13	4	7	6	3	6	11	12	13	8	6	8	8	8	17
.02 TO .09 IN	13	9	13	11	8	14	15	11	7	12	10	13	13	9	12	15	13	8	8	14	15	14	12	12	21
.10 TO .24 IN	4	6	6	4	5	3	2	4	7	7	3	3	2	4	6	6	2	4	4	1	1	3	5	5	19
.25 TO .49 IN	1	3	1	1	1	2			3	1		3	3	4					3				1		17
.50 TO .99 IN			1	1							1														15
1.00 TO 1.99 IN																									12
2.00 IN AND OVER																									2
TOTAL	52	57	59	53	56	58	58	60	55	54	56	52	56	51	51	55	46	49	50	52	55	56	60	56	137

D PERCENTAGE FREQUENCIES OF CEILING–VISIBILITY:

VISIBILITY (MILES)	CEILING (FEET)									
	0	100-200	300-400	500-900	1000-1900	2000-2900	3000-4900	5000-9500	OVER 9500	TOT.
0 TO 1/8	+	+							+	+
3/16 TO 3/8	+	1	+	+	+		+		+	1
1/2 TO 3/4		+	1	+	+				+	1
1 TO 2 1/2		+	2	1	+	+	+	+	2	6
3 TO 6		+	1	3	2	1	1	1	9	16
7 TO 15		+	+	2	4	3	5	6	55	75
20 TO 30										
35 OR MORE										
TOTAL	+	2	3	6	6	4	6	8	66	100

E PERCENTAGE FREQUENCIES OF SKY COVER, WIND, AND RELATIVE HUMIDITY:

HOUR OF DAY	CLOUDS SCALE 0-10			WIND SPEED (M.P.H.)				RELATIVE HUMIDITY (%)					
	0-3	4-7	8-10	0-3	4-12	13-24	25-& OVER	0-29	30-49	50-69	70-79	80-89	90-100
00	48	10	43	5	60	34	2		3	32	19	22	24
01	47	12	41	5	62	30	3		3	30	21	19	27
02	49	7	44	6	63	29	2		2	29	19	22	28
03	50	7	44	4	63	31	3		1	25	22	23	28
04	49	7	44	5	57	35	2		2	25	24	20	30
05	43	10	47	6	61	31	3		1	23	23	22	31
06	38	15	47	5	58	34	2		1	23	22	25	29
07	39	11	51	6	51	41	3		2	31	19	23	25
08	41	10	49	5	49	43	3		7	39	18	14	21
09	39	12	49	4	45	46	5	+	19	39	12	11	19
10	36	15	49	3	42	50	5	1	29	33	10	11	16
11	34	17	49	2	39	55	5	2	36	29	9	8	15
12	36	16	47	3	38	54	6	5	37	28	7	9	14
13	38	14	49	1	38	55	6	9	35	27	6	11	12
14	40	12	48	2	33	60	5	9	33	29	6	12	11
15	41	14	45	1	34	61	4	9	30	30	9	9	13
16	43	11	46	1	41	55	4	6	30	28	11	10	14
17	43	13	44	3	46	47	3	3	26	30	13	15	14
18	47	10	43	5	48	45	3	2	20	32	16	14	17
19	46	8	45	5	50	44	2	1	17	30	18	16	18
20	46	8	45	5	51	40	3	+	13	31	19	15	21
21	44	12	44	4	52	43	2	1	9	32	19	17	22
22	45	9	46	4	55	39	2		7	33	18	19	23
23	46	9	45	4	59	35	3		6	33	19	19	24
AVG	43	11	46	4	50	43	3	2	16	30	16	16	21

BOSTON, MASS.
Logan Int.AP

NOVEMBER
7200 Obs.

A TEMPERATURE AND WIND SPEED–RELATIVE HUMIDITY OCCURRENCES:

WIND	0-4 M.P.H.						5-14 M.P.H.						15-24 M.P.H.						25 M.P.H. AND OVER						TOTAL OBS.
REL HUMID / TEMP (°F)	UNDER 30	30-49%	50-69%	70-79%	80-89%	90-100%	UNDER 30	30-49%	50-69%	70-79%	80-89%	90-100%	UNDER 30	30-49%	50-69%	70-79%	80-89%	90-100%	UNDER 30	30-49%	50-69%	70-79%	80-89%	90-100%	
74/ 70							2	10	2		1		3	4	5						2	3			32
69/ 65			1				1	22	10	1	7	1		14	5	11	21	7		1	7	6	7	1	123
64/ 60	1	1	1	2				37	53	14	17	48		15	46	22	33	28			1	14	10	13	356
59/ 55		1	3	3	2	5	7	54	101	43	58	110	18	63	42	30	35	52	4	4	3	2	4	5	649
54/ 50		1	21	3	14	26		95	191	78	96	209	3	105	80	14	44	65	2	9	5	8	6	16	1091
49/ 45	1	3	28	21	11	22	3	106	308	164	172	178	8	105	165	42	54	56	2	11	8	3	6	40	1517
44/ 40		3	15	27	25	21	1	105	276	165	114	78		107	185	82	59	34	3	21	18	17	14	25	1395
39/ 35		5	22	11	27	12	7	105	263	110	123	63	1	122	159	35	30	34		10	19	1		3	1162
34/ 30		8	4	5	1	6	5	78	156	33	13	9	12	69	114	16	4	5		9	15			2	564
29/ 25		2	6	1				13	57	2	1	4	1	26	66	1		2	1	7	10	2	4	2	208
24/ 20		1						2	11				1	28	18				1	12	4				78
19/ 15								6	3					12	2					1	1				25
TOTAL	2	25	101	73	80	92	26	633	1431	610	602	700	47	670	887	253	280	283	13	85	93	56	51	107	7200

B PERCENTAGE FREQUENCIES OF WIND DIRECTION AND SPEED:

DIRECTION	HOURLY OBSERVATIONS OF WIND SPEED IN MILES PER HOUR: 0-3	4-7	8-12	13-18	19-24	25-31	32-38	39-46	47 OVER	TOTAL	AV SPEED
N	+	1	2	2	+	+	+			5	12.2
NNE	+	1	1	1	+	+				3	12.6
NE	+	1	1	1	+	+	+	+	+	3	15.5
ENE	+	1	1	1	1	+	+	+	+	3	15.6
E	+	+	1	+	+	+	+	+		2	17.3
ESE	+	1	1	1	+	+	+			3	11.3
SE	+	1	1	+	+					3	9.2
SSE	+	+	1	1	+	+				2	11.2
S	+	1	2	1	+	+	+			4	10.9
SSW	+	1	3	2	1	+	+	+	+	7	13.9
SW	+	1	3	5	2	1	+	+		12	14.4
WSW	+	1	3	3	1	+	+			8	13.2
W	+	1	3	4	2	+	+	+		10	14.1
WNW	+	1	3	5	2	1	+			13	14.6
NW	+	1	4	5	2	+	+	+		12	13.9
NNW	+	1	2	3	1	+	+			8	13.7
CALM	1									1	
TOTAL	3	13	32	34	13	4	1	+	+	100	13.5

C OCCURRENCES OF PRECIPITATION AMOUNTS:

INTENSITIES	FREQUENCY OF OCCURRENCE FOR EACH HOUR OF THE DAY — A.M. HOUR ENDING AT: 1	2	3	4	5	6	7	8	9	10	11	NOON	P.M. HOUR ENDING AT: 1	2	3	4	5	6	7	8	9	10	11	MID	NO OF DAYS WITH
TRACE	32	33	28	28	23	33	32	23	31	29	34	35	28	20	29	26	30	27	28	34	38	40	33	22	54
.01 IN	8	5	11	6	6	6	5	9	5	7	8	8	7	10	5	7	6	10	9	9	6	10	15	14	17
.02 TO .09 IN	17	15	15	12	14	11	17	13	14	16	15	11	17	13	10	11	9	10	10	10	10	11	13	17	27
.10 TO .24 IN	9	8	6	8	9	6	2	6	9	4	3	5	4	5	7	5	5	2	3	1	3	2	5	8	19
.25 TO .49 IN	1	3	2	1		1	3	2		1	2	3	3	2	1			2		2		2	1	3	12
.50 TO .99 IN				1					1										1						21
1.00 TO 1.99 IN																									9
2.00 IN AND OVER																									4
TOTAL	67	64	62	56	52	57	59	53	60	57	62	62	59	50	52	49	50	51	51	56	57	65	67	64	163

D PERCENTAGE FREQUENCIES OF CEILING–VISIBILITY:

VISIBILITY (MILES)	CEILING (FEET): 0	100-200	300-400	500-900	1000-1900	2000-2900	3000-4900	5000-9500	OVER 9500	TOT.
0 TO 1/8	+	+		+	+					+
3/16 TO 3/8	+	+	+		+					+
1/2 TO 3/4		+	+	+	+	+	+	+	+	1
1 TO 21/2		+	1	1	1	+	+	+	1	5
3 TO 6		+	+	2	2	1	1	1	9	17
7 TO 15			+	2	4	3	6	9	53	76
20 TO 30										
35 OR MORE										
TOTAL	+	1	2	5	7	4	7	10	63	100

E. PERCENTAGE FREQUENCIES OF SKY COVER, WIND, AND RELATIVE HUMIDITY:

HOUR OF DAY	CLOUDS SCALE 0-10: 0-3	4-7	8-10	WIND SPEED (M.P.H.): 0-3	4-12	13-24	25- & OVER	RELATIVE HUMIDITY (%): 0-29	30-49	50-69	70-79	80-89	90-100
00	43	6	51	3	52	41	5		7	37	18	18	20
01	40	9	52	4	50	42	4	+	7	40	15	15	22
02	38	11	52	5	51	40	4		7	39	13	20	21
03	39	9	52	5	49	41	5		6	37	18	18	22
04	39	8	53	5	47	44	4		5	37	17	19	22
05	37	9	54	5	46	44	4		4	38	17	21	21
06	34	12	54	5	49	41	5		4	36	19	21	20
07	32	11	57	5	45	46	4		5	37	19	18	20
08	33	12	55	4	45	46	5		8	42	17	14	19
09	30	17	53	3	39	53	5		16	43	11	14	16
10	29	10	60	4	37	52	7	1	26	37	12	9	14
11	30	13	57	4	37	48	11	2	36	29	11	10	12
12	32	12	56	2	37	51	10	3	40	27	9	8	12
13	29	15	56	+	39	50	10	5	44	24	10	4	13
14	30	14	56	1	37	55	7	7	40	26	7	8	12
15	33	13	54	1	38	55	6	6	38	28	8	10	9
16	35	11	54	+	43	52	5	3	35	30	9	11	12
17	38	11	51	2	46	49	4	1	34	29	12	12	12
18	44	10	46	3	46	48	4	1	27	34	11	12	15
19	44	9	48	1	48	47	4	+	21	36	16	13	14
20	43	11	47	2	46	47	4		18	37	16	14	15
21	42	10	48	3	50	42	6		17	38	16	12	17
22	43	8	49	2	50	44	5		14	38	15	18	16
23	44	9	47	3	52	39	6		10	39	16	18	17
AVG	37	11	53	3	45	47	6	1	20	35	14	14	16

BOSTON, MASS.
Logan Int. AP

DECEMBER
7440 Obs.

A TEMPERATURE AND WIND SPEED–RELATIVE HUMIDITY OCCURRENCES:

WIND	0-4 M.P.H.						5-14 M.P.H.						15-24 M.P.H.						25 M.P.H. AND OVER						TOTAL OBS.
REL HUMID / TEMP (°F)	UNDER 30	30-49%	50-69%	70-79%	80-89%	90-100%	UNDER 30	30-49%	50-69%	70-79%	80-89%	90-100%	UNDER 30	30-49%	50-69%	70-79%	80-89%	90-100%	UNDER 30	30-49%	50-69%	70-79%	80-89%	90-100%	
64/ 60							1		2	3	5		4		4	8	6	3	1				3	4	44
59/ 55							2	3	4	9	34	22	1	12	6	8	25	35		2	1		13	9	186
54/ 50		1	3			1		16	29	12	35	34		25	21	23	28	21		7	5	1	4	9	275
49/ 45		4	4	4	1	9		23	119	55	48	87	6	49	77	27	20	37	1	18	13	5	1	20	628
44/ 40	2	10	25	15	10	7	2	48	228	152	87	74	1	96	166	67	61	60		19	16	8	5	12	1171
39/ 35		8	31	16	17	9	5	58	290	141	124	115	5	93	222	70	47	90	1	10	24	5	11	32	1424
34/ 30		7	24	15	17	13	5	128	305	97	57	77	3	141	200	47	40	65	1	31	15	1	3	29	1321
29/ 25		11	23	2	2	4	2	124	242	45	35	12	8	177	162	11	20	16	4	31	22	1	7	12	973
24/ 20	3	9	12	3	13	2	13	105	180	19	6		9	156	130	4	6	2		17	6	2	5	1	703
19/ 15			5	1	2		2	43	76	18	10	1	4	95	75	14	11			19	3	3	2		384
14/ 10				1			1	30	42	4	2		6	53	41	1	3		1	11	8		8		212
09/ 05								16	8	1			6	18	26	1	4		10	2	4	1			97
04/ 00													2	7					3	2					14
-01/ -05														7						1					8
TOTAL	5	50	127	57	62	45	33	594	1525	556	443	422	55	929	1130	281	271	329	22	170	117	27	62	128	7440

B PERCENTAGE FREQUENCIES OF WIND DIRECTION AND SPEED:

DIRECTION	HOURLY OBSERVATIONS OF WIND SPEED (IN MILES PER HOUR) 0-3	4-7	8-12	13-18	19-24	25-31	32-38	39-46	47 OVER	TOTAL	AV SPEED
N	+	1	2	2	1	+	+			6	13.2
NNE	+	1	1	1	+	+	+			3	13.7
NE	+	+	1	+	+	+	+			2	13.6
ENE	+	+	+	+	+	+	+	+		2	16.6
E	+	+	+	1	+	+	+	+		2	16.0
ESE	+	+	1	+	+	+	+			1	11.5
SE	+	+	1	1	+	+				2	12.0
SSE	+	1	1	+	+	+	+	+		2	12.3
S	+	1	2	1	+	+	+	+		4	11.5
SSW	+	1	2	2	1	+	+			6	12.7
SW	+	1	4	6	2	+	+	+		13	14.2
WSW	+	1	3	5	2	+				10	14.5
W	+	1	2	4	2	1	+			9	15.3
WNW	+	1	2	5	3	2	+	+		14	16.8
NW	+	1	3	5	3	1	+		+	13	15.7
NNW	+	1	2	4	2	+				9	14.8
CALM	+									+	
TOTAL	3	10	26	37	17	6	1	+	+	100	14.5

C OCCURRENCES OF PRECIPITATION AMOUNTS:

INTENSITIES	FREQUENCY OF OCCURRENCE FOR EACH HOUR OF THE DAY — A M HOUR ENDING AT 1	2	3	4	5	6	7	8	9	10	11	NOON	P M HOUR ENDING AT 1	2	3	4	5	6	7	8	9	10	11	MID	NO OF DAYS WITH
TRACE	26	29	29	29	22	26	26	28	31	34	33	26	27	22	19	27	23	34	25	24	30	32	34	30	71
01 IN	6	8	11	12	13	7	8	5	9	7	4	4	7	6	7	8	11	6	11	8	10	8	9	7	11
02 TO 09 IN	17	19	18	15	17	18	16	16	13	13	13	18	14	19	14	14	13	13	15	15	15	13	15	18	24
10 TO 24 IN	11	6	8	7	4	4	4	7	5	4	5	5	7	5	9	5	4	5	5	4	3	8	5	5	17
25 TO 49 IN	1	1		1	2	1	2										2	2	1	1	1		1	2	15
50 TO 99 IN																			1						18
1 00 TO 1 99 IN																									16
2 00 IN AND OVER																									
TOTAL	61	63	66	64	58	56	56	56	58	58	55	53	55	52	49	54	53	60	58	52	59	61	64	62	172

D PERCENTAGE FREQUENCIES OF CEILING–VISIBILITY:

VISIBILITY (MILES)	CEILING (FEET) 0	100-200	300-400	500-900	1000-1900	2000-2900	3000-4900	5000-9500	OVER 9500	TOT
0 TO 1/8		+	+	+	+		+		+	+
3/16 TO 3/8		+	+	+		+				+
1/2 TO 3/4		+	+	+	+	+		+	+	1
1 TO 21/2		+	1	1	1	+	+	+	1	5
3 TO 6		+	+	3	2	1	1	1	8	17
7 TO 15			+	1	3	4	6	8	55	76
20 TO 30										
35 OR MORE										
TOTAL		1	2	6	6	5	7	9	65	100

E PERCENTAGE FREQUENCIES OF SKY COVER, WIND, AND RELATIVE HUMIDITY:

HOUR OF DAY	CLOUDS SCALE 0-10: 0-3	4-7	8-10	WIND SPEED (M. P. H.): 0-3	4-12	13-24	25-& OVER	RELATIVE HUMIDITY (%): 0-29	30-49	50-69	70-79	80-89	90-100
00	46	5	49	3	38	52	7	+	15	45	13	10	16
01	45	6	50	2	38	52	8		15	43	15	10	16
02	44	8	48	4	40	50	7		14	43	14	13	16
03	43	11	46	2	36	55	7		13	43	13	15	15
04	43	6	51	4	36	53	6		14	41	15	14	16
05	46	5	48	3	42	50	5	+	12	41	19	13	15
06	43	9	48	4	41	52	3		12	44	15	15	15
07	40	7	53	3	44	49	4		12	44	15	14	15
08	36	9	55	4	38	55	4		14	43	15	15	13
09	36	8	56	2	40	54	5	2	19	45	11	11	12
10	36	8	56	4	32	55	9	2	25	43	11	9	10
11	34	10	56	2	32	57	9	3	32	38	8	10	9
12	34	14	52	2	32	55	12	4	38	31	8	10	9
13	33	13	54	2	31	58	9	5	42	30	7	8	8
14	36	15	50	2	31	58	9	5	41	29	9	7	9
15	38	13	50	3	30	61	7	5	42	26	10	8	9
16	37	10	53	2	34	58	6	5	35	28	14	10	9
17	40	9	51	2	38	54	6	2	33	33	11	10	11
18	43	8	50	3	37	53	8	1	31	34	12	10	12
19	43	10	47	2	37	53	7	1	27	37	14	9	13
20	44	6	50	1	35	58	6	+	22	42	13	11	12
21	41	8	51	1	38	52	9	1	21	43	11	13	12
22	42	8	50	2	38	51	8	1	18	44	12	12	13
23	44	4	52	3	35	54	8	1	16	44	13	14	13
AVG	40	9	51	3	36	54	7	2	23	39	12	11	12

BOSTON, MASS.
Logan Int. AP

ANNUAL
87,672 Obs.

A TEMPERATURE AND WIND SPEED–RELATIVE HUMIDITY OCCURRENCES:

WIND	0-4 M.P.H.						5-14 M.P.H.						15-24 M.P.H.						25 M.P.H. AND OVER						TOTAL OBS.
REL. HUMID. / TEMP. (°F)	UNDER 30	30-49%	50-69%	70-79%	80-89%	90-100%	UNDER 30	30-49%	50-69%	70-79%	80-89%	90-100%	UNDER 30	30-49%	50-69%	70-79%	80-89%	90-100%	UNDER 30	30-49%	50-69%	70-79%	80-89%	90-100%	
104/100								+						+											+
99/ 95							1	4	+				1	4	+					+					10
94/ 90			+				2	12	3				1	15	4					1	+				39
89/ 85		1	1				3	38	24	+			3	32	21	+			1	2	1				127
84/ 80	+	1	2	+			5	56	70	12	1		6	39	40	5			+	3	2	+			245
79/ 75	+	2	4	4	2		5	68	122	58	31	5	8	44	50	16	6	2	2	4	1	1	1		433
74/ 70	+	2	9	9	9	6	9	66	158	96	94	50	10	45	50	26	22	6	2	3	2	1	1	1	676
69/ 65	+	2	8	11	14	25	8	52	176	101	129	111	8	44	55	21	23	14	2	5	3	2	2	2	819
64/ 60	+	2	9	10	14	21	8	52	139	100	121	121	13	45	64	23	22	22	2	5	3	2	3	4	805
59/ 55	+	2	6	8	13	23	10	55	122	79	97	127	12	49	60	29	29	34	2	6	4	2	6	8	783
54/ 50	+	2	12	6	10	15	6	53	122	72	81	108	9	55	67	33	36	45	2	7	5	4	5	10	765
49/ 45	+	2	8	7	7	12	5	55	130	74	81	89	8	53	77	31	35	45	2	9	8	2	3	14	756
44/ 40	+	3	10	9	8	10	3	50	138	89	76	80	5	56	104	44	39	52	2	11	13	6	4	20	831
39/ 35	+	3	11	7	11	12	4	46	141	78	81	78	5	64	116	38	40	60	1	13	13	3	6	21	851
34/ 30	+	4	9	6	7	10	3	55	138	50	36	51	5	70	94	23	20	37	2	14	13	2	3	19	669
29/ 25	+	4	7	3	2	2	4	44	87	25	20	11	4	64	77	12	11	8	2	18	12	3	4	8	431
24/ 20	+	2	3	1	2	+	4	32	49	9	10	2	4	46	55	5	7	2	2	10	6	1	2	+	254
19/ 15	+	1	2	1	+		2	22	20	4	4	+	4	36	29	5	4	+	+	10	3	1	1		149
14/ 10			+	+			+	10	9	1	3	+	2	20	15	4	1		1	5	3	+	1		75
09/ 05								4	4	+	1		1	10	9	1	1		2	1	1	+			35
04/ 00								2	2	+			+	3	2	1	+		1	1					11
-01/-05									+					1	1					+					3
-06/-10														+	+					+					1
-11/-15															+										+
TOTAL	3	34	100	80	100	136	81	775	1653	849	864	833	112	795	992	315	296	328	28	128	93	30	40	107	8767

In Tables A and C, occurrences are for the average year (10-year total divided by 10). Values are rounded to the nearest whole, but not adjusted to make their sums exactly equal to column or row totals. "+" indicates more than 0 but less than 0.5.

B PERCENTAGE FREQUENCIES OF WIND DIRECTION AND SPEED:

DIRECTION	HOURLY OBSERVATIONS OF WIND SPEED (IN MILES PER HOUR) 0-3	4-7	8-12	13-18	19-24	25-31	32-38	39-46	47 OVER	TOTAL	AV SPEED
N	+	1	2	2	+	+	+	+		5	12.0
NNE	+	1	1	1	+	+	+	+	+	3	13.3
NE	+	1	1	1	1	+	+	+	+	4	14.2
ENE	+	1	1	1	1	+	+	+	+	4	14.1
E	+	1	2	2	+	+	+	+	+	5	12.9
ESE	+	1	2	2	+	+	+			5	12.0
SE	+	1	2	1	+	+	+		+	4	10.8
SSE	+	1	1	1	+	+	+	+	+	3	10.4
S	+	1	2	1	+	+	+	+	+	4	11.0
SSW	+	1	3	2	1	+	+	+	+	7	13.0
SW	+	1	4	5	1	+	+	+		12	13.3
WSW	+	1	3	3	1	+	+	+		7	13.0
W	+	1	2	3	1	+	+	+		8	14.2
WNW	+	1	3	4	2	1	+	+		11	14.6
NW	+	1	3	4	2	1	+	+	+	11	15.2
NNW	+	1	2	3	1	+	+	+		7	13.8
CALM	1									1	
TOTAL	3	12	33	35	12	4	1	+	+	100	13.3

C OCCURRENCES OF PRECIPITATION AMOUNTS:

INTENSITIES	FREQUENCY OF OCCURRENCE FOR EACH HOUR OF THE DAY. A.M. HOUR ENDING AT												P.M. HOUR ENDING AT												NO OF DAYS WITH
	1	2	3	4	5	6	7	8	9	10	11	NOON	1	2	3	4	5	6	7	8	9	10	11	MID	
TRACE	34	38	39	39	39	41	38	41	40	39	41	37	35	35	37	36	33	35	35	34	38	37	36	34	62
01 IN	10	10	12	12	13	10	11	11	10	10	10	9	10	10	9	11	12	11	12	12	10	12	11	12	16
02 TO 09 IN	17	15	16	17	15	17	20	16	16	17	15	16	16	16	14	16	17	18	17	18	15	15	16	17	38
10 TO 24 IN	6	7	7	5	6	5	4	5	6	5	5	5	5	5	6	6	5	4	5	3	4	5	5	6	24
25 TO 49 IN	1	2	1	1	1	1	2	1	1	1	1	1	1	1	1	1	1	1	1	1	1	1	1	1	21
50 TO 99 IN	+	+	+	1	+	+	+	+	+	+	+	+	+	+	1	+	+	+	+	+	+		+		21
1 00 TO 1 99 IN								+	+	+	+	+	+	+	+	+		+	+	+					11
2 00 IN AND OVER																									2
TOTAL	68	72	74	73	73	74	75	75	74	71	72	68	68	67	67	70	68	71	70	67	69	69	69	69	196

D PERCENTAGE FREQUENCIES OF CEILING–VISIBILITY:

VISIBILITY (MILES)	CEILING (FEET) 0	100-200	300-400	500-900	1000-1900	2000-2900	3000-4900	5000-9500	OVER 9500	TOT.
0 TO 1/8	+	+	+	+	+		+		+	+
3/16 TO 3/8	+	+	+	+	+	+	+		+	1
1/2 TO 3/4	+	+	+	+	+	+	+	+	+	1
1 TO 2 1/2		+	1	2	1	+	+	+	1	5
3 TO 6		+	1	3	2	1	1	1	8	16
7 TO 15	+	+	+	2	3	3	5	8	56	76
20 TO 30									+	+
35 OR MORE									+	+
TOTAL	+	1	2	6	6	3	6	10	65	100

E PERCENTAGE FREQUENCIES OF SKY COVER, WIND, AND RELATIVE HUMIDITY:

HOUR OF DAY	CLOUDS SCALE 0-10 0-3	4-7	8-10	WIND SPEED (M.P.H.) 0-3	4-12	13-24	25-& OVER	RELATIVE HUMIDITY (%) 0-29	30-49	50-69	70-79	80-89	90-100
00	43	8	49	4	53	39	3	+	10	34	18	19	20
01	42	9	49	4	54	38	3	+	9	33	18	19	21
02	43	9	49	4	55	37	3	+	8	32	17	20	23
03	42	9	50	4	56	36	4	+	7	31	18	21	23
04	40	9	51	5	55	37	3	+	7	29	19	21	24
05	38	9	52	5	56	36	3	+	6	29	20	21	24
06	37	10	53	4	54	39	3	+	7	32	20	20	22
07	36	10	54	4	50	42	4	+	10	36	18	18	19
08	36	11	53	4	46	45	5	+	16	38	16	14	15
09	36	11	53	3	43	48	6	1	24	38	12	12	13
10	34	12	53	3	40	51	7	3	30	34	11	10	12
11	32	15	53	2	38	54	7	4	34	32	10	9	11
12	31	17	52	2	35	55	8	6	36	30	10	9	10
13	30	17	53	1	32	59	8	8	36	29	10	9	9
14	31	17	52	1	31	60	8	8	36	28	9	9	9
15	31	17	52	1	32	61	6	8	34	28	10	9	9
16	33	14	53	1	35	58	7	7	32	29	11	10	10
17	35	13	51	2	39	54	5	5	29	30	12	12	11
18	37	12	51	2	44	50	4	4	25	32	14	12	14
19	38	12	50	3	47	46	4	2	21	33	15	15	15
20	40	11	49	3	48	45	4	1	17	35	15	16	16
21	40	11	49	4	49	43	4	1	15	35	15	17	17
22	41	10	49	4	50	43	4	+	13	35	16	18	18
23	42	9	49	4	52	41	4	+	11	35	16	19	19
AVG	37	12	51	3	46	46	5	3	20	32	15	15	16

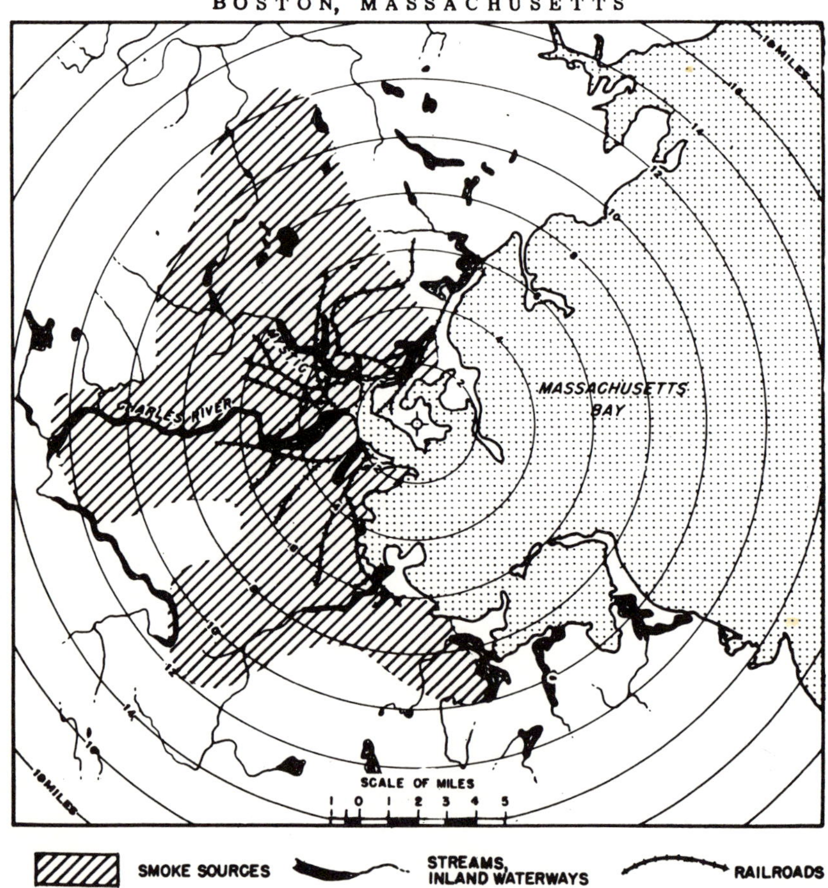
BOSTON, MASSACHUSETTS
18 MILES
16
14
12
10
8
MYSTIC
CHARLES RIVER
MASSACHUSETTS BAY
10
12
14
16
18 MILES
SCALE OF MILES
1 0 1 2 3 4 5
SMOKE SOURCES
STREAMS, INLAND WATERWAYS
RAILROADS

Appendix D
Conferences and Symposiums

WIND ENGINEERING

Wind engineering is a new field of science that has emerged in the past few decades and is continuing to grow, especially through the impetus of national and international conferences. These meetings generate a variety of papers that enhance considerably the state of the art. During February 26–March 1, 1978, the Third U.S. National Conference on Wind Engineering Research was held at the University of Florida in Gainesville. Sponsored by the Wind Engineering Research Council and the National Science Foundation, the conference included six sessions dealing with wind characteristics in general, full-scale measurements of wind, measurements and instrumentation, siting for wind power plants, effect of local wind, effects of turbulence, wind loads, dynamic response of buildings and bridges to wind excitation, etc. A total of 145 papers were presented at the conference. Topics included additional abstracts, practical applications, numerical simulation, wind modeling, wind loads, experimental techniques, dynamic response of real structures, modeling of turbulent flow about obstacles, effects of turbulence, instrumentation, measurements related to boundary layer winds, measurements related to full-scale buildings, siting for wind power, local wind effects, wind characteristics, and full-scale measurements of wind. International conferences on wind engineering are particularly successful in promoting this new field. After four previous conferences, the fifth was held on July 8–14, 1979, at Colorado State University, Fort Collins, Colorado. Sponsoring the conference were AMAX Foundation, Inc., Exxon Research & Engineering Co., Libbey-Owens-Ford Co., Electric Power Research Institute, Nuclear Regulatory Commission, National Science Foundation, International Association for Wind Engineering, Wind Engineering Research Council, Inc., Fluid Mechanics and Wind Engineering Program, and Colorado State University.

The purpose of the conference was to promote international communication of the new developments in knowledge and practice of wind engineering. Its fundamental goal was the exchange of recent research findings by engineers, meteorologists, and fluid dynamicists related to wind effects on building, structures, air pollution, agriculture, and energy production. An equally important objective was the application of new findings to the development of design methods and the codification of information for practicing engineers, architects, and urban planners.

The papers were presented at the conferences included the following topics: Social and economic impact of wind storms, wind characteristics and description, local wind environment, steady wind loads, unsteady wind loads, dynamic response of tall buildings and towers, dynamic response of bridges and roof membranes, physical and mathematical modeling, wind engineering applications, and wind engineering practice.

SEVERE LOCAL STORMS

The field of severe local storms is a specific discipline that, in addition to exploring the structure of violent winds (tornadoes, hurricanes, thunderstorms, etc.), is fundamental for the applied sciences and specifically for structural engineering. Critical wind loadings on buildings, bridges, and other structures require further scientific exploration and quantification of the forces involved. Ninety-eight papers were presented at the Tenth Conference on the topics of Satellite I; Satellite II; radar; probes and sensors; thunderstorm hazards to aircraft; climatology, statistics, and probability; modeling thunderstorms and squall lines; modeling tornadoes; forecasting thunderstorms and severe weather events; tornado forecasting and verification; warning systems (preparedness, dissemination, and public response); tornado case studies; mesoscale aspects of dust storms, thunderstorms, and hailstorms. Topics at the Sixth Conference, presented in a total of 70 papers, included dynamics of convective storms; recognition and forecasting of thunderstorm conditions; electromagnetic sensing of convective storms, waterspouts, tornadoes, hail, and aircraft and severe storms.

The Eleventh Conference on Severe Local Storms was held in Kansas City, Missouri, October 2–5, 1979. The papers will be published in the proceedings of the conference.

TORNADOES

Research work on tornadoes had been mostly restricted within the field of meteorology up to approximately a decade ago. The topic was usually included with other disturbances such as hurricanes and thunderstorms, and papers on tornadoes were usually presented at the severe-local-storms conferences held periodically in the United States. With the impulse of several authors in the field of applied sciences (mostly engineering), the necessity for analyzing tornado forces per se generated a new field that can be called "tornado engineering." As a result, the first conference on tornadoes was held at the University of Wisconsin in May 1970, and a second one at Texas Tech University on June 22–24, 1976, entitled "Symposium on Tornadoes: Assessment of Knowledge and Implications for Man." It is interesting to note that the fields represented by the participants included meteorology, structural and nuclear engineering, and even the insurance sector.

However, the dissemination of information on tornadoes must reach the moving forces in the front lines that can apply the engineering principles to the reality of our building environment. Such forces include of course environmental designers, architects, engineers, and builders. Thirty-seven papers were presented on such topics as wind speeds in tornadoes; tornado vortex models; tornado prediction, detection, and assessment of risk; tornado-structure interaction and engineering implications; tornado-generated missiles and their effects; applications of tornado technology; and new perspectives.

Appendix E
Conversion Tables

LENGTH

TO CONVERT FROM	TO	MULTIPLY BY
Centimeters	Ångström units	1×10^8
"	Feet	0.032808399
"	Feet (U.S. Survey)	0.032808333
"	Hands	0.098425197
"	Inches	0.39370079
"	Links (Gunter's)	0.049709695
"	Links (Ramden's)	0.032808399
"	Meters	0.01
"	Microns	10000
"	Miles (naut., Int.)	5.3995680×10^{-6}
"	Miles (statute)	6.2137119×10^{-6}
"	Millimeters	10
"	Millimicrons	1×10^7
"	Mils	393.70079
"	Picas (printer's)	2.3710630
"	Points (printer's)	28.452756
"	Rods	0.0019883878
"	Wave length of orange-red line of krypton 86	16507.6373
"	Wave length of red line of cadmium	15531.6413
"	Yards	0.010936133
Feet	Centimeters	30.48
"	Chains (Gunter's)	0.01515151
"	Fathoms	0.166666
"	Feet (U.S. Survey)	0.99999800
"	Furlongs	0.00151515

TO CONVERT FROM	TO	MULTIPLY BY
"	Inches	12
"	Meters	0.3048
"	Microns	304800
"	Miles (naut., Int.)	0.00016457883
"	Miles (statute)	0.000189393
"	Rods	0.060606
"	Ropes (Brit.)	0.05
"	Yards	0.333333
Feet (U.S. Survey)	Centimeters	30.480061
"	Chains (Gunter's)	0.015151545
"	Chains (Ramden's)	0.010000020
"	Feet	1.0000020
"	Inches	12.000024
"	Links (Gunter's)	1.5151545
"	Links (Ramden's)	1.0000020
"	Meters	0.30480061
"	Miles (statute)	0.00018939432
"	Rods	0.060606182
"	Yards	0.33333400
Inches	Ångström units	2.54×10^8
"	Centimeters	2.54
"	Chains (Gunter's)	0.00126262
"	Cubits	0.055555
"	Fathoms	0.013888
"	Feet	0.083333
"	Feet (U.S. Survey)	0.083333167
"	Links (Gunter's)	0.126262
"	Links (Ramden's)	0.083333
"	Meters	0.0254
"	Mils	1000
"	Picas (printer's)	6.0225

TO CONVERT FROM	TO	MULTIPLY BY
"	Points (printer's)	72.27000
"	Wave length of orange-red line of krypton 86	41929.399
"	Wave length of the red line of cadmium	39450.369
"	Yards	0.027777
Kilometers	Åstronomical units	6.68878×10^{-9}
"	Centimeters	100000
"	Feet	3280.8399
"	Feet (U.S. Survey)	3280.833
"	Light years	1.05702×10^{-13}
"	Meters	1000
"	Miles (naut., Int.)	0.53995680
"	Miles (statute)	0.62137119
"	Myriameters	0.1
"	Rods	198.83878
"	Yards	1093.6133
Meters*	Ångström units	1×10^{10}
"	Centimeters	100
"	Chains (Gunter's)	0.049709695
"	Chains (Ramden's)	0.032808399
"	Fathoms	0.54680665
"	Feet	3.2808399
"	Feet (U.S. Survey)	3.280833
"	Furlongs	0.0049709695
"	Inches	39.370079
"	Kilometers	0.001
"	Links (Gunter's)	4.9709695
"	Links (Ramden's)	3.2808399

* The meter is defined as 1,650,763.73 wavelengths of the orange-red light from krypton-86, measured in a vacuum.

TO CONVERT FROM	TO	MULTIPLY BY
"	Megameters	1×10^{-6}
"	Miles (naut., Brit.)	0.00053961182
"	Miles (naut., Int.)	0.00053995680
"	Miles (statute)	0.00062137119
"	Millimeters	1000
"	Millimicrons	1×10^{9}
"	Mils	39370.079
"	Rods	0.19883878
"	Yards	1.0936133
Miles (naut., Brit.)	Cable lengths (Brit.)	8.4444
"	Fathoms	1013.333
"	Feet	6080
"	Meters	1853.184
"	Miles (Adm., Brit.)	1
"	Miles (naut., Int.)	1.0006393
"	Miles (statute)	1.151515
Miles (naut., Int.)	Cable lengths	8.4390493
"	Fathoms	1012.6859
"	Feet	6076.1155
"	Feet (U.S. Survey)	6076.1033
"	Kilometers	1.852
"	Leagues (naut., Int.)	0.333333
"	Meters	1852
"	Miles (geographical)	1
"	Miles (naut. Brit.)	0.99936110
"	Miles (statute)	1.1507794
Miles (statute)	Centimeters	160934.4
"	Chains (Gunter's)	80
"	Chains (Ramden's)	52.8
"	Feet	5280

TO CONVERT FROM	TO	MULTIPLY BY
"	Feet (U.S. Survey)	5279.9894
"	Furlongs	8
"	Inches	63360
"	Kilometers	1.609344
"	Light years	1.70111×10^{-13}
"	Links (Gunter's)	8000
"	Meters	1609.344
"	Miles (naut., Brit.)	0.86842105
"	Miles (naut., Int.)	0.86897624
"	Myriameters	0.1609344
"	Rods	320
"	Yards	1760
Milliliters	Cu. cm	1.000028
"	Cu. inches	0.06102545
"	Drams (U.S., fluid)	0.2705198
"	Gills (U.S.)	0.008453742
"	Liters	0.001
"	Minims (U.S.)	16.23119
"	Ounces (Brit., fluid)	0.03519609
"	Ounces (U.S., fluid)	0.03381497
"	Pints (Brit.)	0.001759804
"	Pints (U.S., liq.)	0.002113436
Millimeters	Ångström units	1×10^{7}
"	Centimeters	0.1
"	Decimeters	0.01
"	Dekameters	0.0001
"	Feet	0.0032808399
"	Inches	0.039370079
"	Meters	0.001
"	Microns	1000

TO CONVERT FROM	TO	MULTIPLY BY
"	Mils	39.370079
"	Wave length of orange-red line of krypton 86	1650.76373
"	Wave length of red line of cadmium	1553.16413
Rods	Centimeters	502.92
"	Chains (Gunter's)	0.25
"	Chains (Ramden's)	0.165
"	Feet	16.5
"	Feet (U.S. Survey)	16.499967
"	Furlongs	0.025
"	Inches	198
"	Links (Gunter's)	25
"	Links (Ramden's)	16.5
"	Meters	5.0292
"	Miles (statute)	0.003125
"	Perches	1
"	Yards	5.5
Yards	Centimeters	91.44
"	Chains (Gunter's)	0.4545454
"	Chains (Ramden's)	0.03
"	Cubits	2
"	Fathoms	0.5
"	Feet	3
"	Feet (U.S. Survey)	2.9999940
"	Furlongs	0.00454545
"	Inches	36
"	Meters	0.9144
"	Poles (Brit.)	0.181818

TO CONVERT FROM	TO	MULTIPLY BY
AREA		
Acres	Sq. cm	40468564
"	Sq. ft.	43560
"	Sq. ft. (U.S. Survey)	43559.826
"	Sq. inches	6272640
"	Sq. kilometers	0.0040468564
"	Sq. links (Gunter's)	1×10^{5}
"	Sq. meters	4046.8564
"	Sq. miles (statute)	0.0015625
"	Sq. perches	160
"	Sq. rods	160
"	Sq. yards	4840
Sq. centimeters	Ares	1×10^{-6}
"	Circ. mm	127.32395
"	Circ. mils	197352.52
"	Sq. chains (Gunter's)	2.4710538×10^{-7}
"	Sq. chains (Ramden's)	1.0763910×10^{-7}
"	Sq. decimeters	0.01
"	Sq. feet	0.0010763910
"	Sq. ft. (U.S. Survey)	0.0010763867
"	Sq. inches	0.15500031
"	Sq. meters	0.0001
"	Sq. mm	100
"	Sq. mils	155000.31
"	Sq. rods	3.9536861×10^{-6}
"	Sq. yards	0.00011959900
Sq. feet	Acres	2.295684×10^{-5}
"	Ares	0.0009290304

TO CONVERT FROM	TO	MULTIPLY BY
"	Sq. cm	929.0304
"	Sq. chains (Gunter's)	0.00022956841
"	Sq. ft. (U.S. Survey)	0.99999600
"	Sq. inches	144
"	Sq. links (Gunter's)	2.2956841
"	Sq. meters	0.09290304
"	Sq. miles	3.5870064×10^{-8}
"	Sq. rods	0.0036730946
"	Sq. yards	0.111111
Sq. feet (U.S. Survey)	Acres	$2.29569330 \times 10^{-5}$
"	Sq. centimeters	929.03412
"	Sq. chains (Ramden's)	0.00010000040
"	Sq. feet	1.0000040
Sq. inches	Circ. mils	1273239.5
"	Sq. cm	6.4516
"	Sq. chains (Gunter's)	1.5942251×10^{-6}
"	Sq. decimeters	0.064516
"	Sq. feet	0.0069444
"	Sq. ft. (U.S. Survey)	0.0069444167
"	Sq. links (Gunter's)	0.01594225
"	Sq. meters	0.00064516
"	Sq. miles	$2.4909767 \times 10^{-10}$
"	Sq. mm	645.16
"	Sq. mils	1×10^{6}
Sq. kilometers	Acres	247.10538
"	Sq. feet	1.0763910×10^{7}
"	Sq. ft. (U.S. Survey)	1.0763867×10^{7}
"	Sq. inches	1.5500031×10^{9}
"	Sq. links (Gunter's)	24.710538
"	Sq. links (Ramden's)	10.763910

TO CONVERT FROM	TO	MULTIPLY BY
"	Sq. miles	3.8610216×10^{-7}
"	Sq. mm	1×10^{6}
"	Sq. rods	0.039536861
"	Sq. yards	1.1959900
Sq. miles	Acres	640
"	Hectares	258.99881
"	Sq. chains (Gunter's)	6400
"	Sq. feet	2.7878288×10^{7}
"	Sq. ft. (U.S. Survey)	2.78288×10^{7}
"	Sq. kilometers	2.5899881
"	Sq. meters	2589988.1
"	Sq. rods	102400
"	Sq. yards	3.0976×10^{6}
"	Sq. meters	1×10^{6}
"	Sq. miles	0.38610216
"	Sq. yards	1.1959900×10^{6}
Sq. links (Ramden's)	Acres	2.2956841×10^{-5}
"	Sq. feet	1
Sq. meters	Acres	0.00024710538
"	Ares	0.01
"	Hectares	0.0001
"	Sq. cm	10000
"	Sq. feet	10.763910
"	Sq. inches	1550.0031
"	Sq. kilometers	1×10^{-6}
Sq. millimeters	Circ. mm	1.2732395
"	Circ. mils	1973.5252
"	Sq. cm	0.01
"	Sq. inches	0.0015500031
"	Sq. meters	1×10^{-6}

TO CONVERT FROM	TO	MULTIPLY BY
Sq. mils	Circ. mils	1.2732395
"	Sq. cm	6.4516×10^{-6}
"	Sq. inches	1×10^{-6}
"	Sq. mm	0.00064516
Sq. rods	Acres	0.00625
"	Ares	0.2529285264
"	Hectares	0.002529285264
"	Sq. cm	252928.5264
"	Sq. feet	272.25
"	Sq. ft. (U.S. Survey)	272.24891
"	Sq. inches	39204
"	Sq. links (Gunter's)	625
"	Sq. links (Ramden's)	272.25
"	Sq. meters	25.29285264
"	Sq. miles	9.765625×10^{-6}
"	Sq. yards	30.25
Sq. yards	Acres	0.00020661157
"	Ares	0.0083612736
"	Hectares	8.3612736×10^{-5}
"	Sq. cm	8361.2736
"	Sq. chains (Gunter's)	0.0020661157
"	Sq. chains (Ramden's)	0.0009
"	Sq. feet	9
"	Sq. ft. (U.S. Survey)	8.9999640
"	Sq. inches	1296
"	Sq. links (Gunter's)	20.661157
"	Sq. links (Ramden's)	9
"	Sq. meters	0.83612736

TO CONVERT FROM	TO	MULTIPLY BY
"	Sq. miles	$3.228305785 \times 10^{-7}$
"	Sq. perches (Brit.)	0.033057851
"	Sq. rods	0.033057851

VELOCITY

TO CONVERT FROM	TO	MULTIPLY BY
Centimeters/sec	Feet/min	1.9685039
"	Feet/sec	0.032808399
"	Kilometers/hr	0.036
"	Kilometers/min	0.0006
"	Knots (Int.)	0.019438445
"	Meters/min	0.6
"	Miles/hr	0.022369363
"	Miles/min	0.00037282272
Feet/hour	Cm./hr	30.48
"	Cm./min	0.508
"	Cm./sec	0.0084666
"	Feet/min	0.0166666
"	Inches/hr	12
"	Kilometers/hr	0.0003048
"	Kilometers/min	5.08×10^{-6}
"	Knots (Int.)	0.0001645788
"	Miles/hr	0.000189393
"	Miles/min	3.156565×10^{-6}
"	Miles/sec	5.2609428×10^{-8}
Feet/minute	Cm./sec	0.508
"	Feet/sec	0.0166666
"	Kilometers/hr	0.018288
"	Meters/min	0.3048
"	Meters/sec	0.00508
"	Miles/hr	0.01136363
Feet/second	Cm./sec	30.48
"	Kilometers/hr	1.09728
"	Kilometers/min	0.018288

TO CONVERT FROM	TO	MULTIPLY BY
"	Meters/min	18.288
"	Miles/hr	0.68181818
"	Miles/min	0.01136363
Inches/hr	Cm/hr	2.54
"	Feet/hr	0.0833333
"	Miles/hr	1.578282×10^{-5}
Inches/min	Cm./hr	152.4
"	Feet/hr	5
"	Miles/hr	0.000946969
Kilometers/hr	Cm./sec	27.7777
"	Feet/hr	3280.8399
"	Feet/min	54.680665
"	Knots (Int.)	0.53995680
"	Meters/sec	0.277777
"	Miles (statute)/hr	0.62137119
Kilometers/(hr. × sec.)	Cm./(sec. × sec.)	27.7777
"	Ft./(sec. × sec.)	0.91134442
"	Meters/(sec. × sec.)	0.277777
Kilometers/min	Cm./sec	1666.666
"	Feet/min	3280.8399
"	Kilometers/hr	60
"	Knots (Int.)	32.397408
"	Miles/hr	37.282272
"	Miles/min	0.62137119
Knots (Int.)	Cm./sec	51.4444
"	Feet/hr	6076.1155
"	Feet/min	101.26859
"	Feet/sec	1.6878099
"	Kilometers/hr	1.852
"	Meters/min	30.8666
"	Meters/sec	0.514444
"	Miles (naut., Int.)/hr	1
"	Miles (statute)/hr	1.1507794

TO CONVERT FROM	TO	MULTIPLY BY
Meters/hr.	Feet/hr	3.2808399
"	Feet/min	0.054680665
"	Knots (Int.)	0.00053995680
"	Miles (statute)/hr	0.00062137119
Meters/min	Cm./sec	1.666666
"	Feet/min	3.2808399
"	Feet/sec	0.054680665
"	Kilometers/hr	0.06
"	Knots (Int.)	0.032397408
"	Miles (statute)/hr	0.037282272
Meters/sec	Feet/min	196.85039
"	Feet/sec	3.2808399
"	Kilometers/hr	3.6
"	Kilometers/min	0.06
Miles/hr	Cm./sec	44.704
"	Feet/hr	5280
"	Feet/min	88
"	Feet/sec	1.466666
"	Kilometers/hr	1.609344
"	Knots (Int.)	0.86897624
"	Meters/min	26.8224
"	Miles/min	0.0166666
Miles/hr. × min.)	Cm./(sec. × sec.)	0.7450666
Miles/(hr. × sec.)	Cm./(sec. × sec.)	44.704
"	Ft./(sec. × sec.)	1.466666
"	Kilometers/(hr. × sec.)	1.609344
"	Meters/(sec. × sec.)	0.44704
Miles/min	Cm./sec	2682.24
"	Feet/hr	316800
"	Feet/sec	88
"	Kilometers/min	1.609344
"	Knots (Int.)	52.138574
"	Meters/min	1609.344

PRESSURE

TO CONVERT FROM	TO	MULTIPLY BY
Atmospheres	Bars	1.01325
"	Cm. of Hg (0°C.)	76
"	Cm. of H_2O (4°C.)	1033.26
"	Dynes/sq. cm	1.01325×10^6
"	Ft. of H_2O (39.2°F.)	33.8995
"	Grams/sq. cm	1033.23
"	In. of Hg (32°F.)	29.9213
"	Kg./sq. cm	1.00323
"	Mm. of Hg (0°C.)	760
"	Pounds/sq. inch	14.6960
"	Tons (short)/sq. ft.	1.05811
"	Torrs	760
Bars	Atmospheres	0.986923
"	Baryes	1×10^6
"	Cm. of Hg (0°C.)	75.0062
"	Dynes/sq. cm	1×10^6
"	Ft. of H_2O (60°F.)	33.4883
"	Grams/sq. cm	1019.716
"	In. of Hg (32°F.)	29.5300
"	Kg./sq. cm	1.019716
"	Millibars	1000
"	Pounds/sq. inch	14.5038
Cm. of Hg (0°C.)	Atmospheres	0.013157895
"	Bars	0.0133322
"	Dynes/sq. cm	13332.2
"	Ft. of H_2O (4°C.)	0.446050
"	Ft. of H_2O (60°F.)	0.446474
"	In. of Hg (0°C.)	0.39370079
"	Kg./sq. meter	135.951
"	Pounds/sq. ft	27.8450
"	Pounds/sq. inch	0.193368
"	Torrs	10
Cm. of H_2O (4°C.)	Atmospheres	0.000967814
"	Dynes/sq. cm	980.638
"	Pounds/sq. inch	0.0142229

TO CONVERT FROM	TO	MULTIPLY BY
Dynes/sq. cm	Atmospheres	9.86923×10^{-7}
"	Bars	1×10^{-6}
"	Baryes	1
"	Cm. of Hg (0°C.)	7.50062×10^{-5}
Feet of air (1 atm., 60°F.)	Atmospheres	3.6083×10^{-5}
"	Ft. of Hg (32°F.)	0.00089970
"	Ft. of H_20 (60°F.)	0.0012244
"	In. of Hg (32°F.)	0.0010796
"	Pounds/sq. inch	0.00053027
Feet of Hg (32°F.).	Cm. of Hg (0°C.)	30.48
"	Ft. of H_2O (60°F.)	13.6085
"	In. of H_2O (60°F.)	163.302
"	Ounces/sq. inch	94.3016
"	Pounds/sq. inch	5.89385
Feet of H_2O (4°C.)	Atmospheres	0.0294990
"	Cm. of Hg (0°C.)	2.24192
"	Dynes/sq. cm	29889.8
"	Grams/sq. cm	30.4791
"	In. of Hg (32°F.)	0.882646
"	Kg./sq. meter	304.791
"	Pounds/sq. inch	0.433515
Gram/sq. cm	Pounds/sq. inch	0.000341717
Inches of Hg (32°F.)	Atmospheres	0.0334211
"	Bars	0.0338639
"	Dynes/sq. cm	33863.9
"	Ft. of air (1 atm., 60°F.)	926.24
"	Ft. of H_2O (39.2°F.)	1.132957
"	Grams/sq. cm	34.5316
"	Kg./sq. meter	345.316
"	Mm. of Hg (60°C.)	25.4
"	Ounces/sq. inch	7.85847
Inches of Hg (32°F.)	Pounds/sq. ft	70.7262
Inches of Hg (60°F.)	Atmospheres	0.0333269
"	Dynes/sq. cm	39768.5

TO CONVERT FROM	TO	MULTIPLY BY
"	. Grams/sq. cm	34.4343
"	. Mm. of Hg (60°F.)	25.4
"	. Ounces/sq. inch	7.83633
"	. Pounds/sq. ft	70.5269
Inches of H_2O (4°C.)	Atmospheres	0.0024582
"	Dynes/sq. cm	2490.82
"	In. of Hg (32°F.)	0.0735539
"	Kg./sq. meter	25.3993
"	Ounces/sq. ft	83.2350
"	Ounces/sq. inch	0.578020
"	Pounds/sq. ft	5.20218
"	Pounds/sq. inch	0.03612628
Kilograms/sq. cm ...	Atmospheres	0.967841
" ...	Bars	0.980665
" ...	Cm. of Hg (0°C.)	73.5559
" ...	Dynes/sq. cm	980665
" ...	Ft. of H_2O (39.2°F.)	32.8093
" ...	In. of Hg. (32°F.)	28.9590
" ...	Pounds/sq. inch	14.223343
Kilograms/sq. meter .	Atmospheres	9.67841×10^{-5}
" .	Bars	9.80665×10^{-5}
" .	Dynes/sq. cm	98.0665
" .	Ft. of H_2O (39.2°F.)	0.00328093
" .	Grams/sq. cm	0.1
" .	In. of Hg (32°F.)	0.00289590
" .	Mm. of Hg (0°C.)	0.0735559
" .	Pounds/sq. ft	0.20481614
" .	Pounds/sq. in	0.0014223343
Kilograms/sq. mm ..	Pounds/sq. ft	204816.14
" ..	Pounds/sq. in	1422.3343
" ..	Tons (short)/sq. in	0.71116716
Kilogram sq. cm	Pounds sq. ft	0.0023730360
"	Pounds sq. in	0.34171719
"	Miles/hr	60

TO CONVERT FROM	TO	MULTIPLY BY
Millibars	Atmospheres	0.000986923
"	Bars	0.001
"	Baryes	1000
"	Dynes/sq. cm	1000
"	Grams/sq. cm	1.019716
"	In. of Hg (32°F.)	0.0295300
"	Pounds/sq. ft	2.088543
"	Pounds/sq. inch	0.0145038
Millimeters of Hg (0°C)	Atmospheres	0.0013157895
"	Bars	0.00133322
"	Dynes/sq. cm	1333.224
"	Grams/sq. cm	1.35951
"	Kg./sq. meter	13.5951
"	Pounds/sq. ft	2.78450
"	Pounds/sq. inch	0.0193368
"	Torrs	1
Pounds/sq. ft	Atmospheres	0.000472541
"	Bars	0.000478803
"	Cm. of Hg (0°C.)	0.0359131
"	Dynes/sq. cm	478.803
"	Ft. of air (1 atm., 60°F.)	13.096
"	Grams/sq. cm	0.48824276
"	In. of Hg (32°F.)	0.0141390
"	In. of H_2O (39.2°F.)	0.192227
"	Kg./sq. meter	4.8824276
"	Mm. of Hg (0°C.)	0.359131
Pounds/sq. inch	Atmospheres	0.0680460
"	Bars	0.0689476
"	Cm. of Hg (0°C.)	5.17149
"	Cm. of H_2O (4°C.)	70.3089
"	Dynes/sq. cm	68947.6
"	Grams/sq. cm	70.306958
"	In. of Hg (32°F.)	2.03602
"	In. of H_2O (39.2°F.)	27.6807
"	Kg./sq. cm	0.070306958

TO CONVERT FROM	TO	MULTIPLY BY
"	Mm. of Hg (0°C.)	51.7149
Tons (long)/sq. ft	Atmospheres	1.05849
"	Dynes/sq. cm	1.07252×10^6
"	Grams/sq. cm	1093.6638
"	Pounds/sq. ft	2240
Tons (short)/sq. ft	Atmospheres	0.945082
"	Dynes/sq. cm	957.605
"	Grams/sq. cm	976.486
"	Pounds/sq. inch	13.8888
Tons (long)/sq. in	Atmospheres	152.423
"	Dynes/sq. cm	1.54443×10^8
"	Grams/sq. cm	157487.59
Tons (short)/sq. in	Dynes/sq. cm	1.37895×10^8
"	Kg./sq. mm	1406.139
"	Pounds/sq. inch	2000
Torrs (*or* Tors)	Millimeters of Hg (0°C.)	1

POWER

TO CONVERT FROM	TO	MULTIPLY BY
B.t.u./hr	Cal., *kg.*/hr	0.251996
"	Ergs/sec	2.928751×10^6
"	Foot-pounds/hr	777.649
"	Horsepower	0.000392752
"	Horsepower (boiler)	2.98563×10^{-5}
"	Horsepower (electric)	0.000392594
"	Horsepower (metric)	0.000398199
"	Kilowatts	0.000292875
"	Lb. ice melted/hr	0.0069714
"	Tons of refrig. (U.S. comm.)	8.32789×10^{-5}
"	Watts	0.292875
B.t.u./min	Cal., *kg.*/min	0.251996
"	Ergs/sec	1.75725×10^8
"	Foot-pounds/min	777.649

TO CONVERT FROM	TO	MULTIPLY BY
"	Horsepower	0.0235651
"	Horsepower (boiler)	0.00179138
"	Horsepower (electric)	0.0235556
"	Horsepower (metric)	0.0238920
"	Joules/sec	17.5725
"	Kg.-meters/min	107.514
"	Kilowatts	0.0175725
"	Lb. ice melted/hr	0.41828
"	Tons of refrig. (U.S. comm.)	0.00499673
"	Watts	17.5725
B.t.u. (mean)/min	B.t.u. (mean)/hr	60
"	Cal., *kg.* (mean)/hr	15.1197
"	Cal., *kg.* (mean)/min	0.251996
"	Ergs/sec	1.75978×10^{8}
"	Foot-pounds/min	778.768
"	Horsepower	0.0235990
"	Horsepower (boiler)	0.00179396
"	Horsepower (electric)	0.0235895
"	Horsepower (metric)	0.0239264
"	Joules/sec	17.5978
"	Kg.-meters/min	107.669
"	Kilowatts	0.0175978
"	Lb. ice-melted/hr	0.41888
Foot-pounds/hr	B.t.u./min	2.14321×10^{-5}
"	B.t.u. (mean)/min	2.14013×10^{-5}
"	Cal., gm./min	0.00540080
"	Cal., *gm.* (mean)/min	0.00539304
"	Ergs/min	2.25970×10^{5}
"	Foot-pounds/min	0.0166666
"	Horsepower	5.050505×10^{-7}
"	Horsepower (metric)	5.12055×10^{-7}
"	Kilowatts	3.76616×10^{-7}
"	Watts	0.000376616
"	Watts (Int.)	0.000376554
Foot-pounds/min	B.t.u. sec	2.14321×10^{-5}

TO CONVERT FROM	TO	MULTIPLY BY
"	B.t.u. (mean)/sec	2.14013×10^{-5}
"	Cal., *gm.*/sec	0.00540080
"	Cal., *gm.* (mean)/sec	0.00539304
"	Ergs/sec	2.25970×10^{5}
"	Foot-pounds/sec	0.0166666
"	Horsepower	3.030303×10^{-5}
"	Horsepower (metric)	3.07233×10^{-5}
"	Joules/sec	0.0225970
"	Joules (Int.)/sec	0.0225932
"	Kilowatts	2.25970×10^{-5}
"	Watts	0.0225970
Horsepower	B.t.u. (mean)/hr	2542.48
(mechanical)	B.t.u./min	42.4356
"	B.t.u. (mean)/sec	0.706243
"	Cal., *gm.*/hr	6.41616×10^{5}
"	Cal., *gm.* (IST.)/hr	6.41196×10^{5}
"	Cal., *gm.* (mean)/hr	6.40693×10^{5}
"	Cal., *gm.*/min	10693.6
"	Cal., *gm.* (IST.)/min	10686.6
"	Cal., *gm.* (mean)/min	10678.2
"	Ergs/sec	7.45700×10^{9}
"	Foot-pounds/hr	1980000
"	Foot-pounds/min	33000
"	Foot-pounds/sec	550
"	Horsepower (boiler)	0.0760181
"	Horsepower (electric)	0.999598
"	Horsepower (metric)	1.01387
"	Joules/sec	745.700
"	Kilowatts	0.745700
"	Kilowatts (Int.)	0.745577
"	Tons of refrig. (U.S., comm.)	0.21204
"	Watts	745.700
Horsepower (boiler)	B.t.u. (mean)/hr	33445.7
"	Cal., *gm.*/min	140671.6
"	Cal., *gm.* (mean)/min	140469.4
"	Cal., *gm.* (15°C.)/min	140611.1

TO CONVERT FROM	TO	MULTIPLY BY
"	.. Cal., *gm.*, (20°C.)/min	140742.2
"	.. Ergs/sec	9.80950×10^{10}
"	.. Foot-pounds/min	434107
"	.. Horsepower	13.1548
"	.. Horsepower (electric)	13.1495
"	.. Horsepower (metric)	13.3372
"	.. Horsepower (water)	13.1487
"	.. Joules/sec	9809.50
"	.. Kilowatts	9.80950
"	.. Lb. H_2O evap. per hr. from and at 212°F	34.5
Horsepower (electric)	B.t.u./hr	2547.16
"	B.t.u. (IST.)/hr	2545.50
"	B.t.u. (mean)/hr	2543.50
"	Cal., *gm.*/sec	178.298
"	Cal., *kg.*/hr	641.874
"	Ergs/sec	7.46×10^9
"	Foot-pounds/min	33013.3
"	Foot-pounds/sec	550.221
"	Horsepower	1.00040
"	Horsepower (boiler)	0.0760487
"	Horsepower (metric)	1.0142777
"	Horsepower (water)	0.999942
"	Joules/sec	746
"	Kilowatts	0.746
"	Watts	746
Horsepower (metric)	. B.t.u./hr	2511.31
"	. B.t.u. (IST.)/hr	2509.66
"	. B.t.u. (mean)/hr	2507.70
"	. Cal., *gm.*/hr	6.32838×10^5
"	. Cal., *gm.* (IST.)/hr	6.32425×10^5
"	. Cal., *gm.* (mean)/hr	6.31929×10^5
"	. Ergs/sec	7.35499×10^9
"	. Foot-pounds/min	32548.6
"	. Foot-pounds/sec	542.476
"	. Horsepower	0.986320
"	. Horsepower (boiler)	0.0749782

TO CONVERT FROM	TO	MULTIPLY BY
"	Horsepower (electric)	0.985923
"	Horsepower (water)	0.985866
"	Kg.-meters/sec	75
"	Kilowatts	0.735499
"	Watts	735.499
Horsepower (water)	Foot-pounds/min	33015.2
"	Horsepower	1.00046
"	Horsepower (boiler)	0.0760531
"	Horsepower (electric)	1.00006
"	Horsepower (metric)	1.01434
"	Kilowatts	0.746043
Kilogram-meters/sec	Watts	9.80665
Kilowatts	Horsepower (boiler)	0.101942
"	Horsepower (electric)	1.34048
"	Horsepower (metric)	1.35962
"	Joules/hr	3.6×10^6
"	Joules (IST.)/hr	3.59941×10^6
"	Joules/sec	1000
"	Kg.-meters/hr	3.67098×10^5
"	Kilowatts (Int.)	0.999835
"	Watts (Int.)	999.835
Kilowatts (Int.)	B.t.u./hr	3414.99
"	B.t.u. (IST.)/hr	3412.76
"	B.t.u. (mean)/hr	3410.08
"	B.t.u. (mean)/min	56.8347
"	B.t.u. (mean)/sec	0.947244
"	Cal., *gm.* (mean)/hr	859326
"	Cal., *gm.* (mean)/min	14322.1
"	Cal., *kg.*/hr	860.563
"	Cal., *kg.* (IST.)/hr	860
"	Cal., *kg.* (mean)/hr	859.326
"	Cu. cm.-atm./hr	3.55351×10^7
"	Cu. ft.-atm./hr	1254.91
"	Ergs/sec	1.000165×10^{10}
"	Foot-poundals/min	1.42406×10^6
"	Foot-pounds/min	44261.0

TO CONVERT FROM	TO	MULTIPLY BY
"	Foot-pounds/sec	737.684
"	Gram/cm./sec	1.01988×10^7
"	Horsepower	1.34124
"	Horsepower (boiler)	0.101959
"	Horsepower (electric)	1.34070
"	Horsepower (metric)	1.35985
"	Joules/hr	3.60059×10^6
"	Joules (Int.)/hr	3.6×10^6
"	Kg.-meters/hr	367158
"	Kilowatts	1.000165
Kilowatts	B.t.u./hr	3414.43
"	B.t.u. (IST.)/hr	3412.19
"	B.t.u. (mean)/hr	3409.52
"	B.t.u. (mean)/min	56.8253
"	B.t.u. (mean)/sec	0.947088
"	Cal., *gm.* (mean)/hr	859184
"	Cal., *gm.* (mean)/min	14319.7
"	Cal., *gm.* (mean)/sec	238.662
"	Cal., *kg.* (mean)/hr	859.184
"	Cal., *kg.* (mean)/min	14.3197
"	Cal., *kg.* (mean)/sec	0.238662
"	Cu. ft.-atm./hr	1254.70
"	Ergs/sec	1×10^{10}
"	Foot-poundals/min	1.42382×10^6
"	Foot-pounds/hr	2.65522×10^6
"	Foot-pounds/min	44253.7
"	Foot-pounds/sec	737.562
"	Gram-cm./sec	1.019716×10^7
"	Horsepower	1.34102
Watts	B.t.u./hr	3.41443
"	B.t.u. (mean)/hr	3.40952
"	B.t.u. (mean)/min	0.0568253
"	B.t.u./sec	0.000948451
"	B.t.u. (mean)/sec	0.000947088
"	Cal., *gm.*/hr	860.421
"	Cal., *gm.* (mean)/hr	859.184
"	Cal., *gm.* (20°C.)/hr	860.853

TO CONVERT FROM	TO	MULTIPLY BY
"	Cal., *gm.*/min	14.3403
"	Cal., *gm.* (IST.)/min	14.3310
"	Cal., *gm.* (mean)/min	14.3197
"	Cal., *kh.*/min	0.0143403
"	Cal., *kg.* (IST.)/min	0.0143310
"	Cal., *kg.* (mean)/min	0.0143197
"	Ergs/sec	1×10^7
"	Foot-pounds/min	44.2537
"	Horsepower	0.00134102
"	Horsepower (boiler)	0.000101942
"	Horsepower (elec.)	0.00134048
"	Horsepower (metric)	0.00135962
"	Joules/sec	1
"	Kilowatts	0.001
"	Liter-atm./hr	35.5282
Watts (Int.)	B.t.u./hr	3.41499
"	B.t.u. (mean)/hr	3.41008
"	B.t.u./min	0.569165
"	B.t.u. (mean)/min	0.0568347
"	Cal., *gm.*/hr	860.563
"	Cal., *gm.* (mean)/hr	859.326
"	Cal., *kg.*/min	0.0143427
"	Cal., *kg.* (IST.)/min	0.0143333
"	Cal., *kg.* (mean)/min	0.0143221
"	Ergs/sec	1.000165×10^7
"	Joules (Int.)/sec	1
"	Watts	1.000165
Horsepower (water)	Foot-pounds/min	33015.2
"	Horsepower	1.00046
"	Horsepower (boiler)	0.0760531
"	Horsepower (electric)	1.00006
"	Horsepower (metric)	1.01434
"	Kilowatts	0.746043
Kilogram-meters/sec	Watts	9.80665
Kilowatts	Horsepower (boiler)	0.101942
"	Horsepower (electric)	1.34048

TO CONVERT FROM	TO	MULTIPLY BY
"	Horsepower (metric)	1.35962
"	Joules/hr	3.6×10^6
"	Joules (IST.)/hr	3.59941×10^6
"	Joules/sec	1000
"	Kg.-meters/hr	3.67098×10^5
"	Kilowatts (Int.)	0.999835
"	Watts (Int.)	999.835
Kilowatts (Int.)	B.t.u./hr	3414.99
"	B.t.u. (IST.)/hr	3412.76
"	B.t.u. (mean)/hr	3410.08
"	B.t.u. (mean)/min	56.8347
"	B.t.u. (mean)/sec	0.947244
"	Cal., *gm.* (mean)/hr	859326
"	Cal., *gm.* (mean)/min	14322.1
"	Cal., *kg.*/hr	860.563
"	Cal., *kg.* (IST.)/hr	860
"	Cal., *kg.* (mean)/hr	859.326
"	Cu. cm.-atm./hr	3.55351×10^7
"	Cu. ft.-atm./hr	1254.91
"	Ergs/sec	1.000165×10^{10}
"	Foot-poundals/min	1.42406×10^6
"	Foot-pounds/min	44261.0
"	Foot-pounds/sec	737.684
"	Gram/cm./sec	1.01988×10^7
"	Horsepower	1.34124
"	Horsepower (boiler)	0.101959
"	Horsepower (electric)	1.34070
"	Horsepower (metric)	1.35985
"	Joules/hr	3.60059×10^6
"	Joules (Int.)/hr	3.6×10^6
"	Kg.-meters/hr	367158
"	Kilowatts	1.000165
Kilowatts	B.t.u./hr	3414.43
"	B.t.u. (IST.)/hr	3412.19
"	B.t.u. (mean)/hr	3409.52
"	B.t.u. (mean)/min	56.8253
"	B.t.u. (mean)/sec	0.947088

TO CONVERT FROM	TO	MULTIPLY BY
"	Cal., *gm.* (mean)/hr	859184
"	Cal., *gm.* (mean)/min	14319.7
"	Cal., *gm.* (mean)/sec	238.662
"	Cal., *kg.* (mean)/hr	859.184
"	Cal., *kg.* (mean)/min	14.3197
"	Cal., *kg.* (mean)/sec	0.238662
"	Cu. ft.-atm./hr	1254.70
"	Ergs/sec	1×10^{10}
"	Foot-poundals/min	1.42382×10^{6}
"	Foot-pounds/hr	2.65522×10^{6}
"	Foot-pounds/min	44253.7
"	Foot-pounds/sec	737.562
"	Gram-cm./sec	1.019716×10^{7}
"	Horsepower	1.34102
Watts	B.t.u./hr	3.41443
"	B.t.u. (mean)/hr	3.40952
"	B.t.u. (mean)/min	0.0568253
"	B.t.u./sec	0.000948451
"	B.t.u. (mean)/sec	0.000947088
"	Cal., *gm.*/hr	860.421
"	Cal., *gm.* (mean)/hr	859.184
"	Cal., *gm.* (20°C.)/hr	860.853
"	Cal., *gm.*/min	14.3403
"	Cal., *gm.* (IST.)/min	14.3310

TO CONVERT FROM	TO	MULTIPLY BY
"	Cal., *gm.* (mean)/min	14.3197
"	Cal., *kh.*/min	0.0143403
"	Cal., *kg.* (IST.)/min	0.0143310
"	Cal., *kg.* (mean)/min	0.0143197
"	Ergs/sec	1×10^7
"	Foot-pounds/min	44.2537
"	Horsepower	0.00134102
"	Horsepower (boiler)	0.000101942
"	Horsepower (elec.)	0.00134048
"	Horsepower (metric)	0.00135962
"	Joules/sec	1
"	Kilowatts	0.001
"	Liter-atm./hr	35.5282
Watts (Int.)	B.t.u./hr	3.41499
"	B.t.u. (mean)/hr	3.41008
"	B.t.u./min	0.569165
"	B.t.u. (mean)/min	0.0568347
"	Cal., *gm.*/hr	860.563
"	Cal., *gm.* (mean)/hr	859.326
"	Cal., *kg.*/min	0.0143427
"	Cal., *kg.* (IST.)/min	0.0143333
"	Cal., *kg.* (mean)/min	0.0143221
"	Ergs/sec	1.000165×10^7
"	Joules (Int.)/sec	1

Appendix F
Main Types of Clouds

GENERA

Cirrus. Detached clouds in the form of white, delicate filaments or white or mostly white patches or narrow bands. These clouds have a fibrous (hairlike) appearance, a silky sheen, or both.

Cirrocumulus. Thin, white patch, sheet, or layer of cloud without shading, composed of very small elements in the form of grains, ripples, etc., merged or separate, and more or less regularly arranged; most of the elements have an apparent width of less than 1° (approximately the apparent width of the little finger at arm's length).

Cirrostratus. Transparent, whitish cloud veil of fibrous (hairlike) or smooth appearance, totally or partly covering the sky, and generally producing halo phenomena.

Altocumulus. White or gray, or both white and gray, patch, sheet, or layer of cloud, generally with shading, composed of laminae, rounded masses, rolls, etc., which are sometimes partly fibrous or diffuse and which may or may not be merged; most of the regularly arranged small elements usually have an apparent width of between 1° and 5° (approximately the apparent width of three fingers at arm's length).

Altostratus. Grayish or bluish cloud sheet or layer of striated, fibrous, or uniform appearance, totally or partly covering the sky, and having parts thin enough to reveal the sun at least vaguely, as through ground glass. Altostratus does not show halo phenomena.

Nimbostratus. Gray cloud layer, often dark, the appearance of which is rendered diffuse by more or less continuously falling rain or snow, which in most cases reaches the ground. It is thick enough throughout to blot out the sun. Low, ragged clouds frequently occur below the layer, with which they may or may not merge.

Stratocumulus. Gray or whitish, or both gray and whitish, patch, sheet, or layer of cloud which almost always has dark parts, composed of tessellations, rounded masses, rolls, etc., which are nonfibrous (except for virga) and which may or may not be merged; most of the regularly arranged small elements have an apparent width of more than 5°.

From World Meteorological Organization. 1969. *International Cloud Atlas.* pp. 9–14.

Stratus. Generally gray cloud layer with a fairly uniform base, which may give drizzle, ice prisms, or snow grains. When the sun is visible through the cloud, its outline is clearly discernible. Stratus does not produce halo phenomena except, possibly, at very low temperatures. Sometimes Stratus appears in the form of ragged patches.

Cumulus. Detached clouds, generally dense and with sharp outlines, developing vertically in the form of rising mounds, domes, or towers, of which the bulging upper part often resembles a cauliflower. The sunlit parts of these clouds are mostly brilliant white; their base is relatively dark and nearly horizontal. Sometimes Cumulus is ragged.

Cumulonimbus. Heavy and dense cloud, with a considerable vertical extent, in the form of a mountain or huge towers. At least part of its upper portion is usually smooth, fibrous, or striated, and nearly always flattened; this part often spreads out in the shape of an anvil or vast plume. Under the base of this cloud, which is often very dark, there are frequently low ragged clouds either merged with it or not, and precipitation sometimes in the form of virga.

SPECIES

Following are the definitions of the various species. The cloud genera in which these species generally occur are also mentioned.

Fibratus. Detached clouds or a thin cloud veil, consisting of nearly straight or more or less irregularly curved filaments which do not terminate in hooks or tufts. This term applies mainly to Cirrus and Cirrostratus.

Uncinus. Cirrus often shaped like a comma, terminating at the top in a hook, or in a tuft the upper part of which is not in the form of a rounded protuberance.

Spissatus. Cirrus of sufficient optical thickness to appear grayish when viewed toward the sun.

Castellanus. Clouds that present, in at least some portion of their upper part, cumuliform protuberances in the form of turrets which generally give the clouds a crenellated appearance. The turrets, some of which are taller than they are wide, are connected by a common base and seem to be arranged in lines. The castellanus character is especially evident when the clouds are seen from the side. This term applies to Cirrus, Cirrocumulus, Altocumulus, and Stratocumulus.

Floccus. A species in which each cloud unit is a small tuft with a cumuliform appearance, the lower part of which is more or less ragged and often accompanied by virga. This term applies to Cirrus, Cirrocumulus and Altocumulus.

Stratiformis. Clouds spread out in an extensive horizontal sheet or layer. This term applies to Altocumulus, Stratocumulus, and, occasionally, to Cirrocumulus.

Nebulosus. A cloud like a nebulous veil or layer, showing no distinct details. This term applies mainly to Cirrostratus and Stratus.

Lenticularis. Clouds having the shape of lenses or almonds, often very elongated and usually with well-defined outlines; they occasionally show irisation. Such clouds appear most often in cloud formations of orographic origin, but may also occur in regions without marked orography. This term applies mainly to Cirrocumulus, Altocumulus, and Stratocumulus.

Fractus. Clouds in the form of irregular shreds, which have a clearly ragged appearance. This term applies only to Stratus and Cumulus.

Humilis. Cumulus clouds of only a slight vertical extent; they generally appear flattened.

Mediocris. Cumulus clouds of moderate vertical extent, the tops of which show fairly small protuberances.

Congestus. Cumulus clouds which are markedly spouting and are often of great vertical extent; their bulging upper part frequently resembles a cauliflower.

Calvus. Cumulonimbus in which at least some protuberances of the upper part are beginning to lose their cumuliform outlines but in which no cirriform parts can be distinguished. Protuberances and sproutings tend to form a whitish mass, with more or less vertical striations.

Capillatus. Cumulonimbus characterized by the presence, mostly in its upper portion, of distinct cirriform parts of clearly fibrous or striated structure, frequently having the form of an anvil, a plume, or a vast, more or less disorderly mass of hair. Cumulonimbus capillatus is usually accompanied by a shower, or by a thunderstorm, often with squalls and sometimes with hail; it frequently produces very well defined virga.

VARIETIES

The different varieties of clouds that have been established are defined below. The varieties *intortus, vertebratus, undulatus, radiatus, lacunosus,* and *duplicatus* refer to the arrangement of the cloud elements; the varieties *translucidus, perlucidus,* and *opacus* refer to the degree of transparency of the clouds. The cloud genera in which the varieties are encountered are added.

Intortus. Cirrus, the filaments of which are very irregularly curved and often seemingly entangled in a capricious manner.

Vertebratus. Clouds, the elements of which are arranged in a manner suggestive of vertebrae, ribs, or a fish skeleton. This term applies mainly to Cirrus.

Undulatus. Clouds in patches, sheets, or layers, showing undulations. These undulations may be observed in fairly uniform cloud layers or in clouds composed of elements, separate, or merged. Sometimes a double system of undulations is in evidence. This term applies mainly to Cirrocumulus, Cirrostratus, Altocumulus, Altostratus, Stratocumulus, and Stratus.

Radiatus. Clouds showing broad parallel bands or arranged in parallel bands, which, owing to the effect of perspective, seem to converge toward a point on the horizon or, when the bands cross the whole sky, toward two opposite points on the horizon, called radiation points. This term applies mainly to Cirrus, Altocumulus, Altostratus, Stratocumulus, and Cumulus.

Lacunosus. Cloud patches, sheets, or layers, usually rather thin, marked by more or less regularly distributed round holes, many of them with fringed edges. Cloud elements and clear spaces are often arranged in a manner suggestive of net or a honeycomb. This term applies mainly to Cirrocumulus and Altocumulus; it may also apply, though very rarely, to Stratocumulus.

Duplicatus. Superposed cloud patches, sheets, or layers, at slightly different levels, sometimes partly merged. This term applies mainly to Cirrus, Cirrostratus, Altocumulus, Altostratus, and Stratocumulus.

Translucidus. Clouds in an extensive patch, sheet, or layer, the greater part of which is sufficiently translucent to reveal the position of the sun or moon. (The varieties translucidus and opacus are mutually exclusive.) This term applies to Altocumulus, Altostratus, Stratocumulus, and Stratus.

Perlucidus. An extensive cloud patch, sheet, or layer, with distinct but sometimes very small spaces between the elements. The spaces allow the sun, the moon, the blue of the sky, or overlying clouds to be seen. (The variety perlucidus may be observed in combination with the varieties translucidus or opacus.) This term applies to Altocumulus and Stratocumulus.

Opacus. An extensive cloud patch, sheet, or layer, the greater part of which is sufficiently opaque to mask completely the sun or moon. This term applies to Altocumulus, Altostratus, Stratocumulus, and Stratus.

SUPPLEMENTARY FEATURES AND ACCESSORY CLOUDS

The definitions of the supplementary features and of the accessory clouds follow. The cloud genera in which they are usually encountered are also mentioned.

Supplementary Features

Incus. The upper portion of a Cumulonimbus spread out in the shape of an anvil with a smooth, fibrous, or straited appearance.

Mamma. Hanging protuberances, like udders, on the under surface of a cloud. This supplementary feature occurs mostly with Cirrus, Cirrocumulus, Altocumulus, Altostratus, Stratocumulus, and Cumulonimbus.

Virga. Vertical or inclined trails of precipitation (fallstreaks) attached to the undersurface of a cloud, which do not reach the earth's surface. This supplementary feature occurs mostly with Cirrocumulus, Altocumulus, Altostratus, Nimbostratus, Stratocumulus, Cumulus, and Cumulonimbus.

Praecipitatio. Precipitation (rain, drizzle, snow, ice pellets, hail, etc.) falling from a cloud and reaching the earth's surface. (Although precipitation is normally considered a hydrometeor, it is treated here as a supplementary feature because it appears as an extension of the cloud.) This supplementary feature is mostly encountered with Altostratus, Nimbostratus, Stratocumulus, Stratus, Cumulus, and Cumulonimbus.

Arcus. A dense, horizontal roll with more or less tattered edges, situated on the lower front part of certain clouds and having, when extensive, the appearance of a dark, menacing arch. This supplementary features occurs with Cumulonimbus and, less often, with Cumulus.

Tuba. Cloud column or inverted cloud cone, protruding from a cloud base; it constitutes the cloudy manifestation of a more or less intense vortex. This supplementary feature occurs with Cumulonimbus and, less often, with Cumulus.

Accessory Clouds

Pileus. An accessory cloud of small horizontal extent, in the form of a cap or hood above the top or attached to the upper part of a cumuliform cloud, which often penetrates it. Several pilei may fairly often be observed in superposition. Pileus occurs principally with Cumulus and Cumulonimbus.

Velum. An accessory cloud veil of a great horizontal extent, close above or attached to the upper part of one or several cumuliform clouds, which often pierce it. Velum occurs principally with Cumulus and Cumulonimbus.

Pannus. Ragged shreds sometimes constituting a continuous layer, situated below another cloud and sometimes attached to it. This accessory cloud occurs mostly with Altostratus, Nimbostratus, Cumulus, and Cumulonimbus.

Bibliography

WIND FORCES ON BUILDINGS

Ackeret, J. Der winddruck auf schornsteine mit kreisquerschnitt. *Schweiz. Bauzeitung* **108:** 25 (1936).

Akins, R. E., J. A. Peterka, and J. E. Cermak. Mean Pressure Coefficients for Model Buildings in a Turbulent Boundary Layer. Paper presented at Third U.S. Nat. Conf. Wind Engr. Res., Gainesville, Fla. Feb.–Mar. 1978.

Anderson et al. Lateral forces of earthquake and wind. *Proc. Separate No. 66, ASCE* **77** (April 1951).

Arnstein, K. and W. Klemperer. Wind pressures on the Akron Airship Dock. *J. Aeronaut. Sci.* **3:** 88 (1936).

ASCE. Wind forces on structures. *Trans ASCE* **126:** 1124–1198 (1962).

ASCE Sub-Committee No. 31. Wind bracing in steel buildings. Final report, Committee on Steel. *Trans. ASCE* **105:** 1713 (1940).

Bailey, A. Wind pressure on buildings. *J. Inst. Civ. Engrs.*, Select. Engrg. Paper No. 139 (1933).

——— and N. D. G. Vincent. Wind pressure on buildings including the effect of adjacent buildings. *Proc. Inst. Civ. Engrs., London* **20** (1942).

Baker, J. F. and E. F. Williams. The Effect of Wind Loads on Frames with Semi-rigid Connections. Final report, Steel Struct. Res. Comm., London (1936).

Base, T. E. and Badr, H. M. Airload Prediction Using Turbulence Models. Paper presented at Third U.S. Nat. Conf. Wind Engr. Res., Gainesville, Fla., Feb.–Mar. 1978.

Berneburg, H., H. Fiedler, and E. Mercker. Pressure Distribution on the Top of a Cube in Crossflow. Paper presented at Third U.S. Nat. Conf. Wind Engr. Res., Gainesville, Fla., Feb.–Mar. 1978.

Biron, F. L'action du vent sur les constructions. *Science of Indust. Constr.* **18**(16): 131 (1934).

Blanjean, L. L'action du vent sur les constructions. *Ossature Metallique*, Brussels, Belgium, February 1949.

Brekka, G. N. Wind pressures in various areas of the United States. *Building Materials and Structures Report 152*, National Bureau of Standards, pp. 4, 7 (April 1959).

Cissel, H. and L. M. Legatski. Aerodynamic Characteristics of Circular Arch Roof Structures. Eng. Res. Prof. M-518, University of Michigan, Ann Arbor, MI (1944).

Cohen, E. and H. Parrin. Design of multi-level guyed towers; wind loading. *Proc. Paper No. 1355, ASCE* **83** (ST5), September 1957.

———, J. W. Vellozzi, and S. Suh. Calculation of wind forces and pressures on antennas. *Ann. NY Acad. Sci.* **116**(1): 161–221 (June 1964).

———. Wind Loads on Towers. American Meteorological Society, *Meteorological Monographs.* **4:** 22 (1960).

Collings, G. F. Determining basic wind loads. *Proc. Paper No. 825, ASCE,* **81** (November 1955).

Corke, T. C. Sensitivity of Wind Loads on a Building Model. Paper presented at Third U.S. Nat. Conf. Wind Engr. Res., Gainesville, Fla., Feb.–Mar. 1978.

Davenport, A. G. Gust loading factors. *J. Struct. Div., ASCE* **93** (ST3): 11–34 (June 1967).

———. Note on the random distribution of the largest value of a random function with application to gust loading. *Proc. Inst. Civil Engnr., London* **28** (1964).

———. The application of statistical concept to the wind loading of structures. *Proc. Inst. Civ. Engnrs., London* **19** (1961).

———. The response of line like structures to a gusty wind. *Proc. Inst. Civ. Engnrs., London* **23** (1962).

Dryden, H. L. and G. C. Hill. Wind Pressure on a Model of the Empire State Building. Research Paper No. 545, National Bureau of Standards, 1933.

———. Wind Pressures on a Model of a Mill Building. U.S. Bureau of Standards Research Paper No. 301, 1931.

———. Wind Pressure on Circular Cylinders and Chimneys. Research Paper No. 221, National Bureau of Standards, 1930.

———. Wind Pressures on Structures. Sci. Paper 523, National Bureau of Standards, 1926.

Ferrington, H. Wind loads on buildings. *Struct. Engnr.* **21:** 497 (1943) and **22:** 15 (1944).

Giovannozzi, R. L'Azione del vente sulle construzioni. *Aerotecnica* **16:** 413 (1936).

Harris, R. I. Measurements of wind structure at heights up to 598 ft above ground level. *Proc. Symp. Wind Effects on Buildings and Structures.* Loughborough University, U.K. (1968).

Hirt, C. W. and L. R. Stein. Numerical Simulation of Wind Forces on Structures. Paper presented at Third U.S. Nat. Conf. Wind Engr. Res., Gainesville, Fla., Feb.–Mar. 1978.

Holdredge, E. W. and B. Reed. *Pressure Distribution on Buildings,* College Station, Texas, Texas Engineering Experiment Station, 1956.

Holmes, J. D. and R. J. Best. Model Tests of Turbulent Pressures on Isolated Houses. Paper presented at Third U.S. Nat. Conf. Wind Engr. Res., Gainesville, Fla., Feb.–Mar. 1978.

Horne, M. R. Wind loads on structures. *J. Inst. Civ. Engrs.* **33:** 155 (January 1950).

Howe, J. W. Wind pressure on elementary building forms evaluated by model tests. *Civ. Engnrng.* 42 (May 1952).

Irminger, J. O. V. and C. Nøkkentved. *Wind Pressures on Buildings.* Copenhagen, Denmark, 1930, 1936.

Jeary, A. P., B. E. Lee, and P. R. Sparks, The Determination of Wind Induced Modal Forces. Paper presented at Third U.S. Nat. Conf. Wind Engr. Res., Gainesville, Fla., Feb.–Mar. 1978.

Kareem, A. and J. E. Cermak. Dynamic Pressure Fluctuations on a Square Prism. Paper presented at Third U.S. Nat. Conf. Wind Engr. Res., Gainesville, Fla., Feb.–Mar. 1978.

Katzmayr, R. Wind pressure on engineering structures. *Oesterr. Ing. Arch. Verein, Vienna* **82**(21–22): 175 (1930) and (23–24): 191 (1930).

Klemin, A., E. B. Schaefer, and L. H. Burer, Jr. Aerodynamics of the Perisphere and Trylon at the World's Fair. *Trans. ASCE* **104:** 1449 (1939).

Lee, J. S., C. T. Crowe, and J. A. Roberson. Lateral force on H-sections in cross flows. Paper presented at Third U.S. Nat. Conf. Wind Engr. Res., Gainesville, Fla., Feb.–Mar. 1978.

Modi, V. J. and R. T. Mitry. Blockage effects for cylinders at low Reynolds numbers. Paper presented at Third U.S. Nat. Conf. Wind Engr. Res., Gainesville, Fla., Feb.–Mar. 1978.

Newberry, C. W., et al. The Nature of Gust Loading on Tall Buildings. BRS Current Paper 66/68, Watford, Herts, U.K., Building Research Station, August 1968.

Nøkkentved, C. *Wind Pressure on Buildings.* Publ. Int. Assoc. Bridge and Structural Engrg., Zurich, Switzerland. Vol. I, p. 365, 1932; Vol. II, p. 257, 1934.

Pagon, W. W. *Wind forces on structures: Plate girders and trusses. Proc. Paper No. 1711, ASCE* **84** (July 1958).

———. Closure to *Wind forces on structures: Plate girders and trusses. Proc. ASCE* **85** (September 1959).

Pris, R. L'Action du vent sur les batiments et constructions. *Trav. Architect, Construct.* **20:** 71 (1936).

Rathbun, J. C. Wind forces on a tall building. *Trans. ASCE* **105:** 1 (1940).

Roberson, J. A. and J. A. Crowe. Pressure Distribution on Model Buildings at Small Angles of Attack in Turbulent Flow. Paper presented at Third U.S. Nat. Conf. Wind Engr. Res., Gainesville, Fla., Feb.–Mar. 1978.

Scruton, C. and C. W. Newberry. On the estimation of wind loads for buildings and structural design. *Proc. Inst. Civil Engnrs.* **25:** 97–126 (June 1963).

———, R. E. Whitbread, and T. M. Charlton. An aerodynamic investigation of a tower block. NPL/Aero/1032, 1964.

Sherlock, R. H. Gust factors for the design of buildings. Publications Vol. 8, Internatl. Assoc. for Bridge and Structural Engrg., Zurich, Switzerland, 1947.

Sourochnikoff, B. Wind stresses in semi-rigid connections of steel framework. *Trans. ASCE* **115:** 382 (1950).

Stathopoulos, T., A. G. Davenport, and D. Surry. The Assessment of Effective Wind Loads Acting on Flat Roofs. Paper presented at Third U.S. Nat. Conf. Wind Engr. Res., Gainesville, Fla., Feb.–Mar. 1978.

Surry, D., T. Stathopoulos, L. Apperley, and A. G. Davenport. Current Research on Wind Loads on Low Rise Buildings. Paper presented at Third U.S. Nat. Conf. Wind Engr. Res., Gainesville, Fla., Feb.–Mar. 1978.

Sylvester, H. Wind Loads on Airship Hangars. 6th Natl. Aeronautical Meeting of ASME, 1932.

———. Wind pressure distribution on sharp-edged bodies. *Bygningsstatiske Meddelelser* **8:** 41 (1936).

Thompson, R. A. Protection of Small Buildings Against High Velocity Winds. Bull. 28, Fla. Engrg. and Industrial Experiment Sta., Gainesville, Fla., 1949.

Tyler, L. H. *The Big Storm.* New Haven, Connecticut: The City Printing Co., 1938.

van Koten, H. Some measurements on tall buildings in the Netherlands. Inst. T.N.O., B.B.C., Delft, 1967.

Vellozzi, J. W. and E. Cohen. Dynamic Response of Tall Flexible Structures to Wind Loading. NBS Building Sciences Series 30, National Bureau of Standards, U.S. Department of Commerce, Washington, D.C., Government Printing Office, 1970.

———. Gust response factors. *J. Struct. Div. ASCE* **94:** 1295–1313 (June 1968).

Vickery, B. J. On the Reliability of Gust Loading Factors. NBS Building Science Series 30, National Bureau of Standards, U.S. Department of Commerce, Washington, D.C., Government Printing Office, 1970.

———, and P. N. Georgiou. Wind Loads on Multiple Frames. Paper presented at Third U.S. Nat. Conf. Wind Engr. Res., Gainesville, Fla., Feb.–Mar. 1978.

ATMOSPHERIC CIRCULATION

Recent Studies

Barry, R. G. and R. J. Chorley. *Atmosphere, Weather and Climate.* London: Methuen, 1971.

Blumenstock, D. I. *The Ocean of Air.* New Brunswick, NJ: Rutgers University Press, 1959.

Fisher, R. M. *How About the Weather?* New York: Harper & Brothers, 1951.

Fujita, T. T. Thunderstorm Downbursts. Third U.S. National Conference Wind Engineering Research, University of Florida, Gainesville, Fla., Feb. 26–Mar. 1, 1978.

Hare, F. K. *The Restless Atmosphere.* Hutchinson: London, 1966.

Humphreys, W. J. *Physics of the Air.* New York: McGraw-Hill Co., 1940.

Lilly, D. K. and J. B. Klemp. Downslope Mountain Windstorms. Third U.S. National Conference Wind Engineering Research, University of Florida, Gainesville, Fla., Feb. 26–Mar. 1, 1978.

Panofsky, H. A. New Results on Wind Fluctuations, under Convective Conditions. Third U.S. National Conference Wind Engineering Research, University of Florida, Gainesville, Fla., Feb. 26–Mar. 1, 1978.

SethRaman, S. Turbulence during High-Wind Episodes. Third U.S. National Conference Wind Engineering Research, University of Florida, Gainesville, Fla., Feb. 26–Mar. 1, 1978.

Simiu, E., J. Bietry, and J. J. Filliben. Sampling Errors and Extreme Wind Estimates. Third U.S. National Conference Wind Engineering Research, University of Florida, Gainesville, Fla., Feb. 26–Mar. 1, 1978.

Tannehill, I. R. *Weather Round the World.* Princeton, NJ: Princeton University Press, 1951.

Wenstrom, W. H. *Weather and the Ocean of Air.* Boston: Houghton Mifflin Co., 1942.

Ancient Writings

Aristotle. *Meterologica.* Cambridge, Mass.: Loeb Classical Library, Harvard University Press, 1952.

Bacon, Sir Francis. *Naturall and Experimental History of Winds.* Translated by R. G., Gent., 1653.

Brooks, C. E. P. *Climate in Everyday Life.* New York: Philosophical Library, 1951.

Brooks, C. F. *Why the Weather?* New York: Harcourt, Brace & Co., 1924.

Crombie, A. C. *Medieval and Early Modern Science.* New York: Doubleday & Co., 1959.

Davis, W. M. *Elementary Meteorology.* Boston: Ginn & Co., 1894.

De la Rue, E. A. *Man and the Winds.* New York: Philosophical Library, 1955.

Department of Agriculture. *Climate and Man.* Washington, D.C.: Government Printing Office, 1941.

Espy, J. P. *The Philosophy of Storms.* 1841.

Ferrel. *A Popular Treatise on the Winds.* 1872.

Gautier, E.-F. *Le Sahara.* Paris: Payot, 1923.

Hann, Julius. *Handbook of Climatology.* 1903.

Hedin, Sven Anders. *Through Asia.* 1899.

———. *The Gobi Desert.* 1931.

Heninger, S. K., Jr. *A Handbook of Renaissance Meteorology.* Durham, North Carolina: Duke University Press, 1960.

Huntington, Ellsworth. *Civilization and Climate.* New Haven, Conn.: Yale University Press, 1924.

Inwards, Richard. *Weather Lore.* New York: John F. Rider, 1950.

Kendrew, W. G. *The Climates of the Continents.* New York: Oxford University Press, 1953.

Knox, Alexander. *Climate of the Continent of Africa.* New York: Cambridge University Press, 1911.

Koeppe and de Long. *Weather and Climate.* New York: McGraw-Hill Co., 1958.
Krick and Fleming. *Sun, Sea and Sky.* Philadelphia: J. B. Lippincott Co., 1954.
Wolf, A. *A History of Science, Technology and Philosophy in the 16th and 17th Centuries.* New York: Harper & Brothers, 1959. 2 vols.

WIND LOADS

Ackeret, J. Der winddruck auf schornsteine mit kreisquerschnitt. *Schweiz. Bauzeitung* **108:** 25 (1936).
Alton, J. Masts and guys under wind action. *Canadian Eng. J.* 1069 (1951).
Anderson, A. W., et al. Lateral forces of earthquake and wind. *Proc. Separate No. 66, ASCE* **77** (April 1951).
Arnstein, K. and W. Klemperer. Wind pressures on the Akron Airship Dock. *J. Aeronaut. Sci.* **3:** 88 (1936).
ASCE. Wind-bracing in Steel Buildings. Fifth progress report, Sub-Committee No. 31, Committee on Steel, Structural Division. *Proc. ASCE* **62:** 397 (March 1936).
———. Wind Bracing in Steel Buildings. Final Report, ASCE Sub-Committee No. 31, Committee on Steel. *Trans. ASCE* **105:** 1713 (1940).
Bailey, A. Wind pressure on buildings. *J. Inst. Civ. Engrs.* Select. Engrg. Paper No. 139 (1933).
——— and N. D. G. Vincent. Wind pressure on buildings including effects of adjacent buildings. *J. Inst. Civ. Engrs.* **20:** 243 (1943).
Baines, W. D. Effects of Velocity Distribution on Wind Loads and Flow Patterns on Buildings. N.P.L. Symposium, Paper 6.
Baker, J. F. and E. F. Williams. The Effect of Wind Loads on Frames with Semirigid Connections. Final report, Steel Struct. Res. Comm., London, England, 1936.
Bamber, D. *The Design of Towers, Masts and Pylons.* Richmond, Surrey, England: DATA, 1957.
Biron, F. L'Action du vent sur les constructions. *Science et Indust. Constr.* **18**(16): 131 (1934).
Blanjean, L. L'action du vent sur les constructions. *Ossature Metallique,* Brussels, Belgium, February 1949.
Chien, N., Y. Feng, H. Wang, and T. Liao. *Wind Tunnel Studies of Pressure Distribution on Elementary Building Forms.* Iowa Inst. of Hydr. Research, State Univ. of Iowa, Ames, Iowa, 1951.
Cissel, J. H. and L. M. Legatski. Aerodynamic characteristics of circular arch roof structures. Eng. Res. Prof. M-518, University of Michigan, Ann Arbor, Michigan, 1944.
Cohen, E. and H. Perrin. Design of multi-level guyed towers. *J. Struct. Div., Proc. Amer. Soc. Civ. Eng.,* Paper 1356 (1957).

Collins, G. F. Determining basic wind loads. *Proc. Paper No. 825, ASCE* **81** (November 1955).
Danish Standard. Normer for bygningskonstruktioner 1. Belastningsforskrifter. June 1948.
Davenport, A. G. *Wind Loads on Structures.* Div. of Building Research Paper No. 88, Ottawa, 1960.
Dryden, H. L. and G. C. Hill. *Wind Pressure on Model of the Empire State Building.* Research Paper No. 545, National Bureau of Standards, 1933.
———. *Wind Pressures on a Model of a Mill Building.* U.S. Bureau of Standards, Research Paper No. 301, 1931.
———. *Wind Pressure on Circular Cylinders and Chimneys.* Research Paper No. 221, National Bureau of Standards, 1930.
———. *Wind Pressure on Structures.* Sci. Paper 523, National Bureau of Standards, 1926.
Ferrington, H. Wind loads on buildings. *Struct. Engnr.* **21:** 497 (1943) and **22:** 15 (1944).
Giovannozzi, R. L'Azione del vente sulle construzioni. *Aerotecnica* **16:** 413 (1936).
Harris, C. L. Influence of neighboring structures on the wind pressure on tall buildings. *Pa. State Coll. Engrg. Experiment Sta., Ser. Bull.* 43 (1934); also *Natl. Bur Stds., J. Research* **12**(1): 103 (1934).
Horne, M. R. Wind loads on structures. *J. Inst. Civ. Engrs.* **33:** 155 (January 1950).
Howe, J. W. Wind pressure on elementary building forms evaluated by model tests. *Civ. Engnrng.* 42 (May 1952).
Hull, F. H. Stability analysis of multi-level guyed towers. *J. Struct. Div., Proc. Amer. Soc. Civ. Eng.* Paper 3091 (1962).
Irminger, J. O. V. and C. Nøkkentved. *Wind Pressures on Buildings.* Copenhagen, Denmark, 1930, 1936.
Ishizaki, H. *Effects of Wind Pressure Fluctuations on Structures.* International Research Seminar, Wind Effects on Buildings and Structures, N.R.C., Ottawa, 1967.
Katzmayr, R. Wind pressure on engineering structures. *Oesterr. Ing. Arch. Verein, Vienna* **82**(21–22): 175 (1930) and **82**(23–24): 191 (1930).
Klemin, A., E. B. Schaefer, and L. H. Burer, Jr. Aerodynamics of the Perishpere and Trylon at the World's Fair. *Trans. ASCE* **104:** 1449 (1939).
Nøkkentved, C. *Wind Pressure on Buildings.* Publ. Int. Assoc. Bridge and Structural Engrg., Zurich, Switzerland. Vol. I, p. 365, 1932; Vol. II, p. 257, 1934.
Pris, R. L'Action du vent sur les batiments et constructions. *Trav. Architect. Constr.* **20:** 71 (1936).
Rathbun, J. C. Wind forces on a tall building. *Trans. ASCE* **105:** 1 (1940).
Raymer, W. G., H. L. Nixon, and J. F. M. Maybrey. Tests on Callender's television mast in the duplex wind tunnel. NPL/Aero/268, 1954.
Scruton, C. and C. W. Newberry. On the estimation of wind loads for building and structural design. *Proc. Inst. Civ. Eng.* **25:** 97–126 (1963).

Smith, A. Windloads on buildings. *J. West. Soc. Engnrs.* (1912, 1914).

Sourochnikoff, B. Wind stresses in semi-rigid connections of steel framework. *Trans. ASCE* **115:** 382 (1950).

Sylvester, H. *Wind Loads on Airship Hangars.* 6th Natl. Aeronautical Meeting of ASME, 1932.

———. Wind pressure distribution on sharp-edged bodies. *Bygningsstatiske Meddelelser* **8:** 41 (1936).

Thompson, R. A. *Protection of Small Buildings against High Velocity Winds.* Bull. 28, Fla. Engrg. and Industrial Experiment Sta., Gainesville, Fla., 1949.

Tyler, L. H. (ed.). *The Big Storm.* New Haven, Conn.: The City Printing Co., 1938.

Van Erp, J. W. T. Wind-load standards in Europe. *Proc. Separate No. 42, ASCE* **76** (November 1950).

WIND-BUILDING INTERACTION IN THE URBAN ENVIRONMENT

Abbey, R. F., Jr. Characterization of Building Wake Effects. Presented at Third U.S. Nat. Conf. Wind Engineering Res., Gainesville, Fla., Feb.–Mar. 1978.

Beranek, W. J. and H. van Koten. Wind Environment around Buildings. Presented at Third U.S. Nat. Conf. Wind Engr. Res., Gainesville, Fla., Feb.–Mar. 1978.

Davenport, A. G., M. Hogan, and B. J. Vickery. An Analysis of Records of Wind Induced Building Movement and Column Strain Taken at the John Hancock Center (Chicago). Univ. of Western Ontario, Engr. Science Research Report BLWT-10-70, Sept. 1970.

Enrich, T. and J. Perlman. Dizzy heights. *Wall Street Journal,* p. 1 (Jan. 10, 1973).

———. Taming the tower. *Time,* p. 84 (Oct. 11, 1976).

Feasey, R. and D. H. Freeston. Wind pressure measurements on a multi-storey building. Presented at Third U.S. Nat. Conf. Wind Engr. Res., Gainesville, Fla., Feb.–Mar. 1978.

Flynn, M. D. The Relationship of Wind Engineering Research to the Architectural Design Process. Presented at Third U.S. Nat. Conf. Wind Engr. Res., Gainesville, Fla., Feb.–Mar. 1978.

Gandemer, J. Discomfort due to wind at approaches to buildings: Aerodynamic study of speed field in building groups. ADYM 12–73.

———. Wind Environment around Buildings: Aerodynamic Concepts. International Conf. on Wind Effects on Buildings and Structures, London, Heathrow, Sept. 1975.

——— and G. Barnaud. Simulation of dynamic properties of wind by neutral stability in the CSTB wind tunnel with limited bed. In French.

——— and A. Guyot. Methodological guide and practical plans for including phenomena

due to wind in designing urban groupings.
——— and H. Maille. Method of quantitative visualization applied to the study of air flow in groups of buildings.
——— and G. Narnaud. Discomfort due to wind at approaches to buildings: Aerodynamic study of speed fields in building groups, complementary study. ADYM 1–75.
Gapp, P. Tower finally stops shedding glass. *Chicago Tribune,* p. 9 (Sept. 22, 1976).
Hautoy, C. The CSTB wind tunnel with limited bed. ADYM 12–73.
Hunt, J. C. R. Wind tunnel experiments on the effects of wind on people. BRE (Sept. 1974).
———, E. C. Poulton, and J. C. Mumford. The effects of wind on people: New criteria based on wind tunnel experiments. *Bldg. Sci.*
Isyumov, N., and A. G. Davenport. The Ground Level Wind Environment in Built-up Areas. International Conf. on Wind Effects on Buildings and Structures, London, Heathrow, Sept. 1975.
Lawson, T. V. The Wind Environment of Buildings: A Logical Approach to the Establishment of Criteria. University of Bristol, Aeronautical Department, Eng. TVL/7301, 1973.
Mehta, K. C. Wind Loads on Flat Roof Area through Full Scale Experiment. Presented at Third U.S. Nat. Conf. Wind Engr. Res., Gainesville, Fla., Feb.–Mar. 1978.
Melbourne, N. H. and P. N. Joubert. Problems of Wind Flow at the Base of Tall Buildings and Structures. Tokyo, Conferences on Wind Effects on Buildings and Structures, 1971.
Penwarden, A. D. Acceptable wind speeds in towns. *Bldg. Sci.* **8**(2): Sept. 1973.
Sacre, C. Wind in the planetary limited bed. CLI 10–73.
Simiu, E., and D. W. Lozier. The buffeting of tall structures by strong winds. U.S. Dept. of Commerce, National Bureau of Standards, October 1975.
Vaicaitis, R. Transmission of wind-induced noise. Presented at Third U.S. Nat. Conf. Wind Engr. Res., Gainesville, Fla., Feb.–Mar. 1978.
Williams, D. Measurements of wind effects on tall buildings. Presented at Third U.S. Nat. Conf. Wind Engr. Res., Gainesville, Fla., Feb.–Mar. 1978.

WIND-TUNNEL WORK

Bradshaw, P. Simple wind tunnel degisn. NPL/Aero/1258, 1968.
Chien, N., Y. Feng, H. Wang, and T. Liao. Wind Tunnel Studies of Pressure Distribution on Elementary Building Forms. Iowa Inst. of Hydr. Research, State Uni. of Iowa, Ames, Iowa, 1951.
Counihan, J. The Structure and the Wind Tunnel Simulation of Rural and Adiabatic Boundary Layers. Symp. External Flows, University of Bristol, 1972.

Durgin, F. H. Some methods and techniques used at the Wright Brothers wind tunnel. Presented at Third U.S. Nat. Conf. Wind Engr. Res., Gainesville, Fla., Feb.–Mar. 1978.

Euromech 50. Symp. Wind Tunnel Simulation of the Atmospheric Boundary Layers. Tech. Univ. W. Berlin, 1974.

Fox, N. L. and B. Dayman, Jr. Preliminary Report on Paraboloidal Reflector Antenna Wind-Tunnel Tests. Report JPL, CP3, Calif., 1962.

Godshall, F. A. Artificial Development of an Atmospheric Boundary Layer in a Wind Tunnel. Ph.D. thesis, 117 pp. Maryland University, College Park.

Gould, R. W. F. and W. G. Raymer. Six-component balance on the 7′ × 7′ wind tunnel. NPL/Aero/1004, September 1962.

Johnson, F. and D. Batty. Maps for models. Span. (GB), 8(4), pp. 3–7, January 1972.

Maher, F. J., D. Frederick, E. R. Estes, and D. B. Steinman. Wind tunnel tests of suspension bridge section models. *Bull., Va. Polytech. Inst. Engr. Experiment Sta.* **41**(6), Series No. 69, 1948.

Maskell, E. C. A Theory of Blockage Effects on Bluff Bodies and Stalled Wings in a Closed Wind Tunnel. Paper No. Aero 2685, Royal Aircraft Estab., Farnborough, Hants.

Morgan, D. R. and M. D. Chutchlow. The design of a three component dynamic wind tunnel balance. The Marconi Co., Ltd., 1971.

NPL Report. (unspecified) Simulation of the Atmospheric Boundary Layer in a Wind Tunnel. Teddington, U.K., 1967.

Owen & Zienkiewicz. Production of uniform shear flow in a wind tunnel. *J. Fluid Mech.* **2:** 521–531 (1957).

Petersen, H. A type of wind tunnel for simulating phenomena in the natural wind. Advisory Group for Aeronautical Research and Development, Paris, 13 pp., October 1960.

Scruton, C. The use of wind tunnels in industrial aerodynamic research. Agard Report 309, October 1960.

White, K. C. Wind tunnel testing to determine the environmental wind conditions for the proposed centre development at Corby. Paper 34, *Proc. Symp. Wind Effects on Buildings and Structures,* Univ. Loughborough, 1968.

MYTHOLOGY AND HISTORY

Anesaki, M. *History of Japanese Religion.* London: Kegan Paul, 1930.

Baker, B. The forth bridge. *Engineering* **38:** 213 (August 1884).

Bixby, W. H. Wind pressure in engineering construction. *Engineering News-Record* **33:** 175 (1895).

Dowson, J. *Classical Dictionary of Hindu Mythology.* London: Kegan Paul, 1961.
Drioton, E. *The Religion of the Ancient East.* London: Burrs & Oates, 1959.
Dryden, H. L. and G. C. Hill. Wind Pressure on Structures. Sci. Paper 523, National Bureau of Standards, 1926.
Duchemin, N. V. Recherches experimentales sur les lois de la resistance des fluides. *Memorial de l'Artillerie* **5:** 65 (1842).
Eiffel, A. G. *Nouvelles Recherches sur la Resistance de l'Air et la Pression du Vent sur les Hangars de Dirigeables.* Paris, 1914. Translated by J. C. Hunsaker. New York: Houghton-Mifflin Co.
Farquharson, F. B. "Aerodynamic stability of suspension bridges," Univ. of Washington Bull. No. 116, Part I (1949); Part II (1950); Part III (1952); Part IV (1954); Part V (1954).
Flachsbart, O. and H. Winter. Modellversuche ueber die belastung von gitterfachwerken durch windkraefte. *Der Stahlbau* No. 9; p. 65; No. 10 (1934), p. 73; No. 8 (1935), p. 65; Trans. by Sandia Corp. AFSWP-464.
Fleming, R. *Wind Stresses in Buildings.* New York: John Wiley & Sons, 1930.
Galileo. Discorsi a due nuove scienze. Bologna, 1655.
Gurney, O. R. *The Hittites.* Harmondsworth, England: Penguin Books, 1953.
Harvey, E. D. *The Mind of China.* Yale University Press, 1933.
Hooke, S. H. *Babylonian and Assyrian Religion.* London: Hutchinson, 1953.
James, E. O. *Myth and Ritual in the Ancient Near East.* London: Thames & Hudson, 1958.
Kramer, S. N. *History Begins at Sumer.* London: Thames & Hudson, 1958.
Newton, Sir Isaac. *Philosophiae Naturalis Principia Mathematica.* 1687.
Pagon, W. W. Aerodynamics and the civil engineer. *Engineering News-Record* **112:** 348 (1934); **113:** 41, 456, 814 (1934); **114:** 582, 665, 742 (1935); **115:** 601 (1935).
Parrot, A. *Sumer.* London: Thames & Hudson, 1960.
Prandtl, L. and A. Betz. Ergebnisse der aerodynamischen versuchsanstalt (A VA) zu goettingen. Vol. I (1920); Vol. II (1923); Vol. III (1927) by O. Flachsbart; Vol. IV (1932), Berlin.
Reincourt, A. de. *The Soul of China.* London: Jonathan Cape, 1959.
Rose, H. J. *Ancient Roman Religion.* London: Hutchinson, 1949.
Sanders, N. K. (trans.). *The Epic of Gilgamesh.* Harmondsworth: Penguin Classics, 1960.
Smeaton, J. *An Experimental Investigation Concerning the Natural Powers of Water and Wind.* Phil. Trans. Roy. Soc., London, 1759.
Warner, R. *Men and Gods.* Harmondsworth: Penguin Books, 1952.
Watson, W. *China.* London: Thames & Hudson, 1961.
Weber, M. *The Religion of India.* London: Allen & Unwin, 1958.
Werner, E. T. C. *Myths and Legends of China.* London: Harrap, 1922.

VIBRATIONS

Anderson, A. N. Lateral forces of earthquake and wind. *Proc. Amer. Soc. Civ. Engnrs.* **77** (1951).

Anthony, K. C. *The Background to the Statistical Approach.* Paper 2, The Modern Design of Wind Sensitive Structures, Constr. Ind. Res. Inf. Assn., 1971.

Bishop, R. E. D. and D. C. Johnson. *Mechanics of Vibrations.* Cambridge University Press.

Chiu, A. N. L. and S. K. Takahashi. Dynamic response measurements of a guyed tower. Presented at Third U.S. Nat. Conf. Wind Engr. Res., Gainesville, Fla., Feb.–Mar. 1978.

Crandall, S. H. Random vibration. Cambridge, Mass.: MIT Press, 1963.

Davenport, A. G. Gust loading factors. *J. Struct. Div. Am. Soc. Civ. Engrs.* (June 1967).

———. The Treatment of Wind Loading on Tall Buildings. Symp. Tall Buildings, University of Southampton, 1966.

Den Hartog, J. P. *Mechanical Vibrations.* New York: McGraw-Hill, 1947.

Dickey, W. L. and G. B. Woodruff. The vibrations of steel stacks. *Trans. ASCE* **121:** 1054–1087 (1956).

Dockstader, E. A., W. F. Swiger, and E. Ireland. Resonant vibrations of steel stacks. *Trans. ASCE* **121** (1956).

Eaton, K. J. Damage Due to Tornadoes in SE England, 25 January 1971. Bldg. Res. Stn. Curr. Pap. CP 27/71, 1971.

Edwards, B. and A. E. Bryson. Active flutter control using generalized unsteady aerodynamic theory. *J. Guidance and Control, AIAA* (Jan. 1978).

Farquharson, F. B., G. S. Vincent, et al. Aerodynamic stability of suspension bridges with special reference to the Tacoma Narrows bridge. *Bull. 16, Univ. Washington Engng. Stn.* 1949–54.

——— and R. E. McHugh, Jr. Wind tunnel investigation of conductor vibration with use of rigid models. *Power Apparatus and Systems, AIEE* 871–878 (October 1956).

Frazer, R. A. and C. Scruton. A Summarized Account of the Severn Bridge Aerodynamic Investigation Report. NPL Aerodyn. Rep. 222, London, HMSO, 1952.

Harris, C. M. and C. E. Crede. *Shock and Vibration Handbook.* New York: McGraw-Hill, 1961.

Housner, G. W. and G. A. Brady. Natural periods of vibration of buildings. *J. Eng. Mech. Div., ASCE* **89**(EM4): 1963.

Hunt, J. The effect of wind upon people. Minutes, 39th Int. Fluid Mech. Res. Mtg., NPL, 1974.

Hurty, W. C. and M. F. Rubenstein. *Dynamics of Structures.* New Jersey: Prentice-Hall, 1964.

Jayachandran, P. and W. E. Saul. Dynamic response of tall buildings to stochastic wind. Presented at Third U.S. Nat. Conf. Wind Engr. Res., Gainesville, Fla., Feb.–Mar. 1978.

Johns, D. J. and R. J. Allwood. Wind-Induced Ovaling Oscillations of Circular Cylindrical Shell Structures. Paper 28, Loughborough Conf., 1968.

Klein, R. E., C. Cusano, and J. J. Stukel. Investigation of a Method to Stabilize Wind Induced Oscillations in Large Structures. Paper No. 72-WA/AUT-11, presented at the 1972 ASME Winter Annual Meeting, New York, NY, November 1972.

Klein, R. E. Methods for Vibrational Energy Dissipation. Presented at Third U.S. Nat. Conf. Wind Engr. Res., Gainesville, Fla., Feb.–Mar. 1978.

Klein, R. E. and H. Salhi. Optimal Vibration Control in Structures. Presented at Third U.S. Nat. Conf. Wind Engr. Res., Gainesville, Fla., Feb.–Mar. 1978.

Kramer, C. and H. J. Gerhardt. Wind Effects on Flat Roofed Buildings. Presented at Third U.S. Nat. Conf. Wind Engr. Res., Gainesville, Fla., Feb.–Mar. 1978.

Lee, J. S., C. T. Crowe, and J. A. Roberson. H-Section Oscillation in Turbulent Crossflows. Presented at Third U.S. Nat. Conf. Wind Engr. Res., Gainesville, Fla., Feb.–Mar. 1978.

Lin, Y. K. and S. T. Ariaratnam. Self-excited bridge motion in turbulent wind. Presented at Third U.S. Nat. Conf. Wind Engr. Res., Gainesville, Fla., Feb.–Mar. 1978.

Novak, M., J. F. Howell, and H. Tanaka. Vortex induced oscillation of cylinders. Presented at Third U.S. Nat. Conf. Wind Engr. Res., Gainesville, Fla., Feb.–Mar. 1978.

Nøkkentved, C. Vibrations Produced by Wind. Dansk Selskab for Bygningsstatik, Copenhagen, Denmark, 1941.

Ozker, M. S. and J. C. Smith. Factors influencing the dynamic behavior of tall stacks under the action of wind. *Trans. ASME* **78**(6): 1956.

Pagon, W. W. Vibration problems in tall stacks solved by aerodynamics. *Engineering News-Record,* July 12, 1934.

Parkinson, G. V. and V. J. Modi. Recent Research on Wind Effects on Bluff Two-Dimensional Bodies. Paper 18, Proc. Int. Res. Semin. Wind Effects on Buildings and Structures, Ottawa, 1967.

Penzien, J. Wind induced vibration of cylindrical structures. *Proc. Paper No. 1141, ASCE* **83** (EM1): January 1957.

Price, P. Suppression of the fluid-induced vibration of circular cylinders. *Proc. Paper No. 1030, ASCE* **82**(EM3): July 1956.

Report of the Special Committee of the Board of Directors. Failure of the Tacoma Narrows bridge. *Proc. Am. Soc. Civ. Engrs.,* June 1944.

Richards, D. J. Aerodynamic Properties of the Severn Crossing Conductor. NPL Symp. 16: Wind Effects on Buildings and Structures, 1963, London, HMSO, 1965.

Robson, J. D. *An Introduction to Random Vibration.* Edinburgh University Press, 1963.

Rojiani, K. and Y. Wen. Evaluation of Safety of Buildings under Wind Loadings. Presented at Third U.S. Nat. Conf. Wind Engr. Res., Gainesville, Fla., Feb.–Mar. 1978.

Roshko, A. The Flow Past a Circular Cylinder at Very High Reynolds Numbers. Cal. Inst. Tech., Pasadena, November 1960.

Scruton, C. Aerodynamics of structures. *Proc. International Research Seminar on Wind Effects on Buildings and Structures, Ottawa, 1967.*

———. On the wind excited oscillations of stacks, towers and masts. *Proc. Symp. Wind Effects on Buildings and Structures.* NPL, Teddington, U.K., 1963.

———. Note on the aerodynamic stability of truncated circular cones and tapered stacks. NPL/Aero/305.

——— and A. R. Flint. Wind-excited oscillations of structures. *Proc. Inst. Civ. Engnrs.* **27:** 673–702 (1964).

——— and D. E. J. Walshe. A means for avoiding wind excited oscillations of structures with circular or nearly circular cross sections. NPL/Aero/335, October 1957.

———, L. Woodgate, and A. J. Alexander. The aerodynamic investigation for the proposed Runcorn-Widnes suspension bridge. NPL Aerodyn. Rep. 29, 1955.

Soong, T. T. Active Control against Wind Induced Motion. Presented at Third U.S. Nat. Conf. Wind Engr. Res., Gainesville, Fla., Feb.–Mar. 1978.

Vickery, B. J. Fluctuating lift and drag on a long cylinder. NPL/Aero/1146, 1965.

——— and D. E. Walshe. An aerodynamic investigation for a proposed multi-flue smoke stack at Fawley power station. NPL Aerodyn. Rep. 1132, 1965.

Vincent, G. S. Golden Gate Bridge vibration studies. *Trans. Am. Soc. Civ. Engrs.* **127**(II): 1962.

Walsh, D. E. J. *Wind Excited Oscillations of Structures.* H. M. Stationery Office, London, 1972.

Ward, H. S. and R. Crawford. Wind-Induced Vibrations and Building Modes. National Research Council of Canada, Ottawa, 1965.

Wardlaw, R. L., K. R. Cooper, and H. P. A. H. Irwin. Vibration absorbers for bridge truss members. Presented at Third U.S. Nat. Conf. Wind Engr. Res., Gainesville, Fla., Feb.–Mar. 1978.

Weaver, W. Experimental Investigation of Wind-Induced Vibration in Antenna Members. Group Report No. 75–4 Lincoln Lab. Mass. Inst. Tech., Cambridge, Mass., August 26, 1959.

———. Wind-induced vibrations in antenna members. *ASCE J. Eng. Mech.* **87**(EM1): Feb. 1961.

Woodgate, L. and J. F. M. Mabrey. Further experiments on the use of helical strakes for avoiding wind excited oscillations of structure with circular or near circular cross section. NPL/Areo/381, June 1959.

Wootton, L. P. The oscillations of model circular stacks due to vortex shedding at Reynolds numbers in the range 10^5 to 3×10^6. Paper 18, *Proc. Symp. Wind Effects on Buildings and Structures,* Loughborough, 1968.

——— and C. Scruton. Aerodynamics Stability. Paper 5, The Modern Design of Wind Sensitive Structures, Constr. Ind. Res. In. Assn., 1971.

BUILDING MOTION

Cohen, H. L. and R. C. Anderson. Optimal Instrument to Measure Building Sway. Presented at Third U.S. Nat. Conf. Wind Engr. Res., Gainesville, Fla., Feb.–Mar. 1978.

Citko, S. G. and R. A. Parmelee. Across-wind response of high-rise buildings. Presented at Third U.S. Nat. Conf. Wind Engr. Res., Gainesville, Fla., Feb.–Mar. 1978.

Dalgliesh, W. A. Wind-induced motion of a tall office building. Presented at Third U.S. Nat. Conf. Wind Engr. Res., Gainesville, Fla., Feb.–Mar. 1978.

Friedmann, P. Design, Testing and Correlation of Aeroelastic Models. Presented at Third U.S. Nat. Conf. Wind Engr. Res., Gainesville, Fla., Feb.–Mar. 1978.

Kareem, A. and J. E. Cermak. Reduction of Wind Induced Motion of Buildings. Presented at Third U.S. Nat. Conf. Wind Engr. Res., Gainesville, Fla., Feb.–Mar. 1978.

——— and J. A. Peterka. Wind-Induced Response of High Rise Buildings. Presented at Third U.S. Nat. Conf. Wind Engr. Res., Gainesville, Fla., Feb.–Mar. 1978.

WIND'S DISPERSIVE ACTION

Fanaki, F. H. Wind as an air pollutant dispersive mechanism. Presented at Third U.S. Nat. Conf. Wind Engr. Res., Gainesville, Fla., Feb.–Mar. 1978.

Hoydysh, W., F. Sabetta, and R. Piva. A Combined Scale Model and Numerical Simulation of Pollutant Dispersion. Presented at Third U.S. Nat. Conf. Wind Engr. Res., Gainesville, Fla., Feb.–Mar. 1978.

Meroney, R. N., D. E. Neff, and J. E. Cermak. Transient Behavior of Dense Gas Plumes. Presented at Third U.S. Nat. Conf. Wind Engr. Res., Gainesville, Fla., Feb.–Mar. 1978.

Petersen, R. L. and J. E. Cermak. Numerical and Wind-Tunnel Modeling of Plume Rise. Presented at Third U.S. Nat. Conf. Wind Engr. Res., Gainesville, Fla., Feb.–Mar. 1978.

Ruscheweyh, H. and K. Fischer. A Wind Tunnel Investigation of Ground Level Concentration of a Plume in the Effect Range of a Power Plant with Large Natural Draught Cooling Towers. Presented at Third U.S. Nat. Conf. Wind Engr. Res., Gainesville, Fla., Feb.–Mar. 1978.

Skinner, G. T. and G. R. Ludwig. Atmospheric Boundary Layer Diffusion Modeling. Presented at Third U.S. Nat. Conf. Wind Engr. Res., Gainesville, Fla., Feb.–Mar. 1978.

Way, J. L., H. M. Nagib, A. C. Crawford, and T. C. Corke. Thermal Tracers Applied to Dispersion Studies. Presented at Third U.S. Nat. Conf. Wind Engr. Res., Gainesville, Fla., Feb.–Mar. 1978.

TURBULENCE

Bradshaw, P. *An Introduction to Turbulence and Its Measurement.* Oxford: Pergamon Press, 1971.

Cotes, D. Law of the wake in a turbulent boundary layer. *J. Fluid Mech.* (July 1956).

Connell, J. R. Measured wind and turbulence at 0 to 500m above the surface of a lake. Presented at Third U.S. Nat. Conf. Wind Engr. Res., Gainesville, Fla., Feb.–Mar. 1978.

Gibson, C. H. and R. B. Williams. Turbulence Structure in the Atmospheric Boundary Layer over Open Ocean. Paper 5, AGARD Conf. Proc. CP48, 1969.

Perkin, R. L. Space-Time Structure of Boundary Layer Winds. Presented at Third U.S. Nat. Conf. Wind Engr. Res., Gainesville, Fla., Feb.–Mar. 1978.

Tieleman, H. W. and S. E. Mullins. Atmospheric Turbulence Measurements at Wallops Island. Presented at Third U.S. Nat. Conf. Wind Engr. Res., Gainesville, Fla., Feb.–Mar. 1978.

SITING FOR WIND POWER UTILIZATION

Cheng, E. D. H. Tradeoff considerations between wind power potential and site selection. Presented at Third U.S. Nat. Conf. Wind Engr. Res., Gainesville, Fla., Feb.–Mar. 1978.

Corotis, R. B. A probabilistic basis for wind power siting. Presented at Third U.S. Nat. Conf. Wind Engr. Res., Gainesville, Fla., Feb.–Mar. 1978.

Eskinazi, S. and P. J. Brennan. Siting and concept design considerations for hydro pump back wind energy conversion system. Presented at Third U.S. Nat. Conf. Wind Engr. Res., Gainesville, Fla., Feb.–Mar. 1978.

Howell, W. E. Environmental Impact of Large Windpower Farms. Presented at Third U.S. Nat. Conf. Wind Engr. Res., Gainesville, Fla., Feb.–Mar. 1978.

Ossenbruggen, P. and G. Pregent. Coastal and offshore winds. Presented at Third U.S. Nat. Conf. Wind Engr. Res., Gainesville, Fla., Feb.–Mar. 1978.

Snow, J. W., M. Garstang, R. Pielke, and H. Cooper. The wind power potential of the coastal zone. Presented at Third U.S. Nat. Conf. Wind Engr. Res., Gainesville, Fla., Feb.–Mar. 1978.

WINDS–NATURE

Donely, P. Effective Gust Structure at Low Altitudes. NACA, TR 692, 1940.

Ekman, V. W. On the influence of the earth's rotation on ocean currents. *Arkiv. Mat. Astr. Fysik* (1905).

Gentry, R. C. Discussion of "Nature of Wind" by R. N. Sherlock. *Proc. ASCE* **85**(ST3). March 1959.

Hidy, G. M. *The Winds: The Origins and Behavior of Atmospheric Motion.* Princeton, N.J.: D. Van Nostrand Co., 1967.

Hughes, L. A. On the low-level wind structure of tropical storms. *J. Meteorol.* **9:** 422 (1952).

Humphreys, W. J. *Physics of the Air,* 3rd Ed. New York: McGraw-Hill, 1940.

Jensen, M. *Model-law for phenomena in natural wind. Ingenioren* **2**(4), 1959.

Lettau, H. *Atmosphaerische Turbulenz.* Akad. Verlagsges, Leipzig, 1939.

Nee, V. W., B. Cheung, V. Liu, A. Szewczyk, and K. T. Yang. Physical Modeling of the Atmospheric Surface Layer. Presented at Third U.S. Nat. Conf. Wind Engr. Res., Gainesville, Fla., Feb.–Mar. 1978.

O'Neill, P. G. G. Experiments to simulate a natural wind gradient. In the CAT, NPL/Aero/ 313, 1956.

Riehl, H. *Tropical Meteorology.* New York: McGraw-Hill, 1954.

Rossby, C. G. A generalization of the theory of the mixing length with applications to atmospheric and oceanic turbulence. *Meteorological Papers* **1**(4): 33. Mass. Inst. of Tech., Cambridge, Mass., 1932.

Roy, J. Properties of spectra of atmospheric turbulence at 100 meters. *Quart., Met. Soc.* **80:** 546 (1954).

Sacre, C. Experimental and theoretical study of the airflow over hill. Presented at Third U.S. Nat. Conf. Wind Engr. Res., Gainesville, Fla., Feb.–Mar. 1978.

Shellard, H. C. Extreme wind speeds over the U.K. for periods ending 1959. *Met. Mag.* **91:**39–47 (1962).

Sherlock, R. H. Closure to "Nature of the Winds." *ASCE* **86** (July 1960).

Tannehill, I. R. *Hurricanes.* Princeton, N.J.: Princeton University Press, 1938.

Teunissen, H. W. Measurement and Analysis of Wind Structure. Presented at Third U.S. Nat. Conf. Wind Engr. Res., Gainesville, Fla., Feb.–Mar. 1978.

Thom, H. C. S. Distributions of extreme winds in the United States. *ASCE* **86** (April 1960).

———. New distribution of extreme winds in the United States. To be published.

HURRICANES

Arakawa, H. and K. Suda. Analysis of winds, wind waves, and swells over the sea to the east of Japan during the typhoon of September 26, 1935. *Monthly Weather Review* **81:** 31–37 (1953).

Ballenzweig, E. M. *Seasonal Variation in the Frequency of North Atlantic Tropical Cyclones Related to the General Circulation.* National Hurricane Research Project, Report No. 9, 1957.

Bangs, N. H. Effects of the 1926 Florida hurricane upon engineer designed buildings. *Bull. Amer. Meteorol. Soc.* **10:** 46–47 (1929).

Beerbower, G. M. Hurricane effects on buildings at Hollywood, Florida. *Engineering News Record* **97:** 752 (1926).

Boughner, C. C. Hurricane Hazel. *Weather* **10:** 200–205 (1955).

Braham, R. R. An Exploratory Experiment in Hurricane Seeding. Proceedings, Technical Conference on Hurricanes, Miami Beach, Fla., November 20–22, 1958.

Brooks, C. F. Hurricane hazard in New England. *Geographical Review* (January 1939).

Bunting, D. C. A comparison of six great Florida hurricanes. *Weatherwise* **8:** 64–82 (June 1955).

Cline, I. M. Tropical cyclones. New York: Macmillan Company (1926).

Colón, J. A. A study of hurricane tracks for forecasting purposes. *Monthly Weather Review* **81:** 53–66 (1953).

———. On the formation of Hurricane Alice, 1955, with notes on other cold-season tropical storms. *Monthly Weather Review* **84:** 1–14 (1956).

Conner, W. C., R. H. Kraft, and D. L. Harris. Empirical methods for forecasting the maximum storm tide due to hurricanes and other tropical storms. *Monthly Weather Review* **85:** 113–116 (1957).

Deppermann, C. E. *Some Characteristics of Philippine Typhoons.* Manila, P.I.: Bureau of Printing, 1939.

Douglas, M. S. *Hurricane.* New York: Rinehart and Company, 1958.

Drumm, W. M. *The Pioneer Forecasters of Hurricanes.* Havana, Cuba, 1905.

Duane, J. E., Jr. The hurricane of September 2, 1935, at Long Key, Florida. *Bull. Amer. Met. Soc.* **16:** 238–239 (1935).

Dunn, G. E. Cyclogenesis in the tropical Atlantic. *Bull. Amer. Met. Soc.* **21:** 215–229 (1940).

———. Areas of hurricane development. *Monthly Weather Review* **84:** 47–51 (1956).

———. Tropical cyclones. In *Compendium of Meteorology.* Boston, Mass.: Amer. Met. Soc., 1951.

———, W. R. Davis, and P. L. Moore. Hurricanes of 1955. *Monthly Weather Review* **83:** 315–326 (1955).

Dunn and Miller. *Atlantic Hurricanes.* Baton Rouge, Louisiana: Louisiana State University Press, 1960.

Fassig, O. L. *Hurricanes of the West Indies.* Bull. No. 13, U.S. Weather Bureau, Washington, D.C., 1913.

Fawbush, E. J., R. C. Miller, and L. G. Starrett. An empirical method of forecasting tornado development. *Bull. Amer. Met. Soc.* **32:** 1–9 (1951).

Fletcher, R. D. Computation of maximum winds in hurricanes. *Bull. Amer. Met. Soc.* **36:** 246–250 (1955).

Frank, N. L. Hurricane Disaster Potential Along U.S. Coastlines. Presented at Third U.S. National Conference Wind Engineering Research, University of Florida, Gainesville, Fla., Feb. 26–Mar. 1, 1978.

Garriott, E. B. Forecasts and warnings. *Monthly Weather Review* **28:** 371–377 (1900).

———. The West Indian hurricane of September 1906. *Monthly Weather Review* **34:** 416–417 (1906).

———. Weather, forecasts and warnings for the month. *Monthly Weather Review* **37:** 829–831 (1909).

Gentry, R. C. Wind velocities during hurricanes. *Paper No. 2731, Trans. ASCE* 120, 169 (1955).

———, P. L. Moore, and A. M. Marshall. *Hoist Hurricane Warnings.* Miami Springs, Fla.: Weather Vane Publications, 1950.

Haggard, W. H. Where hurricanes begin. *Weatherwise* **12:** 145–150 (August 1959).

Hales, J. V. *An Evaluation of the Effects of Cloud Seeding in Southern Utah.* Bull. No. 46, Utah University, 1955.

Harris, D. L. Meteorological aspects of storm surge generation. Unpublished manuscript, U.S. Weather Bureau, Washington, D.C., 1957.

Haurwitz, B. The height of tropical cyclones and of the eye of the storm. *Monthly Weather Review* **63:** 45–49 (1935).

Higgs, R. L. Severe floods of October 12–15, 1954, in Puerto Rico. *Monthly Weather Review* **82:** 301–304 (1954).

Hoover, R. A. Empirical relationships of the central pressures in hurricanes to the maximum surge and storm tide. *Monthly Weather Review* **85:** 167–174 (1957).

Horiguti, Y. *Memoirs of the Imperial Marine Observatory, Kobe, Japan.* Nos. 2 and 3, 1926–1927.

Hubert, L. F. Frictional filling of hurricanes. *Bull. Amer. Meteorol. Soc.* **36:** 440–445 (1955).

———. A case study of hurricane formation. *J. Meteorol.* **12:** 486–492 (1955).

Hubert, L. F. and G. B. Clark. The hurricane surge. U.S. Weather Bureau, Washington, D.C., 34 pp., 1955.

——— and O. Berg. A rocket portrait of a tropical storm. *Monthly Weather Review* **83:** 119–124, 1955.

Hughes, L. A. On the low-level wind structure of tropical storms. *J. Meteorol.* **9:** 422–428 (1952).

Hurricanes and tropical storms 1887–1956. *Weatherwise* **10:** 132, 133, 139 (August 1957).

Jordan, C. L. Estimating the central pressures of tropical cyclones from aircraft data. National Hurricane Research Project, Report No. 10, 1957.

———. *Mean Soundings for the Hurricane Eye.* National Hurricane Research Project, Report No. 13, 1957.

Kessler, Edwin III. Eye region of Hurricane Edna, 1954. *J. Meteorol.* **15:** 264–270 (1958).

Klein, W. H. and J. S. Winston. The path of the Atlantic hurricane of September 1947 in relation to the hemispheric circulation. *Bull. Amer. Meteorol. Soc.* **28:** 447–452 (1947).

Knox, J. L. The storm Hazel. *Bull. Amer. Met. Soc.* **36:** 239–246 (1955).

Krueger, A. F. The weather and circulation of October 1954. *Monthly Weather Review* **82:** 296–300, 1954.

Krueger, D. W. *A Relation between the Mass Circulation through Hurricanes and Their Intensity.* University of Chicago, Dept. of Met., 1957.

Laban, J. Sequence of weather events, Hurricane Janet, September 26–27, Swan Island, West Indies. *Weatherwise* **9:** 121–124 (1955).

Landers, P. G. Hurricane winds for design. Third U.S. National Conference Wind Engineering Research, University of Florida, Gainesville, Fla., Feb. 26–Mar. 1, 1978.

Landers, P. G. and R. D. Mazzagatti. An Analysis of Hurricane Winds in Florida. Third U.S. National Conference Wind Engineering Research, University of Florida, Gainesville, Fla., Feb. 26–Mar. 1, 1978.

Ligda, M. G. H. Analysis of Small Precipitation Areas and Bands in the Hurricane of August 23–28, 1949. Mass. Inst. of Tech., Tech. Note No. 3, Cambridge, Mass., 1955.

Malkin, W. and Galway, J. G. Tornadoes associated with hurricanes. *Monthly Weather Review* **81:** 299–303 (1953).

Malkus, J. On the Thermal Structure of the Hurricane Core. Proceedings, Technical Conference on Hurricanes, Miami Beach, Fla., November 20–22, 1958.

———. On the structure and maintenance of the mature hurricane eye. *J. Meteorol.* **15:** 337–349 (1958).

——— and H. Riehl. On the Dynamics and Energetics of the Hurricane area. Proceedings, Technical Conference on Hurricanes, Miami Beach, Fla., November 20–22, 1958.

McDonald, W. F. On a hypothesis concerning normal development and disintegration of tropical hurricanes. *Bull. Amer. Meteorol. Soc.* **23:** 73–78 (1942).

Merceret, F. J. The Distribution of Turbulence in Hurricanes. Third U.S. National Conference Wind Engineering Research, University of Florida, Gainesville, Fla., Feb. 26–Mar. 1, 1978.

Miller, B. I. On the maximum intensity of hurricanes. *J. Meteorol.* **15:** 185–195 (1958).

———. The three dimensional wind structure around a tropical cyclone. National Hurricane Research Project, Report No. 15, 1958.

———. The use of mean layer winds as a hurricane steering mechanism. National Hurricane Research Project, Report No. 18, 1958.

———. Rainfall rates in Florida hurricanes. *Monthly Weather Review* **86:** 258–264 (1958).

——— and P. L. Moore. A Comparison of Hurricane Steering Levels. Proceedings, Technical Conference on Hurricanes, Miami Beach, Fla., November 20–22, 1958.

Mitchell, C. L. *West Indian Hurricanes and Other Tropical Cyclones of the North Atlantic Ocean. Monthly Weather Review.* Supplement No. 24, 1924.

Mook, C. P. and P. H. Kutschenreuter. Hurricane Rains and Floods of August 1955 Carolinas to New England. U.S. Weather Bureau Tech. Paper No. 26, Washington, D.C., 1956.

———, E. W. Hoover, and R. A. Hoover. Analysis of the movement of the hurricane off the East Coast of the U.S., October 12–14, 1947. *Monthly Weather Review* **85:** 243–248 (1957).

Moore, R. L. Forecasting the motion of tropical cyclones. *Bull. Amer. Meteorol. Soc.* **27:** 410–415 (1946).

Namias, J. Long range factors affecting the genesis and paths of tropical cyclones. *Proceedings, UNESCO Symposium on Typhoons,* Tokyo, November 9–12, 1954, pp. 213–219, 1955.

———. Tropical cyclones related to the general circulation. *Trans. NY Acad. Sci.* (1955).

———. Forms of the general circulation related to hurricane genesis and path. *Proc. Tech. Conf. on Hurricanes,* Miami Beach, Fla., November 20–22, 1958.

——— and C. R. Dunn. The weather and general circulation of August 1955. *Monthly Weather Review* **83:** 163–170 (1955).

Norton, G. Hurricane forecasting. Unpublished manuscript, Weather Bureau Office, Miami, Fla., 53 pp., c1947.

———. *Florida Hurricanes.* By R. W. Gray, revised by G. Norton. Washington, D.C.: U.S. Weather Bureau Publications, 6 pp., 1949.

Olascoaga, M. J. Some aspects of Argentine rainfall. *Tellus* **2:** 312–318, 1950.

Palmén, E. On the formation and structure of tropical hurricanes. *Geophysica* (3): 26–38 (1948).

———. Dynamics of Hurricanes. Lecture at Caribbean Hurricane Seminar, Ciudad Trujillo, D.R., February 18, 1956.

Piddington, H. *The Sailor's Hornbook.* London, 1851.

Redfield, A. C. and A. R. Miller. Memorandum on Water Levels Accompanying Atlantic Coast Hurricanes. U.S. Weather Bureau, Washington, D.C., 1955.

Reid, W. F. Forecasts and warnings. *Monthly Weather Review* **45:** 457–461 (1917).

Riehl, H. Waves in the Easterlies and the Polar Front in the Tropics. Dept. Meteor., University of Chicago, Misc. Reports No. 17, 1945.

———. Aerology of tropical storms. In *Compendium of Meteorology.* Boston, Mass.: Amer. Met. Soc., pp. 902–913, 1951.

———. Tropical meteorology. New York: McGraw-Hill Book Company, 1954.

———, W. H. Haggard, and R. W. Sanborn. On the prediction of 24-hour hurricane movement. *J. Meteorol.* **13:** 415–420 (1956).

——— and E. Palmén. Budget of angular momentum and energy in tropical cyclones. *J. Meteorol.* **14:** 150–159 (1957).

———. On the Production of Kinetic Energy from Condensation Heating. National Hurricane Research Project, Report No. 22, 25 pp., 1958.

Riehl, H. Comments on the formation of hurricanes. *Proc. Tech. Conf. on Hurricanes,* Miami Beach, Fla., November 20–22, 1958.

Rockney, V. D. *Hurricane Detection by Radar and Other Means.* Tropical Cyclone Symposium, Brisbane, Australia, 19 pp., 1956.

Rossby, C. G. On the mechanism for the release of potential energy in the atmosphere. *J. Meteorol.* **6:** 163–180 (1949).

Schoner, R. W. Frequency and distribution of areal rainfall averages associated with tropical storms entering the Texas coast. Unpublished manuscript, U.S. Weather Bureau, Washington, D.C., 9 pp., 1957.

——— and S. Molansky. *Rainfall Associated with Hurricanes.* National Hurricane Research Project, Report No. 3, 305 pp., 1956.

Senn, H. V., H. W. Hiser, and R. C. Bourett. *Studies of Spiral Bands as Observed on Radar.* University of Miami, Coral Gables, Fla., 21 pp., 1957.

Senn, H. V. and H. W. Hiser. The origin and behavior of hurricane spiral bands as observed on radar. *Proc. Seventh Weather Radar Conference,* Miami Beach, Fla., November 17–20, 1958.

Sherman, L. On the propagation of hurricanes. *Trans. Amer. Geophys. Union* **31:** 531–535 (1950).

Simpson, R. H. On the movement of tropical cyclones. *Trans. Amer. Geophys. Union* **27:** 641–645 (1946).

———. Exploring the eye of Typhoon Marge, 1951. *Bull. Amer. Meteorol. Soc.* **33:** 286–298 (1952).

———. Hurricanes. *Scientific American* (June 1954).

———. On the structure of tropical cyclones as studied by aircraft reconnaissance. *UNESCO Symp. on Typhoons,* Tokyo, November 9–12, 1954, pp. 129–150, 1955.

——— and H. Riehl. Mid-tropospheric ventilation as a constraint on hurricane development and maintenance. *Proc. Tech. Conf. on Hurricanes,* Miami Beach, Fla., November 20–22, 1958.

Staff, National Hurricane Research Project. *Details of Circulation in the High Energy Core of Hurricane Carrie.* National Hurricane Research Project, Report No. 25, 15 pp., 1958.

Sverdrup, H. U. *Oceanography for Meteorologists.* New York: Prentice-Hall 1942.

Tannehill, I. R. Sea swells in relation to the movement and intensity of tropical storms. *Monthly Weather Review* **64:** 231–238 (1936).

———. *Hurricanes, Their Nature and History.* Princeton, N.J.: Princeton University Press, 1938.

———. *Hurricanes.* Princeton, N.J.: Princeton University Press, 1950.

———. *Hurricane Hunters.* New York: Dodd, Mead, and Company, 1954, 1956.

Tepper, M. A theoretical model for hurricane radar bands. *Proc. Seventh Weather Radar Conf.,* Miami Beach, Fla., November 17–20, 1958.

U.S. Army Corps of Engineers. Hurricanes affecting the Texas coast from Galveston to the Rio Grande. Tech. Memo. No. 78, March 1956.

U.S. Navy Bureau of Aeronautics. *Marine Climatic Atlas of the World. Vol. 1, North Atlantic Ocean.* 1955.

———. *Intensification of Tropical Cyclones, Atlantic and Pacific Areas.* Fourth Research Report, Project AROWA, Norfolk, Va., 28 pp., 1956.

U.S. Weather Bureau. *Atlas of Climatic Charts of the Oceans.* Washington, D.C., 1938.

Veigas, K. W. and R. G. Miller. Probabilistic Prediction of Hurricane Movement by Synoptic Climatology. Technical Conference on Hurricanes, Miami Beach, Fla., November 20–22, 1958.

Vinẽs, B. Investigation of the Cyclonic Circulation and the Translatory Movement of West Indian Hurricanes. U.S. Weather Bureau, Washington, D.C., 1898.

Visher, S. *Tropical Cyclones of the Pacific.* Bernice P. Bishop Museum, Bull. No. 20, Honolulu, T. H., 1925.

Weems, J. E. *A Weekend in September.* New York: Henry Holt and Co., 1957.

Wexler, H. Structure of hurricanes as determined by radar. *Annals of New York Academy of Science* **48:** 821–844 (1947).

Willett, H. C. *A Study of Tropical Hurricanes along the Atlantic and Gulf Coasts of the United States.* New York: Inter-regional Insurance, 1955.

Winston, J. S. The weather and circulation of August 1954, including a discussion of Hurricane Carol in relation to the planetary waves. *Monthly Weather Review* **82:** 228–236 (1954).

Yeh, T. C. The motion of tropical storms under the influence of a superimposed southerly current. *J. Meteorol.* **7:** 108–113 (1950).

WIND EFFECTS ON BRIDGES

ASCE. Aerodynamic stability of bridges. Report of the Advisory Board on the Investigations of Suspension Bridges, *ASCE* **120:** 721 (1955).

Baird, R. C. Wind-induced vibrations of a pipeline suspension bridge, and its cure. *ASME* Paper No. 54-PET-12, July 1955.

———. Wind-induced vibrations of a pipe-line suspension bridge and its cure. *Trans. ASME* **77** (August 1955).

Biggs, J. M. Wind loads on truss bridges. *Trans. ASCE* **119:** 879 (1954).

———, S. Namyet, and J. Adachi. Wind loads on girder bridges. *Trans. ASCE* **121:** 101 (1956).

Farquharson, F. B. *Aerodynamic Stability of Suspension Bridges.* Pt. III, Univ. of Washington, Bull. No. 116, 1952; Pt. IV, 1954.

NPL/Aero/222. A Summarized Account of the Severn Bridge Aerodynamic Investigation. H.M.S.O., 1952.

NPL/Aero/1052. An Investigation of the Aerodynamic Stability of the Towers Proposed for the River Severn Suspension Bridge. 1963.

Scruton, C. Aerodynamic buffeting of bridges. *The Engineer* **199:** 654–667 (May 13, 1955).

——— and D. E. J. Walshe. The Aerodynamic Investigation for the Proposed Tamar Suspension Bridge. NPL/Aero/311, 1956.

Selberg, A. Aerodynamic Effects on Suspension Bridges. N.P.L. Symposium, Paper 11 (ref. 11).

———. Dampening effects on suspension bridges. *Int. Assoc. Bridge & Struct. Eng.* **10.**

Smith, F. C. and G. S. Vincent. Aerodynamic stability of suspension bridges. Pt. II, Mathematical Analyses, Univ. of Washington. Bull. No. 116, 1950.

Vincent, G. S. *Aerodynamic Stability of Suspension Bridges.* Pt. V, Univ. of Washington, Bull. No. 116, 1954.

———. *Investigation of Wind Forces on Highway Bridges.* Highway Research Board Special Report 10, No. 272, 1953.

———. Summary of Laboratory and Field Studies in the U.S.A. on Wind Effects on Suspension Bridges. N.P.L. Symposium, Paper 20 (ref. 11).

Walshe, D. E. J. Use of models to predict the oscillatory behavior of suspension bridges in wind. N.P.L. Symposium, Paper 17 (ref. 11).

Walshe, D. E. J. and M. A. Packer. Investigation of the Aerodynamic Stability of the Forth Suspension Bridge at Various Stages of Construction. NPL/Aero/1058, 1963.

Williams, D. H., H. L. Nixon, and W. C. Skelton. *Experiments on a Static Model of a Section of the Severn Bridge.* Natl. Physical LAb., England, Aero/209.

———. *Tests of Plate Girder Bridges in the Duplex Wind Tunnel.* Natl. Physical Lab., England, Aero/216.

Wind Resistance of Lattice Girder Bridges. London, England: Inst. of Civ. Engrs., 1948.

Wyatt, T. A. Effect of Wind on Slender Long-Span Bridges. Paper 36, Loughborough Conference.

WIND EFFECTS ON SUSPENDED LINES

Barray, A. Action of sleet on the conductors of overhead lines. *C.I.G.R.E.* VIII (1935).

Bleich, F. Dynamic instability of truss-stiffened suspension bridges under wind action. *Trans. ASCE* **114** (1949).

———, C. B. McCullough, R. Rosecrans, and G. S. Vincent. *The Mathematical Theory of Vibration in Suspension Bridges.* Bur. of Pub. Rds., 1950.

Carroll, J. S. Laboratory studies of conductor vibrations. *Trans. AIEE* **55:** 543 (1936).

Davis, D. A., D. J. Richardson, and R. A. Scriven. Investigation of Conductor Oscillation on the 275 kV Crossing over the Rivers Severn and Wye. Paper 4102P, Proc. Inst. Elect. Engrs., 1962.

Edwards, A. T. and A. Madeyski. Progress report on the investigation of galloping transmission line conductors. *Trans. AIEE,* Paper No. 55–213, 1956.

Farquharson, F. B. Aerodynamic Stability of Suspension Bridges. Univ. of Washington Engrg. Experiment Sta., Bull 116, Part III, 1952.

Hogg, A. D. and A. T. Edwards. The Status of the Conductor Galloping Problem in Canada. N.P.L. Symposium, Paper 12, H.M.S.O., 1965.

Krishnasamy, S. G. Wind and Ice Loads on Overhead Transmission Lines. Presented at Third U.S. Natl. Conf. Wind Engr. Research, Gainesville, Fla., Feb.–Mar. 1978.

Manuzio, C. Wind Effects on Suspended Cables. International Research Seminar on Wind Effects on Buildings, Ottawa 1967.

Richardson, A. S., J. R. Martucelli, and W. S. Price. Research Study on Galloping of Electric Power Transmission Lines. N.P.L. Symposium, H.M.S.O., 1965.

Sewell, P. and P. N. Taylor. The Aerodynamic Effects of Wind on Stranded Cables. University of Bristol, Aero. Rep. No. 58, 1961.

Simpson, A. and T. V. Lawson. Oscillations of "Twin" Power Transmission Lines. Loughborough Symposium, Paper 25, 1968.

Steinman, D. B. Aerodynamic theory of bridge oscillations. *Trans. ASCE* **115:** 1180 (1950).

Wind Effects on Bridges and Other Flexible Structures. Notes on Applied Science, No. 11, Natl. Physical Lab., England, 1955.

WIND-INDUCED COLLAPSES

Page, J. K. Field Investigations of Wind Failures in the Sheffield Gales of 1962. Loughborough University, 1968.

Report of the Committee of Inquiry into Collapse of Cooling Towers at Ferrybridge, Central Electricity Generating Board, November 1965.

Tacoma report. Aerodynamic Stability of Suspension Bridges, Part I, Univ. of Washington, Eng. Corp. Stn., Bulletin No. 116.

WIND-RELATED CATASTROPHES

Abercrombie, G. F. December fog in London and the emergency bed service. *Lancet* **1:** 234–235 (1953).

Firket, J. Fog along the Meuse Valley. *Trans. Faraday Soc.* **32:** 1192–1197 (1936).

Glasser, M., L. Greenburg, and F. Field. Mortality and morbidity during a period of high levels of air pollution. *Arch. Environ. Health* **15:** 684–694 (1967).

Logan, W. P. D. Mortality in the London fog incident, 1952. *Lancet* **1:** 336–338 (1953).

Ochsner, A. The hazards of air pollution, fact or fiction? *Medical Tribune* (November 30, 1967).

Snodderly, D. M., Jr. Biomedical and social aspects of air pollution. In *Advances in Environmental Science and Technology,* Vol. 3. New York: John Wiley & Sons, 1974.

INTERIOR VENTILATION

Building Digest No. 49, Central Building Research Institute, Roorkee, India, Jan. 1967.

Givoni, B. Basic Study of Ventilation Problems in Housing in Hot Countries. Research Report to Ford Foundation, Building Research Station, Technion, Haifa, 1962.

———. Laboratory study of the effect of window size and location on indoor air motion. *Architect. Sci. Rev.* **8:** 42–46 (June 1965).

———. Ventilation Problems in Hot Countries. Research Report to Ford Foundation, Building Research Station, Technion, Haifa, 1968.

———. Man, Climate and Architecture. Building Research Station, Technion, Israel Institute of Technology.

Hollemen, T. R. Air Flow through Conventional Window Openings. Research Report No. 33, Texas Engineering Experiment Station, College Station, Texas, 1951.

van Straaten, J. E. *Thermal performance of buildings.* New York: Elsevier Publishing, 1967.

Wise, A. E. E., D. E. Sexton, and M. S. T. Lillywhite. Studies of air flow round buildings. *Architects' J.* **141:** 1185–1189 (May 19, 1965).

SHELTERBELT AND WINDBREAKS

Bates, C. G. The Windbreak as a Farm Asset. Fmrs. Bull. U.S. Dept. Agric. No. 1405 (rev.), 1944.

Caborn, J. M. Shelterbelts and Microclimate. *For. Comm. Bull.* No. 29, H.M.S.O., 1957.

———. *Shelterbelts and Windbreaks.* London: Faber & Faber, 1965.

Cadman, W. A. Shelterbelts for Western Hill Farms. For. Comm. For. Record No. 22, H.M.S.O., 1963.

Eimern, J. van (Ed.) Windbreaks and Shelterbelts. World Meteorological Organization Tech. Note No. 59, Geneva, 1964.

Jensen, M. *Shelter Effect.* Copenhagen: Danish Technical Press, 1954.

Kreutz, W. *Der windschutz.* Dortmund, West Germany: Ardey Verlag, 1952.

WIND-CARRIED POLLUTANTS AND HUMAN COMFORT

American Conference Government Industrial Hygienists. Threshold Limit Values for 1967. Cincinnati, Ohio, 1967.

Barrett, C. F. Dispersion of Pollution Experiments on Models and at Full Scale. Symp. External Flows, Univ. Bristol, 1972.

Bersin, T. Schä digungen von enzymsystemen durch toxische verbindungen der autoabgase. *Vitalstoffe und Zivilisationskrankheiten* **11**(55): 207–210 (1966).

California State Dept. Publ. Health. Clean Air for California. Berkeley, Calif., 1955.

Davenport, S. J. and G. G. Morgis. *Air Pollution; A bibliography.* U.S. Bureau of Mines Bull. 537, 448 pp., 1954.

Dubrovskaya, F. I. Hygiene evaluation of pullution of atmospheric air of a large city with SO_2. In: V. A. Ryazanov, ed., *Limits of Allowable Concentrations of Atmospheric Pollutants* (translated by B. S. Levine). Washington, D.C.: U.S. Dept. Commerce, 1957.

Glasser, M., et al. Mortality and morbidity during a period of high levels of air pollution. *Arch. Environ. Health* **15:** 684–694 (Dec. 1967).

Goldsmith, J. R. Effects of air pollution on human health. In A. C. Stern, ed., *Air Pollution,* 2nd ed., vol. 1, pp. 547–615. New York: Academic Press, 1968.

Greenburg, L., et al. Air pollution, influenza, and mortality in New York City. *Arch. Environ. Health* **15:** 430–438 (1967).

———. Report of an air pollution incident in New York City, November 1953. *Publ. Health Rept.* **77** (1962).

Heimann, H. Effects of air pollution on human health. In *World Health Organization,* ed., Air Pollution, Monogr. Ser. No. 46, pp. 159–220, 1961.

Henschler, A., et al. Olfactory threshold of some important gases and manifestations in man by low concentrations. *Arch. Gewerbepathol. und Gewerbehyg.* **17:** 547–570 (1960).

Holland, G. J., et al. Air pollution simulation and human performance. *Amer. J. Publ. Health* **58:** 1684–1691 (Sept. 1968).

Hueper, W. C., et al. Carcinogenic bioassays on air pollutants. *Arch. Pathol.* **74:** 89–116 (1962).

Hueter, F. G., et al. Biological effects of atmospheres contaminated by auto exhaust. *Arch. Environ. Health* **12:** 553–560 (May 1966).

Kotin, P. and H. L. Falk. Atmospheric factors in pathogenesis of lung cancer. In A. Haddow, ed., *Advances in Cancer Research,* vol. 7, pp. 475–514. New York: Academic Press, 1963.

Lawther, P. J. Cancerogen in der städischen luft *Zeitschrift für Bakteriologie* **1,** Abt. 176, Originale, pp. 187–193, 1959.

——— et al. Carbon monoxide in town air. An interim report. *Ann. Occup. Hyg.* **5:** 241–248 (1962).

Los Angeles County Air Pollution Control. District Quart. Contaminant Reports, 1955–1962.

Lynn, D. A., et al. The November–December 1962 Air Pollution Episode in the Eastern U.S. USDHEW, PHS Publ. No. 999-AP-7, Cincinnati, Ohio, Sept. 1964.

Manos, N. E. Comparative Mortality among Metropolitan Areas in the U.S., 1949–1951. PHS. Publ. No. 562, 1957.

Martin, A. E. Mortality and morbidity statistics and air pollution. *Proc. Roy. Med. Soc.* **57:** 969–975 (1964).

McCarroll, T. Measurements of morbidity and mortality related to air pollution. *J Air Poll. Contr. Assoc.* **17**(4): 203–209 (1967).

Meyers, F. H. and C. H. Hine. Some Experience of NO_2 in Animals and Man. Presented at 5th Air Poll. Med. Res. Conf., Los Angeles, 1961.

Nat. Tub. Resp. Dis. Assoc. *Air Pollution Primer.* New York: Nat. Tub. Resp. Dis. Assoc., 1969.

Pemberton, J. and C. Goldberg. Air pollution and bronchitis. *Brit. Med. J.* **4887:** 567–570 (1954).

Remmers, J. E. and O. J. Balchum. Effects of Los Angeles Urban Pollution upon Respiratory Function of Emphysematous Patients. Presented at AMA Air Poll. Med. Res. Conf., Los Angeles, 1966.

Russell, W. T. The relative influence of fog and low temperature on the mortality from respiratory disease. *The Lancet* 1128–1130 (Nov. 1926).

Schrenk, H. H., et al. Air pollution in Donora, Pa., epidemiology of the unusual smog episode of October, 1948. *Publ. Health Bull.* 306 (1949).

Scott, J. A. The London fog disaster. *Proc. 20th Ann. Clean Air Conf. Glasgow* 25–27 (1953).

Speizer, F. E. An epidemiological appraisal of the effects of ambient air on health: Particulates and oxides of sulphur. *JAPCA* **19**(9): 647–655 (Sept. 1969).

Stanford Res. Inst. Literature Review of Metropolitan Air Pollutant Concentrations—Preparation, Sampling and Essay of Synthetic Atmospheres. Final Rpt., Menlo Park, Calif., 1956.

Sterling, T. D., et al. Measuring the effect of air pollution on urban morbidity. *Arch. Environ. Health* **18:** 485–494 (April 1969).

Stokinger, H. E. and D. L. Coffin. Biologic effects of air pollutants. In: A. C. Stern, ed. *Air Pollution,* 2nd ed., vol. 1, pp. 445–546. New York: Academic Press, Inc. (1968).

Tomono, Y. Effects of SO_2 on human pulmonary functions. *Japan J. Ind. Health* **3:** 77–85 (1961).

Waller, R. E. Air pollution as an etiological factor in lung cancer. Publ. of Acta Union Int. contre le Cancer, 15, p. 437, 1959.

——— and B. T. Commins. Episodes of high pollution in London 1952–66. *Int. Clean Air Congr. Proc.* **1:** 228–231 (London, October 1966).

Wayne, W. S., et al. Oxidant air pollution and athletic performance. *J. Am. Med. Assoc.* **199**(12): 151–154 (March 1967).

Wynder, E. L. and E. C. Hamond. A study of air pollution carcinogenesis. *Cancer* **15:** 79–92 (1967).

Zeidberg, L. D., et al. Nashville air pollution study, I. SO_2 and bronchial asthma. A preliminary report. *Amer. Rev. Resp. Dis.* **84:** 489–503 (1961).

WIND-FIRE INTERACTION

Blackadar, A. K. Boundary layer wind maxima and their significance for the growth of nocturnal inversions. *Bull. Amer. Meteorol. Soc.* **38:** 283–290 (1957).

Byram, G. M. Atmospheric Conditions Related to Blowup Fires. U.S. Forest Serv. Southeast. Forest Expt. Sta. Paper 35, 1954.

——— and R. M. Nelson. The possible relation of air turbulence to erratic fire behavior in the southeast. *Fire Control Notes* **12:** 1–8 (1951).

Davis, K. P. *Forest Fire—Control and Use.* New York: McGraw-Hill, 1959.

Graham, H. E. Fire whirlwind formation as favored by topography and upper winds. *Fire Control Notes* **18:** 20–24 (1957).

Hissong, J. E. Whirlwinds at oil-tank fire, San Luis Obispo, Calif. *Monthly Weather Review* **54:** 161–163 (1926).

Jones, D. R. Note on observed vertical wind shear at low levels over the ocean. *Bull. Amer. Meteorol. Soc.* **34** 393–396 (1953).

Musham, H. A. The great Chicago fire, October 8–10, 1871. In *Papers in Illinois History and Transactions for 1940.* Springfield, Ill.: Illinois State Historical Society, 1941.

Schaefer, V. J. The relationship of jet streams to forest wildfires. *J. Forestry* **44:** 419–425 (1957).

Small, R. T. The relationship of weather factors to the rate of spread of the Robie Creek fire. *Monthly Weather Review* **85:** 1–8 (1957).

WIND-DIFFUSED ODORS IN THE ENVIRONMENT

Albin, R. C. Swine fecal odor as affected by feed additives. Texas Tech University, personal communication, 4 pp., 1973.

American Public Health Association. *Standard Methods for the Examination of Water and Wastewater.* 13th Edition, American Public Health Assn., Washington, D.C., 1971.

American Conference of Governmental Industrial Hygienists. *Threshold Limit Values.* 1967.

American Society for Testing and Materials. *Manual on Sensory Testing Methods.* ASTM STP No. 434, 1968.

Amoore, J. E. Stereochemical theory of olfaction. *Nature* **198:** 274 (1963).

———, G. Palmer, and E. Wauke. Molecular shape and odor: pattern analysis by PAPA. *Nature* **216:** 1084 (1967).

Baker, R. A. Response parameters including synergism-antagonism in aqueous odor measurements. *Ann. N.Y. Acad. Sci.* **116**(2): 495 (1964).

Barneby and Cheney. Scentometer: An Instrument for Field Odor Measurement. Instruction sheet 9–68.

Barth, C. Why Does It Smell So Bad? Amer. Soc. Agric. Engnrg., paper 70–416, July 1970.

Barth, C. L., D. T. Hill, and L. B. Polkowski. Correlating OII and Odorous Components in Stored Dairy Manure. Amer. Soc. Agric. Engnrg. Paper 72–950, December 1972.

Bell, R. J. Aeration of liquid poultry manure: A stabilization process or an odor control measure. *Poultry Sci.* **50:** 155–158 (1971).

Benham, C. L. The Effectiveness of Caustic Soda in Preventing the Development of Smell in Poultry Droppings in Slurry Pits. Report from National Agricultural Advisory Service, Shardlow Hall, Shardlow, Derby, England, 1967.

Bethea, R. M. and R. S. Narayan. Identification of beef cattle feedlot odors. *Trans. Amer. Soc. Agric. Engnrg.* **15:** 1135–1137 (1972).

Burnett, W. E. Qualitative determination of the odor quality of chicken manure. In: Odors, Gases, and Particulate Matter from High Density Poultry Management Systems as They

Relate to Air Pollution. Final Report Cornell Univ. Agricultural Engineering Dept. Contract No. C-1101, pp. 2–17, 1969.

——— and N. C. Dondero. The control of air pollution (odors) from animal wastes—evaluation of commercial odor control products by an organoleptic test. Amer. Soc. Agric. Engnrg. Paper 68–909, December 1968.

———. Microbiological and chemical changes in poultry manure associated with decomposition and odor generation. In: *Animal Waste Management,* Cornell University Conf. on Agric. Waste Management, pp. 271–291, 1969.

Carlson, D. A. and R. C. Gumerman. Hydrogen sulfide and methyl mercaptans removals with soil columns. *Proc. 21st Industr. Waste Conf.,* Purdue Univ. Engin., Lafayette, Indiana, 1966.

Carlson, D. A. and C. P. Leiser. Soil beds for the control of sewage odors. *J. Water Poll. Contr. Fed.* **34:** 829–840 (1966).

Charles, D. R. and C. G. Payne. The influence of graded levels of atmospheric ammonia on chickens. *Brit. Poultry Sci.* **7:** 177–198 (1966).

Chemical Rubber Publishing Co. *Handbook of Chemistry and Physics,* 13th ed., pp. 1642–1645. Chemical Rubber Publishing Co., 1955–56.

Chittenden, J. A. Control of odors from anaerobic lagoons treating meat-packing wastes. *Proc. 8th Nat. Symp. Food Processing* 38–61 (1977).

Converse, J. C. Odor control and degradation of swine manure with minimum aeration. Ph.D. thesis, Dept. Agric. Engr. Univ. of Illinois, 198 pp., 1970.

Davies, J. T. A theory of the quality of odours. *J. Theoret. Biol.* **8:** 1 (1965).

——— and F. H. Taylor. A model system for the olfactory membrane. *Nature* **174:** 693 (1954).

Day, D. L. Liquid hog manure can be deodorized by treatment with chlorine or lime. *Illinois research* (*Summer*) **127:** 16 (1966).

———, E. L. Hansen, and S. Anderson. Gases and odors in confinement swing buildings. *Trans. Amer. Soc. Agric. Engnrg.* **8:** 118–121 (1965).

Dean, J. A., ed. *Lange's Handbook of Chemistry.* 11th ed. New York: McGraw-Hill, 1973.

Deibel, R. H. Biological aspects of the animal waste disposal problem. In: Nyle C. Brady, ed. *Agriculture and the Quality of Our Environment.* Amer. Assoc. for Advancement of Science, pp. 395–399, 1967.

Dyson, G. M. The scientific basis of odour. *Chem. & Ind.* **57:** 647 (1938).

Faith, W. L. Odor control in cattle feed yards. *J. Air Poll. Contr. Assoc.* **14:** 459–460 (1964).

Feedlot Management. Sagebrush for odor control: In the feed or in the manure? *Feedlot Management* **14**(5): 74 (1972).

Fosnaugh, J. and E. R. Stephens. Identification of Feedlot Odors. Final report, Dept. HEW, PHS Grant No. UI 00531–02, Statewide Air Pollution Research Center, University of California, Riverside, 1969.

Friedman, H. H., D. A. Mackay, and H. L. Rosano. Odor measurement possibilities via energy changes in cephalin monolayers. *Ann. N.Y. Acad. Sci.* **116**(2): 602 (1964).

Frus, J. D., T. E. Hazen, and J. R. Miner. Chemical oxygen demand as a numerical measure of odor level. *Trans. Amer. Soc. Agric. Engnrg.* **14:** 837–840 (1971).

Gloyna, E. F. and E. Espino. Sulfide production in waste stabilization ponds. *Proc. Amer. Soc. Civil Engnrs., J. San. Engr. Div.* **SA3:** 607 (1969).

Guadnagni, D. G., R. G. Buttery, S. Okano, and H. K. Burn. Additive effect of sub-threshold concentrations of some organic compounds associated with food aromas. *Nature* **200:** 1288 (1963).

Gumerman, R. C. and D. A. Carlson. Chemical aspects of odor removal in some soil systems. *Animal Waste Management,* Cornell Univ. Conf. on Agri. Waste Management, pp. 292–302, 1969.

Hartung, L. D., E. G. Hammond, and J. R. Miner. Identification of carbonyl compounds in a swing building atmosphere. *Livestock Waste Management and Pollution Abatement,* ASAE Publication PROC-271, pp. 105–106, 1971.

Hashimoto, A. Aeration under caged laying hens. *Trans. Amer. Soc. Agric. Engnrg.* **15:** 1119–1123 (1971).

——— and D. C. Ludington. Ammonia desorption from concentrated chicken manure slurries. *Livestock Waste Management and Pollution Abatement,* ASAE Publication PROC-271, pp. 117–121, 1971.

Hollenbach, R. C. Manure odor abatement using hydrogen peroxide. Personal communication, FMC Corp. Rep. No. 5638-R, Princeton, New Jersey, 1971.

Huey, N. A., L. C. Broering, G. A. Jutze, and C. W. Gruber. Objective odor pollution control investigations. *J. Air Poll. Contr. Assoc.* **10:** 441–444 (1960).

Hutchinson, G. L. and F. G. Viets., Jr. Nitrogen enrichment of surface water by absorption of ammonia volatilized from cattle feedlots. *Science* **166:** 515 (1969).

Ingram, S. H., R. C. Albin, C. D. Jones, A. M. Lenon, L. F. Tribble, L. B. Porter, and C. T. Gaskins. Swine fecal odor as affected by feed additives (abstr.) *J. Animal Sci.* **36**(1): 207 (1973).

Jacobs, M. B. *The Chemical Analysis of Air Pollutants.* New York: Interscience Publications, 1960.

Jones, D. D., D. L. Day, and A. D. Dale. *Aerobic Treatment of Livestock Wastes.* Agricultural Experiment Station Bulletin 737, University of Illinois, p. 4, April 1971.

Jones, F. N. and M. H. Waskow. On the intensity of odor mixtures. *Ann. NY Acad. Sci.* **116:** 484 (1964).

Kendall, D. A. and A. J. Neilson. Correlation of subjective and objective odor responses. *Ann. NY Acad. Sci.* **116:** 567 (1964).

Kibbell, W. H., Jr., C. W. Raleigh, and J. A. Shephard. Hydrogen peroxide for pollution control. *Proc. 27th Purdue Industr. Waste Conf.*, Purdue University, May 24, 1972.

Koelliker, J. D. and J. R. Miner. Desorption of ammonia from anaerobic lagoons. *Trans. Amer. Soc. Agric. Engnrg.* **16**(1): 148–151 (1973).

Kowalki, O. L., O. A. Hougen, and K. M. Watson. Transfer Coefficients of Ammonia in Absorption Towers. Winsconsin Eng. Expt. Sta. Bull. 68, 1925.

Lebeda, D. L. and D. L. Day. Waste-caused air pollutants are measured in swine buildings. *Illinois Res.* (*Fall*) 15 (1965).

Leonardos, G., D. Kendall, and N. Barnard. Odor threshold determinations of 53 odorant chemicals. *J. Air. Poll. Contr. Assoc.* **19**(2): 91 (1969).

Lillie, R. J. *Air Pollutants Affecting the Performance of Domestic Animals: A Literature Review.* Agriculture Handbook No. 380, p. 109, 1970.

Ludington, D. C., A. T. Sobel, and A. G. Hashimoto. Odors and gases liberated from diluted and undiluted chicken manure. Amer. Soc. Agri. Engin. Paper 69–462, June 1969.

Luebs, R. E., A. E. Laag, and K. R. Davis. Ammonia and related gases emanating from a large dairy area. *Cali. Agri.* **27**(2): 10–12 (1973).

McCalla, T. M. and L. F. Elliott. The role of microorganisms in the management of animal wastes on beef cattle feedlots. In: *Livestock Waste Management and Pollution Abatement,* ASAE Publication PROC-271, pp. 132–134, 1971.

Merkel, J. A., T. E. Hazen, and J. R. Miner, 1969. Identification of gases in a confinement swine building atmosphere. *Trans. ASAE* **12**(3): 310.

Miner, R. J. *Control of Odors from Anaerobic Lagoons Treating Food Processing Wastewaters.* EPA-600/2–78–151, July 1978.

——— (ed.) *Farm Animal-Waste Management.* N.C. Regional Publication 206, p. 12, 1971.

——— and T. E. Hazen. Ammonia and amines: Components of swine building odor. *Trans. Amer. Soc. Agric. Engnrg.* **12:** 772–774 (1969).

Moncrieff, R. W. An instrument for measuring and classifying odors. *J. Appl. Physiol.* **16:** 742 (1961).

———. *Odour Preferences.* New York: John Wiley, 1966.

———. *The Chemical Senses.* London: Leonard Hill, 1967.

Moulton, D. G. Physiological aspects of olfaction. *J. Food Sci.* **30:** 908 (1965).

Moum, S. G., W. Seltzer, and T. M. Goldhaft. A simple method of determining concentrations of ammonia in animal quarters. *Poultry Sci.* **46:** 347–348 (1969).

O'Neil, E. T. Manure odor abatement with hydrogen peroxide. Personal communication, FMC Corporation Report No. 4760-R, Princeton, New Jersey, 1972.

Paine, M. D. *Feedlot Odor.* Cooperative Extension Project GPE-7, Pub. GPE-7800, Oklahoma State University, 1973.

Paulson, D. J. Commercial feedlots—nuisance, zoning and regulation. *Washburn Law J.* **6:** 493–507 (1967).

Pomeroy, R. D. and H. Cruse. Hydrogen sulfide odor threshold. *J. Amer. Water Works Assoc.* **61**(12): 677 (1969).

Posselt, H. S. and A. H. Reidies. Odor abatement with potassium permanganate solutions. *Ind. Engin. Chem. Product Research & Development* **4**(3): 48–50 (1965).

Rands, M. B. and D. E. Cooper. Development and operation of a low cost anaerobic plant for meat wastes. In: *Proc. 21st Purdue Industrial Waste Conf.,* Lafayette, Indiana, pp. 613–638, 1966.

Rosano, H. L. and S. Q. Scheps. Adsorption-induced electrode potential in relation to olfaction. *Ann. NY Acad. Sci.* **116:** 590 (1964).

Rosen, A. A., J. B. Peter, and F. M. Middleton. Odor thresholds of mixed organic chemicals. *J. Water Poll. Contr. Fed.* **34:** 7 (1962).

Rowe, N. R. Odor control with activated charcoal. *J. Air Poll. Contr. Assoc.* **13:** 150–153 (1963).

Sawyer, C. N. and P. L. McCorty. *Chemistry for Sanitary Engineers.* Second edition. New York: McGraw-Hill, 1967.

Seltzer, W., S. G. Moum, and T. M. Goldhaft. A method for the treatment of animal wastes to control ammonia and other odors. *Poultry Sci.* **48:** 1912 (1969).

Sheehy, J. P., W. C. Achinger, and R. A. Simon. *Handbook of Air Pollution.* Environment Health Series, Air Pollution, 99-AP-44, A-17, pp. 14–19 (undated).

Sobel, A. T. Measurement of the odor strength of animal manures. *Animal Waste Management,* Cornell Univ. Conf. on Agric. Waste Management, pp. 260–270, 1969.

———. Olfactory measurement of animal manure odors. Agricultural Waste Mnagement and Associated Odor Control, Cornell University, AWM 71–04, 1971.

Stahl, W. H., ed. *Compilation of Odor and Taste Threshold Values Data.* ASTM Data Series DS 48, Philadelphia, pp. 104–130, 1973.

Stern, A. G. *Air Pollution.* Vol. II. New York: Academic Press, 1968.

Stombaugh, D. P., H. S. Teague, and W. L. Roller. Effects of atmospheric ammonia on the pig. *J. Animal Sci.* **28:** 44–47 (1969).

Stone, H. Behavioral aspects of absolute and differential olfactory sensitivity. *Ann. NY Acad. Sci.* **116:** 527 (1964).

Swets, J. A. Is there a sensory threshold? *Science* **134:** 168 (1961).

Theimer, E. T. and J. T. Davies. Olfaction, musk odor and molecular properties. *J. Agric. Food Chem.* **14:** 6 (1963).

Venstrom, D. and J. E. Amoore. Olfactory threshold in relation to age, sex, and smoking. *J. Food Sci.* **33:** 264 (1968).

Waldo, D. A. An Evaluation of an Environetics Controliner. Unpublished special report, Advisor D. H. Vanderholm, Department of Agricultural Engineering, University of Illinois, 16 pp., 1976.

White, R. K., E. P. Taiganides, and C. D. Cole. Chromatographic identification of malodors from dairy animal waste. *Livestock Waste Management and Pollution Abatement.* ASAE Publication PROC-271, pp. 110–113, 1971.

Willrich, T. L. Manufacturers of odor control chemicals for use in controlling manure odors. Oregon State University, personal communication, 2 pp., 1973.

——— and J. R. Miner. Litigation experiences of five livestock and poultry producers. *Livestock Waste Management and Pollution Abatement,* ASAE Publication PROC-271, pp. 99–101, 1971.

Wright, R. H. Odor and chemical constitution. *Nature* **173:** 831 (1954).

———. Why is an odour? *Nature* **209:** 551 (1966).

Yushok, W. and F. E. Bear. Poultry manure: Its preservation, deodorization and disinfection. *NJ Agric. Expt. Sta. Bull.* No. 707, Rutgers University, New Brunswick, New Jersey, 1948.

WIND ENERGY STORAGE SYSTEMS

Allison, H. J. *A Wind Energy Storage Technique Utilizing a Hydrogen-Oxygen Electrolysis Cell System.* Frontiers of Power Technology Conference, Oklahoma State University, October 29, 1968.

———. Energy conversion to storage research at Oklahoma State University. *Proc. 2nd Ann. Conf. Energy Conversion and Storage,* Engineering Research, Oklahoma State University, 1964.

———. Research on energy storage and conversion at Oklahoma State University. *Advanced Battery Technology* **3** (9). (1967)

———. Energy storage system concepts, limitations, and efficiencies. *Proc. Conf. Energy Conversion and Storage,* School of Electrical Engineering, Oklahoma State University, 1963.

———, R. Ramakumar, and W. L. Hughes. *Energy Storage Research at Oklahoma State University.* Midwest Power Symposium, Ames, Iowa, October 1970.

Bruckner, A., Fabrycky, W. J. Economic optimization of energy conversion with storage,

Third Annual Conference on Energy Conversion and Storage, Stillwater, Okla., October 1965.

Easter, B. H. Computer aided design of power systems using energy storage, Oklahoma State Univ., Stillwater, Okla.

Energy Storage and Conversion. Annual report, Engineering Research, Oklahoma State University, June 30, 1965.

Energy Storage and Conversion. Quarterly report, Engineering Research, Oklahoma State University. March 1965; September 1965; December 31, 1965; March 1966; June 1966; August 1966; December 1966; March 1967; June 1967.

Hughes, W. L. A report on some of the experimental work on energy conversion and storage at Oklahoma State University and A quantitative evaluation of power density and storage capacity for solar and wind energy. *Proc. Conference on Energy Conversion and Storage,* Oklahoma State University, School of Electrical Eng., October 1963.

———. Uses of energy storage systems. *Proc. Conf. Energy Conversion and Storage,* School of Electrical Engineering, Oklahoma State University, October 1963.

———, R. G. Ramakumar, and H. J. Allison. An economical energy storage system utilizing hydrogen-oxygen regenerative fuel cells. IEFE Summer Power Meeting, New York City, January 1970.

———, K. A. McCollom, H. J. Allison, R. Ramakumar, B. H. Easter, and W. E. Knabach. *Energy Conversion and Storage Research Report.* Engineering Research, August 31, 1968.

———, C. M. Summers, H. J. Allison, and W. J. Fabrycky. *Energy Storage . . . Key to Our Economic Future.* Engineering Research, Oklahoma State University, 1961.

———, C. M. Summers, and H. J. Allison. *Feasibility of Energy Storage by Water Electrolysis and High Pressure Hydrogen Storage.* Engineering Research Report, Oklahoma State University, 1962.

Hughes, W. L., C. M. Summers, K. A. McCollom, H. J. Allison, H. T. Fristoe, W. J. Fabrycky, A. B. Bruckner, and J. B. West. *Energy Storage and Conversion.* Quarterly report, Engineering Research, Oklahoma State University, December 1964.

Mackenthun, W. Neuwerk windmill power generation plant. *Elektrizitsetswirtschaft (Frankfurt am Main)* **11:** 322–325 (1951).

McCollum, K. A. Use of energy storage with unconventional energy sources to aid developing countries. *Proc. Second Intersociety Energy Conversion Engineering Conference (IECEC),* Miami Beach, Florida, August 1967.

Rabenhorst, D. W. *Superflywheel Energy Storage System.* NASA, Lewis Res. Center, Wind Energy Conversion Systems, pp. 137–145, December 1933.

Ramakumar, R., K. A. McCollom, H. J. Allison, and W. L. Hughes. Wind energy storage
Barber, E. M. On-Site Wind vs. Central Station Power Generation, Nomograph for Economic Comparison. Yale University, School of Architecture, 1973.
and conversion system for use in underdeveloped countries. *Proc. 4th Intersociety Energy Conversion Eng. Conf.,* Washington, D.C., pp. 606–13, September 22–26, 1969.

Schwartz, H. J. *Batteries for Storage of Wind-Generated Energy.* NASA, Lewis Res. Center, Wind Energy Conversion Systems, pp. 146–151, December 1973.

Shefter, Y. I. Semiautomatic wind-electric stations with inertia storage batteries. *Vestnik sel'skokhozyaystvennoy nauk* 12 (1958).

Summers, C. M. A quantitative evaluation of power density and storage capacity for solar and wind energy. *Proc. Conf. Energy Conversion & Storage,* School of Electrical Engineering, Oklahoma State University, October 1963.

———. A Solar to Electrical Energy System for Intermediate Energy Storage. IEEE Winter Power Meeting, New York, February 1967.

———. The basic technical problems associated with a solar-to-electrical system with intermediate energy storage. *Proc. Conf. Energy Conversion & Storage,* School of Electrical Engineering, Oklahoma State University, October 1963.

Szego, G. C. *Energy Storage by Compressed Air.* NASA, Lewis Res. Center, Wind Energy Conversion Systems, pp. 152–154, December 1973.

Zlokovic, V. S. Wind-power plant and energy storage. *Elektroprivreda* (*Yogoslavia*) **22:** 377 (1969).

WIND POWER (GENERAL)

A National Plan for Energy Research, Development & Demonstration: Creating Energy Choices for the Future. ERDA-48, vol. 1, June 28, 1975.

An Assessment of Solar Energy as a National Energy Resource. NSF/NASA Solar Energy Panel, Dec. 1972, 85 pp.

Adrianov, V. Stability of a synchronous generator driven by wind-power and working on a power system. *Elektrichestvo,* No. 10, October 1949, pp. 26–32.

——— and D. N. Bystritski. Parallel operation of a wind power station with a powerful grid. *Elektrichestvo,* No. 5, 1951; pp. 8–12.

——— and A. L. Pokatoer. Regulation of the output as a wind power station. *Elektrichestvo,* No. 6, 1952, pp. 19–24.

Ardman, H. A novel proposal for the fuel of the future. *American Legion magazine,* v. 95, July 1973, pp. 4–9, pp. 39–40.

Asta, A. Experience in the utilization of wind power for electric power generation. *Ricerca Scientifica,* Vol. 23, No. 14, April 1953, pp. 537–558.

Back to the windmill to generate power. *Business Week,* Mayll, 1974, N. 2330, pp. 140–142.

Barton, T. H. and K. Repole. A Simple Electric Transmission for A Free-Running Windmill. Report Number T. 68, Brace Research Institute, 1970.

Bartria, J. New designs of wind-power generators. *Elektrotech, Obzar,* Vol. 38, November 1949, pp. 528–585.

Baumeister, T. and L. Marks. *Standard Handbook for Mechanical Engineers.* 7th Edition. New York: McGraw-Hill Book Company, 1967.

Bennett, R. and J. Elton. *Hystory of Corn Milling,* vol. 2, Watermills and Windmills (1898).

Bettiqnies, C. Wind energy—its utilization in isolated & Arctic regions (1/4 to 100 kW). *Northern Engineer,* v. 5, Winter 1973–74, 13–17.

Brand, D. It's an ill wind, etc.: Energy crisis may be good for Windmills. *Wall Street Journal,* Jan. 11, 1974, p. 1, 16.

———. Power pioneers. *Wall Street Journal,* March 18, 1975, p. 1, 29.

Brown, C. A. New interest in an old power resource. *Cooperation Canada,* July–Aug. 1974, 14–19.

Bruekner, A. Taking Power Off the Wind. New Scientist City University of U.K. March 28, 1974, V. 61, N. 891, pp. 812–815.

Can we harness pollution-free electric power from windmills? *Popular Science,* November 1972.

Clark, W. Interest in wind is picking up as fuels dwindle. *Smithsonian,* v. 4, Nov. 1973, 70–71, 73–75, 77–79.

Clews, H. *Electric Power from the Wind.* Solar Wind Publications, Happytown, Maine, 1972.

Climatic Atlas of the United States. (Gives wind averages, the strongest wind, and wind direction. The atlas and a monthly wind report for each state are available from Environmental Data Service, National Climatic Center, Federal Building, Asheville, North Carolina 28801.)

Chilcott, R. E. *The Design Development and Testing of a Low Cost 10 hp Windmill Prime Mover.* Report Number MT 7, Brace Research Institute, 1969.

Collins, J. *Wind Power.* Washington Reference Section, Science and Technology Division, Library of Congress. Science and Technology Division, LC science tracer bullet, TB 73–17.

De Korne, J. B. The answer is blowin' in the wind. *Mother Earth News,* N. 24, Nov. 1973, 67–75.

Development, testing and operation of a 200 kW wind power station in Denmark. Report of the Windpower Committee of the Association of Danish Electricity Undertakings (OEF), Cophagen, 1962.

Energy From the Wind. Technical Note No. 4, World Meteorology Organization, 1954, WMO, No. 32, T.P. 10, Geneva, Switzerland.

Fan Blade Windmill. (US/AID) Vita Publication, Number 11133.3.

Fateev, E. M. and I. V. Rozhdestvenskii. Achievements of Soviet wind power engineering. *Vestnik Mashinostroeniya,* No. 9, 1952, pp. 24–28. (From a translation into German, *Energietechnik,* Vol. 3, February 1953, pp. 53–56.)

Freese, S. *Windmills and Millwrighting.* London, 1968.

Fuller, B. R. Energy through wind power. *New York Times,* Jan. 17, 1974, p. 39.

Ganger, B. *Elastic Power Transmission in Wind Power Installations.* H. V. Institute of Karlsruhe Technical High School, March 1974, 12 pp.

Ghosh, P. K. Low Drag, Laminar Flow Aerotoil Section of Windmill Blades. Number CP 20, Brace Research Institute, May 1969.

Gimpel, G. and A. Stodhart. *Windmills for Electricity Supply in Remote Areas.* 1958, pp. 24.

Golding, E. W. Large-Scale Generation of Electricity by Wind Power—Preliminary Report. 1949, pp. 15.

———. *The Utilization of Wind Power in Desert Areas.* 1953, pp. 11.

———. *The Combination of Local Energy Resources to Provide Power Supplies in Underdeveloped Areas.* 1956, pp. 21.

———. *The Generation of Electricity by Wind Power.* Philosophical Library, Inc., New York, 1955.

——— and A. H. Stodhart. *The Selection and Characteristics at Wind-Power Sites.* 1952, pp. 32.

———. *The Use of Wind Power in Denmark.* 1954, pp. 16, illustration sheet.

Hartley, William and Ellen. The wind shifts to windmills. *Popular Mechanics* (November 1974)

Hearings Before the Subcommittee on Energy Research, Development and Demonstration of the Committee on Science and Technology U.S. House of Representatives, Feb. 21, 1975, ERDA—Authorization—Part IV 1976 and Transition Period Solar and Physical Research, pp. 272–375; pp. 543–574.

Helical Sail Windmill. Number 11131.1, VITA Publication, Mount Rainier, Maryland.

Heronemus, W. E. Power from the offshore winds. *Proc. 8th Annual Marine Technology Society Conference,* Washington, D.C., 1972.

———. Testimony prepared for the atomic safety and licensing board hearings in the matter of Long Island Lighting Company proposed Shoreham Nuclear Power Station. AEC docket No. 50–322. In Extention of remarks of Mike Gravel. Congressional record (daily ed.), v. 119, Apr. 30, 1973: E 2666-E 26669.

———. The U.S. energy crisis: some proposed gentle solutions. In extention of remarks of Mike Gravel. Congressional record (daily ed.), V. 118, Feb. 9, 1972, E 1043-E 1049.

———. Using two renewables. *Oceans,* v. 17, summer 1974, 20–27.

Hicks, N. Energy crisis impels many to study and erect windmills as power source. *New York Times,* May 20, 1974, p. 33.

The Homemade Windmills. Bulletin of the U.L.S. Agricultural Experiment Station of Nebraska. The University of Nebraska. Bulletin Number 59. Volume XI, Article V. 1897.

How To Construct a Cheap Wind Machine for Pumping Water. Brace Research Institute, 1973.

Hydrogen Economy Miami Energy (THEME). Conference, Miami Beach, Fla. 1974. Conference proceedings. Coral Gables, Fla., University of Miami, 1974. 1 v. (various proceedings).

Jacobs, M. The Playboy interview: Marcellus Jacobs. *Mother Earth News,* No. 24, Nov. 1973, 52–58.

Johnson, C. C., R. T. Smith, and R. K. Swanson. Wind power development and applications. *Power Engineering,* v. 78, Oct. 1974, 50–53.

Jousse, M. *L'Art de Charpenterie.* 2nd ed. (1702).

Juul, J. Investigation of the possibilities of utilization of wind-power. *Elektroteknikeren,* vol. 45, October 22, 1949, pp. 607–635.

Kadivar, M. S. Potential for Wind Power Development. Report Number MT 10, Brace Research Institute, 1970.

Kaspir, F. The synchronous generator and its use in wind-power station. *Elektrotech. Obzor,* vol. 34, no. 4, pp. 60–63, no. 5–6, pp. 81–89, 1945.

Kroms, A. Wind power stations working in connection with existing power systems. *A.S.E. Bull.,* Vol. 45, no. 5, March 6, 1954, pp. 135–144.

Lacroix, G. Electrical problems raised by the utilization of wind power. *Bull. Soc. Franc. Elect.* Vol. 10, April 1950, pp. 211–215.

Lerza, C. How to kick the fossil fuel habit. *Env. Action,* Feb. 2, 1974, V. 5, N. 18, pp. 3–12.

Lilley, G. M. and W. J. Rainbird. A preliminary report on the design and performance of ducted windmills. 1957, 65 pp., 7 illustration sheets.

Lines, V. *The English Windmill.* illus. New York: Augustus M. Kelly, 1968.

Linperch, P. *Architectura Mechanica of MooleBoek* (1727).

Low Cost Windmill for Developing Nations. VITA Publication Number 20, 1970.

Meyer, H. *Wind Energy.* Domebook 2.

———. Wind generators. *Popular Science,* November 1972, p. 103.

Mogilnitskii, I. D. On the regulation of slow-running wind-power generators. Paper of the Lenin Academy of Agricultural Sciences, No. 5, 1950, pp. 36–40.

Moore, J. G. Utilization of solar energy in northeastern Ohio. In extension of remarks of Charles A. Vanik. Congressional record daily ed., v. 119, Oct. 29, 1973, E 6810-E 6812.

Morrison, J. G. The development of a method for measurement of strains in the blades of a windmill rotor. 1957, pp. 28.

Nakra, H. L. A Report on Preliminary Testing of a Lubing Windmill Generator. Number T. 75, Brace Research Institute, pp. 5.

Other technical concepts are exciting but the roads to power are long. *Coal Age,* Apr. 1974, V. 79, N. 4, pp. 106–111.

Power From the Wind. P. C. Putnam, New York, 1948.

Preston, D. J. One man's answer to the energy crisis. *American forests,* v. 80, Feb. 1974, 20–33, 64–66.

Prinsen Molen Committee. Research Inspired by the Dutch Windmill (1958).

Proceedings of a Conference on Energy Conversion and Storage. Oklahoma State University, October 28, 1963.

Proceedings of the New Delhi Symposium on Wind and Solar Energy, 1956.

Puthoff, R. L. and Sirocky, P. Status Report of 100 kW Experimental Wind Turbine Generator Project. Lewis Research Center, NASA TMX—71758, June 1975.

Putnam, C. *Power from the Wind.* D. Van Nostrand Co., New York, 1948.

Rabenhorst, D. W. Superflywheel. *Proceedings of the Solar Heating and Cooling for Buildings Workshop,* Washington, D.C., March 21–23, pp. 60–68.

Reed, J. W. Wind power climatology. Sandie Lab., New Mexico. *Weatherwise,* Dec. 1974, V. 27, N. 6, pp. 36.

Reyynolds, J. *Windmills and Watermills.* Praeger Publishing Co., N.Y. 1970.

Ronse, A. *De windmolens.* 1934.

Rosenbrock, H. H. An extension of the momentum theory of wind turbines. 1951, p. 10, 2 illustration sheets.

———. The design and development of three new types of gust anemometer. 1951, p. 37.

———. Vibration and stability problems in large wind turbines having hinged blades. 1955, p. 53.

——— and J. R. Tagg. Wind and gustmeasuring instrument developed for a wind-power survey. 1951, 10 pp.

Rosseler, G. Electrical Power supply by wind power to the 10 cm Schoneberg (Eitel) Radio Link and experience gathered with its operation. *Nachirichtentech.* Z (*N.T.Z.*), Vol. 12, July 1959, pp. 352–360.

Savonius Rotor. *Windmill,* VITA Publication, Number 11132.1.

Santa Barbara Fan Blade Windmill. VITA Publication Number 11133.2.

Schenfer, K. and A. Lvanor. Lines of development of rural wind-power plants. *Elektrichestvo,* No. 5, May 1941, pp. 21–22.

Schwartz, M. Can windmills supply farm power? *Organic Gardening,* Jan. 1974, V. 21, N. 1, pp. 155–158.

Seidel, G. R. Modern wind-power utilization. *Der Elektrotechniker,* Vol. 4, March 1952, pp. 61–66.

Sektorov, V. R. The present state of planning and erection of large experimental wind-power stations. *Elektrichestvo,* No. 2, 1933, pp. 9–13.

Sencenbaugh, J. I built a wind charger for $400.00. *Mother Earth News,* No. 20.

Simonds, M. H. and A. Boclek. Performance Test of A Savonius Rotor. Report No. T-10, Brace Research Institute, 1964.

Smeaton. *An Experimental Enquiry Concerning the Natural Powers of Wind and Water* (1974).

Souci, G. Pulling power out of thin air: pinwheels of the high seas and high peaks? *Audubon,* v. 76, May 1974, 81–88.

South, P. and R. S. Ragni. A Wind Tunnel Investigation of a 14/1 Diameter Vertical Axis Windmill. National Research Council of Canada, Laboratory Technical Report LTR-LA-105, Sept. 1972.

———. The performance and Economics of the Vertical-Axis Wind Turbine Developed at the National Research Council. Ottawa, Canada. American Society of Agricultural Engineers, 1973.

Stabb, D. Wind. *Architectural Design.* March, 1972.

Sterne, L. H. G. and G. C. Rose. The aerodynamics of windmills used for the generation of electricity. 1951, 12 pp.

Stodhart, A. H. The economic value of hydrogen produced by wind power. 1954, 8 pp.

Stokhuyzen, F. The Dutch windmill. (1962). Sun, sea, wind, geysers-new energy from old sources? *U.S. News & World Report,* v. 78, Jan. 27, 1975, 37–39.

Tagg, J. R. Wind data related to the generation of electricity by wind power. 1957, 52 pp.

———. Wind driven generators. The difference between the estimated output and the actual energy obtained. 1960, 7 pp. 2 illus.

Thomas, P. H. Electric Power From the Wind. U.S. Federal Power Commission, 1945.

Thomas, R. L. Wind energy conversion; Energy environment productivity. *Proceedings of the First Symposium on RANN,* Washington, D.C., Nov. 18–20, 1973, pp. 39–41.

——— and J. E Sholes. Preliminary Results of the Large Experimental Wind Turbine Phase of the National Wind Energy Program. Lewis Research Center, NASA TMX—71796, October 1975.

U.N. Conference on New Sources of Energy. Volume 7: *Wind Power.* Rome 1961. Published by the U.N., 1963.

U.S. Congress. House. Committee on Science and Astronautics. Subcommittee on Energy. Wind energy, Hearings, 93d Cong., 2d sess. May 21, 1974. Washington, U.S. Govt. Print. Off., 1974. p. 386 No. 49.

U.S. energy resources; underdeveloped, over-regulated, wastefully used. *Coal Age,* v. 79, Apr. 1974, 69–116. A joint report by the editors of: *Coal Age* and *Engineering and Mining Journal.* Also appears in *Engineering and Mining Journal,* April 1974.

Vezzani, R. Experimental data on models of devices for accumulating wind energy in space. *Elettrotecnia,* Vol. 35, Dec. 1948, pp. 488–493.

Villecco, M. Wind power. *Architecture Plus,* May–June 1974, V. 2, N. 3, pp. 64–78.

Villers, D. E. The testing of wind-driven generators operating in parallel with a network. 1957, 22 pp.

Wade, N. Windmills: the resurrection of an ancient energy technology. *Science,* v. 184, June 7, 1974: 1055–1058.

Wailes. *The English Windmill* (1954, reprinted 1967).

Walker, J. G. The automatic operation of a medium-sized wind driven generator running in isolation. Preliminary report, 1960, 14 pp.

Wax, M. P. An experimental study of wind structure (with reference to the design and operation of wind-driven generators). 1956, 24 pp.

Wentink, T., Jr. Wind power for Alaska: an impossible dream? *Northern Engineer,* v. 5, Winter 1973–1974, 8–12.

Wind energy conversion system: workshop proceedings (Washington, U.S. Govt. Print. Off.) 1973 (i.e., 1974) 258 p. NSF/RA/W-73-006.

Wind Energy Systems Workshop, June 11–13, 1973. Washington, D.C.

Wind Power. FPC Bureau of Power Staff Report, Sept. 1973.

Wind Power. Public Interest Report 1974.

Wolff, A. R. *The Windmill as a Prime Mover* (1885).

Tornado Bibliography

Abbey, R. F., Jr. 1975a. Research efforts in severe storms applied to nuclear reactors. *Preprints, Second U.S. National Conf. on Wind Engineering Research,* Fort Collins, Colorado, June 23–25.

———. 1975. Establishment of maximum regional tornado criteria for nuclear power plants. *Preprints, Ninth Conf. Severe Local Storms,* Norman, Oklahoma, October 21–23, 368–375. American Meteorological Society, Boston, Mass.

———. 1975. Establishment of regional tornado criteria for power plants. *Preprints, Ninth Conf. Severe Local Storms,* Oct. 21–23, Norman, Okla.

———. Tornado research supported by the USNRC (invited paper). Third U.S. National Conference Wind Engineering Research, University of Florida, Gainesville, Fla., Feb. 26–Mar. 1, 1978.

——— and T. T. Fujita. 1973. Use of tornado path lengths and gradations of damage to assess tornado intensity probabilities. *9th Conf. Severe Local Storms. Amer. Meteorol. Soc.,* 286–292.

——— and T. T. Fujita. 1975. Use of tornado path lengths and gradations of damage to assess tornado intensity probabilities. *Preprints, Ninth Conf. Severe Local Storms,* Norman, Oklahoma, October 21–23, 286–293. American Meteorological Society, Boston, Mass.

———. 1976. Risk probabilities associated with tornado windspeeds.

——— and T. T. Fujita. 1975. Use of tornado path lengths and gradations of damage to assess tornado intensity probabilities. *Preprints, Ninth Conf. Severe Local Storms,* Oct. 21–23, Norman, Okla.

———, R. L. Schwiesow, and J. H. Golden. 1975. Use of Doppler Lidar to Measure Waterspout Windspeeds. Paper given at 7th International Laser Radar Conference, Menlo Park, Calif., Nov. 4–7, 1975.

Abdullah, A. J. 1955. Some aspects of the dynamics of tornadoes. *Mon. Wea. Rev.,* **83**(4): 83–94.

Achtemeier, G. L. Some Observations of Splitting Thunder over Iowa on August 25–26, 1965. *Department of Meteorology Report No. 69–4,* Florida State University, 1969, 17 pp.

Agee, E. M. 1969. Tornado project activities, Purdue University. *Bull. Amer. Meteorol. Soc.* **50** 806.

———. 1970. Tornado project activities, Purdue University. *Bull. Amer. Meteorol. Soc.* **51** (951).

———. 1970. The climatology of Indiana tornadoes. *Proc. Ind. Acad. Sci. for 1969* **79:** 299–308.

———, C. Church, C. Morris, and J. Snow. 1975. Some synoptic aspects and dynamic features of vortices associated with the tornado outbreak of 3 April 1974. *Mon. Wea. Rev.* **103:** 318–333.

Akins, R. E., J. A. Peterka, and J. E. Cermak. n.d. Mean force and moment coefficients for buildings in turbulent boundary layers. *J. Industr. Aerodynam.*

Allen, C. R. 1975. Geological criteria for evaluating seismicity. *Geol. Soc. Am. Bull.* **86:** 1041–1057.

Almuti, A. 1974. Engineering report on the Xenia tornado on April 3, 1974. Bechtel Corporation Report, Ann Arbor, Michigan.

American Meteorological Society. 1975. Presentation to the Subcommittee on the Environment and the Atmosphere (HR). Presentation during the Ninth Severe Local Storms Conference, American Meteorological Society, Norman, OK, October 1975.

American National Standards Institute. 1972. Building Code Requirements for Minimum Design Loads in Buildings and Other Structures. ANSI A58.1–1972.

American Nuclear Society. 1976. Standard for Estimating Tornado, Hurricane, and Other Extreme Wind Parameters at Nuclear Reactor Sites.

———. 1976. Committee ANS 2.3. Guidelines for determining tornadoes, hurricanes and wind at power reactor sites. Interim report.

American Peoples Encyclopedia. 1965. New York: Grolier Incorporated.

American Society of Civil Engineers. Task Committee on Wind Forces. 1961. Final report. Paper No. 3269 on wind forces. *Trans. ASCE* 1124–97.

———. 1961. Wind forces on structures. *Trans. ASCE* **125**(II): 1124–1198.

Amirikian, A. 1950. Design of Protective Structures: A New Concept of Structural Behavior. Bureau of Yards and Docks, Dept. of the Navy, p. 51, Washington, D.C.

ANSI. 1972. *Building Code Requirements for Minimum Design Loads in Buildings and Other Structures.* ANSI A58.1, American National Standards Institute, New York.

Asp, M. O. 1950. Tornadoes in Oklahoma, 1875–1949. *Monthly Weather Review* 23–26 (February).

———. 1956. Geographical distribution of tornadoes in Arkansas. *Monthly Weather Review* **84:** 143–145.

———. 1963. History of tornado observations and data sources. *Key to Meteor. Records Doc.* No. 3.131, U.S. Weather Bureau, Washington, D.C.

Auer, A. H. 1967. Tornadoes in northeastern Colorado. *Monthly Weather Review* **95:** 32–34.

Baier, J. 1896. Low pressure in the St. Louis tornado. *Monthly Weather Review* 332 (September).

Bailey, A. 1933. *Wind Pressures on Buildings.* Inst. of Civil Engineering, selected Eng. Paper No. 189.

——— and N. D. G. Vincent. 1943. Wind pressures on buildings including the effects of adjacent buildings. *J. Inst. Civ. Engnrng.* **20:** 243–275.

Baines, W. D. 1963. Effects of Velocity Distribution on Wind Loads and Flow Patterns on Buildings. *Proc. Conf. Wind Effects Buildings and Structures,* National Physical Laboratory,

Teddington, England, 198–225.
Banks, D. C. 1963. A Study of Bearing Capacity in Sands under Dynamic Loadings. M.S. thesis, Georgia Institute of Technology, Atlanta, Georgia.
Barber, R. B. 1973. Steel Rod/Concrete Slab Impact Tests (Experimental Simulation), Bechtel Power Corporation, San Francisco, California.
Barcilon, A. I. 1967. Vortex decay above a stationary boundary. *J. Fluid Mech.* **27:** 155–175.
———. 1967. A theoretical and experimental model for a dust devil. *J. Atmos. Sci.* **24:** 453–466.
——— and P. G. Drazin. 1972. Dust devil formation. *Geophys. Fluid Dynam.* **4:** 147–158.
Barnes, S. L. 1970. Some aspects of a severe, right moving thunderstorm deduced from mesonetwork rawinsonde observations. *J. Atmos. Sci.* **27:** 634–648.
———. 1974. Morphology of Two Tornadic Storms: An Analysis of NSSL Data on April 30, 1970. *Papers on Oklahoma Thunderstorms, April 29–30, 1970.* NOAA Technical Memorandum ERL NSSL-69, 1974, pp. 125–39.
Bates, F. C. 1962. Tornadoes in the Central United States. *Trans. Kansas Acad. Sci.* **65**(3): 215–246.
———. 1967. A major hazard to aviation near severe thunderstorms. Aviation Safety Monogr. No. 1, Dept. of Geosciences, St. Louis Univ., 36 pp.
———. 1968. A theory and model of the tornado. *Proc. Int. Conf. Cloud Physics,* Toronto, Canada, Aug. 26–30, 559–563.
——— and A. E. Swanson. 1967. Tornado Design Considerations for Nuclear Power Plants. Paper presented at Annual Meeting of the Am. Nuc. Soc., Chicago, Ill., November 6–9, 28 pp.
Battan, L. J. 1959. Duration of tornadoes. *Bull. Am. Meteor. Soc.* **40:** 340–342.
———. 1961. *The Nature of Violent Storms.* Garden City, N.Y.: Doubleday & Co.
Beebee, R. G. 1958. Tornado proximity soundings. *Bull. Amer. Meteorol. Soc.* **38:** 195–201.
———. 1960. The Life Cycle of the Dallas Tornadoes. Research Paper No. 41. U.S. Weather Bureau, Washington, D.C. pp. 3–9.
Beeth, D. R. and S. H. Hobbs. 1975. Analysis of Tornado Generated Missiles. Topical Report B&R-001, Brown and Root, Inc., Houston, Texas.
Bellamy-Knights, P. G. 1970. An unsteady two-cell vortex solution of the Navier-Stokes equations. *JFM* **41:** 673–687.
———. 1971. Unsteady multicellular viscous vortices. *JFM* **50:** 1–16.
———. 1974. An axisymmetric boundary layer solution for an unsteady vortex above a plane. *Tellus* **26:** 318–324.
Benjamin, J. R. and C. A. Cornell. 1970. *Probability, Statistics, and Decision for Civil Engineers.*

New York: McGraw-Hill, Inc.

Banjamin, T. B. 1962. Theory of the vortex breakdown phenomenon. *J. Fluid Mech.* **14:** 593–629.

Bergman, K. H. 1970. On the dynamic stability of convective atmospheric vortices. US Army Electronics Command, Fort Monmouth, N.J., ECOM 68G6–1, 169 pp.

Berry, R. E. and J. W. Reed. 1972. A discussion of the tornado environment and occurrence climatology. *Sandia Lab. Rept.* SC-M-72, Albuquerque, New Mexico.

Biggs, J. M. 1964. *Introduction to Structural Dynamics.* New York: McGraw-Hill Book Company, p. 69.

Biggs, W. G. and P. J. Waite. 1970. Can TV really detect tornadoes? *Weatherwise* **23:** 120–25.

Bigler, S. G. 1957. *Tornado Damage Surveys.* A & M project 156, scientific report No. 1. Department of Oceanography and Meteorology, Texas A & M University, College Station, Texas.

Bigler, V. J. 1975. The Xenia tornado at its beginning. *Weatherwise* **28:** 78–79.

Binnie, A. M. and G. A. Hookings. 1948. Laboratory experiments on whirlpools. *Proc. Royal Soc.* **A, 194:** 348–415.

Blanchette, V. G. 1971. An experiment with tornado-like winds. *Scientific American* **225**(4): 111–112.

Blechman, J. B. 1975. The Wisconsin tornado event of April 21, 1974. Observations and theory of secondary vortices. Report No. 75–3 (M.S. thesis), Dept. of Meteorology, Univ. of Wisconsin, 130 pp. Published in *Preprints, Ninth Conf. on Severe Local Storms,* Norman, Okla, pp. 344–349.

———, C. Anderson, and E. Lovell. 1974. Windspeed estimates from analysis of fence post damage by Wisconsin tornadoes of April 21, 1974. Personal correspondence from J. B. Blechman (U. of Wisc.) to J. E. Minor (Texas Tech Univ.), dated August 8, 1974.

Bleeker, W. and A. Delver. 1951. Some new ideas on the formation of windspouts and tornadoes. *Archiv. f. Meteor. Geophys. u. Bioklim.* **A4,** 220–237.

Bode, L., L. M. Leslie, and R. K. Smith. 1975. A numerical study of boundary effects on concentrated vortices with applications of tornadoes and waterspouts. *QJRMS* **101:** 313–324.

Bodewadt, V. T. 1940. Die Drehstromung uber festem Grunde. *Zeits. Ang. Math. Mech.* **20:** 241–253.

Booker, C. A. 1953. Tower damage provides key to Worcester Tornado data. *Electric World* **170:** 22–24.

———. 1954. On transmission towers destroyed by the Worcester, Massachusetts, tornado of June 9, 1953. *Bull. Amer. Meteorol. Soc.* 225, 229 (May).

Borghi, S. and N. Minafra. 1972. La tromba d'aria abbattutasi su Venezia la sera dell'11–9–1970. Indagine su alcuni fattori concomitanti alla sua formazione. *Riv. Met. Aer.* **XXXII** (2, Roma 1972): 133–145. [The tornado that struck Venice the evening of September 11, 1970. Investigation of some of the factors that contributed to the formation of the storm.]

Botchlor, G. K. 1967. *Fluid Dynamics.* Cambridge, England: Cambridge University Press.

Bradbury, D. L. and T. T. Fujita. 1966. Features and Motions of Radar Echoes on Palm Sunday, 1965. *SMRP Research Paper* no. 51, University of Chicago.

Bradshaw, H. and V. Bradshaw. How you can 'see' tornadoes on TV. *Popular Mechanics* **131:** 93–96 (March 1969).

Bradshaw, P. 1973. The effects of streamline curvature on turbulent flow. AGARDograph No. 169.

Brinkmann, W. A. R. 1975. Severe local storm hazard in the United States: a research assessment. *Program on Technology, Environment and Man Monograph* No. NSF-RA-E–75–011, Inst. of Behavioral Sci., Univ. of Colorado, Boulder, Colorado.

British Standards Institution. 1972. Code of Basic Data for the Design of Buildings, Chapter V. Loading, Part 2. Wind Loads. BSI CP3: Chapter V: Part 2. September 1972, London W1A 2BS.

Brook, M. 1967. Electrical currents accompanying tornado activity. *Science* **157:** 1434–1436.

———. n.d. *Why the Weather?* New York: Harcourt, Brace & Co.

Brooks, E. M. 1949. The tornado cyclone. *Weatherwise* **2:** 23–33.

———. 1951. Tornadoes and related phenomena. *Compendium of Meteorology.* Boston, Mass.: American Meteorological Society, pp. 673–80.

———. 1958. *Quantitative Models of Wind Velocity Components in a Tornado Vortex.* Institute of Technology, St. Louis University Report No. 1, June.

Brown, C. W. 1933. A study of the time-, areal- and type-distribution of tornadoes in the United States. *Trans. Am. Geophys. Union* **14:** 100–106.

——— and W. O. J. Roberts. 1935. The distribution and frequency of tornadoes in the United States from 1880 to 1931. *Trans. Am. Geophys. Union* **16:** 151–158.

———. 1937. The areal frequency of tornadoes in the United States by counties, 1880–1931. *Trans. Am. Geophys. Union* **18:** 144–146.

Brown, R. A. 1969. New England tornadoes: climatological survey from first settlement through 1968. *Preprints, Sixth Conf. on Severe Local Storms,* Chicago, Illinois, April 8–10, 238–243. Pub. by Am. Meteor. Soc., Boston, Mass.

——— and L. Lemon. 1976. Single Doppler radar vortex recognition Part II: Tornadic vortex signatures. *Preprints, 17th Conf. Radar Meteorology,* October 26–29, 1976, Seattle, Wash. Printed by the Amer. Meteor. Soc., Boston, Mass.

———, D. W. Burgess, and K. C. Crawford. 1973. Twin tornado cyclones within a severe thunderstorm: single Doppler radar observations. *Weatherwise* **26**(2): 63–69, 71.

———, D. W. Burgess, L. R. Lemon, D. Sirmans, et al. 1975. NSSL dual-Doppler radar measurements in tornadic storms: A preview. *Bull. Amer. Meteorol. Soc.* **56:** 524–526.

Browning, K. A. 1964. Airflow and precipitation trajectories within severe local storms which travel to the right of winds. *J. Atmos. Sci.* **21:** 634–39.

———. 1965. Some inferences about the updraft within severe local storms. *J. Atmos. Sci.* **22**(6): 667–69.

——— and T. T. Fujita. 1965. *A Family of Severe Local Storms—A Comprehensive Study of the Storms in Oklahoma on May 26, 1963. Part I, Air Force Cambridge Research Laboratory,* Bedford, Mass.

Budney, L. J. 1965. Unique damage patterns caused by a tornado in dense woodlands. *Weatherwise* **18**(2): 75–77, 86.

Burdette, E. G., R. A. James, and C-n Sun. 1974. *The Generation of Missiles by Tornadoes.* Division of Engineering Design Report TVA-TR-74-1, Tennessee Valley Authority, Oak Ridge, Tennessee.

Burgess, D. W. and R. A. Brown. 1976. Tornado warning with single Doppler radar. *Preprints, Symp. Tornadoes,* Lubbock, Tex., June 22–24, 1976.

———, L. R. Lemon, and R. A. Brown. 1975. Tornado characteristics revealed by Doppler radar. *Geophy. Res. Lett.* **2:** 183–184.

———, L. D. Hennington, R. J. Doviak, and P. S. Ray. 1976. Multimoment Doppler display for severe storm identification.

Burgers, J. M. 1948. A mathematical model illustrating the theory of turbulence. *Advan. Appl. Mech.* **1:** 197–199.

Burggraf, O. R. and M. R. Foster. 1975. *Theoretical Study of Vortex-Breakdown in Tornado-like Vortices.* Final report NOAA Grant 04–3–022–37, Ohio State University.

———, K. Stewartson, and R. Belcher. 1971. Boundary layer induced by a potential vortex. *Phys. Fluids* **14:** 1821–1833.

Burley, M. W. and P. J. Waite. 1965. Wisconsin tornadoes. *Trans. Wisc. Acad. of Sci., Arts & Lett.* **54:** 1–35.

Byers, H. R. 1942. *Non-Frontal Thunderstorms. Miscellaneous Report, No. 3,* Dept. of Meteorology, University of Chicago, 22 pp.

——— and L. J. Battan. 1949. Some effects of vertical wind shear on thunderstorm structure. *Bull. Am. Meteorol. Soc.* **30:** 168–75.

——— and R. R. Braham. 1949. *The Thunderstorm.* U.S. Government Printing Office, Washington, D.C., 287 pp.

Canipe, Y. J., J. E. Vogel, and R. A. Clark. 1973. Detailed Analysis of Tornado-Producing Thunderstorms Using Digital Radar. *Preprints, Eighth Conf. Severe Local Storms,* American Meteorological Society, Boston, Mass., pp. 57–60.

Carr, J. A. Preliminary report on the tornadoes of March 21–22, 1952. *Monthly Weather Review,* 50–58 (March 1952).

Carrier, G. F. 1970. Singular perturbation theory and geophysics. *SIAM Rev.* **12:** 175–193.

———. 1971. Swirling flow boundary layers. *J. Fluid Mech.* **49:** 133–144.

———. 1971. The intensification of hurricanes. *J. Fluid Mech.* **49:** 145–158.

———, A. L. Hammond, and O. D. George. 1971. A model of the mature hurricane. *J. Fluid Mech.* **47:** 147–170.

Casagrande, A. and W. L. Shannon. 1949. Strength of soils under dynamic loads. *Trans. American Society of Civil Engineers* **114:** 755–72.

Castro, I. P. and A. G. Robins. 1975. *The Effect of a Thick Incident Boundary Layer on the Flow Around a Small Surface Mounted Cube.* Central Electricity Generating Board Report R/M/N 795.

Cermak, J. E. 1952. *Application of Model Techniques to Mass Transfer Studies.* Centennial of Engineering, Chicago, Illinois, Technical Report CER53JEC2, Fluid Dynamics and Diffusion Laboratory, Colorado State University, Fort Collins, Colorado.

———. 1958. *Wind Tunnel for the Study of Turbulence in the Atmospheric Surface Layer.* Technical Report CER58JEC42, Fluid Dynamics and Diffusion Laboratory, Colorado State University, Fort Collins, Colorado.

———. 1970. Determination of wind loading on structural models in wind-tunnel simulated winds. *Proc. Symp. Wind Effects on High-Rise Buildings.* Northwestern University, Evanston, Illinois, 61–88.

———. 1970. Separation-induced pressure fluctuations on buildings. *Proc. USA-Japan Research Seminar on Wind Loads on Structures,* University of Hawaii, 55–70.

———. 1971. Laboratory simulation of the atmospheric boundary layer. *AIAA J.* **9:** 1746–1754.

———. 1975. Applications of Fluid Mechanics to Wind Engineering—a Freeman Scholar Lecture. *Trans. ASME, J. Fluids Eng.* **97:** 9–38.

———. 1976. Aerodynamics of buildings. *Ann. Rev. Fluid Mech.* **8:** 75–106.

———, W. Z. Sadeh, and G. Hsi. 1969. Fluctuating moments on tall buildings produced by wind loading. *Proc. Technical Meeting Concerning Wind Loads on Buildings and Structures,* National Bureau of Standards, Gaithersburg, Maryland, 45–59.

Chang, C. C. 1966. First real man-made tornado is generated in laboratory "cage" by space scientists at the Catholic University of America. News item released by Office of Public Information of the Catholic University of America.

———. 1969. Recent laboratory model study of tornadoes. *6th. Conf. Severe Local Storms,* Am. Meteor. Soc., pp. 244–252.

———. 1971. Tornado wind effects on buildings and structures with laboratory simulation. *3rd. Intern. Conf. Wind Effects on Buildings & Structures, Tokyo, Japan* **II:** 8.1–10.

———. 1971. What we learn from the tornado of Lubbock, Texas, U.S.A. May 11, 1970. *3rd. Intern. Conf. Wind Effects on Buildings & Structures, Tokyo, Japan* **III:** 8.1–10.

——— and C. Park. 1973. New laboratory model of a tornado-like vortex. *8th. Conf. Severe Local Storms,* Amer. Meteor. Soc., pp. 197–198.

Changnon, S. A. and R. G. Semonin. 1966. A great tornado disaster in retrospect. *Weatherwise* **19:** 56–65.

Charba, J. P. 1975. Operation scheme for short range forecasts of severe local weather. *Preprints, Ninth Conf. Severe Local Storms,* Amer. Meteor. Soc., Boston, Mass., pp. 51–57.

——— and L. Sasaki. 1968. Structure and Movement of the Severe Thunderstorms of April 3, 1964, as Revealed from Radar and Surface Mesonetwork Data Analysis. *Technical Memorandum* ERLTM-NSSL41, 47 pp.

——— and M. Livingston. 1973. Preliminary results on short range forecasting of severe storms from surface predictors. *Preprints, Eighth Conf. Severe Local Storms,* Amer. Meteor. Soc., Boston, Mass., pp. 226–231.

Chaussee, D. S. 1972. Numerical solution of axisymmetric vortex formation normal to a solid boundary. Ph.D. thesis, Iowa State Univ.

———, D. N. Yarger, and J. D. Iverson. 1973. *The Use of Observational Data in the Numerical Simulation of an Axisymmetrical Vortex.* Eighth Severe Local Storms Conf., Am. Meteorol. Soc.

Chi, S. W. and T. Costopolous. 1975. *Wind Loadings on Ground Structures in Intense Atmospheric Vortices.* Annual Report prepared for The National Science Foundation, Grant No. GK-41469.

——— and W. J. Glowacki. 1974. Applicability of mixing length theory to turbulent boundary layers beneath intense vortices. *ASME J. Applied Mech.* **41:** 15–19.

——— and J. Jih. 1974. Numerical modeling of the three-dimensional flows in the ground boundary layer of a maintained axisymmetric vortex. *Tellus* **26:** 444–455.

———, S. J. Ying, and C. C. Chang. 1969. The ground turbulent boundary layer of a stationary tornado-like vortex. *Tellus* **21:** 693–700.

Chien, N., Y. Feng, H. J. Wang, and T. T. Siao. 1951. *Wind-Tunnel Studies of Pressure Distribution on Elementary Building Forms.* Iowa Institute of Hydraulic Research.

Chrzanowski, P., J. M. Young, and H. L. Marrett. 1960. *Infrasonic Pressure Waves from Tornadic Storms.* National Bureau of Standards Report No. 7035.

Clarke, R. H. 1962. *Severe Local Wind Storms in Australia.* Division of Meteorological Physics, Tech. Paper No. 13, C.S.I.R.O., Melbourne, Australia.

Cochrane, H. C. 1975. *National Hazards and Their Distributive Effects.* Institute of Behavioral Science, University of Colorado, Boulder, Co.

Colgate, S. A. 1967. Tornadoes: Mechanism and control. *Science* **157:** 1431–1434.

———. 1968. Electrical heating of a tornado vortex. *J. Geophys. Res.* **73:** 6121.

———. 1975. Comment on "On the relation of electrical activity to tornadoes" by R. P. Davies-Jones and J. H. Golden. *J. Geophys. Res.* **80:** 4556.

Cook, A. W. 1953. Summary of tornadoes in Colorado, Wyoming, and New Mexico, 1916–51. *Mon. Wea. Rev.* **81:** 74–76.

Cook, R. K. 1969. *Atmospheric Sound Propagation.* Atmospheric exploration by remote probes. Final Report of the Panel on Remote Atmospheric Probing to the Committee on Atmos. Sci., National Research Council, Vol. 2, pp. 633–669.

Costello, J. R. 1975. Personal correspondence to J. R. McDonald of Texas Tech University; U.S. Nuclear Regulatory Commission, Washington, D.C.

Costen, R. C. 1970. Equation for vortex motion including effects of buoyancy and sources with applications to tornadoes. NASA Tech. Note D-5964.

Court, A. 1970. Tornado incidence maps. *ESSA Tech. Memo.* ERLTM-NSSL 49, National Severe Storms Laboratory, Norman, Ok.

———. 1976. Tornado damage probabilities. *Proc. Symposium on Tornadoes: Assessment of Knowledge and Implications for Man,* Lubbock, Texas, June 22–24.

Cressman, G. P. 1969. Killer storms. *Bull. Amer. Meteorol. Soc.* **50:** 850–855.

Croghan, S. 1977. "Tornado Damage Evaluations of School Buildings." Unpublished M.S. thesis, Dept. of Civil Engineering, Texas Tech University, Lubbock, Texas.

Crowe, C. T. et al. 1974. Are the codes safe for wind pressures? *J. Struct. Div., ASCE* **100** (ST8): 1745–1747.

Crutcher, H. L. et al. 1975. A Preliminary Note on Return Periods and Their Misuse. Available from NOAA National Climatic Center, Asheville, North Carolina, 10 pp.

Crumlish, J. D. and G. F. Wirth. 1967. A preliminary study of engineering seismology benefits. Washington, D.C., U.S. Department of Commerce, Environmental Science Services Administration.

Culver, C., H. S. Lew, G. C. Hart, and C. Pinkham. 1975. Natural hazard evaluation of existing buildings. National Bureau of Standards BSS 61.

Cunny, R. W. and R. C. Sloan. 1961. *Dynamic Loading Machine and Results of Preliminary Small-Scale Footing Tests.* ASTM Symposium on Soil Dynamics, Special Technical Publication, No. 305, pp. 65–77.

Dames & Moore. 1972. Tornado Wind Damage Probability and Recurrence for West Valley, New York. U.S. Nuc. Reg. Comm. Docket No. 50–201.

———. 1975. A meteorological and engineering approach to the regionalization of tornado wind criteria and nuclear power plants.

Danielsen, E. F. 1975. A conceptual theory of tornadogenesis: Based on macro-, meso- and microscale processes. *Ninth Conf. on Severe Local Storms, AMS* 376–383.

Darkow, G. L. 1968. The total energy environment of severe storms. *J. Applied Meteorology* **7**(2): 199–205.

———. 1969. An analysis of over sixty tornado proximity soundings. *Preprints of Papers Presented at Sixth Conf. Severe Local Storms,* American Meteorological Society, Boston, Mass., 218–221.

———. 1971. Periodic tornado production by long-lived parent thunderstorms. *Preprints of Papers Presented at Seventh Conf. Severe Local Storms,* American Meteorological Society, Boston, Mass., 214–217.

Davenport, A. G. 1961. The application of statistical concepts to the wind loading structures. *Proc. Inst. Civil Engr.* **19:** 449–471.

———. 1966. The treatment of wind loading on tall buildings. *Proc. Symp. Tall Buildings,* University of Southampton, England.

———. 1975. Perspectives on the full-scale measurements of wind effects. *J. Industr. Aerodynam.* **1**(1): 23–54.

——— and N. Isyumov. 1967. The application of the boundary layer wind tunnel to the prediction of wind loading. *Proc. International Research Seminar on Wind Effects on Buildings and Structures,* Ottawa, Canada, University of Toronto Press, 201–230.

———. 1967. A Wind-Tunnel Study for the United States Steel Building. Technical Report BLWT–5–67, The University of Western Ontario, London, Canada.

David, C. L. 1973. An objective method for estimating the probability of severe thunderstorms using predictions from the NMC (PE) numerical prediction model and observed surface data. *Preprints of Eighth Conf. Severe Local Storms,* Amer. Meteor. Soc., Boston, Ma., pp. 223–225.

———. 1976. A study of upper air parameters at the time of tornadoes. *Mon. Wea. Rev.* **104:** 546–551.

Davies-Jones, R. P. 1973. The dependence of core radius on swirl ratio in a tornado simulator. *J. Atmos. Sci.* **30**(7): 1427–1430.

———. 1976. Laboratory simulations of tornadoes. *Proc. Symp. on Tornadoes: Assessment of Knowledge and Implications for Man,* Lubbock, Tex., June 24–26. Published by Texas Tech Univ., pp. 151–174.

——— and J. H. Golden. 1975. On the relation of electrical activity to tornadoes. *J. Geophys. Res.* **80:** 1614–1616.

———. 1975. Reply to Comments on "On the relation of electrical activity to tornadoes" by S. A. Colgate. *J. Geophys. Res.* **80:** 4557.

———. 1975. Reply to Comments on "On the relation of electrical activity to tornadoes" by B. Vonnegut. *J. Geophys. Res.* **80:** 4561.

——— and E. Kessler. 1974. Tornadoes. *Weather and Climate Modification,* chapter 16, edited by W. N. Hess. New York: John Wiley & Sons.

——— and G. T. Vickers. 1971. Numerical simulation of convective vortices. NOAA Tech. Memo. ERL NSSL-57, NSSL, Norman, Okla.

———, D. W. Burgess, and L. R. Lemon. 1975. Analysis of the 4 June 1973 Norman tornadic storm. *Preprints Ninth Conf. Severe Local Storms.* American Meteorological Society, Boston, Mass., pp. 384–388.

———. 1976. An atypical tornado producing cumulonimbus. *Weather.*

——— et al. 1973. Psychological response to tornadoes. *Science* **180:** 544.

Defense Civil Preparedness Agency. 1973. Environmental Protection, DCPA, U.S. Department of Defense TR-39, Washington, D.C., December 1973.

———. 1975. Interim guidelines for building occupant protection from tornadoes and extreme winds. TR-83A, Defense Civil Preparedness Agency, U.S. Dept. of Defense, Washington, D.C.

De Beer, E. E. and A. Vesić. 1958. Etude expérimentale de la capacité portante du sable sous des fondations directes établies en surface. *Annales de Travaux Publics de Belgique* **59:** 3–51.

Deissler, R. G. and D. R. Boldman. 1974. Tornado-like gravity-driven vortex model, NASA Tech. Note D-7738.

Dennett, J. T. 1972. Fingerprinting the funnel. NOAA, pp. 48–51.

Denton, H. R. 1975. Probabilistic Approach to Tornado Missile Veolcities. Communication to F. Schroeder, Acting Director, Division of Technical Review, NRR, U.S. Nuclear Regulatory Commission, Washington, D.C.

Dergarabedian, P. and F. E. Fendell. 1967. Parameters governing the generation of free vortices. *Phys. Fluids* **10:** 2293–2299.

———. 1970. *Estimation of Maximum Wind Speeds in Tornadoes and Hurricanes.* Rept., TRW S, Redondo Beach, Calif.

———. 1970. On estimation of maximum wind speeds in tornadoes and hurricanes. *J. Astronaut. Sci.* **17:** 218–236.

———. 1971. A method for rapid estimation of maximum tangential wind speed in tornadoes. *Monthly Wea. Rev.* **99:** 143–145.

———. 1970. Estimation of maximum wind speeds in tornadoes. *Tellus* **22:** 511–516.

———. 1970. The surface frictional layer under a hurricane vortex. *J. Astronaut. Sci.* 9–34.

———. 1973. One- and two-cell tornado structure and funnel-cloud shape. *J. Astronaut. Sci.* **21:** 26–31.

———. 1975. Tornado dynamics. *Second U.S. National Conference on Wind Engineering Research (22–25 June 1975)*. Conference Preprints. Fort Collins, Colorado.

Desio, A. 1922. Su un turbine atmosferico che investi Roma el 1749. [About the tornado that struck Rome in 1749]. *Rivista Geografica, Italiana* **30:** 150–162.

Dessens, H. J. J. 1960. Severe hailstorms are associated with very strong winds between 6,000 and 12,000 M. *The Physics of Precipitation, Geophysical Monograph,* no. 15. Washington, D.C., 1960, pp. 333–38.

Dessens, J. 1972. Influence of ground roughness on tornadoes: a laboratory simulation. *J. Appl. Meteorol.* **11:** 72–75.

———. 1975. The effect of surface friction on the convective process in tornadoes. *Proc. Ninth Conf. Severe Local Storms,* Norman, Ok., Oct. 21–23, Amer. Meteor. Soc., pp. 364–367.

Devine, J. C. 1965. "Shell Structure of the Visible Tornado Vortex." M.S. thesis, St. Louis University.

Dikkers, et al., 1971. Hurricane Camile: August 1969. NBS Technical Note 569, National Bureau of Standards, U.S. Department of Commerce, Washington, D.C.

Dines, W. H. 1896. Experiments illustrating the formation of the tornado cloud. *Quart. J. Roy. Meteorol. Soc.* **22:** 71–73.

Dinwiddie, F. B. 1959. Waterspout-tornado structure and behavior at Nags Head, N.C., August 12, 1952. *Monthly Weather Review,* 239–50.

Doan, P. L. 1969. *Tornadoes and Tornado Effect Considerations for Nuclear Power Plant Structures Including the Spent Fuel Storage Pool.* United Engineers & Constructors Rept. UE&C APED-6807, Philadelphia, Pa.

———. 1970. Tornado considerations for nuclear power plant structures. *Nuc. Safety,* **11:** 296–308.

Dodd, T. 1975. Disaster. *Family Safety* (Summer).

Donaldson, C. duP., 1972. Calculation of turbulent shear flows for atmospheric and vortex motions. *AIAA J.* **10:** 4–12.

———. 1973. The relationship between eddy transport and second-order closure models for stratified media and for vortices. *Free Turbulent Shear Flows,* NASA SP-321, **I:** 233–255.

——— and R. D. Sullivan. 1960. Behavior of the solutions of the Navier-Stokes equations for a complete class of three-dimensional viscous vortices. *Proc. of 1960 Heat Transfer and Fluid Mechanics Inst.,* Stanford Univ. Press, pp. 16–30.

———. 1960. Examination of the solutions of the Navier-Stokes equations for a class of three-dimensional vortices. Part I: Velocity distributions for steady motion. AFOSR Report No. TN 60–1227, 82 pp.

Donaldson, R. J., Jr. 1962. *Radar Observations of a Tornado Thunderstorm in Vertical Section.* National Severe Storm Project Report, No. 8, 21 pp.

———. 1965. Methods for identifying severe thunderstorms by radar: A guide and bibliography. *Bull. Amer. Meteor. Soc.* **46:** 174–193.

———. 1970. Vortex signature recognition by a Doppler radar. *J. Appl. Meteor.* **9:** 661–670.

———, G. Armstrong, A. C. Chemal, and M. J. Draus. 1969. Doppler radar investigation of air flow and shear within severe thunderstorms. *Preprints, Sixth Conference on Severe Local Storms.* American Meteorological Society, Boston, Mass., pp. 146–54.

Donaldson, R. L., Jr. 1973. Doppler radar evidence for anticyclonic circulation in a severe convective storm. *Preprint Volume, Eighth Conference on Severe Local Storms,* American Meteorological Society, Boston, Mass., pp. 48–50.

Doviak, R. J., D. Burgess, L. Lemon, and D. Sirmans. 1974. Doppler velocity and reflectivity structure observed within a tornadic storm. *J. Rech. Atmos.* **8:** 235–243.

Dunlap, J. A. and K. Wiedner. 1971. Nuclear power plant tornado design considerations. *J. Power Div., Proc. Am. Soc. Civil Eng.,* **97:** 407–417.

Dryden, J. L. and G. C. Hill. 1926. *Wind Pressures on Structures.* U.S. National Bureau of Standards, Science paper 523, vol. 3.

Eagleman, J. R. 1967. Tornado damage patterns in Topeka, Kansas, June 8, 1966, *Monthly Weather Review* **95**(6): 370–74.

———. 1971. A double vortex thunderstorm model. *Preprints, Seventh Conf. Severe Local Storms.* American Meteorological Society, Boston, Mass. pp. 177–78.

——— and V. U. Muirhead. 1971. Observed damage from tornadoes and safest location in houses. *Preprints, Seventh Conf. Severe Local Storms.* American Meteorological Society, Boston, Mass., pp. 171–77.

——— and N. Willems. 1970. *Thunderstorms, Tornadoes and Damage to Buildings.* Research Report, HEW contract No. EC00303, University of Kansas, 253 pp.

———. 1971. *Thunderstorms, Tornadoes and Damage to Buildings.* Research Report, Contract No. EC00303. University of Kansas, Lawrence, Kan., 290 pp.

———. 1972. *Thunderstorms, Tornadoes and Damage to Buildings.* Environmental Publications, Lawrence, Kan. 279 pp.

———. 1975. *Thunderstorms, Tornadoes and Building Damage.* Lexington, Mass: D. C. Heath and Co., 317 pp.

Eaton, K. J. 1971. Damage due to Tornadoes in SE England, 25 January 1971. Bldng. Res. Stn. curr. Pap. CP27/71, 1971.

Einstein, H. A. and H. Li. 1955. Steady vortex flow in a real fluid. *Houille Blanche* **10:** 483.

Ekman, V. W. 1905. On the influence of the earth's rotation on ocean currents. *Arkiv Mathematik, Astr. och. Fysik* **2:** 1–52.

Emmons, H. W. and S. J. Ying. 1967. The fire whirl. *Eleventh Symp. Combustion.* Published by the Combustion Institute, Pittsburgh, Pa., pp. 475–486.

Environmental Science Services Administration. 1966. *Tornadoes.* Washington, D.C., 15 pp.

———. 1969. *Lightning.* PI 660024 Public Information Office, U.S. Dept. of Commerce.

Eom, J. K. 1975. Analysis of the int gravity wave occurrence of 19 April in the Midwest. *Monthly Weather Review* **103:** 226.

Eskridge, R. E. and P. Das. 1976. Effect of a precipitation-driven downdraft on a rotating wind field: a possible trigger mechanism for tornadoes? *J. Aeronaut. Sci.* **33:** 70–84.

Evesson, D. T. 1969. Tornado occurrences in New South Wales. *Australian Meteorological Magazine* **17**(3): 143–165. Also Bureau of Meteorology, working paper No. 130 65/18 of August, Sydney, Australia, 1970.

Fankhauser, J. C. 1971. Thunderstorms—Environment Interaction Determined by aircraft and radar observation. *Monthly Weather Review* **99**(3): 171–92.

Fawbush, E. J. and R. C. Miller. 1952. A mean sounding representative of the tornado air mass. *Bull. Amer. Meteorol. Soc.* **33**(7): 303–7.

———. 1954. The types of air masses in which North American tornadoes form. *Bull. Amer. Meteorol. Soc.,* **35:** 154–65.

Fedyayevsky, K. R. and S. M. Belotserkovsky. 1954. The aerodynamic forces acting on buildings in squalls. *Izv. A.N., U.S.S.R., Otd. tekhn* **6:** 13–24.

Ferrel, W. 1889. *A Popular Treatise on the Winds.* New York: John Wiley & Sons, 505 pp.

Finley, J. P. 1884. Report on the character of six hundred tornadoes. *Prof. Papers of the Signal Service* No. VII, available NOAA library, Silver Spring, Md.

Fisher, W. E. 1962. Experimental studies of dynamically loaded footings on sand. *ASTIA Technical Bulletin,* No. AD-290731, 43 pp.

Fitzjarrald, D. E. 1973. A laboratory simulation of convective vortices. *J. Atmos. Sci.* **30:** 894–902.

Flora, S. D. 1953. *Tornadoes of the United States.* Norman, Oklahoma: Univ. of Oklahoma Press. 194 pp.

Florida Power & Light Co. 1969. *A Comparative Study of Florida's Most Severe Tornadoes with those in other Parts of the Continental U.S.* USAEC Docket no. 50–335, Amendment 4.

Frankenfield, H. C. 1896. The tornado of May 27, at St. Louis, Missouri. *Monthly Weather Review* 77–81 (March).

Franz, H. W. 1967. Ein Theoretisches Trombenmodell. Bericht Nr. 2 Zugleich Abschluss-berischt zu dem mit Mitteln der Deutschen Forschungsgemeinschaft geforderten Forschungsvorhaben WI 150/5 "Versuch Einer Theorie der Tromben" (Einschl. WI 150/7), 60 pp.

———. 1969. Die Zellstruktur stationarer Konvektionswirbel (On the cellular structure of steady convective vortices). *Beitr. Phy. Atm.* **42:** 36–66.

Fred, D. 1975. "The Influence of Topography on Tornadoes." M.S. thesis, Kent State Univ., Kent, Ohio.

Friday, E. W. Jr. 1969. Behavior of discrete convective elements in a rotating fluid. Final Report NSF Grant GA-1328, Univ. of Okla.

Friedman, D. G. 1975. *Severe Storm Climatology of Kansas.* The Travelers Insurance Co., Hartford, Ct.

——— and P. A. Shortell. 1967. *Prospective Weather Hazard Rating in the Midwest with Special Reference to Kansas and Missouri.* The Travelers Insurance Co., Hartford, Ct.

Fritz, C. E., "Disaster, Contemporary Social Problems," New York, Harcourt, Brace and World, Inc., 651–694, 1961.

Fujita, T. T., "Results of Detailed Synoptic Studies of Squall Lines," *Tellus,* 1955, vol. 7, pp. 405–436.

———. 1960. A detailed analysis of the Fargo tornadoes of June 20, 1957. U.S. Weather Bureau Res. Pap. No. 42, Washington, D.C., 67 pp.

———. 1960. Structure of convective storms. *Physics of Precipitation, Proc. Cloud Physics Conf., Woods Hole, Mass.*

———. 1963. Analytical mesometeorology: A review. *Meteorological Monograph, Severe Local Storms, Amer. Meteorolg. Soc.* **5**(27): 77–122.

———. "Estimated wind speeds of the Palm Sunday tornadoes," Satellite and Mesometeorology Research Project Paper 53, University of Chicago, Department of Geophysical Sciences, April, 1967.

Fujita, T., 1967. *Estimation of Tornado Wind Speed from Characteristic Ground Marks.* Satellite and Mesometeorology Research Project Research Paper 69. Chicago: U. of Chicago, Dept. of Geophysical Sciences.

Fujita, T. T. 1970. The Lubbock tornadoes: a study of suction spots. *Weatherwise* **23:** 160–173.

———. 1970. Estimate of areal probability of tornadoes from inflationary reporting of their frequencies. *SMRP Res. Paper* No. 89, Univ. of Chicago, Chicago, Ill.

———. 1970. Estimate of maximum wind speeds of tornadoes in three northwestern states. *SMRP Res. Paper* No. 92, Univ. of Chicago, Chicago, Illinois.

———. 1971. Proposed characterization of tornadoes and hurricanes by area and intensity. *SMRP Res. Paper* No. 91, Univ. of Chicago, Chicago, Illinois.

———. 1972. F-scale classification of 1971 tornadoes. *SMRP Res. Paper* No. 100, Univ. of Chicago, Chicago, Illinois.

———. 1972. Proposed mechanism of suction spots accompanied by tornadoes. *SMRP Res. Paper* No. 102, Univ. of Chicago, Illinois.

———. 1972. Estimate of maximum windspeeds of tornadoes in southernmost Rockies. *SMRP Res. Paper* No. 105, Univ. of Chicago, Chicago, Illinois.

———. 1973. Experimental classification of tornadoes in F P P scale. *SMRP Res. Paper* No. 98, Univ. of Chicago, Illinois.

———. 1973. Tornadoes around the world. *Weatherwise* **26:** 56–62, 78–83.

———. 1974. Jumbo tornado outbreak of 3 April 1974. *Weatherwise* **27:** 116–126.

———. 1975. Superoutbreak tornadoes of April 3–4, 1974. Final edition; map available from the Univ. of Chicago, Ill.

———. 1975. New evidence from April 3–4, 1974 tornadoes. *Preprints, Ninth Conf. on Severe Local Storms,* Norman, Ok., October 21–23, 248–255. American Meteorological Society, Boston, Mass.

———. 1976. Graphic examples of tornadoes. *Bull. Amer. Meteorol. Soc.* **57,** 401–412.

———. 1976. U.S. Tornadoes, 1930–1974. Map available from the Univ. of Chicago, Ill.

———. 1977. Lamont-Argonne tornado of June 13, 1976. *SMRP Res. Paper* No. 144, Univ. of Chicago, Chicago, Ill.

Fujita, T. T. and Hector Grandoso. 1968. Split of a thunderstorm into anticyclonic and cyclonic storms and their motion as determined from numerical model experiments. *J. Atmos. Sci.* **23**(3): 416–39.

——— and A. D. Pearson. 1973. Results of F P P classification of 1971 and 1972 tornadoes. *Preprints, Eighth Conf. Severe Local Storms,* Denver, Colorado. October 15–17, 142–145. Amer. Meteorol. Soc., Boston, Mass.

———, D. L. Bradbury, and C. F. Van Thullenar. 1970. Palm Sunday tornadoes of April 11, 1965. *Monthly Weather Review* **98**(1): 29–69.

——— and D. M. Ludlum. 1975. Long-term fluctuation of tornado activities. *Preprints, Ninth Conf. Severe Local Storms,* Norman, Ok., October 21–23, 417–423. Amer. Meteorol. Soc., Boston, Mass.

———, J. J. Tecson, and L. A. Schaal. 1971. Preliminary results of tornado watch experiment 1971. *Preprints, Seventh Conf. Severe Local Storms,* Kansas City, Mo., October 5–7, 255–261. Amer. Meterol. Soc., Boston, Ma.

——— et al. 1976. Photogrammetric analyses of tornadoes. *Proc. Symp. Tornadoes: Assessment of Knowledge & Implications for Man,* Lubbock, Texas, June 22–24.

——— et al. 1960. A detailed analysis of the Fargo tornadoes of June 20, 1957. U.S. Weather Bureau Research Paper No. 42.

Fulks, J. R. 1962. On the Mechanics of the tornado. *NOAA Tech. Memo.* ERLTM-NSSL 4, NSSL, Norman, Okla., 33 pp.

Fultz, D. 1951. Experimental analogies to atmospheric motions. *Compendium of Meteorology,* Amer. Meteorol. Soc., Boston, 1235–1248.

Gallimore, R. G., Jr. 1968. "Topographic Influence on Tornado Tracks and Frequencies in Arkansas and Wisconsin." M.S. thesis, Univ. of Wisconsin, Madison, Wis.

——— and H. H. Lettau. 1970: Topographic influence on tornado tracks and frequencies in Wisconsin and Arkansas. *Trans. Wisc. Acad. Sci., Arts & Lett.* **58:** 101–127.

Galway, J. G. 1967. SELS forecast verification 1952–1966. Preprints, *Fifth Conf. Severe Local Storms,* St. Louis, Mo., October 19–20, 140–145. Am. Meteorol. Soc., Boston, Mass.

———. 1975. Relationship of tornado deaths to severe weather watch areas. *Mon. Wea. Rev.* **103:** 737–741.

Galway, J. L. 1956. The Lifted Index as a Predictor of Latent Instability. *Bull. Amer. Meteorol. Soc.* **36:** 528–29.

Garson, R. C. et al. 1974. *Tornado Design Winds Based on Risk.* Structures Publication No. 396, MIT Department of Civil Engineering, Cambridge, Mass.

———, J. Morlá Catalán, and C. A. Cornell, 1975. Tornado risk evaluation using wind speed profiles. *J. Struct. Div., ASCE* **101:** 1167–1171.

———. 1975. Tornado design winds based on Risk. *J. Struct. Div., ASCE* **101.**

Gates, D. M. 1972. *Man and His Environment: Climate.* The University of Michigan, Michigan.

Glaser, A. H. 1960. An observational deduction of the structure of a tornado vortex. In *Cumulus Dynamics* (Proc. 1st Conf. Cumulus Convection, 1959, Portsmouth, N.H.), Pergamon Press, pp. 157–166.

Glasstone, S. (ed.). 1962. *Effects of Nuclear Weapons.* U.S. Dept. of Defense, U.S. Atomic Energy Commission, Washington, D.C., 730 pp.

Golden J. H. 1973. Some statistical aspects of waterspout formation. *Weatherwise* **26:** 108–117.

———. 1975. *A U.S. East and Gulf Coast Waterspout Climatology.* Summary rept. prepared for U.S. Nuc. Reg. Comm., Washington, D.C.

Golden, J. H. 1976. Windspeeds in Tornadoes. *A Symposium on Tornadoes: Assessment of Knowledge and Implications for Man,* held at Texas Tech University, Lubbock, Texas, June 22–24, 1976.

Golden, J. H. and R. P. Davies-Jones. 1975. Photogrammetric windspeed analysis and damage interpretation of the Union City, Oklahoma tornado, May 24, 1973. *Proc. Second U.S. National Conf. Wind Engineering Research,* Ft. Collins, Colo., June 23–25 (published by Wind Engineering Research Council), II–2–1 to II–2–4.

Golden, J. H. and D. Purcell. 1974. Photogrammetric velocity and morphological analysis of the Union City, Oklahoma, tornado, May 24, 1973. Paper presented at Fall Annual AGU Meeting, San Francisco, Calif., Dec. 12–17, 1974. Abstract in *EOS* **56**(12).

———. 1975. Photogrammetric velocities for the Great Bend, Kansas, tornado-accelerations and asymmetrics. *Ninth Conf. Severe Local Storms,* Amer. Meteorol. Soc., pp. 336–343.

Goldstik, M. A. 1960. A paradoxical solution of the Navier Stokes equations. *PMM* **24:** 610–621.

Goolsby, D. E. 1974. *In-Residence Shelters for Extreme Winds.* Department of Civil Engineering, Texas Tech University, Lubbock, Texas, p. 8.

Gordon, A. H. 1951. Waterspouts. *Weather* **6:** 6–12, 364–371.

Gordon, H. 1971. Comment on a unique tornado report. *Monthly Weather Review* **99**(8): 649.

Granger, R. A. 1966. Steady three-dimensional vortex flow. *J. Fluid Mech.* **25:** 557–576.

———. 1972. A steady axisymmetric vortex flow. *Geophys. Fluid Dynam.* **3:** 45–88.

Grant, H. L., R. W. Stewart, and A. Moilliet. 1962. Turbulence Spectra from a Tidal Channel. *J. Fluid Mech.* **12**(2): 241.

Gray, W. M. 1969. Hypothesized importance of vertical wind shear in tornado genesis. *Preprints, Papers presented at Sixth Conf. Severe Local Storms,* Amer. Meteorol. Soc. Boston, Mass., 230–237.

———. 1971. Research methodology, observations, and ideas on tornado genesis. *Preprints, Papers Presented at Seventh Conf. Severe Local Storms,* Amer. Meteorol. Soc., Boston, Mass., 292–298.

Greenspan, H. P. 1968. *The Theory of Rotating Fluids.* Cambridge University Press, U.K. 327 pp.

Greneker, E. F., C. S. Wilson, and J. I. Metcalf. 1975. Remote detection of severe storms and tornadoes in Georgia. *Final Tech. Rept. Project* E220–903/E230–902. Engineering Experiment Station, Georgia Inst. of Tech., Atlanta, Geogia.

Grimm, J. 1883. *Teutonic Mythology.* (English trans. 1966), Dover Publications, N.Y. (3 vols).

Gupta, Y. M. and L. Seama. 1975. *Dynamic Behavior of Reinforced Concrete under Missile Impact Loading.* ASCE Speciality Conference on Structural Design of Nuclear Plant Facilities. Vol. 1-A, New Orleans, Louisiana, pp. 637–662.

Gutman, L. N. 1957. Theoretical model of a waterspout. *Bull. Acad. Sci.,* USSR (Geophysical Series). Pergamon Press translation, N.Y. Vol. I, pp. 87–103.

———. 1969. *Introduction to the Nonlinear Theory of Mesoscale Meteorological Processes.* Gidrometeorologicheskoe Izdatel'stvo, Leningrad. Translated from Russian, Israel Program for Sci. Translations, Jerusalem, 1972.

Gwaltney, R. C. 1968. *Missile Generation and Protection in Light-Water-Cooled Reactor Plants.* ORNL-NSIC-22, Oak Ridge National Laboratory, Oak Ridge, Tennessee.

Haddon, V. D. 1960. The use of wind tunnel models for determining the wind pressure on buildings. *Civil Engnrg. & Public Works Rev.* **55**(645): 500.

Haglund, C. T. 1969. *A Study of a Severe Storm of 16 April 1967.* NSSL Tech. Memo No. 44, U.S. Department of Commerce, ESSA Research Laboratories.

Hall, M. G. 1966. The structure of concentrated vortex cores. *Prog. Aeronaut. Sci.* **7:** 53–110.

———. 1972. Vortex breakdown. *Ann. Rev. Fluid Mechan.* **4:** 195–218.

Hall, R. S. 1951. Inside a Texas tornado. *Weatherwise* **4**(3): 54–57, 65.

Hamel, G. 1916. Spiralformige Bewegung Zaher Flussigkeiten. *Jahresher d. Dt. Mathematikev-Vereinigung* **25:** 34–60.

Handa, K. N. and B. L. Clarkson. 1968. Response of tall structures to atmospheric turbulence. *Proc. Wind Effects on Buildings and Structures,* Loughborough University of Technology, England, vol. 1, Paper 14.

Harding, E. T. 1965. Part 1—Hurricanes. *Heavy Weather Guide,* U.S. Naval Institute, Annapolis, Maryland, 3–67.

Hardy, R. N. 1971. The Cyprus waterspouts and tornadoes of 22 December 1969. *Meteorolog. Mag.* **100:** 74–82.

Harkness, G. S. 1894: The tornado at Little Rock, Arkansas, October 2, 1894, editorial in *Monthly Weather Review,* Oct. 1894, pp. 413–414.

Harlow, F. H. and L. R. Stein. 1974. Structural analysis of tornado-like vortices. *J. Atmos. Sci.* **31:** 2081–2098.

——— and J. E. Welch. 1965. Numerical calculation of time-dependent viscous incompressible flow of fluid with a free surface. *Phys. Fluids* **8:** 2182–2189.

Harris, C. M. and C. E. Crede. 1961. *Shock and Vibration Handbook.* New York: McGraw-Hill Book Co.

Harrison, H. T. and W. I. Orendorff. 1941. Prefrontal Squall Lines. United Air Lines, Meteorology Dept., Circulation Memo #29, pp. 1–8.

Harrold, T. W. 1966. A Note on the Development and Movement of Storms over Oklahoma on May 7, 1965. *National Severe Storms Laboratory Technical Memorandum, No. 29.*

Hart, G. C. 1976. Estimation of structural damage due to tornadoes. *Proc. Symp. Tornadoes: Assessment of Knowledge and Implications for Man,* Lubbock, Texas, June 22–24.

Hatton, L. 1975. Stagnation point flow in a vortex core. *Tellus* **27:** 269–279

Haurwitz, B. 1935. The height of tropical cyclones and the "eye" of the storm. *Monthly Weather Review* **63:** 45–49.

Hazen, H. A. 1890. Tornadoes. A prize essay. *Amer. Meteorol. J.* **7:** 205–229.

———. 1890: *The Tornado.* N.D.C. Hodges, N.Y.

———. 1890: Annual Report of the Chief Signal Officer of the Army to the Secretary of War for the Year 1890. G.P.O., Wash., D.C.

Hebb, L. E. 1975. Wind Probabilities in DCPA Region Six. *Proc. Second U.S. National Conf. Wind Engineering Research,* Ft. Collins, Colo., June 1975, II–12–1 to II–12–4.

Heller, L. W. 1964. *Failure Modes of Impact-Leaded Footings on Dense Sand.* Technical Report,

R-281, U.S. Naval Civil Engineering Laboratory, Port Hueneme, California, 31 pp.

Hellmann, G. 1917. Die ältesten Untersuchunger über Windhosen. *Beitr. Geschichte Meteor.* **2:** 239–334.

Helmholtz, H. von. 1858. Crelle's Jour. 55 (also 1867 *Phil. Mag.* (IV) **33:** 485, and Wissenschaftliche Abhand Kungen **1:** 101.)

Hide, R. 1966. On the dynamics of rotating fluids and related topics in geophysical fluid dynamics. *Bull. Amer. Meteorol. Soc.* **47:** 873–885.

——— and C. W. Titman. 1968. Detached shear layers in a rotating fluid. *J. Fluid Mech.* **29:** 39–60.

Hoecker, W. H. Jr. 1959. History and measurement of the two major Scottsbluff tornadoes of 27 June 1955. *Bull. Amer. Meteorol. Soc.* **40:** 117–133.

———. 1960. The dimensional and rotational characteristics of the tornadoes and their cloud system. *The Tornadoes at Dallas, Tex., April 2, 1957.* U.S. Weather Bureau Research Paper 41, U.S. Govt. Printing Office, Washington, D.C., 53–113.

Hoecker, W. H. 1961. Three-dimensional pressure pattern of the Dallas tornado and some resultant implications. *Monthly Weather Review* 533–42.

———. 1960. Wind speed and air flow patterns in the Dallas tornado of April 2, 1957. *Monthly Weather Review* 167–80 (May).

———, R. G. Beebe, D. T. Williams, J. T. Lee, S. G. Bigler, and E. P. Segner, Jr. 1960. The tornadoes at Dallas, Texas, April 2, 1957. U.S. Department of Commerce, Weather Bureau Research Paper No. 41, 175 pp.

Hoerner, S. F. 1965. *Fluid-Dynamic Drag.* Published by the author. Brick Town, New Jersey, p. 4–4.

Hoffman, E. R. and P. N. Joubert. 1963. Turbulent line vortices. *J. Fluid Mech.* **16:** 395–411.

Hoyle, R. J. 1973. *Wood Technology in the Design of Structures,* Mountain Press Publishing Co., Missoula, Mont.

House, D. C. 1963. Forecasting tornadoes and severe thunderstorms. *Meteorol. Monographs* **5:** 141–156.

Howard, L. N. and A. S. Gupta. 1962. On the hydrodynamic and hydromagnetic stability of swirling flows. *J. Fluid Mech.* **14:** 463–476.

Howe, G. M. 1974. Tornado path sizes. *J. Appl. Meteorol.* **13:** 343–347.

Howe, J. W. 1952. Wind pressure on elementary building forms evaluated by model tests. *Civil Engnrng* 42–46, 330–334.

Hsu, C. T. 1973. Laboratory modelling of the tornado suction mechanism. *8th. Conf. Severe Local Storms,* Amer. Meteorol. Soc., 199–202.

Hsu, C. T. and B. Fattahi. 1975. Laboratory Simulation of Tornadoes. *Proc. Second U.S. National Conference on Wind Engineering Research,* Colorado State University, Fort Collins, Colorado.

———. 1975. Tornado funnel formation from a tornado cyclone. *Proc. Ninth Conference on Severe Local Storms,* Norman, Okla, Oct. 21–23, Amer. Meteorol. Soc., pp. 358–363.

Hsu, C. T. and C. Odetunde. 1978. Velocity and pressure distributions of a tornado-like vortex. Third U.S. National Conf. Wind Engineering Research, University of Florida, Gainesville, Fla., Feb. 26–Mar. 1, 1978.

Hsu, C. T. and H. Tesfamariam. 1976. Computer simulation of a tornado-like vortex boundary layer flow. Preprint ERI-76231, Eng. Res. Inst., Iowa State Univ., Ames, Iowa.

———. 1975. Turbulent modeling of a tornado boundary layer flow. Preprint ERI-76126, Eng. Res. Inst., Iowa State Univ., Ames, Iowa.

Huff, F. A., H. W. Hiser, and S. G. Bibler. 1954. Study of an Illinois Tornado Using Radar Synoptic Weather and Field Survey Data. *Report of Investigation Number 22,* Illinois State Water, Survey, Urbana, Ill., 73 pp.

Humphreys, W. J. 1940. *Physics of the Air.* New York: McGraw-Hill.

Inman, R. 1970. Operational objective analysis schemes at the National Severe Storms Forecast Center. Tech. Circular No. 10, NSSL, 50 pp.

Institute for Disaster Research, 1972. *Guideline to Tornado Damage Survey.* Dept. of Civil Engr., Texas Tech Univ., Lubbock, Tex. 6 pp.

Iotti, R. C. 1974. Velocities of Tornado Generated Missiles. Ebasco Services, Inc. Rept. ETR-1003, New York, N.Y.

———. 1975. Velocities of tornado generated missiles. Ebasco Services, Inc. Rept. ETR-100, New York, N.Y.

Jaech, J. L. 1970. Statistical analysis of tornado data for the three northwestern states.

James Cook University. 1972. *Cyclone Althea, Part I—Buildings.* James Cook University of North Queensland, Townsville, Australia.

Jensen, M. 1958. The model law for phenomena in natural wind. *Ingeniøren.* **2**(4): 121–28.

Jet Propulsion Laboratory. 1976. *Wind Field and Trajectory Models for Tornado-Propelled Objects.* Technical Report prepared for Electric Power Research Institute, EPRI 308, No. 1, Palo Alto, Ca.

Jischke, M. C. and M. Parang. 1974. Properties of simulated tornado-like vortices. *J. Atmos. Sci.* **31:** 506–512.

———. 1975. Fluid dynamics of a tornado-like vortex flow. Final Report NOAA Grant N22–200–72(G) and 04–4–022–13, Univ. of Okla.

Jones, H. L. 1951. A sferic method of tornado identification and tracking. *Bull. Amer. Meteor. Soc.* **32:** 380–385.

Joos, L. A. 1960. Time trends in mid-west tornado statistics. Unpublished manuscript available from NOAA library, Washington, D.C. 6 pp.

JPL. 1976. *Wind Field and Trajectory Models for Tornado-Propelled Objects.* EPRI 308, Technical Report 1, Electric Power Research Institute, Palo Alto, Ca.

Keller, D. and B. Vonnegut. 1976. Wind speed estimation based on the penetration of straws and splinters into wood. *Weatherwise* **29** (5).

———. 1976. Wind speeds required to drive straw splinters into wood. Dept. of Atmos. Sci., State University of New York at Albany.

Kelly, D. L., J. T. Schaefer, R. P. McNulty, C. A. Doswell, and R. F. Abbey. 1978. An augmented tornado climatology. *Monthly Weather Review* **106:** 1172–1183.

Kessler, E. 1965. Purposes and program of the U.S. Weather Bureau National Severe Storms Laboratory. *Trans. Amer. Geophys. Union* **46:** 389–397.

———. 1966. A storm's incalculable energy. *Natural History* **75:** 12–17.

———. 1967. Correspondence to Peter A. Morris, Director of the Division of Reactor Licensing of the U.S. Atomic Energy Commission, February 21, 1967. (Copy in possession of M. Melaragno.)

———. 1970. Tornadoes. *Bull. Am. Meteor. Soc.* **51:** 926–936.

——— and J. T. Lee. 1976. Normalized indices of destruction and deaths by tornadoes. *NOAA Tech. Memo.* ERL NSSL-77, National Severe Storms Laboratory, Norman, Okla.

Kiesling, E. W. 1975. Anniversary of tornado is marked with opening of all-weather house. *Arkansas Gazette* (July 13): 150–160.

——— and D. E. Goolsby. 1974. In-Home shelters from extreme winds. *Civil Engineering,* Vol. 44, No. 9, September 1974, 105–107.

Kinzer, G. D. and B. Morgan. Location and movement of lightning centers associated with a tornadic storm. Paper presented at the AGU-AMS Meeting in Washington, D.C., 8–11 April, 1968.

Köppen, W. 1896. Die Windhose vom 5. Juli 1890 bei Oldenburg und die Gewitterboe vom 10. Juli 1896 in Ostholstein. *Ann. Hydro. Marit. Meteor.* **10:** 448.

Kraus, M. J. 1973. Doppler radar observations of the Brookline, Mass., tornado of Aug. 9, 1972. *Bull. Am. Meteor. Soc.,* **54:** 519–524.

Kuhlbrodt, E. 1920. Das Wind system im Fuss der Sachsenwaldtrombe vom 28. Juni 1920. *Ann. Hydro. Marit. Meteor.* **50:** 154–158.

Kuo, H. L. 1966. On the dynamics of convective atmospheric vortices. *Journal of the Atmospheric Sciences.* Vol. 23, No. 1, 25–42.

———. 1967. Note on the similarity solutions of the vortex equations in an unstably stratified atmosphere. *J. Atmos. Sci.* **24:** 95–97.

———. 1971. Axisymmetric flow in the boundary layer of a maintained vortex. *J. Atmos. Sci.* **28:** 20–41.

Lalanne, M. 1839. Sur l'évaluation numerique de la force qui a produit certains effets de rupture á Chatenay. *Comptes Rendus* **9:** 219–223.

Lamb, H. 1932. *Hydrodynamics.* 6th Ed., Dover Publications, New York, 738 pp.

Lamb, H. H. 1957. Tornadoes in England, May 21, 1950. *Geophys. Memoirs* No. 99, Air Ministry Meteorological Office, London, 38 pp.

Lambert, B. K. et al. 1975. Economics of Wind Resistant Construction: Expected Value of Damage vs Additional Construction Costs in DCPA Region Six. Institute for Disaster Research, Texas Tech University June 1975.

League, L. D. 1971. "Wind Flow Simulations Around and within Double Vortex Thunderstorm Cells." M.A. Thesis, University of Kansas, Lawrence, Kan., 90 pp.

Lee, A. J. H. 1973. A General Study of Tornado Generated Missiles, Gilbert Associates, Inc., Reading, Pennsylvania.

———. 1975. Design parameters for tornado-generated missiles. *Topical Rept.* No. GAI-TR-102, Gilbert Associates, Inc., Reading, Pa.

Lee, J. T. 1958. Tornadoes in the United States, 1950–1956. *Mon. Wea. Rev.,* **86:** 219–228.

Lemon, L. R. 1970. Formation and Emergence of an Anticyclonic Eddy within a Severe Thunderstorm as Revealed by Radar and Surface Data. *Preprints, Fourteenth Radar Meteorology Conference,* American Meteorological Society, Boston, Mass., pp. 323–28.

———. 1974. Thunderstorm Wake Vortex Structure and Aerodynamic Origin. *NOAA technical memo ERL NSSL-71,* pp. 17–43.

———, D. W. Burgess, and R. A. Brown. 1975. Tornado production and storm sustenance. Ninth Conf. on Severe Local Storms, A.M.S., 100–104.

Leslie, L. M., B. R. Morton, and R. K. Smith. 1970. On modeling tornadoes. *QJRMS* **96.**

Letzmann, J. 1923. Das Bewegungsfeld im Fuss einer fortschreitenden Windoder Wasserhose. *Acta. Comm. Univ. Dorpat A* **6:** 1–136.

———. 1928. Zur Methodik der Trombenforschung. *Meteor. Zeit.* **45:** 434–439.

———. 1937. Richtlinien zur Erforschung von Tromben, Tornadoes, Wassenhosen und Kleintromben. IMO, Klimat. Komm., Salzburg, Publ. No. 38, 91–110.

——— and A. Wegnener. 1930. Die Druckerniedrigung in Tromben. *Meteor. Zeit.* **47:** 165–169.

Leverson, V. H. and P. C. Sinclair. 1975. Waterspout Wind, Temperature, and Pressure Structure Deduced from Aircraft Measurements. *Preprints of Ninth Conference on Severe Local*

Storms, American Meteorological Society, Norman, OK, October 1975, 350–357.

Lewellen, W. S. 1976. Theoretical models of the tornado vortex. *Proceedings,* Symp. on Tornadoes: Assessment of Knowledge and Implications for Man, Lubbock, Tex., June 24–26 (published by Texas Tech Univ.), 107–143.

Lewis, W. and P. J. Perkins. 1953. Recorded pressure distribution in the outer portion of a tornado vortex. *Mon. Wea. Rev.* **81:** 374–385.

Lilly, D. K. 1969. Tornado dynamics. Research manuscript 69–117, National Center for Atmospheric Research, Boulder, Colorado.

Lincoln Library of Essential Information. 1969. Frontier Press, Buffalo, New York.

Linehan, U. J. 1957. Tornado deaths in the United States. *U.S. Weather Bureau Tech. Paper* No. 30, Washington, D.C.

Logie, J. 1919. Note on tornadoes. *Quart. J. R. Meteor. Soc.* **45:** 317.

Long, R. R. 1956. Sources and sinks at the axis of a rotating liquid. *Quart. J. Mech. Appl. Math.* **9:** 385–393.

———. 1958. Vortex motion in a viscous fluid. *J. Meteor.* **15:** 108–112.

———. 1960. Tornadoes and dust whirls. Johns Hopkins University, Baltimore, Md., Dept. of Mech., Tech. Rept. No. 10 (ONR series) or No. 13 (CWB series), 27 pp.

Lorenz, E. 1967. The nature and theory of the general circulation of the atmosphere. World Meteor. Org., TP 115, 158 pp.

Ludlam, F. H. 1963. Severe Local Storms, A Review. *Meteorological Monographs, Severe Local Storms,* American Meteorological Society, Boston, Mass., vol. 5, no. 27, pp. 1–32.

Ludlum, D. M. 1970. *Early American Tornadoes 1586–1870.* American Meteorological Society, Boston, Massachusetts, 219 pp.

Maddox, R. A. 1973. A study of tornado proximity data and an observationally derived model of tornado genesis. Dept. of Atmospheric Science paper 212, Colorado State University, Fort Collins, Colorado.

Mal'bakhov, V. M. 1972. Investigation of the structure of tornadoes. *Izvestiya, 8,* 17–28; English translation, *Atmopsheric & Oceanic Physics* **8:** 8–14.

Malkus, J. S. 1958. On the structure and maintenance of the mature hurricane eye. *J. Meteorol.* **15:** 337–349.

——— and H. Riehl. 1960. On the dynamics and energy transformation in steady-state hurricanes. *Tellus* **12:** 1–20.

Marchenko, A. S. 1961. Determination of maximum wind velocities in tornadoes and tropical cyclones. *Meteorology and Hydrology* **5:** 11–16 (in Russian).

Markee, E. H., Jr. et al. 1974. Technical basis for interim regional tornado criteria. U.S. Atomic Energy Commission, WASH-1300, Office of Regulation, Washington, D.C.

Martin, J. R. 1940. Tornadoes in the United States, 1916–1937, inclusive. Mimeo. summary in NOAA library, Washington.

Martins, C. 1850. Answeisung Zur Beobachtung der Windhosen oder Tromben. *Ann. Physik* **81:** 444, 467 (fr. French orig.).

Maxworthy, T. 1967. The flow creating a concentration of vorticity over a stationary plate. *JPL Space Programs Summary* 37–44, IV, 243.

———. 1972. On the structure of concentrated, columnar vortices. *Astro. Acta.* 363–374.

———. 1973. Vorticity source for large scale dust devils and other comments on naturally occurring columnar vortices. *JAS* **30:** 1717–1722.

McDonald, J. R. 1970. *Response of Twenty Story Building to the Lubbock Tornado.* Storm Research Report TTUSRRO1, Texas Tech University, Lubbock, Texas, 60 p.

———. 1971. *The Hereford Tornado, April 9, 1971.* Department of Civil Engineering, Texas Tech University, Lubbock, Texas, 32 p.

McDonald, J. R., K. C. Mehta, and J. E. Minor. 1974. Tornado-Resistant Design of Nuclear Power-Plant Structures. *Nuclear Safety* 15–4, 432–439.

McDonald, J. R., K. C. Mehta, and J. E. Minor. 1972. *Development of a Design Basis Tornado for the Rocky Flats Site, Colorado.* rept. for C. F. Braun & Co., Engineers, under Agreement 4410–210–980–108.

———. 1973. Tornado generated missiles. *Specialty Conf.* on Structural Design of Nuclear Plant Facilities, Chicago, Ill., December 17–18, Vol. II, 543–556.

———. 1974. *Development of Design Basis Tornadoes for the Plantex Plant Site.* rept. for USAEC under contract No. DA-11-173-AMC-487(a).

———. 1975. *Development of a Windspeed Risk Model for the Argonne National Laboratory Site.* rept. for ANL under purchase order No. 823905.

———. 1975. *Development of a Design Basis Tornado and Structural Design Criteria for the Nevada Test Site, Nevada.* rept. for Lawrence Livermore Laboratory, Univ. of California, under purchase order No. 5062405.

———. 1975. *Tornado Risks and Design Windspeeds for the Paducah Plant Site.* Rept. for Union Carbide Corp. under subcontract No. 4077.

———. 1975. *Tornado Risks and Design Windspeeds for the Oak Ridge Plant Site.* Rept. for Union Carbide Corp. under subcontract No. 4077.

———. 1975. *Tornado Risks and Design Windspeeds for the Portsmouth Plant Site.* Rept. for Union Carbide Corp. under subcontract No. 4077.

———. 1975. *Development of Design Basis Tornadoes and Design Manual for the Pantex Plant Site.* Rept. for USAEC under contract No. DA-11-173-AMC-487(A).

———. 1975. *Development of Windspeed Risk Models for the Savannah River Plant Site.* Rept. for E. I. Du Pont de Nemours Co. under purchase order No. AXC 672-W.

———. 1975. *Development of a Design Basis Tornado and Structural Design Criteria for Lawrence Livermore Laboratory's Site 300, California.* Rept. for Lawrence Livermore Lab., Univ. of California, under purchase order No. 5062405.

———. 1975. Development of a windspeed risk model for the Argonne National Laboratory Site. draft report submitted to ANL by Texas Tech Univ. Dept. of Civil Eng., Lubbock, Texas.

McKay, G. A. and A. B. Lowe. 1960. The tornado in western Canada. *Bull. Am. Meteor. Soc.* **41:** 1–8.

McLaughlin, J. M. 1970. Design of nuclear power plants for tornadoes. *Proceed.* Conf. on Tornado Phenomenology and Related Protective Design Measures, Univ. of Wisconsin, Madison.

Meaden, G. T. 1976. Tornadoes in Britain: their intensities and distribution in space and time. *J. Meteor.* **1:** 242–251.

Mehta, K. C. 1976. Windspeed estimates: engineering analyses. *Proceedings,* Symp. on Tornadoes: Assessment of Knowledge and Implications for Man, Lubbock, Tex., June 24–26 (published by Texas Tech Univ.).

Mehta, K. C., et al. 1975. *Engineering Aspects of the Tornadoes of April 3–4, 1974.* Published by the Committee on Natural Disasters, National Academy of Sciences, Washington, D.C., 110 pp.

Mehta, K. C., J. R. McDonald, J. E. Minor, and A. J. Sanger. 1971. Response of Structural Systems to the Lubbock Storm, SRR 03, Texas Tech University, Lubbock, Texas (NTIS Accession No. PB-204-938).

Mehta, K. C., J. E. Minor, and J. P. McDonald. 1976. Windspeed Analyses of April 3–4, 1974 Tornadoes. *J. Struct. Div., ASCE,* Vol. 102, No. St9, September 1976, 1709–1724.

Mehta, K. C., J. E. Minor, J. R. McDonald, B. R. Manning, J. J. Abernethy, and U. F. Koehler. 1975. Engineering aspects of the tornadoes of April 3–4, 1974. Report available from Committee on Natural Disasters of the National Academy of Sciences, 2101 Constitution Ave., Washington, D. C., 20418.

Melaragno, M. G. *Tornado Forces and Their Effects on Buildings,* Kansas State Printing Service. Manhattan, Kan., 1968, 51 pp.

———. Outdoor Tornado Shelters for Residential Areas, *K-State Printing Service,* Kansas State University, Manhattan, Kan., 1968, 36 pp.

———. Structural design and Australian tornadoes. Architectural Science Review, Vol. 19, No. 4, pp. 78–82, December, 1976.

———. Tornado forces and their effects on buildings. 2nd ed., Office of Civil Defense, 51 pp., January, 1970.

———. 1968. Outdoor tornado shelters for residential areas. Kansas State University, 36 pp.
———. Dwelling structures for the Great Plains. 1st ed. *Building Science,* Vol. 7, pp. 87–92, 1972.
———. Some aspects of tornado protection in the design of structures. *Ingegneria Civile,* No. 43, 1973.
———. Better buildings in tornadoes and hurricane zones. *The Florida Architect,* Vol. 23, No. 2, pp. 19–21, March–April, 1973.
———. New buildings can survive tornadoes. *Texas Architect,* Vol. 23, No. 3, May–June, 1973.
———. Design for tornadoes. Engineering, Ipe. (in print), Paper No. 765 accepted by the Institution of Structural Engineers. (England).
Miller, B. I. 1958. On the maximum intensity of hurricanes. *J. Meteorol.* **15:** 184–195.
———. 1967. Characteristics of hurricanes. *Science* **157:** 184–195.
Miller, J. E. 1955. A tornado model and the fire whirlwind. *Weatherwise* **8:** 88–91.
Miller, R. C. 1967. Notes on analysis and severe-storm forecasting procedures of the military weather warning center. Tech. rept. 200, USAF Air Weather Service, Kansas City, Missouri.
Miller, R. C. 1972. Notes on analysis and severe storms forecasting procedures at the Air Force Global Weather Central. *AWS Technical Report 200* (*Rev*)
Minor, J. E. 1975. Concretely Speaking. Editorial, *Concrete Construction,* Vol. 20, No. 10, October 1975 (back cover).
——— et al. 1973. Designing Critical Facilities for Extreme Winds. *Proceedings of the Conference: Designing to Survive Disaster.* IIT Research Institute, Chicago, November 1973, 97–114.
——— et al. 1972. Failures of Structures due to Extreme Winds. *J. Struct. Div., ASCE,* Vol. 98, No. St11, November 1972.
——— et al. 1972. Impact of the Lubbock Storm on Regional Systems. Technical Report to the Defense Civil Preparedness Agency, TTU SRR 05, Texas Tech University, Lubbock, TX, June 1972.
——— et al. 1976. Air Flow Around Buildings as Reflected in Failure Modes. *ASHRAE Transactions,* Vol. 82, Part I, July 1976.
———. 1976. Applications of tornado technology in professional practice. *Proceedings,* Symp. on Tornadoes: Assessment of Knowledge and Implications for Man, Lubbock, Tex., June 24–26 (published by Texas Tech Univ.), 375–392.
——— and W. L. Beason. 1975. Window glass failures in windstorms. ASCE National Structural Engineering Meeting, New Orleans.
Minor, J. E., J. R. McDonald, and K. C. Mehta. Aspects of tornadoes of engineering interest.

The Third U.S. National Conference Wind Engineering Research, University of Florida, Gainesville, Fla., Feb. 26–Mar. 1, 1978.

Miranda, C. 1975. *A Report on Tornado Damage to the Windsor Curling Club on April 3, 1974.* Consulting Engineer, Box 800, Detroit, Michigan 48221, July 1975.

Modahl, A. C., and W. M. Gray. 1971. Summary of funnel cloud occurrences and comparison with tornadoes. *Mon. Wea. Rev.* **99:** 877–882.

Mohorovicic, A., 1892. Der Tornado bei Novska. *Meteor. Zeit* **4,** p. 320.

Moore, C. V. 1967. The design of barricades for hazardous pressure systems. *Nuclear Engineering and Design 5* 81–97. North-Holland Publishing Company, Amsterdam, Holland.

Moore, H. E. 1958. Tornadoes over Texas, a study of Waco and San Angelo in disaster. Austin, University of Texas Press, 334 pp.

Morton, B. R. 1966. Geophysical vortices. *Progress in Aeronautical Sciences* Vol. 7, Pergamon Press, New York, 145–193.

———. 1966. Geophysical vortices. *Progress in Aeronautical Science.* Vol. 7, Pergamon, New York, New York, 145–194.

———. 1969. The strength of vortex and swirling core flows. *JFM* **38:** 315–333.

MRI. 1965. Wind Effect on a Flat Roof. *Final Report Midwest Research Institute Project No. 2815-P.*

Muirhead, V. U. 1973. Compressible Vortex Flow. *American Institute of Aeronautics and Astronautics Paper No. 73–106.* Eleventh Aerospace Sciences meeting, New York, Jan. 1973.

——— and J. R. Eagleman. Laboratory Compressible Flow Tornado Model. *Preprints Seventh Conference on Severe Local Storms,* American Meteorological Society, Boston, Mass., Oct. 1971, pp. 284–91.

——— and J. R. Eagleman. 1973. Tornado Vortex Air Flow. *Preprints, Eighth Conference on Severe Local Storms.* American Meteorological Society, Boston, Mass., pp. 213–28.

National Research Council of Canada. 1975. National Building Code of Canada, 1975, NRCC No. 13982, National Research Council of Canada, Ottawa, 169 pp.

PUBLICATIONS OF NATIONAL SEVERE STORMS LABORATORY

The NSSL Technical Memorandum, beginning with No. 28, continue the sequence established by the U.S. Weather Bureau National Severe Storms Project, Kansas City, Missouri. Numbers 1–22 were designated NSSP Reports. Numbers 23–27 were NSSL Reports, and 24–27 appeared as subentries of Weather Bureau Technical Notes. These Reports are available from the National Technical Information Service, Operations Division, Springfield, VA 22151, for $3.00 and a microfiche version for $0.95. NTIS numbers are given in parentheses.

No.

1 National Severe Storms Project Objectives and Basic Design. Staff, NSSP. March 1961. (PB-168207)

2 The Development of Aircraft Investigations of Squall Lines from 1956–1960. B. B. Goddard. (PB-168208)

3 Instability Lines and Their Environments as Shown by Aircraft Soundings and Quasi-Horizontal Traverses. D. T. Williams. February 1962. (PB-168209)

4 On the Mechanics of the Tornado. J. R. Fulks. February 1962. (PB-168210)

5 A Summary of Field Operations and Data Collection by the National Severe Storms Project in Spring 1961. J. T. Lee. March 1962. (PB-165095)

6 Index to the NSSP Surface Network. T. Fujita. April 1962. (PB-168212)

7 The Vertical Structure of Three Dry Lines as Revealed by Aricraft Traverses. E. L. McGuire. April 1962. (PB-168213)

8 Radar Observations of a Tornado Thunderstorm in Vertical Section. Ralph J. Donaldson, Jr. April 1962. (PB-174859)

9 Dynamics of Severe Convective Storms. Chester W. Newton. July 1962. (PB-163319)

10 Some Measured Characteristics of Severe Storms Turbulence. Roy Steiner and Richard H. Rhyne. July 1962. (N62-16401)

11 A Study of the Kinematic Properties of Certain Small-Scale Systems. D. T. Williams. October 1962. (PB-168216)

12 Analysis of the Severe Weather Factor in Automatic Control of Air Route Traffic. W. Boynton Beckwith. December 1962. (PB-168217)

13 500-Kc./Sec. Sferics Studies in Severe Storms. Douglas A. Kohl and John E. Miller. April 1963. (PB-168218)

14 Field Operations of the National Severe Storms Project in Spring 1962. L. D. Sanders. May 1963. (PB-168219)

15 Penetrations of Thunderstorms by an Aircraft Flying at Supersonic Speeds. G. P. Roys. Radar Photographs and Gust Loads in Three Storms of 1961 Rough Rider. Paul W. J. Schumacher. May 1963. (PB-168220)

16 Analysis of Selected Aircraft Data from NSSP Operations, 1962. T. Fujita. May 1963. (PB-168221)

17 Analysis of Methods for Small-Scale Surface Network Data. D. T. Williams. August 1963. (PB-168222)

18 The Thunderstorm Wake of May 4, 1961. D. T. Williams. August 1963. (PB-168223)

19 Measurements by Aircraft of Condensed Water in Great Plains Thunderstorms. George P. Roys and Edwin Kessler. July 1966. (PB-173048)

20 Field Operations of the National Severe Storms Project in Spring 1963. J. T. Lee, L. D. Sanders, and D. T. Williams. January 1964. (PB-168224)

21 On the Motion and Predictability of Convective Systems as Related to the Upper Winds in a Case of Small Turning of Wind with Height. James C. Fankhauser. January 1964. (PB-168225)

22 Movement and Development Patterns of Convective Storms and Forecasting the Probability of Storm Passage at a Given Location. Chester W. Newton and James C. Fankhauser. January 1964. (PB-168226)

23 Purposes and Programs of the National Severe Storms Laboratory, Norman, Oklahoma. Edwin Kessler. December 1964. (PB-166675)

24 Papers on Weather Radar, Atmospheric Turbulence, Sferics, and Data Processing. August 1965. (AD-621586)

25 A Comparison of Kinematically Computed Precipitation with Observed Convective Rainfall. James C. Fankhauser. September 1965. (PB-168445)

26 Probing Air Motion by Doppler Analysis of Radar Clear Air Returns. Roger M. Lhermitte. May 1966. (PB-170636)

27 Statistical Properties of Radar Echo Patterns and the Radar Echo Process. Larry Armijo. May 1966. The Role of the Kutta-Joukowski Force in Cloud Systems with Circulation. J. L. Goldman. May 1966. (PB-170756)

28 Movement and Predictability of Radar Echoes. James Warren Wilson. November 1966. (PB-173972)

29 Notes on Thunderstorm Motions, Heights, and Circulations. T. W. Harrold, W. T. Roach, and Kenneth E. Wilk. November 1966. (AD-644899)

30 Turbulence in Clear Air Near Thunderstorms. Anne Burns, Terence W. Harrold, Jack Burnham, and Clifford S. Spavins. December 1966. (PB-173992)

31 Study of a Left-Moving Thunderstorm of 23 April 1964. George R. Hammond. April 1967. (PB-174681)

32 Thunderstorm Circulations and Turbulence from Aircraft and Radar Data. James C. Frankhauser and J. T. Lee. April 1967. (PB-174860)

33 On the Continuity of Water Substance. Edwin Kessler. April 1967. (PB-175840)

34 Note on the Probing Balloon Motion by Doppler Radar. Roger M. Lhermitte. July 1967. (PB-175930)

35 A Theory for the Determination of Wind and Precipitation Velocities with Doppler Radars. Larry Armijo. August 1967. (PB-176376)

36 A Preliminary Evaluation of the F-100 Rough Rider Turbulence Measurement System. U. O. Lappe. October 1967. (PB-177037)

37 Preliminary Quantitative analysis of Airborne Weather Radar. Lester P. Merritt. December 1967. (PB-177188)

38 On the Source of Thunderstorm Rotation. Stanley L. Barnes. March 1968. (PB-178990)

39 Thunderstorm—Environment Interactions Revealed by Chaff Trajectories in the Mid-Troposphere. James C. Fankhauser. June 1968. (PB-179659)

40 Objective Detection and Correction of Errors in Radiosonde Data. Rex L. Inman. June 1968. (PB-180284)

41 Structure and Movement of the Severe Thunderstorms of 3 April 1964 as Revealed from Radar and Surface Mesonetwork Data Analysis. Jess Charba and Yoshikazu Sasaki. October 1968. (PB-183310)

42 A Rainfall Rate Sensor. Brian E. Morgan. November 1968. (PB-183979)

43 Detection and Presentation of Severe Thunderstorms by Airborne and Ground-Based Radars: A Comparative Study. Kenneth E. Wilk, John K. Carter, and J. T. Dooley. February 1969. (PB-183572)

44 A Study of a Severe Local Storm of 16 April 1967. George Thomas Haglund. May 1969. (PB-184970)

45 On the Relationship Between Horizontal Moisture Convergence and Convective Cloud Formation. Horace R. Hudson. March 1970. (PB-191720)

46 Severe Thunderstorm Radar Echo Motion and Related Weather Events Hazardous to Aviation Operations. Peter A. Barclay and Kenneth E. Wilk. June 1970. (PB-192498)

47 Evaluation of Roughness Lengths at the NSSL-WKY Meteorological Tower. Leslie D. Sanders and Allen H. Weber. August 1970. (PB-194587)

48 Behavior of Winds in the Lowest 1500 ft in Central Oklahoma: June 1966–May 1967. Kenneth C. Crawford and Horace R. Hudson. August 1970.

49 Tornado Incidence Maps. Arnold Court. August 1970. (COM-71-00019)

50 The Meteorologically Instrumented WKY-TV Tower Facility. John K. Carter. September 1970. (COM-71-00108)

51 Papers on Operational Objective Analysis Schemes at the National Severe Storms Forecast Center. Rex L. Inman. November 1970. (COM-71-00136)

52 The Exploration of Certain Features of Tornado Dynamics Using a Laboratory Model. Neil B. Ward. November 1970. (COM-71-00139)

53 Rawinsonde Observation and Processing Techniques at the National Severe Storms Laboratory. Stanley L. Barnes, James H. Henderson, and Robert J. Ketchum. April 1971. (Com-71-00707)

54 Model of Precipitation and Vertical Air Currents. Edwin Kessler and William C. Bumgarner. June 1971. (COM-71-00911)

55 The NSSL Surface Network and Observations of Hazardous Wind Gusts. Operations Staff. June 1971. (COM-71-00910)

56 Pilot Chaff Project at the National Severe Storms Laboratory. Edward A. Jessup. November 1971. (COM-72-10106)

57 Numerical Simulation of Convective Vortices. Robert P. Davies-Jones and Glenn T. Vickers. November 1971. (COM-72-10269)

58 The Thermal Structure of the Lowest Half Kilometer in Central Oklahoma: December 9, 1966–May 31, 1967. R. Craig Goff and Horace R. Hudson. July 1972. (COM-72-11281)

59 Cloud-to-Ground Lightning Versus Radar Reflectivity in Oklahoma Thunderstorms. Gilbert D. Kiazer. September 1972. (COM-73-10050)

60 Simulated Real Time Displays of Velocity Fields by Doppler Radar. L. D. Hennington and G. B. Walker. November 1972. (COM-73-10515)

61 Gravity Current Model Applied to Analysis of Squall-Line Gust Front. Jess Charba. November 1972. (COM-73-10410)

62 Mesoscale Objective Map Analysis Using Weighted Time-Series Observations. Stanley L. Barnes. March 1973. (COM-73-10781)

63 Observations of Severe Storms on 26 and 28 April 1971. Charles L. Vlcek. April 1973. (COM-73-11200)

64 Meteorological Radar Signal Intensity Estimation. Dale Sirmans and R. J. Doviak. September 1973. (COM-73-11923/2AS)

65 Radiosonde Altitude Measurement Using Double Radiotheodolite Techniques. Stephan P. Nelson. September 1973. (COM-73-11934/9AS)

66 The Motion and Morphology of the Dryline. Joseph T. Schaefer. September 1973. (COM-74-10043)

67 Radar Rainfall Pattern Optimizing Technique. Edward A. Brandes. March 1974. (COM-74-10906/AS)

68 The NSSL/WKY-TV Tower Data Collection Program: April–July 1972. R. Craig Goff and W. David Zittel. May 1974. (COM-74-11334/AS)

69 Papers on Oklahoma Thunderstorms, April 29–30, 1970. Stanley L. Barnes, Editor. May 1974. (COM-74-11474/AS)

70 Life Cycle of Florida Key's Waterspouts. Joseph H. Golden. June 1974. (COM-74-11477/AS)

71 Interaction of Two Convective Scales Within a Severe Thunderstorm: A Case Study and Thunderstorm Wake Vortex Structure and Aerodynamic Origin. Leslie R. Lemon. June 1974. (COM-74-11642/AS)

72 Updraft Properties Deduced from Rawinsoundings. Robert P. Davies-Jones and James H. Henderson. October 1974. (COM-75-10583/AS)

73 Severe Rainstorm at Enid, Oklahoma—October 10, 1973. L. P. Merritt, K. E. Wilk, and M. L. Weible. November 1974. (COM-75-10583/AS)

74 Mesonetwork Array: Its Effect on Thunderstorm Flow Resolution. Stanley L. Barnes. October 1974. (COM-75-10248/AS)

75 Thunderstorm-Outflow Kinematics and Dynamics. R. Craig Goff. December 1975. (PB-250808/AS)

76 An Analysis of Weather Spectra Variance in a Tornadic Storm. Philippe Waldteufel. May 1976. (PB-258456/AS)

77 Normalized Indices of Destruction and Deaths by Tornadoes. Edwin Kessler and J. T. Lee. June 1976. (PB-260923/AS)

78 Objectives and Accomplishments of the NSSL 1975 Spring Program. K. Wilk, K. Gray, C. Clark, D. Sirmans, J. Dooley, J. Carter, and W. Bumgarner. July 1976. (PB-263813/AS)

79 Subsynoptic Scale Dynamics As Revealed By The Use Of Filtered Surface Data. Charles A. Doswell III. December 1976. (PB-265433/AS)

80 The Union City, Oklahoma Tornado of 24 May 1973. Rodger A. Brown, Editor. December 1976. (PB-269443/AS)

81 Mesocyclone Evolution and Tornado Generation Within the Harrah, Oklahoma Storm. Edward A. Brandes. May 1977. (PB-271675/AS)

Newton, C. W. 1963. Dynamics of severe convective storms. *Meteorological Monographs, Severe Local Storms,* American Meteorological Society, vol. 5, no. 27, pp. 33–58.

Newton, C. W. 1971. Hurricane. *McGraw-Hill Encyclopedia of Science and Technology.* Vol. 6 GAB-HYS, McGraw-Hill, New York, New York, 589–590.

——— and J. C. Fankhauser. 1964. On the Movements of Convective Storms, with Emphasis on Size Discrimination in Relation to Water-budget Requirements. *Journal of Applied Meteorology* vol. 3, no. 6, pp. 651–68.

NOAA. 1974. The widespread tornado outbreak of April 3–4, 1974. *National Disaster Survey Report 74–1,* 42 pp.

NOAA. 1975. *Tornado.* NOAA/PA 70007, National Oceanic andAtmospheric Administration, Department of Commerce, Washington, D.C.

———. n.d. *Tornado Safety Rules.* Public information brochure, NOAA, U.S. Department of Commerce, Washington, D.C.

Notis, C. and J. L. Stanford. 1973. The contrasting synoptic and physical character of northeast and southeast advancing tornadoes in Iowa. *J. Appl. Meteor.* **12:** 1163–1173.

———. 1975. The synoptic and physical character of Oklahoma tornadoes. *Preprints, Ninth Conf. on Severe Local Storms,* Norman, Oklahoma, October 21–23, 409–416. Pub. by Am. Meteor. Soc., Boston, Mass.

———. 1976. The synoptic and physical character of Oklahoma tornadoes. *Mon. Wea. Rev.* **104:** 397–406.

Nunley, R. E. 1971. Living Maps of the Field Plotter. *Commission of College Geography Technical Paper No. 4.* Association of American Geographers, Washington, D.C.

OEP. 1972. Report to the Congress, disaster preparedness. Office of Emergency Preparedness, Executive Office of the President (3 Vols.), 184, 26, and 143 pp.

Orville, H. D., and L. J. Sloan. 1970. A numerical simulation of the life history of a rainstorm. *J. Atmos. Sci.* **27:** 1148–1159.

Outram, T. S. 1904. Storm of August 20, 1904, in Minnesota. *Monthly Weather Review* 365–66 (August).

Paddleford, D. F. 1969. Characteristics of tornado generated missiles. Westinghouse Electric Corp. WCAP-7897, Pittsburgh, Pa., 34 pp.

Palmen, E., and C. W. Newton. 1969. *Atmospheric Circulation Systems.* Academic Press, New York, 471–522.

Pautz, M. E. 1969. Severe Local Storm Occurrences, 1955–1967. *ESSA Technical Memorandum WBTM RCST 13.* Office of Meteorological Operations, Silver Spring, Md. 77 pp.

Peace, R. L., and R. A. Brown. 1968. Comparison of single and double Doppler radar velocity measurements in convective storms. Proc. 13th Radar Meteor. Conf., Amer. Meteor. Soc., Boston, 464–473.

Pearson, A. D. 1971. Statistics on tornadoes that caused fatalities 1960–1970. *Preprints, Seventh Conf. on Severe Local Storms,* Kansas City, Mo., October 5–7, 194–197. Pub. by Am. Meteor. Soc., Boston.

Pearson, A. 1976. Privately communicated. [Also see "NSSFC's tornado data available," *Bull. of the Amer. Meteor. Soc.* **57:** 314.]

Pearson, A. D., J. G. Galway, and R. L. Inman. 1967. Relationship of surface dew point and integrated moisture in the planetary boundary layer. *Preprints, Fifth Conference on Severe Local Storms,* Amer. Meteor. Soc., Boston, 135–139.

Peltier. 1840. *Meteorologie Observations et recherches experimentales sur les causes qui concourent a la formation des trombes.* Paris.

Penner, S. S. 1972. Elementary considerations of the fluid mechanics of tornadoes and hurricanes. *Astronaut. Acta* **17:** 351–362.

Peterson, R. E. (ed.) 1976. *Proceedings, Symp. on Tornadoes: Assessment of Knowledge and Implications for Man.* Lubbock, Tex., June 24–26 (published by Texas Tech Univ.), 696 pp.

Pope, Alan. 1951. *Basic Wind and Airfoil Theory.* 1st ed., McGraw-Hill Publications, New York.

——— and J. J. Harper. 1966. *Low-Speed Wind Tunnel Testing.* John Wiley & Sons. 325 pp.

Prosser, N. E. 1964. Aerial photographs of a tornado path in Nebraska, May 5, 1964. *Monthly Weather Review,* pp. 593–98.

Puppo, A. and P. Longo. 1934. La Tromba del 24 Luglio 1930 nel Territorio di Treviso-Udine, *Mem. R. Off. Cent. Meteor. Geof.,* Ser. 3, **4,** 68 p.

Purdom, J. R. W. 1975. *Tornadic Thunderstorms and GOES Satellite Imagery.* Paper prepared for Ninth Conf. on Severe Local Storms, Norman, Okla., Oct. 21–23, available from Amer. Meteor. Soc., Boston, Mass.

Ray, P. S., R. J. Doviak, G. B. Walker, D. Sirmans, J. Carter, and B. Bumgarner. 1975.

Dual-Doppler observation of a tornadic storm. *J. Appl. Meteor.* **14:** 1521–1530.

Reap, R. M. and D. S. Foster. 1975. New operational thunderstorm and severe storm probability forecasts based on model output statistics (MOS). *Preprints of Ninth Conference on Severe Local Storms,* Amer. Meteor. Soc., Boston, Ma., 58–63.

Redmann, G. H., J. R. Radbill, J. E. Marte, P. Dergarabedian, and F. Fendell. 1976. Wind field and trajectory models for tornado-propelled objects. Tech. report 308–1, Electric Power Research Institute, Palo Alto, Calif.

Reed, J. W. 1970. *Window Damage Study of the Lubbock Tornado.* Technical Memorandum SC-TM-70-535, Sandia Laboratories, Albuquerque, New Mexico.

———. 1971. Some averaged measures of tornado intensity based on fatality and damage reports. *Preprints, Seventh Conf. on Severe Local Storms,* Kansas City, Mo., October 5–7, 187–193. Pub. by Am. Meteor. Soc., Boston, Mass.

Reichelderfer, F. W. 1957. Hurricanes, tornadoes, and other storms. *Annals of Am. Acad. of Political and Social Sci.* **309:** 23–35.

Research on Short-Term Weather Phenomena. 1974. Hearings before the Subcommittee on Space Science and Applications, Committee on Science and Astronautics, U.S. House of Representatives, 93rd Congress, Nov. 6, 7, and 9, 1973, U.S. Government Printing Office, 329 pp.

Reynolds, G. W. 1950. *A Systematic Method for Reporting Tornado Data Applied to a Tornado of 21 May 1949.* Master's Thesis, St. Louis University, p. 53.

———. 1957. A common wind damage pattern in relation to the classical tornado. *Bulletin of the American Meteorological Society,* vol. 38, no. 1, pp. 1–5.

———. 1958. Venting and other building practices as practical means of reducing damage from tornado low pressures. *Bulletin of the American Meteorological Society* 14–20 (January).

Rossman, F. 1951. Über Wolkentromben. *Annalen d. Meteor.* 48–55.

Rossman, F. O. 1953. The Physical Process in the Development of Tornadoes and Consequences Which Influence Their Dissipation. *Proc. Conf. Radio Meteor., Bur. Eng. Res., U. of Texas, Austin,* Session X-7, 10 pp.

Rossow, V. J. 1966. On the origin of circulation for tornado formation. *Proceedings, Twelfth Conference Radar Meteorology,* Norman, Okla., Oct. 17–20, Amer. Meteor. Soc., 183–189.

Rossow, V. J. 1967. A study of an electrostatic-motor drive for tornado vortices. Paper presented at Eighth Symposium on Engineering Aspects of MHD, Stanford University, March 28–30, 1967.

———. 1970. Observations of waterspouts and their parent clouds. *NASA Technical Note* D-5854, National Aeronautics and Space Administration, Washington, D.C., 63 pp.

Rotz, J. V., G. C. K. Yeh, and W. Bertwell. 1974. Tornado and extreme wind design criteria for nuclear power plants. *Topical Rept.* BC-TOP-3-A, Rev. 3, Bechtel Power Corp., San Francisco, Calif.

Ryan, R., and B. Vonnegut. 1969. Generation of a vortex by electrical heating. *Preprints of Papers Presented at the Sixth Conference on Severe Local Storms,* American Meteorological Society, Boston, Mass., 253–256.

———. 1970. Miniature whirlwinds produced in the laboratory by high-voltage electrical discharges. *Science* **168:** 1349–1351.

SAA. 1975. SAA loading code, part 2—wind forces. AS1170, Part 2—1975, Stand. Association of Australia, Sydney, 52 pp.

Sadowski, A. 1965. Potential casualties from tornadoes. Paper presented at 244th National Meeting of the Am. Meteor. Soc., Cloud Physics and Severe Local Storms, Reno, Nevada, October 18–22, 11 pp.

Safford, A. T. 1970. *The Influence of Terrain on the Frequency Distribution of Tornado and Hail Occurrence in the Central Midwest.* M.S. thesis, Saint Louis Univ., St. Louis, Missouri.

Salter, C. A. 1958. Wind Loadings on Flat Roof Buildings. *Engineering,* vol. 186, no. 4832, pp. 508–10.

Samuel, F. J. and C. W. Haman. 1939. *Civil Protection.* London: The Architectural Press. *In* Amirikian, Arsham. 1950. Design of Protective Structures. U.S. Navy, NavDocks P-51.

Sandborn, V. A. and R. D. Marshall. 1965. Local Isotropy in Wind-Tunnel Turbulence. Technical Report, U.S. Army Grant DA-AMC-28-64-G-9, Fluid Dynamics and Diffusion Laboratory, Colorado State University, Fort Collins, Colorado.

Sanders, W. W., Jr., L. F. Greimann, and D. I. McKeown. 1975. Structural damage from Ankeny, Iowa Tornado—June 18, 1974. Engineering Research Institute, Iowa State University, Ames, 15 pp.

Sandstrom, J. W. 1909. Über die Bewegung der Flussigkeiten. *Ann. Hydro. Marit. Meteor.* **37:** 242–254.

Sanger, A. J. and R. R. Minor. 1971. *Observations of the Response of Metal Building Systems to the Lubbock Tornado.* Storm Research Report, TTU SRR 02, Texas Tech University, Lubbock, Texas, 110 pp.

Sangster, W. 1960. A method of representing the horizontal pressure force without reduction of station pressures to sea level. *J. Meteor.* **17:** 166–176.

Sarpkaya, T. 1971. On stationary and traveling vortex breakdowns. *JFM* **45:** 545–559.

Schefer, J. T., D. L. Kelly, and R. F. Abbey, Jr. 1979. Tornado track characteristics and hazard probabilities, *Preprints of 5th International Conf. on Wind Engineering,* 8–14 July 1979, Colo. State U.

Schroeder, T. A. 1971. *The Effect of Topography Upon Tornado Incidence for the Central Midwest.* M.S. thesis, Purdue University, West Lafayette, Indiana.

——— and E. M. Agee. 1971. A regression model for tornado distribution in a synoptically homogeneous region. *Preprints, Seventh Conf. on Severe Local Storms,* Kansas City, Mo., October 5–7, 45–48. Pub. by Am. Meteor. Soc., Boston, Mass.

Schwiderski, E. W. 1968. On the axisymmetric vortex flow over a flat surface. Naval Weapons Lab. Report TR-2210.

Schwiesow, R. L. and R. E. Cupp. 1975. Remote Doppler velocity measurements of atmospheric 'dust devil' vortices. Manuscript submitted to *Applied Optics,* 6 pp.

Seed, H. B. and R. Lundgren. 1954. Investigation of the effect of transient loading on the strength and deformation characteristics of saturated sands. Proc. American Society for Testing Materials, vol. 54 pp. 1288–306.

Seelye, C. J. 1945. Tornadoes in New Zealand. *New Zealand J. of Sci. and Tech.* **27:** 166–174. Also Air Dept.—N.Z. Meteor. Off. Note No. 28, 1946.

Segman, Ralph. 1971. Wet Tornado. NOAA, pp. 45–47.

Segner, P., Jr. 1960. Estimates of minimum wind forces causing structural damage. *The Tornadoes at Dallas, Tex., April 2, 1957,* U.S. Weather Bureau Research Paper 41, U.S. Govt. Printing Office, Washington, D.C., 169–175.

Segner, E. P. 1960. Estimates of minimum wind forces causing structural response to natural wind. *Proceedings,* Wind Effects on Buildings and Structures, Univ. of Toronto Press, **1:** 595–630.

Serrin, J. 1972. The swirling vortex. *Phil. Trans. Roy. Soc. London,* **A, 271:** 325–360.

Shanahan, J. A. 1976. Tornado structure interaction: Engineering implications (in preprints for *Symposium on Tornadoes: Assessment of Knowledge and Implications for Man*).

———. Evaluation of and design for extreme tornado phenomena. *Proceedings,* Symp. on Tornadoes: Assessment of Knowledge and Implications for Man, Lubbock, Tex., June 24–26 (published by Texas Tech Univ.), 251–282.

Shapiro, A. H. 1962: Bathtub vortex. *Nature* **196:** 1080–1081.

Shenkman, S., and K. E. McKee. 1961. Bearing capacities of dynamically loaded footings. ASTM Symposium on Soil Dynamics, Special Technical Publication, No. 305, pp. 78–90.

Sherman, Zachary. 1971. *Residential Buildings Engineered to Resist Tornadoes and Hurricanes.* Paper presented at the XXII International Astronautical Congress in Brussels, Belgium.

———. 1973. Residential buildings engineered to resist tornadoes. *J. Struc. Div., ASCE* **99:** 701–714.

Showalter, A. K. 1943. The Tornado—An analysis of antecedent meteorological conditions. USWB, Washington, D.C. 3–139

———. 1953. A Stability Index for Forecasting Thunderstorms. *Bulletin of American Meteorological Society.* vol. 34, p. 250.

Shuman, R., and J. B. Hovermale. 1968. An operational six-layer primitive equation model. *J. Appl. Meteor.* **7:** 525–547.

Shunk, J. F. 1972. On the dynamic structure of a tornado vortex near the ground. Ph.D. thesis, Texas A&M.

Simpson, J., and V. Wiggert. 1969. Models of precipitating cumulus towers. *Monthly Weather Review* **97:** 471–489.

Sims, J. H., and D. D. Baumann. 1972. The tornado threat: coping styles of the North and South. *Science* **176:** 1386–1392.

Sinclair, P. C. 1964. Some preliminary dust devil measurements. *Monthly Weather Review* **92:** 363–367.

———. 1969. General characteristics of dust devils. *J. Appl. Meteor.,* **8:** 32–45.

———. 1973. The lower structure of dust devils. *J. Atmos. Sci.* **30:** 1599–1619.

Singh, M. P., A. Morcos, and S. L. Chu. 1973. Probabilistic treatment of problems in nuclear power plant design. *Specialty Conf.* on Structural Design of Nuclear Plant Facilities, Chicago, Ill., December 17–18, 263–289. pub. by Am. Soc. Civ. Eng.

Sirmans, D., and R. J. Doviak. 1973. Pulsed-Doppler velocity isotach displays of storm winds in real time. *J. Appl. Meteor.* **12:** 694–697.

Skaggs, R. H. 1969. Analysis and regionalization of the diurnal distribution of tornadoes in the United States. *Mon. Wea. Rev.,* **97:** 103–115.

———. 1970. On tornado probabilities. *Proc. Assn. of Am. Geographers* **2:** 123–126.

Slade, D. H. 1969. Wind Measurements on a Tall Tower in Rough and Inhomogeneous Terrain. *Journal of Applied Meteorology* **8**(2): 293–297.

Smart, H. R., L. K. Stevens, and P. N. Joubert. 1967. Dynamic structural response to natural wind. *Proceedings, Wind Effects on Buildings and Structures.* Univ. of Toronto Press, **1:** 595–630.

Smith, R. C. 1967. Flow in the boundary layer beneath an intense vortex such as a tornado. NCAR pre-publication review copy, National Center for Atmospheric Research, Boulder, Colo., 33 pp.

Smith, R. C. and P. Smith. 1965. Theoretical flow pattern of a vortex in the neighborhood of a solid boundary. *Tellus* **17:** 213–219.

Smith, R. K. 1978. Tornado dynamics. The Third U.S. National Conference Wind Engineering Research, University of Florida, Gainesville, Fla., Feb. 26–Mar. 1, 1978.

Smith, R. L. and D. W. Holmes. 1961. Use of doppler radar in meteorological observations. *Mon. Wea. Rev.* **89:** 1–7.

Smith, T. B., and V. A. Mirabella. 1972. *Characteristics of California Tornadoes.* Rept. for Lawrence Livermore Laboratory, Univ. of California, No. MRI 72 FR-996.

Snow, J. T., and E. M. Agee. 1975. Vortex splitting in the mesocyclone and the occurrence of tornado families. *Ninth Conference on Severe Local Storms,* American Meteorological Society, Boston, Mass., 270–277.

Somes, N.F., R. D. Dikken, and T. H. Boone. 1970. Lubbock tornado—A survey of building damage in an urban area. NBS Report 10254, National Bureau of Standards, U.S. Dept. of Commerce, 69 pp.

Spohn, H. R., and P. J. Waite. 1962. Iowa tornadoes. *Mon. Wea. Rev.* **90:** 398–406.

Starr, V. P. 1974. The tornado mechanism and its possible artificial duplication. *Rivista Italiana di Geofisica* **23:** 267–271.

———, N. E. Gaut, and R. D. Rosen. 1974. An angular momentum theorem for tornadoes and related topics. *Rivista Italiana di Geofisica* **23:** 317–328.

Stephenson, A. E. 1975. *Tornado Vulnerability—Nuclear Production Facilities,* Sandia Laboratories, RS 9333/13.

———. 1975. *Addendum to Tornado Vulnerability—Nuclear Production Facilities,* Sandia Laboratories, R423595.

———. 1976. *Full-Scale Tornado-Missile Impact Tests,* EPRI, Interim Report, RP399, Sandia Laboratories, SAND76–0195.

———. n.d. *Full-Scale Tornado-Missile Impact Tests,* Final Report, Sandia Laboratories.

Stephenson, A. E., G. E. Sliter, and E. Burdette. 1975. "Full-Scale Tornado-Missile Impact Tests Using a Rocket Launcher." *Specialty Conference on the Structural Design of Nuclear Plant Facilities,* Volume 1-A, pp. 611–636, New Orleans, LA.

Stevenson, J. D. 1972. Tornado design of class I structures for nuclear power plants. Proc. of Symposium on Structural Design of Nuclear Power Plant Facilities, 409–417.

Stevenson, J. D., ed. n.d. *Nuclear Reactor Design Manual.* Amer. Soc. Civil Eng.

Stong, C. L. 1963. How to make and investigate vortexes in water and flame. *Scientific American* **209**(4): 133–142.

Suarez, M. A. 1973. Selection of design basis tornado. Paper presented at *ASCE Specialty Conf.* on Structural Design of Nuclear Plant Facilities, Chicago, Ill., Dec. 17–18, 35 pp.

Sullivan, R. D. 1959. A two-cell vortex solution of the Navier-Stokes equation. *J. Aerospace Sci.* **26:** 767–768.

Swanson, A. E., R. E. Stippich, and F. C. Bates. 1967. Presented at the Fifth Conference on Severe Storms, St. Louis, Missouri.

Sykes, W. 1967. Tornado splinter data requested. *Weatherwise* **20:** 271.

Szillinsky, A. 1970. Numerical experiments simulating the penetration of a two-cell tornado-like vortex into the bottom boundary layer. *BPA* **43:** 47–73.

——— and F. Wippermann. 1969. The penetration of tornado-like vortices into the boundary layer—numerical experiments. Final Rept. Technische Hochschule, Darmstadt, West Germany, Inst. fur Meteorologie, AD-691879, AFCRL-69-0321, 106 pp.

Takeda, T. 1969. Numerical simulation of large convective clouds. Stormy Weather Group Scientific Report WM–64 (in two volumes), McGill University Montreal.

Taylor, W. L. 1973. Evaluation of an Electromagnetic Tornado-Detection Technique. *Preprints, Eighth Conference on Severe Local Storms,* American Meteorological Society, Boston, Mass., pp. 165–68.

Tecson, J. J. 1972. Characterization of 1965 tornadoes by their area and intensity. *SMRP Res. Paper* No. 94, Univ. of Chicago, Chicago, Illinois.

Teesdale, L. V. 1928. *Tornado-Resistant Construction and Building Possible by Venting.* Madison Forest Products Laboratory Branch, Madison, Wisconsin.

Tepper, Morris. 1957. On the evaluation of tornado reports and a numerical method for their classification. *Mon. Wea. Rev.* **85:** 159–165.

———. 1958. Tornadoes. *Scientific American.* May 1958. "The tornado at Little Rock, Arkansas, October 2, 1894." Editorial in *Monthly Weather Review,* pp. 413–14, October 1894.

Thom, H. C. S. 1963. Tornado probabilities. *Mon. Wea. Rev.* **91:** 730–736.

———. 1968. New distributions of extreme winds in the United States. *J. Struc. Div., Proc. Am. Soc. Civil Eng.* **94:** 1787–1801.

Thresher, A. A., II. 1971. "A Diagnostic Study of Great Plains Tornado Conditions." M.S. thesis. Naval Postgraduate School, Monterey, Calif.

"Tornado Central USA." NOAA, pp. 18–25, January 1972.

"Tornado facts." Washington, D.C., U.S. Department of Commerce, Environmental Science Services Administration, Weather Bureau, November, 1965 (L.S. 6403 Rev.).

Tornadoes-Hailstones-Thunderclouds. 1906. *Mon. Wea. Rev.* **34:** 30–31.

"Tornado information." Washington, D.C., U.S. Department of Commerce, Weather Bureau, April 1964.

"Tornado preparedness planning." NOAA, P170009, Washington, D.C., October 1970.

"Tornado-resistant buildings possible." Note in *Scientific American,* p. 65, July 1928.

Turner, J. S. 1966. The constraints imposed on tornado-like vortices by the top and bottom boundary conditions. *J. Fluid Mech.* **25**(2): 377–400.

Umenhofer, T. A. 1975. Overshooting top behavior of three tornado-producing thunderstorms. *Preprints Ninth Conference on Severe Local Storms,* American Meteorological Society, Boston, Mass. 96–99.

Uniform Building Code. 1973. International Conference of Building Officials.

Univ. of Wisconsin. 1970. *Proceedings of a Short Topical Conference on Tornado Phenomenology and Protective Design Measures.* The University of Wisconsin,.Madison, April 1970.

USAEC. 1974. Design Basis Tornadoes for Nuclear Power Plants. Regulatory Guide 1.76, Atomic Energy Commission, Washington, D.C., April 1974.

U.S. Department of Commerce. 1971–1975. *Storm Data.* Environmental Data Service, NOAA, Vol. 13–17.

———. *Storm Data.* U.S. Government Printing Office, Washington, D.C., 1950–74.

———. Tornado Occurrences in the United States. *Technical Paper No. 20.* U.S. Government Printing Office, Washington, D.C., 1953, 43 pp.

U.S. Department of Defense. 1962. The Effects of Nuclear Weapons. U.S. Atomic Energy Commission.

U.S. Department of Health, Education, and Welfare. 1979. *Morbidity and Mortality Weekly Report.* Center for Disease Control. Vol. 28, No. 17, May 4, 1979.

U.S. Department of Housing and Urban Development. 1966. HUD recommends ways to beat storm damage. News item released October 7, 1966.

U.S. Weather Bureau. 1954. *Tornado Safety Rules.* A two-page circular.

U.S. Weather Bureau. 1965. Tornado facts. November 1965.

Vaiksnoras, J. V. 1971. *Tornadoes in Tennessee (1916–1970).* Rept. available from Univ. of Tennessee, Institute for Public Service, Knoxville, Tennessee.

Van Tassel, E. L. 1955. The North Platte Valley tornado outbreak of June 27, 1955. *Monthly Weather Review* **83:** 255–264.

Vann, W. P. and M. Seniwongse. Response of tall steel buildings to tornadoes. The Third U.S. National Conference Wind Engineering Research, University of Florida, Gainesville, Fla., Feb. 26–Mar. 1, 1978.

Visconti, I., M. Capelli, and A. e Fagioli. 1969. Un esempio di analisi particolareggiata: la tempesta avvettiva dell'8–6–1964 nell'alto e nel medio Adriatico. (An example of specific analysis of the advective storm that hit the northern and central Adriatic Sea on June 8, 1964.) *CNR-IFA,* STR n. 10, Roma, 1969, p. 21.

Vesíc, A. 1963. Beaning capacity of deep foundations in sand. National Academy of Sciences, National Research Council. *Highway Research Record,* No. 39, pp. 112–53.

Vonnegut, Bernard. 1960. Electrical Theory of Tornadoes. *Journal of Geophysical Research,* vol. 65, pp. 203–12.

———. 1975. Chicken plucking as measure of tornado wind speed. Unpublished typescript, Dept. of Atmos. Sci., State University of New York at Albany.

———. 1975. Comment on "On the relation of electrical activity to tornadoes" by R. P. Davies-Jones and J. H. Golden. *J. Geophys. Res.* **80:** 4559–4560.

——— and James R. Weyer. 1966. Luminous phenomena in nocturnal tornadoes. *Science* 1213–20 (September).

———, C. B. Moore and C. K. Harris. 1960. Stabilization of a high-voltage discharge by a vortex. *J. Meteor.* **17:** 468–471.

———. 1975. Comments on "On the relation of electrical activity to tornadoes" by R. P. Davies-Jones and J. H. Golden. *J. Geophys. Res.* **80:** 4559–4560.

Waite, P. J. and C. E. Lamoureux. 1969. Corn Striations in the Charles City Tornado in Iowa. *Weatherwise,* Vol. 22, 50–59.

Wan, C. A. and C. C. Chang. 1972. Measurement of velocity field in a simulated tornado-like vortex using a three-dimensional velocity probe. *J. Atmos. Sci.* **29:** 116–127.

Walker, G. R. 1972. Cyclone Althea. James Cook Univ. of North Queensland, Townsville, Queensland, Australia, Part I, 211 pp.

———. 1975. Investigation of Wind Design Criteria Using a Statistical Simulation Model. *Proceedings of the International Conference in Wind Effects on Buildings,* London 1975.

———. 1975. Report on Cyclone 'Tracy' Effect on Buildings—December 1974. Melbourne, Australian Department of Housing and Construction, March 1975.

———. 1976. The Rational Design of Low Rise Housing in Tropical Cyclone Prone Areas. *Proceedings of the Annual Engineering Conference,* Institution of Engineers, Townsville, Queensland, Australia, May 1976.

———, J. E. Minor, and R. D. Marshall. 1975. The Darwin cyclone, valuable lesson in structural design. *Civil Engineering,* Vol. 45, No. 12, December 1975, 82–86.

Ward, N. B. 1956. Temperature inversion as a factor in formation of tornadoes. *Bull. Amer. Meteor. Soc.* **37:** 145–151.

———. 1962. The effect of low level wind shear on the formation of atmospheric vortices. *Proceedings of the Second Conference on Severe Storms,* Norman, Okla., Feb. 1962, Amer. Meteor. Soc.

———. 1971. The exploration of certain features of tornado dynamics using a laboratory model. NOAA Technical Memorandum ERLTM-NSSL-57, Norman, Oklahoma, 30 pp., November.

———. 1972. The exploration of certain features of tornado dynamics using a laboratory model. *J. Atmos. Sci.* **29:** 1194–1204.

Wark, D. Q., and D. T. Hilleary. 1969. Atmospheric temperature: successful tests of remote probing. *Science* **165:** 1256–1258.

Webb, E. K. 1963. Sink vortices and whirlwinds. Proc. First Australian Conf. of Hydraulics and Fluid Mech. 1962, 473–483.

Wegener, A. 1917. *Wind-und Wassenhosen in Europa,* F. Vieweg & Sohn, Braunschweig, 301 pp.

———. 1928. Die Windhose in der Oststeiermark vom 23. September 1927. *Meteor. Zeit.* **45:** 41–49.

Weller, N. and P. J. Waite. 1969. The Weller method: tornado detection by television. *Preprints, Sixth Conf. Severe Local Storms,* Amer. Meteor. Soc., Boston, 169–171.

Wen, Y.-K. 1975. Dynamic tornadic wind loads on tall buildings. *J. Struc. Div.,* Proc. Am. Soc. Civil Eng., **101:** 169–185.

———. 1976. Note on analytical modeling in assessment of tornado risks. *Proceedings,* Symposium on Tornadoes: Assessment of Knowledge and Implications for Man, Lubbock, Texas, June 22–24.

———, and A. H.-S. Ang. 1975. Tornado risk and wind effect on structures. *Proc.* Fourth Intern. Conf. on Wind Effect on Buildings and Structures, London, England, September.

———, and S.-L. Chu. 1973. Tornado risks and design wind speed. *J. Struc. Div.,* Proc. Am. Soc. Civil Eng. **99:** 2409–2421.

White, F. M. 1974. *Viscous Fluid Flow.* McGraw-Hill, pp. 423–25.

White, G. F., and J. E. Haas. 1975. *Assessment of Research on Natural Hazards.* MIT Press, Cambridge, Massachusetts, 487 pp.

Whitman, R. V. 1957. The behavior of soils under transient loading. Proc. Fourth International Conference on Soil Mechanics and Foundation Engineering (London), Vol. 1, pp. 207–12.

——— and K. A. Healy. 1962. Shear strength of sands during rapid loadings. *Proc. American Society of Civil Engineers* **88**(SM2): 99–132.

Whitman, R. V., J. M. Biggs, J. E. Brennan, C. A. Cornell, R. L. Neufuille, and E. H. Vanmarcke. 1975. Seismic design decision analysis. *J. Struc. Div., ASCE,* **101:** 1067–1084.

Wiggins, J. H. 1972. Balanced Risk: An Approach to Reconciling Man's Needs with his Environment. J. H. Wiggins Company, Palos Verdes Estates, CA (reprint of a talk presented at the 42nd Annual International Meeting of the Society of Exploration Geophysicists).

———. 1974. Toward a Coherent Natural Hazards Policy. *Civil Engineering,* Vol. 44, No. 4, April 1974, pp. 74–76.

———. 1976. Risk reduction through natural hazards research, an executive summary.

Wilkins, E. M. 1964. The role of electrical phenomena associated with tornadoes. *J. Geophys. Res.* **69:** 2435–2447.

——— and L. T. McConnell. 1968. Threshold conditions for vortex-stabilized electrical discharges in the atmosphere. *J. Geophys. Res.* **73:** 2559–2568.

———, Y. Sasaki, E. W. Friday, J. McCarthy, and J. McIntyre. 1969. Properties of simulated thermals in a rotating fluid. *J. Geophys. Res.* **74:** 4472–4486.

——— and R. H. Schauss. 1971. Interactions between the velocity fields of successive thermals. *Mon. Wea. Rev.* **99:** 215–226.

Wilkins, E. M., Y. Sasaki, and R. H. Schauss. 1971. Vortex formation by successive thermals: a numerical simulation. *MWR* **99:** 577–592.

———. 1974. Vortex formation in a friction layer: a numerical simulation. *MWR* **102:** 99–104.

——— and H. L. Johnson. 1975. Surface friction effects on the thermal convection in a rotating fluid: a laboratory simulation. *Mon. Wea. Rev.* **103:** 305–317.

Williams, R. J. 1976. Surface parameters associated with tornadoes. *Mon. Wea. Rev.* **104:** 540–545.

Wills, T. G. 1969. "Characteristics of the Tornado Environment as Deduced from Proximity Soundings," *Preprints, Sixth Conference on Severe Local Storms,* American Meteorological Society, Boston, Mass., pp. 222–29.

Wilson, J. W., and S. A. Changnon, Jr. 1971. Illinois tornadoes. *Circular* 103, Ill. State Water Survey, Urbana, Ill.

——— and G. M. Morgan, Jr. 1971. Long-track tornadoes and their significance. *Preprints, Seventh Conf. on Severe Local Storms,* Kansas City, Mo., October 5–7, 183–186. Pub. by Am. Meteor. Soc., Boston, Mass.

Wind Effects on Buildings. 1968. Proceedings, International Research Seminar, University of Toronto Press, vol. 1, 772 pp.

Wind Effects on Buildings. 1968. Proceedings, International Research Seminar, University of Toronto Press, vol. 2, 461 pp.

Winston, J. S. 1956. Forecasting tornadoes and severe thunderstorms. Forecasting Guide No. 1, U.S. Weather Bureau, Kansas City, Mo., 34 pp.

Wippermann, F. and A. Szillinsky. 1968. Numerical experiment on the formation of a tornado funnel in the boundary layer. A.F. Cambridge Res. Lab. Rept. 68–0425.

———, L. Berkofsky, and A. Szillinsky. 1969. Numerical experiments on the formation of a tornado funnel under an intensifying vortex. *QJRMS* **95:** 689–702.

Wolford, L. V. 1960. Tornado occurrences in the United States. *Tech. Paper* No. 20, revised, U.S. Weather Bureau, Washington, D.C., 71 pp.

Woodard, J. M. 1964. An investigation of the dynamic bearing capacity of footings on sand. M.S. thesis, Georgia Institute of Technology, Atlanta, Georgia, 57 pp.

Yamazaki, T. 1973. A laboratory experiment on composite tornado-like vortices formed by the interaction of horizontal shear and vertical instability. *Journal of the Faculty of Science,* Hokkaido University (Japan), Series VII, Vol. 4, No. 2, 59–68.

Ying, S. J. and C. C. Chang. 1970. Exploratory model study of tornado-like vortex dynamics. *J. Atmos. Sco.* **27:** 3–14.

Zrnic, D. S. 1974. Doppler techniques for tornado speed measurements. NSSL Memo for the Record, July 26, 1974, 13 pp.

———. 1975. Estimated tornado spectra and maximum velocity statistics. Final Rept. on NOAA Grant No. 04–5–022–17.

——— and W. C. Bumgarner. 1975. Receiver chain and signal processing effects on the Doppler spectrum. *Preprints,* 16th Radar Meteor. Conf., April 22–24, 1975, Houston, Tex. Published by Amer. Meteor. Soc., Boston, Mass.

——— and R. J. Doviak. 1975. Velocity spectra of vortices scanned with a pulse-Doppler radar. *J. Appl. Meteor.* **14:** 1531–1539.

Index